Endophytes for a Growing World

The book brings together papers covering the most recent scientific research from the top endophyte researchers in the world. It presents the state of the art in our knowledge and technical capacity and explores future directions of this work. It is highly relevant and timely because of the need to improve global food security and its sustainability, and also to provide novel bioactive molecules for medicine. There is also a need to protect forestry in a changing and growing world. Endophytes offer huge potential to reduce environmentally damaging agricultural inputs such as fertilisers and pesticides. They are also a largely overlooked group of organisms where much basic science remains to be undertaken. For example, new molecular tools of DNA profiling using high-throughput environmental sequencing are allowing the exploration of a previously largely unknown resource. There is a pressing need to convert scientific research on endophytes into practical application. This book describes how that will be achieved.

TREVOR R. HODKINSON is Professor in Botany and Head of the Botany Molecular Laboratory, Trinity College Dublin, Ireland. His research is mainly focused in fields known as endophyte biology, molecular systematics and genetic resource characterisation. He has specialist knowledge of the grass family and of forest tree genetics/mycorrhizae.

FIONA M. DOOHAN is Professor of Plant Pathology at University College Dublin, Ireland. Her research interests lie in the control of plant diseases through host genetics and via biological control. Her specialist area of knowledge is the control of cereal diseases. She has led several international consortia on this topic and collaborated extensively with industry and academic partners globally.

MATTHEW J. SAUNDERS is an Assistant Professor in Plant Science in the Botany Department, Trinity College Dublin, Ireland. His research involves the area of environmental physiology, with particular emphasis on how plants respond to changes in their physical, chemical and biological environments and how this information can be used to assess the resilience and adaptive capacity of terrestrial ecosystems to global environmental change.

BRIAN R. MURPHY is a Research Fellow in the Botany Department at Trinity College Dublin, Ireland. From an Irish perspective, he is largely responsible for the exciting research and associated publications relating to fungal endophyte application in agriculture. He is recognised as a leading expert in endophyte discovery from wild relatives of crops and their application to increasing stress resistance in cereal crops.

Endophytes for a Growing World

EDITED BY

TREVOR R. HODKINSON

Trinity College Dublin

FIONA M. DOOHAN

University College Dublin

MATTHEW J. SAUNDERS

Trinity College Dublin

BRIAN R. MURPHY

Trinity College Dublin

CAMBRIDGE
UNIVERSITY PRESS

CAMBRIDGE
UNIVERSITY PRESS

University Printing House, Cambridge CB2 8BS, United Kingdom

One Liberty Plaza, 20th Floor, New York, NY 10006, USA

477 Williamstown Road, Port Melbourne, VIC 3207, Australia

314–321, 3rd Floor, Plot 3, Splendor Forum, Jasola District Centre, New Delhi – 110025, India

79 Anson Road, #06-04/06, Singapore 079906

Cambridge University Press is part of the University of Cambridge.

It furthers the University's mission by disseminating knowledge in the pursuit of education, learning, and research at the highest international levels of excellence.

www.cambridge.org
Information on this title: www.cambridge.org/9781108471763
DOI: 10.1017/9781108607667

First published 2019

Printed in the United Kingdom by TJ International Ltd. Padstow Cornwall

A catalogue record for this publication is available from the British Library.

Library of Congress Cataloging-in-Publication Data
Names: Hodkinson, Trevor R., editor.
Title: Endophytes for a growing world / edited by Trevor R. Hodkinson, Fiona M. Doohan, Matthew J. Saunders, Brian R. Murphy
Description: New York, NY : Cambridge University Press, 2019. | Includes index.
Identifiers: LCCN 2018048892 | ISBN 9781108471763 (hardback) | ISBN 9781108458405 (paperback)
Subjects: LCSH: Endophytes. | Plants – Microbiology
Classification: LCC QR351.E52 2019 | DDC 579/.178–dc23
LC record available at https://lccn.loc.gov/2018048892

ISBN 978-1-108-47176-3 Hardback

Contents

List of Contributors *page* vii

Preface xi

Part I Introduction

1 Endophytes for a Growing World **3**

Trevor R. Hodkinson and Brian R. Murphy

Part II Role of Endophytes in Growth and Biotic and Abiotic Stress Resistance

2 Searching for Novel Fungal Biological Control Agents for Plant Disease Control Among Endophytes **25**

David B. Collinge, Hans J. L. Jørgensen, Meike A. C. Latz, Andrea Manzotti, Fani Ntana, Edward C. Rojas and Birgit Jensen

3 Application of Formulated Endophytic Entomopathogenic Fungi for Novel Plant Protection Strategies **52**

Vivien Krell, Desirée Jakobs-Schoenwandt and Anant V. Patel

4 Crossing Frontiers: Endophytic Entomopathogenic Fungi for Biological Control of Plant Diseases **67**

Lorena Barra-Bucarei, Andrés France and Paz Millas

5 Emerging Methods for Biological Control of Barley Diseases Including the Role of Endophytes **93**

Anna K. Høyer, Hans J. L. Jørgensen, Birgit Jensen, Brian R. Murphy and Trevor R. Hodkinson

6 Phosphate Nutrition in Root–Fungus Interactions **120**

Wael Yakti, Diana R. Andrade-Linares, Bernard Ngwene, Michael Bitterlich, Gábor M. Kovács and Philipp Franken

7 From Darkness to Light: Emergence of the Mysterious Dark Septate Endophytes in Plant Growth Promotion and Stress Alleviation **143**

Charlotte Berthelot, Michel Chalot, Corinne Leyval and Damien Blaudez

Part III Diversity and Community Ecology of Endophytes

8 Microbispora Dominate Diversity of Endophytic Actinobacteria from Australian Rice Plants **167**
FITRI WIDIANTINI AND CHRISTOPHER FRANCO

9 Isolation, Diversity and Potential Use of Endophytes in the Biomass and Bioenergy Crop *Miscanthus* **188**
JET BEEKWILDER, BRIAN R. MURPHY, EOIN MAC MATHUNA, AARON BARRY AND TREVOR R. HODKINSON

10 Life Within the Leaf: Ecology and Applications of Foliar Bacterial Endophytes **208**
C. RUTH MCNEES, ISAAC V. GREENHUT, AUDREY D. LAW, MUHAMMAD SALEEM AND LUKE A. MOE

11 Meta-Omics Approach to Unravel the Endophytic Bacterial Communities of *Brassica napus* and Other Agronomically Important Crops in Response to Agricultural Practices **232**
RIDHDHI RATHORE, KIERAN J. GERMAINE, PATRICK D. FORRISTAL, JOHN SPINK AND DAVID N. DOWLING

12 The Influence of Endophytes on Cork Oak Forests Under a Changing Climate **250**
DANIELA COSTA, RUI M. TAVARES, PAULA BAPTISTA AND TERESA LINO-NETO

Part IV Endophytes for Novel Biomolecules and In Vitro Methods

13 Endophytic Fungi: A Quintessential Source of Potential Bioactive Compounds **277**
VINEET MESHRAM AND MAHITI GUPTA

14 Enhancing Secondary Metabolite Production in Medicinal Plants Using Endophytic Elicitors: A Case Study of *Centella asiatica* (Apiaceae) and Asiaticoside **310**
SHUBHPRIYA GUPTA AND PREETI CHATURVEDI

15 *In Vitro* Methods for Plant–Microbe Interaction and Biocontrol Studies in European Ash (*Fraxinus excelsior* L.) **328**
ANINDITA LAHIRI, GERRY C. DOUGLAS, BRIAN R. MURPHY AND TREVOR R. HODKINSON

Part V Application and Commercialisation of Endophytes in Crop Production

16 The Science Required to Deliver *Epichloë* Endophytes to Commerce **343**
LINDA J. JOHNSON AND JOHN R. CARADUS

17 Plant Growth-Promoting Bacteria Field Trials in Europe **371**
KAREN O'HANLON

18 Prospecting Crop Wild Relatives for Beneficial Endophytes **390**
BRIAN R. MURPHY, FIONA M. DOOHAN AND TREVOR R. HODKINSON

Index 411

Colour plates can be found between pages 204 and 205.

Contributors

DIANA R. ANDRADE-LINARES Research Unit for Comparative Microbiome Analyses, Helmholtz Zentrum München, Germany

PAULA BAPTISTA CIMO, School of Agriculture, Polytechnic Institute of Bragança, Portugal

LORENA BARRA-BUCAREI Banco de Recursos Genéticos Microbianos, Instituto de Investigaciones Agropecuarias INIA, Chile, *and* Facultad de Agronomía, Universidad de Concepción, Chile

AARON BARRY Department of Botany, School of Natural Sciences, The University of Dublin, Trinity College Dublin, Ireland

JET BEEKWILDER Department of Botany, School of Natural Sciences, The University of Dublin, Trinity College Dublin, Ireland

CHARLOTTE BERTHELOT Université de Lorraine, Nancy, France

MICHAEL BITTERLICH Leibniz-Institute of Vegetable and Ornamental Crops, Germany

DAMIEN BLAUDEZ Université de Lorraine, Nancy, France

JOHN R. CARADUS Grasslanz Technology Limited, New Zealand

MICHEL CHALOT Université de Lorraine, Nancy, France

PREETI CHATURVEDI Department of Biological Sciences, College of Basic Sciences and Humanities, GB Pant University of Agriculture and Technology, Pantnagar, India

DAVID B. COLLINGE Department of Plant and Environmental Sciences and Copenhagen Plant Science Centre, University of Copenhagen, Frederiksberg, Denmark

DANIELA COSTA BioSystems & Integrative Sciences Institute (BioISI), Plant Functional Biology Centre, University of Minho, Portugal

FIONA M. DOOHAN School of Biology and Environmental Science, University College Dublin, Ireland

GERRY C. DOUGLAS Teagasc Agriculture & Food Development Authority, Forestry Development Unit, Dublin, Ireland

David N. Dowling Envirocore, Dargan Research Centre, Institute of Technology Carlow, Ireland

Patrick D. Forristal Teagasc Crops Research Centre, Carlow, Ireland

Andrés France Banco de Recursos Genéticos Microbianos, Instituto de Investigaciones Agropecuarias INIA, Chile

Christopher Franco Department of Medical Biotechnology, College of Medicine and Public Health, Flinders University, Adelaide, Australia

Philipp Franken Leibniz-Institute of Vegetable and Ornamental Crops, Germany, *and* Humboldt University of Berlin, Plant Physiology Department, Berlin, Germany

Kieran J. Germaine Envirocore, Dargan Research Centre, Institute of Technology Carlow, Ireland

Isaac V. Greenhut Department of Plant & Soil Sciences, University of Kentucky, Lexington, Kentucky, USA

Mahiti Gupta Department of Biotechnology, School of Biotechnology and Biosciences, Lovely Professional University, Phagwara, India

Shubhpriya Gupta Department of Biological Sciences, College of Basic Sciences and Humanities, GB Pant University of Agriculture and Technology, Pantnagar, India

Trevor R. Hodkinson Department of Botany, School of Natural Sciences, The University of Dublin, Trinity College Dublin, Ireland

Anna K. Høyer Department of Botany, School of Natural Sciences, The University of Dublin, Trinity College Dublin, Ireland

Desirée Jakobs-Schoenwandt WG Fermentation and Formulation of Biologicals and Chemicals, Faculty of Engineering and Mathematics, Bielefeld University of Applied Sciences, Bielefeld, Germany

Birgit Jensen Department of Plant and Environmental Sciences and Copenhagen Plant Science Centre, University of Copenhagen, Frederiksberg, Denmark

Linda J. Johnson AgResearch Ltd., Grasslands Research Centre, New Zealand

Hans J. L. Jørgensen Department of Plant and Environmental Sciences and Copenhagen Plant Science Centre, University of Copenhagen, Frederiksberg, Denmark

Gábor M. Kovács Eötvös Loránd University, Institute of Biology, Department of Plant Anatomy, Budapest, Hungary

Vivien Krell WG Fermentation and Formulation of Biologicals and Chemicals, Faculty of Engineering and Mathematics, Bielefeld University of Applied Sciences, Bielefeld, Germany

Anindita Lahiri Department of Botany, School of Natural Sciences, The University of Dublin, Trinity College Dublin, Ireland

MEIKE A. C. LATZ Department of Plant and Environmental Sciences and Copenhagen Plant Science Centre, University of Copenhagen, Frederiksberg, Denmark

AUDREY D. LAW Department of Plant & Soil Sciences, University of Kentucky, Lexington, Kentucky, USA

CORINNE LEYVAL Université de Lorraine, Nancy, France

TERESA LINO-NETO BioSystems & Integrative Sciences Institute (BioISI), Plant Functional Biology Centre, University of Minho, Portugal

EOIN MAC MATHUNA Department of Botany, School of Natural Sciences, The University of Dublin, Trinity College Dublin, Ireland

C. RUTH MCNEES Department of Plant & Soil Sciences, University of Kentucky, Lexington, Kentucky, USA

ANDREA MANZOTTI Department of Plant and Environmental Sciences and Copenhagen Plant Science Centre, University of Copenhagen, Frederiksberg, Denmark

VINEET MESHRAM Department of Plant Pathology and Weed Research, The Volcani Centre, Agriculture Research Organization, Rishon LeZion, Israel

PAZ MILLAS Banco de Recursos Genéticos Microbianos, Instituto de Investigaciones Agropecuarias INIA, Chile

LUKE A. MOE Department of Plant & Soil Sciences, University of Kentucky, Lexington, Kentucky, USA

BRIAN R. MURPHY Department of Botany, School of Natural Sciences, The University of Dublin, Trinity College Dublin, Ireland

BERNARD NGWENE Leibniz-Institute of Vegetable and Ornamental Crops, Germany

FANI NTANA Department of Plant and Environmental Sciences and Copenhagen Plant Science Centre, University of Copenhagen, Frederiksberg, Denmark

KAREN O'HANLON nadicom GmbH, Karlsruhe, Germany

ANANT V. PATEL WG Fermentation and Formulation of Biologicals and Chemicals, Faculty of Engineering and Mathematics, Bielefeld University of Applied Sciences, Bielefeld, Germany

RIDHDHI RATHORE Envirocore, Dargan Research Centre, Institute of Technology Carlow, Ireland, *and* Teagasc Crops Research Centre, Carlow, Ireland

EDWARD C. ROJAS Department of Plant and Environmental Sciences and Copenhagen Plant Science Centre, University of Copenhagen, Frederiksberg, Denmark

MUHAMMAD SALEEM Department of Plant & Soil Sciences, University of Kentucky, Lexington, Kentucky, USA

JOHN SPINK Teagasc Crops Research Centre, Carlow, Ireland

RUI M. TAVARES BioSystems & Integrative Sciences Institute (BioISI), Plant Functional Biology Centre, University of Minho, Portugal

Fitri Widiantini Department of Medical Biotechnology, College of Medicine and Public Health, Flinders University, Adelaide, Australia, *and* Department of Plant Diseases, Faculty of Agriculture, Universitas Padjadjaran, Bandung, Indonesia

Wael Yakti Leibniz-Institute of Vegetable and Ornamental Crops, Germany *and* Humboldt University of Berlin, Plant Physiology Department, Berlin, Germany

Preface

The 18 chapters of this book are based on the theme of an international conference held at Trinity College Dublin (TCD), Ireland, in August 2017. During the 3-day conference, entitled 'Sustain: Endophytes for a Growing World', there were stimulating presentations, posters and discussions covering a broad range of endophyte biology and application; these influenced the shape and content of this volume. Papers were contributed by a number of conference delegates and others were subsequently invited to broaden the book's scope or address particular topics. Two anonymous book proposal reviewers provided valuable content guidance and many anonymous reviewers also helped improve the chapter contributions. We are particularly grateful to the production team at Cambridge University Press, who have been highly supportive and professional. Finally, we thank all 56 contributing authors to the book (from institutions in Australia, Chile, Denmark, France, Germany, Hungary, India, Indonesia, Ireland, Israel, New Zealand, Portugal and the United States), many of whom also peer reviewed other chapters. We urge all readers to encourage national, regional and global policy makers to embrace the potential of endophyte science to address the Global Sustainable Development Goals of the United Nations.

Part I

Introduction

1

Endophytes for a Growing World

Trevor R. Hodkinson and Brian R. Murphy

Abstract

Endophytes are any microbes that can live within plants. We divide them into three major functional groups: endosyms (endosymbionts), endopaths (pathogens) and endosympaths (those that exist in both forms along a mutualism–parasitism continuum). Within these groups, endophytologists recognise harmful pathogenic microbes and a diverse range of beneficial/commensal microbes, including bacteria and archaea, such as diazotrophs, and fungi, such as the vertically transmitted clavicipitaceous endophytes, the generally horizontally transmitted class 2 fungal endophytes, mycorrhizal fungi and dark septate endophytes. This chapter introduces the science of endophyte biology and its application for a world population that is projected to grow to over 9 billion by 2050. It explores the potential of endophytes for improved agricultural and silvicultural sustainability including: yield improvement and nutrition; biocontrol of pests and diseases; and abiotic stress resistance in the context of climate change. It outlines how bioprospectors are using endophytes as sources of novel metabolites for the pharmaceutical and biochemical industries, and describes how endophytes can be used *in vitro* to elicit the increased production of known secondary metabolites from plants.

1.1 Endophytes

The microbiome of plants is complex and dynamic, and because of this plants are increasingly being considered as holobionts commonly inhabited by endophytic microbial communities (Hardoim *et al.*, 2015; Krell *et al.*, 2019, Chapter 3). Indeed, endophytes are ubiquitous and have been found in all species of plants studied to date (Hardoim *et al.*, 2015). The microorganisms that can behave as endophytes are easy to describe taxonomically, where they can be classed as archaea, bacteria, fungi, protozoans and viruses (or taxa of lower taxonomic rank within these). However, in the absence of a really clear-cut and universally accepted definition, it is more difficult to describe what features of these microorganisms qualify them as endophytes.

In broad terms, there are two schools of thought. The first defines an endophyte as any microorganism found inside plant tissue regardless of behaviour, the second prefers a more restricted definition including only those endophytes that are beneficial or neutral for the plant. In this book, we encounter these differences either implicitly or explicitly in almost every chapter. Here, we evaluate each definition and argue that the more general first definition, of any microorganism living within a plant, is most appropriate with more specific subclasses recognised within it.

The word endophyte literally translates as 'inside plant', so if understood solely on this basis then an endophyte is any organism that can exist within plant tissue; this would include beneficial symbionts, neutralists, commensals and pathogens. It excludes epiphytes (Figure 1.1). Such a broad definition would allow for the inclusion of endophytic organisms that can change the type of interaction with the plant along a mutualism–parasitism continuum (Mandyam and Jumpponen, 2015; Fesel and Zuccaro, 2016; Collinge *et al.*, 2019, Chapter 2; Berthelot *et al.*, 2019, Chapter 7; Costa *et al.*, 2019, Chapter 12; Murphy *et al.*, 2019, Chapter 18). A broad definition also allows for sub-classification into functional types (Figure 1.2).

A more restricted definition would consider endophytes as only beneficial or near-neutral plant symbionts and this interpretation is favoured by most researchers (Schulz and Boyle, 2006; Yakti *et al.*, 2019, Chapter 6; Widiantini and Franco, 2019, Chapter 8; Beekwilder *et al.*, 2019, Chapter 9; McNees *et al.*, 2019, Chapter 10; Gupta and Chaturvedi, 2019, Chapter 14). A further refinement is favoured by some researchers who maintain that endophytes are always beneficial for plants; Meshram and Gupta (2019, Chapter 13), for example, consider that 'endophytic microorganisms asymptomatically live together with plants in mutualistic alliance'.

One thing common to most definitions is that the effect of an endophyte on a plant is environmentally and genetically dependent. If we accept this proviso, then a broader definition would again seem to be favoured. Here, the endophytic microorganism will be detrimental or beneficial for the plant under different conditions. This understanding will be familiar to plant pathologists, many of whom consider that there is no such thing as a pathogen, rather merely a pathosystem where all parts of the necessary causal factors of disease need to be in place – the pathogen, a compatible host and suitable environmental conditions (Fang *et al.*, 2013; Collinge *et al.*, 2019, Chapter 2). For example, a host with some degree of resistance, but not

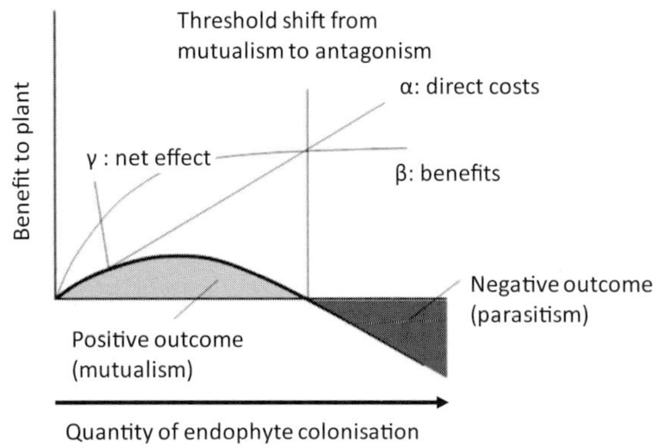

Figure 1.1 Mutualism–parasitism continuum. The dose–response relations in plant–endophyte interaction are shown. Endophyte colonisation has direct metabolic energy costs that should be correlated with their abundance within the host plant (α). Benefits to the plant in contrast are likely to follow a saturation curve (β). The net benefit (γ) follows an optimum curve with positive values at low to intermediate levels indicating mutualistic endophyte densities (light shading). However, negative values indicating antagonistic densities are found at high infection densities. Adapted from Partida-Martínez and Heil (2011) with permission.

immune, will result in an overall lower level of disease. This disease pathosystem model could just as well be applied to endophytes.

Those who favour the more restricted definition of an endophyte as always being beneficial or neutral for the plant would see the environmental and genetic factors in the system as making the endophyte either more or less beneficial for the plant. This definition therefore allows no room for the endophyte to behave as a pathogen and/or to have a detrimental effect on important plant growth parameters. While, in general, many endophytes do not have adverse effects on plants under any circumstances, this is not always the case. Murphy *et al.* (2014b) showed that even an almost universally beneficial endophyte such as the model organism *Serendipita indica* (syn. *Piriformospora indica*) can have negative effects on barley grain yield under severe nitrogen limitation and cold growing conditions. In this and other similar cases, the endophyte may behave as a competitor with the plant for important nutrient resources.

A definition of endophytes that caters for both of these perspectives would necessarily have to be somewhat more complex than either. Endophytes were first described in 1809 by the German botanist Heinrich Friedrich Link, who considered them to be fungal plant parasites. Since then, definitions of endophytes have increased in proportion to the growing number of endophyte studies: Wennstrom (1994) was one of the first to point out how the endophyte definition has been misused and he argued the case for new definitions focusing on particular endophyte behaviours. During the last 25–30 years, nearly all studies that feature endophytes have included a definition of the term, some being highly cited (Wilson *et al.*, 1992; Wilson, 1995; White *et al.*, 1993; Clement *et al.*, 1997; Schulz *et al.*, 1999; Jumpponen, 2001; Stone *et al.*, 2004; Schulz and Boyle, 2006; Rodriguez *et al.*, 2009). Most of these authors describe and define what an endophyte is from a generally plant-beneficial perspective. However, we are now beginning to see definitions that more accurately address the temporal and dynamic nature of the plant–endophyte relationship; for example, from Murphy *et al.* (2014a):

> Endophytes are a class of plant-associated microorganisms (bacteria, fungi and unicellular eukaryotes) that live at least part of their life cycle inter- or intracellularly inside plants, usually without inducing pathogenic symptoms. This can include competent, facultative, obligate, opportunistic and passenger endophytes (a passenger endophyte enters the plant by accident in the absence of selective forces maintaining it in the internal tissue of the plant).

Some also see the plant phenotype as being extended by its microbial inhabitants (Partida-Martínez and Heil, 2011), just like the human phenotype is extended by its own microbiome.

Another complication with the endophyte definition occurs with fungi that form mycorrhizae. Under the broad definition, mycorrhizal fungi can be considered endophytes but a distinction is usually made in the literature. The plant–endophyte relationship is often said to differ from mycorrhizal symbiosis by lacking cellular interfaces where specialised structures develop (e.g. arbuscules) and synchronised development between the plant and the fungal associate (Brundrett, 2006; Berthelot *et al.*, 2019, Chapter 7). As with the term mycorrhizal fungus, the term 'endophyte' is now so firmly entrenched in the literature that it will remain as the word of choice for these types or microorganisms. However, there is a strong case to be made for using this term as a very general description for microorganisms that inhabit plant tissue without resorting to some of the unwieldy definitions discussed above. It is in the nature of any relatively new research field for the terminology and descriptive language to evolve along with new discoveries and insights. Thus, in the spirit of this evolution of language (or linguemes), we could differentiate between the different endophyte lifestyles and variability of behaviour, by suggesting three major subcategories of endophytes (Figure 1.2):

- Endopathogenic microbes (**endopaths**): always detrimental/pathogenic
- Endosymbiotic microbes (**endosyms**): always beneficial or neutral
- Endosympathetic microbes (**endosympaths**): fluctuate between these two different behaviours depending on other factors (facultative endopaths/ facultative endosymbionts).

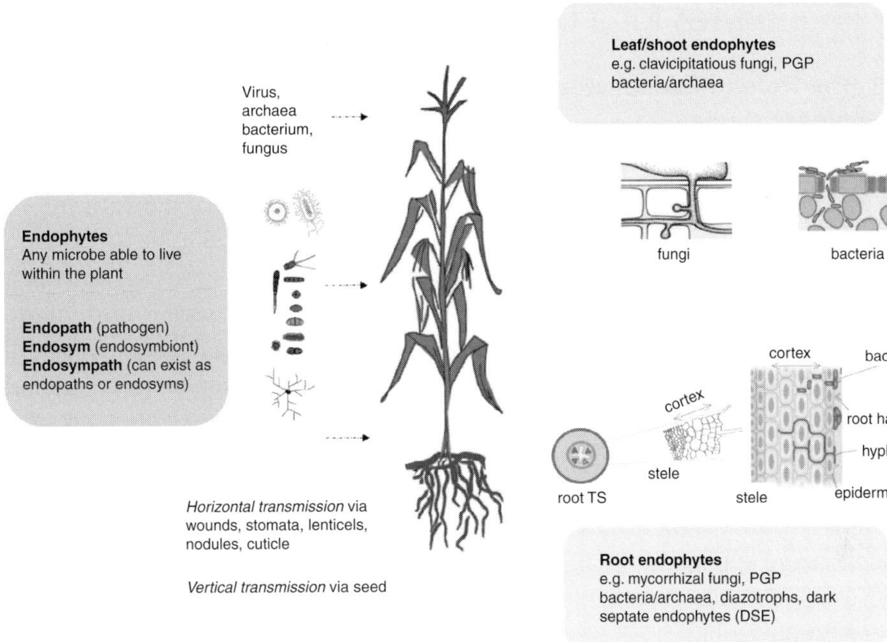

Figure 1.2 Subgroups of endophytes. Endophytes are any microbe that can live within a plant. Subclasses of endophytes can be defined in a hierarchical manner. Endopaths are pathogens, endosymbionts (endosyms) are always mutualistic and endosympaths can exist as both endosymbionts and endopaths. Within endosymbionts, major groups of beneficial microbes can be recognised including plant growth-promoting (PGP) and diazatrophic bacteria/archaea, class 1 (clavicipitatious) fungal endophytes, class 2 fungal endophytes, mycorrhizal fungi and dark septate endophytes (DSEs). They can also be divided by tissue type such as shoot or root. (A black and white version of this figure will appear in some formats. For the colour version, please refer to the plate section.)
Drawn from images in D'Arcy *et al.* (2001), Melotto *et al.* (2008), Hirsch and Mauchline (2012), Gelorini *et al.* (2012), Pickett *et al.* (2014) and Sarah Fogg with permission (www.sarahfogg.co.uk).

The etymology of the word 'sympaths' would also neatly suggest that the endophyte is 'sympathetic' to the environmental conditions (and also either symbionts or pathogens existing on the mutualism–parasitism continuum). These suggested new terms would also have the added advantage that they would be equally applicable to all life forms, not just plants (e.g. human, bird endosympaths). Whatever definition we use for endophytic microorganisms, the dynamic nature of this relatively new field of research will be reflected for some time to come in the equally dynamic and evolving terminology used to describe them. To some researchers it is sufficient for an endophyte to simply be 'what endophytologists study'.

1.2 Endophytes for a Growing World

1.2.1 The Role of Endophytes in Meeting Global Sustainable Development Goals

The rapidly growing world population is projected to reach over 9 billion by 2050 raising serious concerns over food security (FAO *et al.*, 2017). The number of chronically undernourished people in the world is currently estimated to be 815 million (IFPRI, 2018). Environmental damage including pollution, overuse of water, habitat loss, land-use change, biodiversity loss and climate change is also on the rise (Hodkinson *et al.*, 2011). It will be important to maintain high crop yields and productivity while improving sustainability. This is not a trivial challenge for several reasons. Approximately 85% of global water consumption is presently used for agricultural irrigation, of which 15–35% is unsustainable (Rosegrant *et al.*, 2009). Food systems contribute about one-fifth of all greenhouse gas emissions and agriculture is a primary cause of biodiversity loss (Arndt *et al.*, 2016). Furthermore, fertiliser production and application changes both nitrogen and phosphorous cycles, risking expensive, potentially irreversible environmental damage (Rockström *et al.*, 2009).

These problems have been recognised at a policy level by several international initiatives, such as the Strategic Plan for Biodiversity (Convention on Biological Diversity; COP, 2010), the Paris Agreement on Climate Change (2016), the G20 Agriculture Ministers' Action Plan 2017 for protecting water resources, the Global Challenges of the Consultative Group on International Agricultural Research (CGIAR, 2016), and the Sustainable Development Goals (SDGs; including 17 goals and 169 targets) of the United Nations 2030 Agenda for Sustainable Development (United Nations, 2015). Seven of the United Nations SDGs are particularly relevant to the endophyte research community:

- Goal 2: Achieve food security and improved nutrition and promote sustainable agriculture
- Goal 3: Ensure healthy lives and promote well-being for all at all ages
- Goal 6: Ensure access to water and sanitation for all
- Goal 7: Ensure access to affordable, reliable, sustainable and modern energy for all
- Goal 12: Ensure sustainable consumption and production patterns
- Goal 13: Take urgent action to combat climate change and its impacts
- Goal 15: Sustainably manage forests, combat desertification, halt and reverse land degradation, halt biodiversity loss.

There is also a pressing need to discover and produce new and useful biomolecules to provide assistance and relief in all aspects of human life especially those that involve global health (United Nations, 2015). Medicine needs to constantly adapt and cope with new challenges presented by the growing and ageing world population, including increased urbanisation, globalisation and travel (United Nations, 2015). Molecules are needed to tackle drug resistance in bacteria, combat viruses and fight fungal infections (Cannon and Simmons, 2017; Meshram and Gupta, 2019, Chapter 12; Gupta and Chaturvedi, 2019, Chapter 13). Endophytes are proving to be a novel source of metabolites for the pharmaceutical and biochemical industries providing biologically active compounds such as antibiotics, antioxidants, anticancer agents, immunosuppressive compounds, insecticides, plant growth-promoting (PGP) agents and volatile antimicrobial agents representing a wide range of organic molecules including terpenoids, peptides, carbohydrates, aromatics and hydrocarbons (Strobel, 2018; McNees *et al.*, 2019, Chapter 10). Endophytes can also be elicitors and enhancers of secondary metabolite production in plants as outlined in Gupta and Chaturvedi (2019, Chapter 13).

1.2.2 Endophytes in Sustainable Crop Production

Agricultural productivity has improved in pace with population growth particularly since the Green Revolution (Godfray *et al.*, 2010). However, current agricultural and silvicultural production systems need to become more environmentally, economically and socially sustainable to provide ecosystem services and help address problems associated with maintaining productivity including pollution, erosion, soil fertility, greenhouse gas emissions, pests/disease and reduced agroecosystem biodiversity, in line with the United Nations SDGs outlined above. There is a need to minimise chemical applications, better manage water, enhance carbon sequestration, maximise biodiversity and increase resilience to pests and diseases. There is also a need to make sustainable use of biomass and bioenergy crops (Hodkinson *et al.*, 2015; Beekwilder *et al.*, 2019, Chapter 9). To do this, crop management regimes need to manipulate the plant microbial communities, including their endophytes, to best effect (Busby *et al.*, 2017).

Endophytes can contribute to both productivity and sustainability in combination with several other approaches, including plant breeding and improved farm and soil management. The challenge to endophytologists is to widely integrate endophytes into modern agricultural practices in the most efficient and beneficial ways (Le Cocq *et al.*, 2016). To achieve this goal, there is a need to investigate the role of endophytes in the resilience of crops and forest trees to abiotic and biotic stress; understand the biodiversity and community structure of endophytes; and utilise the beneficial attributes of endophytes to improve the sustainability of agroecosystems and forestry.

The potential application of endophytes to agriculture is well documented (Murphy *et al.*, 2014a, b, 2015a, b, c, d, 2017b, 2018; Soares *et al.*, 2016). Some PGP endosyms can improve mineral nutrition and yields (Achatz *et al.*, 2010; Hubbard *et al.*, 2013; Yakti *et al.*, 2019, Chapter 5; Murphy *et al.*, 2019, Chapter 18), confer insect or pathogen resistance (Clement *et al.*, 1997; Vega *et al.*, 2009; Wiewióra *et al.*, 2015; Collinge *et al.*, 2019, Chapter 2) and improve stress tolerance (Waller *et al.*, 2005; Waqas *et al.*, 2012; Bertherlot *et al.*, 2019, Chapter 7). In forestry, the importance of mycorrhizal fungi is well documented but other endophyte groups are understudied and could play important roles in productivity, stress tolerance and the control of pests and diseases such as cork oak decline (Costa *et al.*, 2019, Chapter 12) and ash dieback disease (Lahiri *et al.*, 2019, Chapter 15).

Few of the beneficial endosyms are well characterised from a mechanistic perspective (Finkel *et al.*, 2017). There are some well-known examples, such as alkaloid production in the fungus *Epichloë* (Saikkonen *et al.*, 2016; Johnson and Caradus, 2019, Chapter 16); biological nitrogen fixation by rhizobia in legumes (Desbrosses and Stougaard, 2011); phosphate solubilisation/mobilisation by bacteria and mycorrhizal fungi (Timmusk *et al.*, 2017); hormone production, such as *Azospirillum* induced auxin synthesis in wheat (Dobbelaere *et al.*, 1999), enzymatic effects, such as *Enterobacter* 1-aminocyclopropane-1-carboxylate (ACC) deaminase mediated phytostimulation (Li *et al.*, 2000); and contribution to systemic resistance (Ryu *et al.*, 2004). However, mechanistic studies are still in their infancy and need assessment for the wide range of differing endophyte/plant systems.

Biocontrol offers a promising alternative or supplement to synthetically produced pesticides (Collinge *et al.*, 2019, Chapter 2; Krell *et al.*, 2019, Chapter 3; Barra-Bucarei *et al.*, 2019, Chapter 4; Høyer *et al.*, 2019, Chapter 5). One group of promising biocontrol agents (BCAs) are the endophytic entomopathogenic fungi and most interest has focused on the application of entomopathogenic aerial conidia from *Beauveria* spp. and *Metarhizium* spp. for crop protection. An early study by Bing and Lewis (1991) showed that application of *Beauveria bassiana* to maize and its subsequent colonisation of plant tissue suppressed the European corn borer *Ostrinia nubilalis*. Research on several other entomopathogenic fungal genera, such as *Acremonium*, *Cladosporium*, *Clonostachys*, *Isaria* and *Lecanicillium*, is taking place (reviewed in Krell *et al.*, 2019, Chapter 2 and Barra-Bucarei *et al.*, 2019, Chapter 4). Some of these fungi have demonstrated the dual ability to control both insect pests and plant pathogens (Ownley *et al.*, 2004) and others have been shown to improve other agronomic traits (Murphy *et al.*, 2015a, b).

Some microbes such as the clavicipitatious fungal endosyms of several forage grasses produce alkaloids, phenolics and terpenoids that reduce insect damage and therefore improve yield (Clement *et al.*, 1997; Rasmussen *et al.*, 2012; Żurek *et al.*, 2010; Johnson and Caradus, 2019, Chapter 16). They are often vertically transmitted through seed, and there is evidence that they have co-evolved with the plant host (Saikkonen *et al.*, 2016; Hodkinson, 2018). They colonise the plant from the seed, so are sold as endosym-containing seed (Johnson and Caradus, 2019, Chapter 16).

However, it is necessary to ensure that toxicity is not so high as to poison cattle (White *et al.*, 1987; Porter, 1995).

Some bacterial and fungal endophyte treatments have been demonstrated to improve crop resistance to endopaths (Waller *et al.*, 2005; Murphy *et al.*, 2015b; Høyer *et al.*, 2019, Chapter 5; Widiantini and Franco, 2019, Chapter 8). However, little is known about how endophytic communities can influence pathogen resistance of trees. Lahiri *et al.* (2019, Chapter 15) outline the use of *in vitro* methods for endophyte pathogen interaction studies in ash for the study of dieback disease resistance and Costa *et al.* (2019, Chapter 12) discuss the role of endophytic communities in declining cork oak populations.

Endophytic biocontrol is known to occur via antibiosis (antagonism), parasitism, competition and induced resistance (Hardoim *et al.*, 2015; Ludwig-Müller, 2015; Jensen *et al.*, 2016; Card *et al.*, 2016; Collinge *et al.*, 2019, Chapter 2). Dual culture *in vitro* experimentation (endophyte and pathogen) examining inhibitory activities of the endophyte on the pathogen do not always correlate with *in planta* efficacy, especially since endophytes rely on intimate plant contact. Therefore, it is essential to test for *in planta* biological activity as early in the screening process as is practical (this can be done in pots, *in vitro* or in the field). Furthermore, Høyer *et al.* (2019, Chapter 5) critically evaluated the biocontrol literature in barley and found that few studies have conducted sufficient experimentation to demonstrate biocontrol or its mechanisms.

Endophytes can improve both macro and micronutrient nutrition (Behie and Bidochka, 2014; Timmusk *et al.*, 2017). The role of PGP microbes such as diazotrophic bacteria/archaea (including rhizobia) in N nutrition is well documented (Vessey, 2003; O'Hanlon, 2019, Chapter 17). Phosphate is a non-renewable resource and many natural soils are phosphate deficient. Fungi and PGP bacteria are able to solubilise phosphate from inorganic sources that are not available to plants (Widiantini and Franco, 2019, Chapter 8). They can also free mineral nutrients from organic compounds/sources and some are able to transport phosphate through their hyphae to the roots thereby bridging rhizospheric phosphate depletion zones. Yakti *et al.* (2019, Chapter 6) discuss the role of arbuscular mycorrhizal fungi (Glomeromycotina), fungi from the order Sebacinales (Basidiomycota) and the diverse group of dark septate endophytes (DSEs; Ascomycota) in phosphate nutrition. DSEs have been shown to improve mineral nutrition and degrade complex carbon compounds, produce secondary metabolites such as phytohormones and volatile organic compounds and protect against phytopathogens (Mandyam and Jumpponen, 2005, 2015; Newsham, 2011; Bertherlot *et al.*, 2019, Chapter 7). Furthermore, DSEs have a role in abiotic stress tolerance in trace element/heavy metal contaminated soils and other stressful habitats (Bertherlot *et al.*, 2019, Chapter 7).

1.3 Practical Application and Commercialisation

There are two broad approaches to improve plant–endophyte interactions in crop systems, as outlined below. In the first approach, naturally occurring endophyte communities are manipulated through farm management to allow the natural colonisation of beneficial PGP microbes. In the second approach, endophytes are added directly to the crop via various types of treatments (spraying, drenching, seed coating or vertical transmission in seed) so that key microbes are immediately available to the plant.

1.3.1 Manipulating Endophytic Microbe Communities for Increased Production

Most previous studies have focused on interactions between individual endophytes and single plant species and few have attempted to explore the diversity and potential of endophytes at a community level. In fact, very little is known about the diversity of endophyte communities of most ecosystems and how these symbionts influence plant growth and survival (Schulthess *et al.*, 1998; Wilberforce *et al.*, 2003; Richter-Heitmann *et al.*, 2016). With advances in metagenomic and amplicon barcoding sequencing of bacteria and fungi (Lindahl *et al.*, 2013; Akinsanya *et al.*, 2015), it is now possible to accurately determine the microbiome composition of individual plants and their ecosystems and attempt to develop tools for their manipulation. Manipulating agroecosystems to encourage optimal endophyte levels is a major challenge, but clues can be found by looking at productive semi-natural ecosystems. For example, semi-natural grasslands are remnants of old biodiverse agricultural systems and are likely to provide a rich source of endophytes because of their high diversity of grasses and other organisms. This contrasts to intensively farmed forage or cereal crop systems that tend to be dominated by a few species. Indeed, van der Heijden (2008) found increases in grassland species richness associated with microbial symbionts and concluded that soil microbes must be considered as important drivers of plant diversity and productivity in terrestrial ecosystems.

The soil and surrounding vegetation provides a diverse range of microbes that crop plants can recruit as shoot endosyms and rhizospheric endosyms, including mycorrhizal fungi (Bonfante and Genre, 2010). However, the science behind managing the soil and surrounding vegetation to maximise endosym uptake by crops is still in its infancy (Estendorfer *et al.*, 2017). Several studies have examined the role of management practices on endophytic diversity (Estendorfer *et al.*, 2017). For example, site conditions have been shown to influence endophyte colonisation in grasses (Wiewióra *et al.*, 2011) and Estendorfer *et al.* (2017) showed that land-use intensity strongly influenced the microbiome and endophyte community of the forage grass *Dactylis glomerata*. The role of management in an oil seed rape endophyte community is described in Rathore *et al.* (2019, Chapter 11), who review the role of endophytic bacteria in crop production and the promotion of agricultural practices that maintain the natural diversity of endophytes. They provide evidence to show that agricultural management practices, such as soil tillage, crop rotation and fertiliser and pesticide applications have a major effect on function and structure of the soil, rhizosphere and endophytic bacterial communities.

1.3.2 Develop Endophyte Treatments for Crop Improvement

The alternative to natural colonisation of endosyms is to apply them artificially to crops or to apply biomolecules extracted from them. However, selecting which endophytes to use as treatments for crop improvement is a difficult challenge as microbiome diversity of plants is immense (Murphy *et al.*, 2015d) and their function and mode of action has not been sufficiently characterised (Høyer *et al.*, 2019, Chapter 5). For this reason, there is a need to develop the methodology for selecting endosyms suitable for yield improvement, find the best endosymbiont combinations to use as consortia for crop inoculants and to find the best ways to deliver these endophytes to crops (Murphy *et al.*, 2018).

The endophyte discovery process can be considered as a funnelling approach (Figure 1.3), where selection steps are taken serially to find the most beneficial endophytes. At each step in the process, the number of endophytes is reduced (Figure 1.3). At step 1, the endophytes are targeted from wild, crop wild relative (CWR) or cultivated plants. At step 2, pure endophyte cultures are isolated (only a fraction of all endophytes are culturable). At step 3, the culturable endophytes with high inoculation potential are selected and endophytes with unfavourable characteristics avoided. Positive attributes would include rapid growth in culture, spore formation for crop treatments and persistency. At step 4, vigorous screening and testing is applied. Screening might include bioinformatic and literature search approaches seeking favourable or unfavourable properties (e.g. excluding known endopaths). A functional genomic or proteomic approach might be taken to assess potential functions of the endophytes. Testing usually involves high-throughput phenotyping with plant–microbe interaction studies (e.g. growth experiments *in vitro* or in pots) to establish evidence for endosymbiotic action. Consortia of endophytes can also be tested at this stage. Step 5 involves the field trialling of endosyms previously identified as having highest potential through the serial selection process (Figure 1.3). Field trials are required over many differing environments and seasons (Murphy *et al.*, 2018).

We have used this approach successfully to discover and select endophyte consortia ready for commercial application (reviewed in Murphy *et al.*, 2018). We have developed fungal seed coating systems by screening CWRs at step 1. CWRs have been shown to be rich in endophytes and are likely to be more compatible with the target crop than a randomly chosen source plant (Murphy *et al.*, 2015a, c). Other approaches for sourcing endophtyes and delivering them to the crop (spraying,

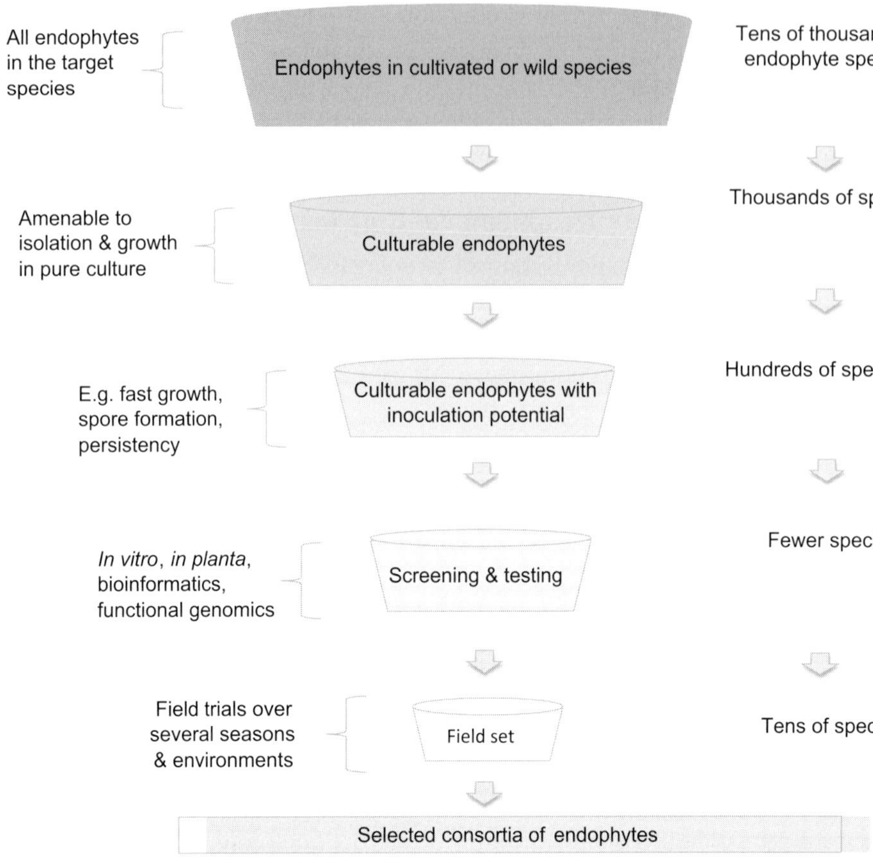

Figure 1.3 Screening microbiome diversity via a serial selection approach. Successively fewer microbes are selected at each step resulting in the final selection of endophyte consortia for inoculation.

drenching, vertical transmission in seed) have also been successful (Hallmann *et al.*, 2007; Parnell *et al.*, 2016). For example, Johnson and Caradus (2019, Chapter 16) describe the delivery systems for vertically transmitted *Epichloë* and O'Hanlon (2019, Chapter 17) discusses the application of PGP bacteria by spraying in commercial trials. The regulatory and commercialisation challenges of getting endophyte products to market are also outlined in the case studies presented by Johnson and Caradus (2019, Chapter 16).

1.4 Conclusions

Endophyte science has the potential to transform agriculture and forestry by improving production and sustainability. Endophytes can also provide a novel source of biomolecules for pharmacy and biochemical industries. This book describes the basic concepts of endophyte biology and outlines how they can be used in various applications. We are sure that they will play a significant role in addressing the 2030 Sustainable Development Goals of the United Nations particularly in the areas or food security, forestry, environmental protection and climate change. We hope that this book goes some way to help achieve that aim.

Acknowledgements

We thank our co-editors Fiona Doohan and Matthew Saunders for useful discussion on the content of this chapter. We thank the Science Foundation Ireland Catalyst Award Scheme for funding the international conference that resulted in the production of this book and chapter. We also thank the Enterprise Ireland Commercialisation Fund (Grant number CF 2017 0625-P) for financial assistance.

References

Achatz, B., von Rüden, S., Andrade, D. *et al.* (2010). Root colonization by *Piriformospora indica* enhances grain yield in barley under diverse nutrient regimes by accelerating plant development. *Plant Soil*, **333**, 59–70.

Akinsanya, M. A., Goh, J. K., Lim, S. P. and Ting, A. S. (2015). Metagenomics study of endophytic bacteria in *Aloe vera* using next-generation technology. *Genomics Data*, **6**, 159–163.

Arndt, C., Msangi, S. and Thurlow, J. (2016). Green Energy: Fueling the Path to Food Security, in 2016 Global Food Policy Report. Washington, DC: International Food Policy Research Institute, pp. 56–65.

Barra-Bucarei, L., France, A. and Millas, P. (2019). Crossing frontiers: endophytic entomopathogenic fungi for biological control of plant diseases. In *Endophytes for a Growing World*, ed. T. R. Hodkinson, F. M. Doohan, M. J. Saunders and B. R. Murphy. Cambridge: Cambridge University Press, Chapter 4.

Beekwilder, J., Murphy, B. R., Mathuna, E. M., Barry, A. and Hodkinson, T. R. (2019). Isolation, diversity and potential use of endophytes in the biomass and bioenergy crop *Miscanthus*. In *Endophytes for a Growing World*, ed. T. R. Hodkinson, F. M. Doohan, M. J. Saunders and B. R. Murphy. Cambridge: Cambridge University Press, Chapter 9.

Behie, S. W. and Bidochka, M. J. (2014). Nutrient transfer in plant–fungal symbioses. *Trends in Plant Science*, **19**, 734–740.

Berthelot, C., Chalot, M., Leyval, C. and Blaudez, D. (2019). From darkness to light: emergence of the mysterious dark septate endophytes in plant growth promotion and stress alleviation. In *Endophytes for a Growing World*, ed. T. R. Hodkinson, F. M. Doohan, M. J. Saunders and B. R. Murphy. Cambridge: Cambridge University Press, Chapter 7.

Bing, L. A. and Lewis, L. C. (1991). Suppression of *Ostrinia nubilalis*

(Huebner) (Lepidoptera: Pyralidae) by endophytic *Beauveria bassiana* (Balsamo) Vuillemin. *Environmental Entomology*, **20**, 1207–1211.

Bonfante, P. and Genre, A. (2010). Mechanisms underlying beneficial plant–fungus interactions in mycorrhizal symbiosis. *Nature Communications*, **1,** 48, 1–11.

Brundrett, M. C. (2006). Understanding the roles of multifunctional mycorrhizal and endophytic fungi. In *Microbial Root Endophytes*, ed. B. Schulz, C. Boyle and T. N. Sieber. Berlin: Springer, pp. 281–298.

Busby, P. E., Soman, C., Wagner, M. R. *et al.* (2017). Research priorities for harnessing plant microbiomes in sustainable agriculture. *PLoS Biol*, **15**, e2001793.

Cannon, P. F. and Simmons, C. M. (2017). Diversity and host preference of leaf endophytic fungi in the Iwokrama Forest Reserve, Guyana. *Mycologia*, **94**, 210–220.

Card, S., Johnson, L. E. B., Teasdale, S. and Caradus, J. (2016). Deciphering endophyte behaviour: the link between endophyte biology and efficacious biological control agents. *FEMS Microbiology Ecology*, **92**, fiw114.

CGIAR (2016). Annual Report 2016: Innovations for Global Food Security. Montpellier, France: CGIAR.

Clement, S. L., Wilson, A. D., Lester, D. G. and Davitt, C. M. (1997). Fungal endophytes of wild barley and their effects on *Diuraphis noxi*a population development. *Entomologia Experimentalis et Applicata*, **82**, 275–281.

Collinge, D. B., Jørgensen, H. J. L., Latz, M. A. C. *et al.* (2019). Searching for novel fungal biological control agents for plant disease control among endophytes. In *Endophytes for a Growing World*, ed. T. R. Hodkinson, F. M. Doohan, M. J. Saunders and B. R. Murphy. Cambridge: Cambridge University Press, Chapter 2.

COP (2010). The Strategic Plan for Biodiversity 2011–2020 and the Aichi Biodiversity Targets (X/2). Conference of the Parties to the Convention on Biological Diversity, Tenth meeting, Nagoya, Japan.

Costa, D., Tavares, R. M., Baptista P. and Lino-Neto, T. (2019). The influence of endophytes on cork oak forests under a changing climate. In *Endophytes for a Growing World*, ed. T. R. Hodkinson, F. M. Doohan, M. J. Saunders and B. R. Murphy. Cambridge: Cambridge University Press, Chapter 12.

D'Arcy, C. J., Eastburn, D. M. and Schumann, G. L. (2001). *Illustrated Glossary of Plant Pathology. The Plant Health Instructor*. doi: 10.1094/PHI-I-2001-0219-01. www.apsnet.org/edcenter/illglossary

Desbrosses, G. J. and Stougaard, J. (2011). Root nodulation: a paradigm for how plant–microbe symbiosis influences host developmental pathways. *Cell Host Microbe*, **10**, 348–358.

Dobbelaere, S., Croonenborghs, A., Thys, A., Vande Broek, A. and Vanderleyden, J. (1999). Phytostimulatory effect of *Azospirillum brasilense* wild type and mutant strains altered in IAA production on wheat. *Plant Soil*, **212**, 153–162.

Estendorfer, J., Stempfhuber, B., Haury, P. *et al.* (2017). The influence of land use intensity on the plant-associated microbiome of *Dactylis glomerata* L. *Frontiers in Plant Science*, **8**, 930.

Fang, X., Jost, R., Finnegan, P. M. and Barbetti, M. J. (2013). Comparative proteome analysis of the strawberry-*Fusarium oxysporum* f. sp. *fragariae* pathosystem reveals early activation of defense responses as a crucial determinant of host resistance. *Journal of Proteome Research*, **12**, 1772–1788.

FAO, IFAD, UNICEF, WFP and WHO (2017). *The State of Food Security and Nutrition in the World 2017. Building Resilience for Peace and Food Security.* Rome: FAO.

Fesel, P. H. and Zuccaro, A. (2016). Dissecting endophytic lifestyle along the parasitism/mutualism continuum in *Arabidopsis*. *Current Opinion in Microbiology*, **32**, 103–112.

Finkel O. M., Castrillo, G., Herrera Paredes, S., Salas González, I. and Dangl, J. L. (2017). Understanding and exploiting plant beneficial microbes. *Current Opinion in Plant Biology*, **38**, 155–163.

G20 (2017). Group of 20 'Towards Food and Water Security: Fostering Sustainability, Advancing Innovation,' G20 Agriculture Ministers' Action Plan 2017. www.g20.utoronto. ca/2017/ 170122-agriculture-action-en.html

Gelorini, V., Verbeken, A., Lens, L. *et al.* (2012). Effects of land use on the fungal spore richness in small crater-lake basins of western Uganda. *Fungal Diversity*, **55**, 125–142.

Godfray, H. C., Beddington, J. R., Crute, I. R. *et al.* (2010). Food security: the challenge of feeding 9 billion people. *Science*, **327**, 812–818.

Gupta, S. and Chaturvedi, P. (2019). Enhancing secondary metabolite production in medicinal plants using endophytic elicitors: a case study of *Centella asiatica* (Apiaceae) and Asiaticoside. In *Endophytes for a Growing World*, ed. T. R. Hodkinson, F. M. Doohan, M. J. Saunders and B. R. Murphy. Cambridge: Cambridge University Press, Chapter 14.

Hallmann, J., Berg, G. and Schulz, B. (2007). *Isolation Procedures for Endophytic Microorganisms*. New York: Springer.

Hardoim, P. R., van Overbeek, L. S., Berg, G. *et al.* (2015). The hidden world within plants: ecological and evolutionary considerations for defining functioning of microbial endophytes. *Microbiology and Molecular Biology Reviews*, **79**, 293–320.

Hirsch, P. R. and Mauchline, T. H. (2012). Who's who in the plant root microbiome? *Nature Biotechnology*, **30**, 961–962.

Hodkinson, T. R. (2018). Evolution and taxonomy of the grasses (Poaceae): a model family for the study of species-rich groups. *Annual Plant Reviews Online*, doi: 10.1002/9781119312994.apr0622.

Hodkinson, T. R., Jones, M. B., Waldren, S. and Parnell, J. A. N., eds. (2011). *Climate Change Ecology and Systematics*. Cambridge: Cambridge University Press.

Hodkinson, T. R., Klaas, M., Jones, M. B., Prickett, R. and Barth, S. (2015). *Miscanthus*: a case study for the utilization of natural genetic variation. *Plant Genetic Resources*, **13**, 219–237.

Hodkinson, T. R. and Murphy, B. R. (2019). Endophytes for a growing world. In *Endophytes for a Growing World*, ed. T. R. Hodkinson, F. M. Doohan, M. J. Saunders and B. R. Murphy. Cambridge: Cambridge University Press, Chapter 1.

Høyer, A. K., Jørgensen, H. J. L., Jensen, B., Murphy, B. R. and Hodkinson, T. R. (2019). Emerging methods for biological control of barley diseases including the role of endophytes. In *Endophytes for a Growing World*, ed. T. R. Hodkinson, F. M. Doohan, M. J. Saunders and B. R. Murphy. Cambridge: Cambridge University Press, Chapter 5.

Hubbard, M., Germida, J. J. and Vujanovic, V. (2013). Fungal endophytes enhance wheat heat and drought tolerance in terms of grain yield and second generation seed viability. *Journal of Applied Microbiology*, **116**, 109–122.

IFPRI (2018). 2018 Global Food Policy Report. Washington, DC: International Food Policy Research Institute. doi: 10.2499/9780896292970.

Jensen, D. F., Karlsson, M., Sarrocco, S. and Vannacci, G. (2016). Biological control using microorganisms as an alternative to disease resistance. In *Plant Pathogen Resistance Biotechnology*, ed. D. B. Collinge. New York and London: Wiley Blackwell, pp. 341–363.

Johnson, L. J. and Caradus, J. R. (2019). The science required to deliver *Epichloë* endophytes to commerce. In *Endophytes for a Growing World*, ed. T. R. Hodkinson, F. M. Doohan, M. J. Saunders and B. R. Murphy. Cambridge: Cambridge University Press, Chapter 16.

Jumpponen, A. (2001). Dark septate endophytes: are they mycorrhizal? *Mycorrhiza*, **11**, 207–211.

Krell, V., Jakobs-Schoenwandt, D. and Patel, A. V. (2019). Application of formulated endophytic entomopathogenic fungi for novel plant protection strategies. In *Endophytes for a Growing World*, ed. T. R. Hodkinson, F. M. Doohan, M. J. Saunders and B. R. Murphy. Cambridge: Cambridge University Press, Chapter 3.

Le Cocq, K., Gurr, S. J., Hirsch, P. R. and Mauchline, T. H. (2016). Exploitation of endophytes for sustainable agricultural intensification. *Molecular Plant Pathology*, **18**, 469–473.

Lahiri, A., Douglas, G. C., Murphy, B. R. and Hodkinson, T. R. (2019). *In vitro* methods for plant-microbe interaction and biocontrol studies in european ash (*Fraxinus excelsior* L.). In *Endophytes for a Growing World*, ed. T. R. Hodkinson, F. M. Doohan, M. J. Saunders and B. R. Murphy. Cambridge: Cambridge University Press, Chapter 15.

Li, J., Ovakim, D. H., Charles, T. C. and Glick, B. R. (2000). An ACC deaminase minus mutant of *Enterobacter cloacae* UW4 no longer promotes root elongation. *Current Microbiology*, **41**, 101–105.

Lindahl, B. D., Nilsson, R. H., Tedersoo, L. *et al.* (2013). Fungal community analysis by high-throughput sequencing of amplified markers: a user's guide. *New Phytologist*, **199**, 288–299.

Ludwig-Müller, J. (2015). Plants and endophytes: equal partners in secondary metabolite production? *Biotechnology Letters*, **37**, 1325–1334.

Mandyam, K. G. and Jumpponen, A. (2005). Seeking the elusive function of the root-colonising dark septate endophytic fungi. *Studies in Mycology*, **53**, 173–189.

Mandyam, K. G. and Jumpponen, A. (2015). Plant-microbe interaction mutualism–parasitism paradigm synthesized from results of root-endophyte models. *Frontiers in Microbiology*, **5**, 1–13.

McNees, C. R., Greenhut, I. V., Law, A. D., Saleem, M. and Moe, L. A. (2019). Life within the leaf: ecology and applications of foliar bacterial endophytes. In *Endophytes for a Growing World*, ed. T. R. Hodkinson, F. M. Doohan, M. J. Saunders and B. R. Murphy. Cambridge: Cambridge University Press, Chapter 10.

Melotto, M., Underwood, W. and He, S. Y. (2008). Role of stomata in plant innate immunity and foliar bacterial diseases.

Annual Review of Phytopathology, **46**, 101–122.

Meshram, V. and Gupta, M. (2019). Endophytic fungi: a quintessential source of potential bioactive compounds. In *Endophytes for a Growing World*, ed. T. R. Hodkinson, F. M. Doohan, M. J. Saunders and B. R. Murphy. Cambridge: Cambridge University Press, Chapter 13.

Murphy, B. R., Doohan, F. M. and Hodkinson, T. R. (2014a). Fungal endophytes of barley roots. *The Journal of Agricultural Science*, **152**, 602–615.

Murphy, B. R., Doohan, F. M. and Hodkinson, T. R. (2014b). Yield increase induced by the fungal root endophyte *Piriformospora indica* in barley grown at low temperature is nutrient limited. *Symbiosis*, **62**, 29–39.

Murphy, B. R., Doohan, F. M. and Hodkinson, T. R. (2015a). Fungal root endophytes of a wild barley species increase yield in a nutrient-stressed barley cultivar. *Symbiosis*, **65**, 1–7.

Murphy, B. R., Doohan, F. M. and Hodkinson, T. R. (2015b). Persistent fungal root endophytes isolated from a wild barley species suppress seed-borne infections in a barley cultivar. *Biocontrol*, **60**, 281–292.

Murphy, B. R., Martin Nieto, L., Doohan, F. M. and Hodkinson, T. R. (2015c). Fungal endophytes enhance agronomically important traits in severely drought-stressed barley. *Journal of Agronomy and Crop Science*, **201**, 419–427.

Murphy, B. R., Martin Nieto, L., Doohan, F. M. and Hodkinson, T. R. (2015d). Profundae diversitas: the uncharted genetic diversity in a newly studied group of fungal root endophytes. *Mycology*, **6**, 139–150.

Murphy, B. R., Doohan, F. M. and Hodkinson, T. R. (2017a). A seed dressing combining fungal endophyte spores and fungicides improves seedling survival and early growth in barley and oat. *Symbiosis*, **71**, 69–76.

Murphy, B. R., Hodkinson, T. R. and Doohan, F. M. (2017b). A fungal endophyte consortium counterbalances the negative effects of reduced nitrogen input on the yield of field-grown spring barley. *The Journal of Agricultural Science*, **155**, 1324–1331.

Murphy, B., Doohan, F. M. and Hodkinson, T. R. (2018). From concept to commerce: developing a successful fungal endophyte inoculant for agricultural crops. *Journal of Fungi*, **4**, 24.

Murphy, B. R., Doohan F. M. and Hodkinson, T. R. (2019). Prospecting crop wild relatives for beneficial endophytes. In *Endophytes for a Growing World*, ed. T. R. Hodkinson, F. M. Doohan, M. J. Saunders and B. R. Murphy. Cambridge: Cambridge University Press, Chapter 18.

Newsham, K. K. (2011). A meta-analysis of plant responses to dark septate root endophytes. *New Phytologist*, **190**, 783–793.

O'Hanlon, K. (2019). Plant growth-promoting bacteria field trials in Europe. In *Endophytes for a Growing World*, ed. T. R. Hodkinson, F. M. Doohan, M. J. Saunders and B. R. Murphy. Cambridge: Cambridge University Press, Chapter 17.

Ownley, B. H., Pereira, R. M., Klingeman, W. E., Quigley, N. B. and Leckie, B. M. (2004). *Beauveria bassiana*, a dual purpose biocontrol organism, with activity against insect pests and plant pathogens. In *Emerging Concepts in Plant Health Management*, ed. R. T. Lartey and A. J. Caesar. Kerala, India: Research Signposts, pp. 255–269.

Ownley, B. H., Gwinn, K. D. and Vega, F. E. (2010). Endophytic fungal entomopathogens with activity against plant pathogens: ecology and evolution. *BioControl*, **55**, 113–128.

Paris Agreement on Climate Change (2016). C.N.92.2016.Treaties-XXVII.7.d. Geneva: United Nations Treaty Collection.

Parnell, J. J., Berka, R., Young, H. A. *et al.* (2016). From the lab to the farm: an industrial perspective of plant beneficial microorganisms. *Frontiers in Plant Science*, **7**, 1110.

Partida-Martinez, L. P. and Heil, M. (2011). The microbe-free plant: fact or artifact? *Frontiers in Plant Science*, **2,** 100, 1–16.

Pickett, J. A., Woodcock, C. M., Midega, C. A. O. and Khan, Z. R. (2014). Push-pull farming systems. *Current Opinion in Biotechnology*, **26**, 125–132.

Porter, J. K. (1995). Analysis of endophyte toxins: fescue and other grasses toxic to livestock. *Journal of Animal Science*, **73**, 871–880.

Rasmussen, S., Parsons, A. J. and Jones, C. S. (2012). Metabolomics of forage plants: a review. *Annals of Botany*, **110**, 1281–1290.

Rathore, R., Germaine, K. J., Forristal, P. D., Spink, J. and Dowling, D. N. (2019). Meta-omics approach to unravel the endophytic bacterial communities of *Brassica napus* and other agronomically important crops in response to agricultural practices. In *Endophytes for a Growing World*, ed. T. R. Hodkinson, F. M. Doohan, M. J. Saunders and B. R. Murphy. Cambridge: Cambridge University Press, Chapter 11.

Richter-Heitmann, T., Eickhorst, T., Knauth, S., Friedrich, M. W. and Schmidt, H. (2016). Evaluation of strategies to separate root-associated microbial communities: a crucial choice in rhizobiome research. *Frontiers in Microbiology*, **7**, 773.

Rockström, J., Steffen, W., Noone, K. *et al.* (2009). Planetary boundaries: exploring the safe operating space for humanity. *Ecology and Society*, **14**, 32.

Rodriguez, R. J., White, J. F., Arnold, A. E. and Redman, R. S. (2009). Fungal endophytes: diversity and functional roles. *The New Phytologist*, **182**, 314–30.

Rosegrant, M. W., Ringler, C. and Zhu, T. (2009). Water for agriculture: maintaining food security under growing scarcity. *Annual Reviews*, **34**, 205–222.

Ryu, C. M., Farag, M., Hu, C. H. *et al.* (2004). Bacterial volatiles induce systemic resistance in Arabidopsis. *Plant Physiology*, **134**, 1017–1026.

Saikkonen, K., Young, C. A., Helander, M. and Schardl, C. L. (2016). Endophytic *Epichloë* species and their grass hosts. *Plant Molecular Biology*, **90**, 665–675.

Soares, M. A., Li, H.-Y., Kowalski, K. P. *et al.* (2016). Evaluation of the functional roles of fungal endophytes of *Phragmites australis* from high saline and low saline habitats. *Biological Invasions*, **18**, 2689–2702.

Schulthess, F. M. and Faeth, S. H. (1998). Distribution, abundances, and associations of the endophytic fungal community of Arizona fescue (*Festuca arizonica*). *Mycologia*, **90**, 569–578.

Schulz, B. and Boyle, C. (2006). What are endophytes? *Microbial Root Endophytes*, **9**, 1–14.

Schulz, B., Rommert, A.-K., Dammann, U., Aust, H.-J. and Strack, D. (1999). The endophyte–host interaction: a balanced antagonism? *Mycological Research*, **103**, 1275–1283.

Stone, J. K., Polishook, J. O. N. D. and White, J. F. (2004). Endophytic fungi. In *Biodiversity of Fungi: Inventory and*

Monitoring Methods, ed. G. M. Mueller, G. F. Bills and M. S. Foster. Burlington, MA: Elsevier Academic Press, pp. 241–270.

Strobel G. (2018). The emergence of endophytic microbes and their biological promise. *Journal of Fungi*, **4**, 57.

Timmusk, S., Behers, L., Muthoni, J., Muraya, A. and Aronsson, A.-C. (2017). Perspectives and challenges of microbial application for crop Improvement. *Frontiers in Plant Science*, **8**, 49.

United Nations (2015). Transforming Our World: The 2030 Agenda For Sustainable Development. *A/RES/*70/1. Geneva: United Nations. sustainabledevelopment.un.org.

van der Heijden, M. G. A., Bardgett, R. D. and Van Straalen, N. M. (2008). The unseen majority: soil microbes as drivers of plant diversity and productivity in terrestrial ecosystems. *Ecology Letters*, **11**, 296–310.

van der Heijden, M. G. A., Dombrowski, N. and Schlaeppi, K. (2017). Continuum of root–fungal symbioses for plant nutrition. *Proceedings of the National Academy of Sciences of the United States of America*, **114**, 11574–11576.

Vega, F. E., Goettel, M. S., Blackwell, M. *et al.* (2009). Fungal entomopathogens: new insights on their ecology. *Fungal Ecology*, **2**, 149–159.

Vessey, J. K. (2003). Plant growth-promoting rhizobacteria as biofertilizers. *Plant and Soil*, **255**, 571–586.

Waller, F., Achatz, B., Baltruschat, H. *et al.* (2005). The endophytic fungus *Piriformospora indica* reprograms barley to salt-stress tolerance, disease resistance, and higher yield. *Proceedings of the National Academy of Sciences of the United States of America*, **102**, 13386–13391.

Waqas, M., Khan, A. L., Kamran M. *et al.* (2012). Endophytic fungi produce gibberellins and indoleacetic acid and promotes host-plant growth during stress. *Molecules*, **17**, 10754–10773.

Wennstrom, A. (1994). Endophyte: the misuse of an old term. *Oikos*, **71**, 535–536.

White, J. F., Cole, G. T. and Morgan-Jones, G. (1987). Endophyte-host associations in forage grasses. VI. A new species of *Acremonium* isolated from *Festuca arizonica*. *Mycologia*, **79**, 148–152.

White, J. F., Morgan-Jones, G. and Morrow, A. C. (1993). Taxonomy, life cycle, reproduction and detection of *Acremonium* endophytes. *Agriculture, Ecosystems and Environment*, **44**, 13–37.

Widiantini, F. and Franco, C. (2019). Microbispora Dominate Diversity of Endophytic Actinobacteria from Australian Rice Plants. In *Endophytes for a Growing World*, ed. T. R. Hodkinson, F. M. Doohan, M. J. Saunders and B. R. Murphy. Cambridge: Cambridge University Press, Chapter 8.

Wiewióra, B., Żurek, G., Pro, M., Żurek, M. and Schmidt, J. (2011). Relations between site conditions and endophyte colonization of grasses in Poland. *Journal of Life Sciences*, **5**, 831–837.

Wiewióra, B., Żurek, G. and Żurek, M. (2015). Endophyte-mediated disease resistance in wild populations of perennial ryegrass (*Lolium perenne*). *Fungal Ecology*, **15**, 1–8.

Wilberforce, E. M., Boddy, L., Griffiths, R. and Griffith, G. W. (2003). Agricultural management affects communities of culturable root-endophytic fungi in temperate grasslands. *Soil Biology and Biochemistry*, **35**, 1143–1154.

Wilson, A. D., Clement, S. L. and Kaiser, W. J. (1992). Endophytic fungi in a *Hordeum* germplasm collection. *Plant Genetic Resources*, **87**, 1–4.

Wilson, D. (1995). Endophyte: the evolution of a term, and clarification of its use and definition. *Oikos*, **73**, 274–276.

Yakti, W., Andrade-Linares, D. R., Ngwene, B. *et al.* (2019). Phosphate nutrition in root-fungus interactions. In *Endophytes for a Growing World*, ed. T. R. Hodkinson, F. M. Doohan, M. J. Saunders and B. R. Murphy. Cambridge: Cambridge University Press, Chapter 6.

Żurek, G., Wiewióra, B., Żurek, M. and Łyszczarz, R. (2017). Environmental effect on *Epichloë* endophyte occurrence and ergovaline concentration in wild populations of forage grasses in Poland. *Plant and Soil*, **410**, 383–399.

PART II

Role of Endophytes in Growth and Biotic and Abiotic Stress Resistance

2

Searching for Novel Fungal Biological Control Agents for Plant Disease Control Among Endophytes

DAVID B. COLLINGE, HANS J. L. JØRGENSEN, MEIKE A. C. LATZ, ANDREA MANZOTTI, FANI NTANA, EDWARD C. ROJAS AND BIRGIT JENSEN

Abstract

There are increasing efforts aiming to utilise endophytes as biological control agents (BCAs) to improve crop production. However, reliability remains a major practical constraint for the development of novel BCAs. Many organisms are adapted to their specific habitat; it is optimistic to expect that a new organism added can find a niche or even out-compete those adapted and already present. Our approach for isolating novel BCAs for specific plant diseases is therefore to look in healthy plants in a habitat where disease is a problem, since we predict that it is more likely to find competitive strains among those present and adapted. *In vitro* inhibitory activities often do not correlate with *in planta* efficacy, especially since endophytes rely on intimate plant contact. They can, however, be useful to indicate modes of action. We therefore screen for *in planta* biological activity as early as possible in the process in order to minimise the risk of discarding valuable strains. Finally, some fungi are endophytic in one situation and pathogenic in another (the mutualism–parasitism continuum). This depends on their biology, environmental conditions, the formulation of inoculum, the health, developmental stage and cultivar of the host plant, and the structure of the plant microbiome.

2.1 Introduction

It is a major challenge to increase crop yields worldwide in a sustainable manner. There is an increasing wish to reduce the use of pesticides, the application of inorganic fertiliser and the need for irrigation, as these factors deplete natural resources and can have negative effects on the environment. We need to meet the challenges of climate change, as well as world population growth and other demographic factors, such as urbanisation and economic growth. The combination of these factors results in increased demand for both water resources and meat, hence of plants as food and fodder. Together, these many factors place high demands on agriculture, motivating research effort to improve yields in a sustainable manner: the global challenge is to produce more food from less agricultural land with reduced inputs.

Plant diseases reduce yield and profitability, both in terms of direct losses, product quality and indirectly through inputs into the crop. However, the grower has many practical tools for combating plant diseases, including cultural practices such as crop rotation, fertilisation, pruning, irrigation and tillage. Chemical control agents are also of seminal importance for many diseases: these can be directly antimicrobial, e.g. fungicides, or act as inducers of plant defence responses and therefore disease resistance, e.g. benzothiadiazole (BTH, Bion®) and probenazole. However, there are increasing problems with fungicide resistance in many important pathogens, which means that some available fungicides have become ineffective for specific diseases (Lucas *et al.*, 2015). Furthermore, the number of approved fungicides is being reduced due to side effects, e.g. accumulation in the environment, conflicts with medicinal use (Berger *et al.*, 2017); at the same time few new products are being developed and approved.

Another important strategy is the use of disease resistance. This can be achieved through conventional plant breeding by introducing resistance genes from within the same plant species (e.g. from landraces) or closely related plant species. Effective sources of disease resistance are often unavailable for conventional plant breeding programmes, especially for necrotrophic pathogens. Resistance can also be achieved through biotechnological approaches, e.g. genetic engineering, to add genes which encode either antimicrobial agents or regulators of defence mechanisms (Collinge *et al.*, 2010, 2016). More recently, an increased understanding of the biology of microbial pathogenicity has led to new breeding technologies, which can manipulate pathogens, for example, by host-induced gene silencing (HIGS) or by using gene editing techniques (site-directed mutation), e.g. CRISPR-Cas9 (Nowara *et al.*, 2010; Collinge, 2018). Even the combination of these approaches is insufficient to control many important diseases of key crops and, furthermore, there is a reluctance to use the new breeding technologies in some parts of the world (Collinge, 2018). There is a need to be innovative in finding new approaches, especially since many important pathogens are well able to adapt to, and can therefore overcome, both disease resistance and pesticides (e.g. *Zymoseptoria tritici, Phytophthora infestans*; Lucas *et al.*, 2015).

An underdeveloped and underutilised means of disease control is biological control. Biological control is defined as direct or indirect inhibition of a disease or the pathogen causing the disease, by another organism (antagonist) or group of organisms. Biological control is seen as an increasingly attractive tool to complement the more traditional control strategies both in organic and conventional agriculture. Furthermore, biological control can be especially useful when used in organic agriculture, as well as in systems where disease resistance is unavailable in the host and where fungicide resistance is prevalent in the pathogen populations. Some biological control organisms (BCAs) can even supplement fertilisers to stimulate plant growth (Coleman-Derr and Tringe, 2014; Jensen *et al.*, 2016a). The exploitation of beneficial microorganisms and the management of the plant microbiome can contribute to these needs, both through biological control such as by stimulation of host growth. Although development of a product may be achieved without understanding how it works, there are many fascinating fundamental questions about the nature and physiological mechanisms underlying plant–microbe interactions which beg study (Latz *et al.*, 2018). There are also potential risks associated with biological control such as the negative consequences of introducing organisms into new habitats and production systems which need to be considered before permitting the use of a new biological control agent. For these reasons, the regulation of commercialisation of novel BCAs is stringent, and therefore expensive (Anon, 2009).

As a division of the plant microbiome, endophytes represent a largely unexploited source of organisms with potential for the biological control of pests and pathogens and for helping plants adapt to abiotic stress. One of the main drawbacks of biological control is inconsistency under variable environments. Since the endophyte is inside the plant, it is potentially better protected from the external environment and, therefore, we consider it likely that at least some endophytes will prove to be more reliable BCAs than epiphytic phyllosphere or rhizosphere organisms. Here, we provide a resumé of current knowledge of the similarities and differences between commensal, mutualistic and pathogenic interactions between plants and microorganisms, with emphasis on different approaches for isolating and studying endophytic fungi. We favour an ecological approach, namely isolating BCAs from plants which are coping with a stressful environment. We isolate from different plant organs (and mutants), to which BCAs can be returned and utilised for plant disease control. Another approach being taken, which we do not address here, is to try to manipulate microbiomes rather than using individual organisms, e.g. by manipulating agricultural practice and choice of cultivar. This is the particular focus of studies using bacterial microbiomes (Finkel *et al.*, 2017), but we predict an increase in studies on fungal microbiomes will follow.

2.2 What Is an Endophyte?

Endophytes (*endo* = within, *phyto* = plant), a term coined by the father of plant pathology Anton de Bary in 1886, are now defined as microorganisms capable of colonising inner parts of plants without causing disease symptoms (Schulz and Boyle, 2005; Kusari *et al.*, 2012; Wani *et al.*, 2015). The plant is not host to a single microorganism, but hosts many different microorganisms simultaneously; collectively, these are termed the endomicrobiome. The endomicrobiome is diverse and comprises, e.g. archaea (Moissl-Eichinger *et al.*, 2018), bacteria (Furnkranz *et al.*, 2012; Aguiar-Pulido *et al.*, 2016) and diverse fungi including both ascomycetes (Newsham, 2011) and basidiomycetes (Weiß *et al.*, 2016; Hardoim *et al.*, 2015), as well as some endophytic oomycetes (Ploch and Thines, 2011). We have much to learn about the individual microorganisms and their interactions in ecological communities of microorganisms comprising the endomicrobiome.

A specific organism might be endophytic in one situation and saprotrophic, epiphytic or even pathogenic in others (Hardoim *et al.*, 2015; Mukherjee *et al.*, 2012). In the latter case, there is the issue of whether it is possible to distinguish whether an organism is endophytic or a quiescent or latent pathogen (Louarn *et al.*, 2013; Zeilinger *et al.*, 2016; Lofgren *et al.*, 2018). Indeed, the term endophyte covers different ecological strategies (latent pathogen, beneficial symbiont, commensal passenger) used by different organisms in different situations (species or growth conditions), which result in what we observe as the endophytic habit (Schulz and Boyle, 2005). It can be difficult to assign a particular interaction one of these specific biological terms and this situation is encompassed by the concept of 'mutualism–parasitism continuum' (Müller and Krauss, 2005; Schulz and Boyle, 2005).

Different strains of specific taxa (e.g. *Fusarium oxysporum*) can be endophytic BCAs or pathogens on the same host. Quite closely related taxa can include both pathogens or endophytes (Fravel *et al.*, 2003; Ma *et al.*, 2010). Some organisms are pathogens of one host, and endophytes in another closely related host species, and can thus exhibit different roles, ranging from latent pathogens or endophyte to pathogen, e.g. *Colletotrichum* spp. and *Fusarium* spp. (De Silva *et al.*, 2017; Lofgren *et al.*, 2018). In contrast, *Ramularia collo-cygni* is an endophyte in wheat but a pathogen in barley under certain physiological conditions or genetic backgrounds (McGrann *et al.*, 2014, 2016). *Fusarium graminearum* is a pathogen of small grain cereals, but apparently an endophyte in other species of grass (Lofgren *et al.*, 2018). What we do not know is whether they are commensals, deriving nutrition or other benefits without causing measurable harm (i.e. quiescent pathogens), and therefore technically not symbionts according to our narrow definition, or whether they are, in fact, beneficial symbionts in these alternative hosts. Furthermore, some pathogens arguably exhibit an endophytic growth habit in the first part of the infection cycle and cause disease later, e.g. *Zymoseptoria tritici* (formerly *Septoria tritici* (anamorph) *Mycosphaerella graminicola* (telomorph)) (Shetty *et al.*, 2003; Sánchez-Vallet *et al.*, 2015). Indeed, it is perhaps useful to exclude the potential pathogens from the definition and confine the term to organisms that, in a specific situation, have a mutualistic or at least commensal relationship with their host. An adapted version of Koch's postulates can be useful to exclude organisms which cause disease when reintroduced into the host (Card *et al.*, 2016).

Genotype can also be important, thus many strains of *Fusarium oxysporum* are endophytic, but possession of some 'B chromosomes' can confer pathogenicity (Ma *et al.*, 2010). Does the plant decide whether the organism is a pathogen or endophyte by the way it reacts to the colonising organism or is the interplay more complex? Interestingly, there is increasing evidence to suggest that some effectors play a role in competitive fungal–fungal interactions (Rovenich *et al.*, 2014). Furthermore, some endophytes can not only induce resistance but, at least in some cases, susceptibility to pathogens (Houterman *et al.*, 2008; Kurose *et al.*, 2012). Thus, the same endophyte may influence different interactions in opposite ways.

2.3 How Do We Find Endophytes?

There are two major approaches to identifying endophytes in a plant: cultivation-dependent (e.g. isolation) and cultivation-independent (e.g. metagenomics) methods. Each approach has advantages and disadvantages (Figure 2.1). It is important that the tissue from which endophytes are isolated is fresh and shows few or no visible disease symptoms. To ensure that organisms isolated are endophytic and not epiphytic, the tissue needs to be surface sterilised. This is especially important for roots and tubers. This can be achieved in several ways (e.g. alcohol, sodium hypochlorite treatment) and there is a compromise between using too harsh a treatment, which penetrates the tissue killing the microorganisms within, and too gentle a treatment, which allows too many epiphytic or rhizospheric organisms to survive (Busby *et al.*, 2016b). Standardising these conditions before working with a new species or tissue is highly recommended as harsh treatments can reduce fungal survival inside the tissue or mild treatments can result in false positives.

Clearly, if an organism is to be used in crop production, it has to be cultivable. Many endophytes are not. In order to increase the chances of isolating, hence cultivating and utilising endophytic microorganisms, the growth media can be supplemented with plant extracts. This has been shown to increase the number of endophytes recovered (Eevers *et al.*, 2015; Murphy *et al.*, 2015). In our experience, plant-extract-complemented media do not improve yield or diversity in the isolation of endophytes from cereals (Latz *et al.*, unpublished). However, different fungi have more or less specific requirements for the media that they will grow on, and, furthermore, the isolation process itself may inhibit the growth of some fungi, since

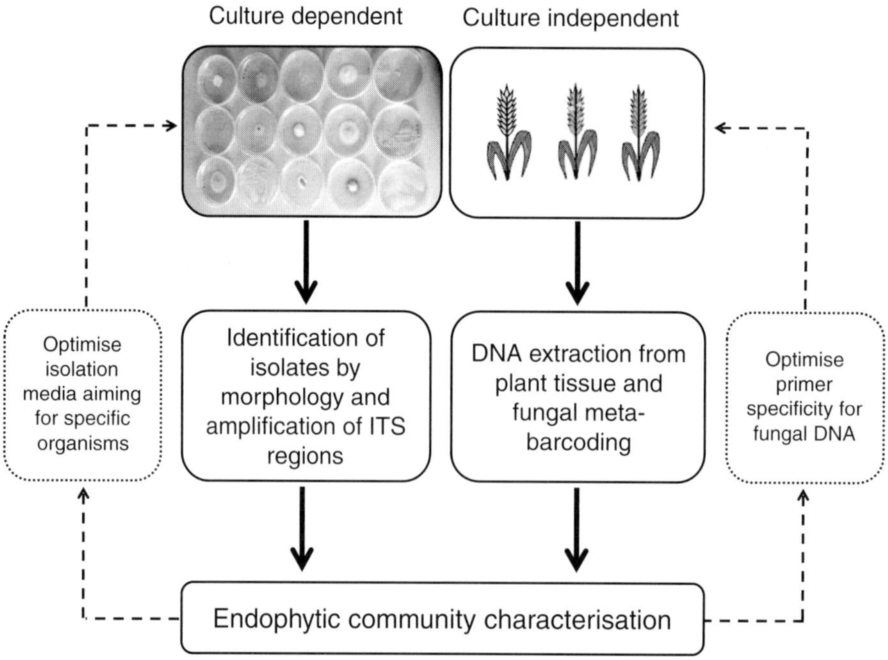

Figure 2.1 Endophytic community characterisation using culture-dependent and culture-independent methods. Both methods can be optimised in order to obtain a more accurate description of fungal microbiome.

antimicrobial compounds can be induced and/or released from the host tissues on damage. Moreover, different microorganisms grow at different rates and some will simply be outcompeted by others. As this is especially a problem with bacteria outcompeting fungi, antibacterial compounds may be incorporated in media when the target endophyte is fungal, although this may not itself be without negative effect. Low nutrient media can favour slow-growing fungi in relation to fast-growing fungi and bacteria. In conclusion, isolation-based approaches have little value in quantitative floristic diversity assessments, but they remain our only door to finding potential biocontrol agents. Additionally, it is important to standardise different techniques and growth media for isolation in each case. Once isolated, the fungi can be identified by morphological traits combined with marker gene sequencing (typically the ribosomal genes).

Metagenomics as a methodology can be used to obtain a more objective and quantitative, although still descriptive, assessment of the endophytic fungal microbiome *in situ* (Diaz *et al.*, 2012; Tian *et al.*, 2015; Abdelfattah *et al.*, 2016; Sapkota *et al.*, 2017). The information obtained may even be used as a guide for choosing specific media for targeted isolation of specific taxa that appear to be enriched in the microbial community. For fungal endophytes (and indeed most eukaryotes), the most common way to perform a fungal community analysis is to amplify parts of the ribosomal genes, for example, the internal transcribed spacer (ITS)-1 (Nicolaisen *et al.*, 2014) or ITS-2 (Hertz *et al.*, 2016; Gdanetz and Trail, 2017), followed by deep-sequencing and bioinformatical analyses in order to investigate the composition of the endophytic communities without the need to isolate living fungi from the host plant. The main challenge with this approach is finding primers which recover most of the fungal diversity without creating phylum biases, while at the same time avoiding targeting the host genome. This has been a constant problem, especially in polyploid species such as wheat. Many pairs of primers have been suggested using *in silico* analysis, but no real comparison has been performed on DNA samples (Toju *et al.*, 2012). Compromises will therefore be necessary!

Shot-gun metagenomics or whole metagenome shotgun sequencing concerns deep sequencing of the mixed genomes of the organisms present in the sampled material (Aguiar-Pulido *et al.*, 2016; Kaul *et al.*, 2016). This approach, initially applied for the study of bacterial endomicrobiomes (Bulgarelli *et al.*, 2012; Lundberg *et al.*, 2012; Hardoim *et al.*, 2015), eliminates the bias unavoidably introduced when selecting amplification primers since the sequencing process itself is objective. The technique is, however, currently more expensive, both in terms of the chemical process and in the need for computing time during bioinformatics analysis; huge amounts of sequence data are generated, which have to be analysed. Thus, the quality of the genome sequence databases is a constraint on the quality of the analysis: although the number of sequenced genomes in the databases increases with time, the extent and quality of annotation of the sequences in them lags, and a large proportion of the data therein are not yet assigned to specific organisms.

A third approach, metatranscriptomics, uses RNA as the substrate for sequencing, i.e. gene transcripts rather than the genome (Aguiar-Pulido *et al.*, 2016; Kaul *et al.*, 2016). This means that the results reflect the physiological activity of organisms and therefore give more information about the actual plant microbiome interactions, and not just an image of the microbial communities within the plant (Kaul *et al.*, 2016). This approach will also indicate which gene families participate in or are necessary for the endophytic lifestyle, for example, genes involved in hormone production or action, effector proteins and specialised metabolites (see below). The value of the analysis depends on the choice of physiological conditions; an important component of the microbiota may have latent infection periods and may therefore be unrepresented in the results.

It should, at least to some extent, be possible to combine these approaches in order to use amplicon sequencing and whole metagenome sequencing (and metatranscriptomics) to identify endophytes with desired traits (functions or e.g. metabolites), and chose, or even design, isolation methods tailored to their real or perceived needs, if such interesting organisms are indeed culturable. Depending on the stamina of the researcher, the classical microbiological approach can easily yield 40–120 fungal taxa per sample, e.g. (Comby *et al.*, 2016; Kernaghan *et al.*, 2017; Kosawang *et al.*, 2018), and amplicon sequencing approaches typically reveals 200 or more within a study (Nicolaisen *et al.*, 2014).

2.4 What Do Endophytes Do to the Plant?

Endophytes can support growth and health of the host plant via different physiological mechanisms. Thus, some fungal endophytes can stimulate abiotic stress tolerance (e.g. drought and salt stress) and plant growth by altering hormone balance (see below) or nutrient acquisition (Zeilinger *et al.*, 2016). Superficially, it may seem surprising that some endophytes do this, but there is no need to look for a more complicated reason for why this is, than that a happier plant can support a higher biomass of a specific endophyte. In other words, in an ecological sense, the endophyte may find a safe niche within the host tissue (protection and reduced competition for nutrients), while improving the plant's performance under specific stress conditions. Survival and proliferation of the endophyte may simply depend on growth and survival of the host. These are by no means universal properties of endophytes, and the majority may only be neutral commensal passengers which do not measurably influence their host. Furthermore, the ability to stimulate plant growth is by no means exclusive to endophytes as plant growth-promoting organisms are also found among rhizosphere microorganisms and mycorrhizal fungi.

2.5 How Do Pathogens and Endophytes Differ?

Evidence coming from overlapping transcription profiles of endophyte- and pathogen-colonised plants indicates a level of similarity between the host response to colonisation and infection, at least at the initial stages of both types of interactions (Guimil *et al.*, 2005; Schäfer *et al.*, 2009; Zamioudis and Pieterse, 2011). In other words, at least some endophytes can induce plant immune responses through microbe-associated molecular patterns (MAMPs) that are recognised by the plant in a similar way to those produced by pathogens (Latz *et al.*, 2018). It has long been speculated that, as for pathogens (Lo Presti *et al.*, 2015), one strategy used by endophytes to avoid detection is to supress plant immunity using effectors (Sánchez-Vallet *et al.*, 2013; Rovenich *et al.*, 2014), and this can even result in induced susceptibility to pathogens (e.g. Kurose *et al.*, 2012; Busby *et al.*, 2016a). So, what are the actual differences between endophytes and pathogens? Is the intention of the pathogen necessarily bad, or is it just a question of compatibility, i.e. expression of genes often associated with pathogenicity such as those encoding cell wall degrading enzymes? The fact that at least some organisms can behave as an endophyte in one situation and as pathogen in others suggests that there is: (1) a continuum from beneficial to harmful (and vice versa); and (2) no simple answer to the question 'How do pathogens and endophytes differ?'

The complexity of the mechanisms of recognition and response to fungi can be illustrated with a pathogen-related example, which may be predicted to apply to endophytes too: chitin is a MAMP where analyses of plant–microorganism interactions have demonstrated several layers of interaction (signal and reaction, counter-reaction, new signal and so on). *Passalora fulva* (formely *Cladosporium fulvum*) is considered to be a pathogen of tomato which displays an initially endophytic lifestyle (traditionally considered to be biotrophic by some authors) in the apoplast without making specific infection structures. Several seminal studies have demonstrated multiple layers of protection used by this pathogen to avoid being recognised by the host or to counter the triggered immune response: (1) Fungal effectors (chitin-binding lectins, including Avr4) coat the hyphal surface and act as inhibitors of the plant defence enzyme chitinase (van den Burg *et al.*, 2006). (2) The effector Avr4 is recognised by the host receptor Cf4 (a resistance gene) triggering immunity (i.e. ETI). The CERK receptor-like protein kinase (3) binds chitin fragments released by host chitinase to enhance the induced immunity, but (4) the pathogen produces the effector Ecp6 to scavenge chitin fragments (de Jonge *et al.*, 2010; Malinovsky *et al.*, 2014).

2.6 Hormones in Plant–Endophyte Interactions

It is well established that hormones, such as abscisic acid (ABA), auxins (e.g. indole acetic acid, IAA), brassinosteroids, cytokinins (many), ethylene, gibberellins, jasmonates (JAs), salicylic acid (SA) and strigolactones are important for plant development and modulation of plant defences against pathogens (Pieterse *et al.*, 2012; De Vleesschauwer *et al.*, 2013; Großkinsky *et al.*, 2016; Ma and Ma, 2016). Pathogens (and presumably also endophytes) modulate hormone levels using several mechanisms. Thus, they can (1) stimulate the host to increase or decrease the production of specific hormones; (2) produce and degrade hormones themselves; (3) insert biosynthetic genes for phytohormones into the host chromosomes; and (4) produce molecules which emulate hormone action. Examples are emerging that many hormones, e.g. auxins (Hilbert *et al.*, 2012), JAs, gibberellins (Schäfer *et al.*, 2009; Jacobs *et al.*, 2011), ABA (Peskan-Berghöfer *et al.*, 2015), ethylene (Khatabi *et al.*, 2012) and SA (Alonso-Ramírez *et al.*, 2014) play a role in regulating the ability of endophytes to colonise the host tissue.

Several hormones are involved in regulating the establishment and maintenance of infections with the model root endophyte *Serendipita indica* (formerly

Piriformospora indica) which can form associations with many important crop plants, e.g. wheat, barley and tomato (Weiß *et al.*, 2016), and model plants, i.e. *Arabidopsis* (Stein *et al.*, 2008) and *Brachypodium* (Ye *et al.*, 2014). *Serendipita indica* is considered to be a root endophyte, although it does induce necrosis at the site of colonisation. *Serendipita indica* has been developed as a model endophyte through the combination of wide host range and its positive effects on many hosts. Experimentally, a fairly well annotated genomic sequence and molecular genetic tools are available, since the fungus can be cultured on artificial media (Zuccaro *et al.*, 2009, 2011). Studies in *Arabidopsis* using the JA-insensitive host mutants *jai1* and *jar1* demonstrated that *S. indica* lost the ability to supress host immunity. On the other hand, mutants involved in gibberellin signalling, a *della* mutant in which gibberellin levels are high and the *ga1-6* mutant (impaired in biosynthesis), exhibited higher and lower levels of infection, respectively, compared to the wild-type plants (Jacobs *et al.*, 2011). The effects of auxin on *S. indica* infection were studied in barley (Hilbert *et al.*, 2012) using gene silencing to compromise auxin production in the fungus: these strains showed reduced rates of early infection of barley roots, but overall levels of infection were unaffected. JA signalling is often related to ethylene signalling and it has been shown that ethylene production is induced by *S. indica* in *Arabidopsis* and barley (Khatabi *et al.*, 2012). *Arabidopsis* mutants with increased ethylene production and ethylene-induced defence mechanisms were more susceptible to *S. indica* infection. *S. indica* treatment of barley resulted in the induction of the expression of genes involved in the metabolism of several hormones including the biosynthesis of auxins, brassinosteroids and gibberellin as well as ABA responsive genes (Schäfer *et al.*, 2009). The possible contribution of hormones produced by the fungus was not taken into account in this study, as described above. However, other studies have demonstrated hormone production by fungi (Hilbert *et al.*, 2012; Khan *et al.*, 2012; Waqas *et al.*, 2015).

The effects of altered hormone levels are both subtle and complex; indeed, the same hormones can stimulate defence or pathogenicity in interactions with different pathogens, and different hormones can modulate each other's effects by complex crosstalk (Pieterse *et al.*, 2012). Most studies address individual hormones in isolation. In some cases, the concerted action (crosstalk) between hormones, e.g. IAA and gibberellins (Waqas *et al.*, 2012), has been demonstrated. Indeed, some endophytes and other beneficial fungi (and bacteria) produce or induce the production of specific plant hormones to modulate this crosstalk for their benefit and often also for the host plants' benefit (Schäfer *et al.*, 2009; Evangelisti *et al.*, 2014; Gutjahr, 2014). Techniques have been developed that allow monitoring of several hormones simultaneously (Kojima *et al.*, 2009; Ionescu *et al.*, 2017). Only by doing this, ultimately, it will be possible to unravel the roles, effects and the mode of action of hormones exploited in the interactions between endophytes and plants under different physiological conditions. We still have much to learn!

2.7 Effectors in Plant–Endophyte Interactions

Effectors are traditionally defined as pathogen proteins introduced into the host cell in order to repress host defence (plant immunity). Indeed many, especially hemibiotrophic/biotrophic filamentous pathogens such as the oomycete *Phytophthora infestans* and ascomycete *Blumeria graminis*, have a huge arsenal of effector proteins for this purpose (Ahmed *et al.*, 2016). Plants need to defend themselves from attack and, taxonomically, endophytes are often related to pathogens, producing the same MAMPs (that can trigger defence, MTI or MAMP-triggered immunity) (Collinge *et al.*, 2016). It is necessary for endophytes to avoid triggering the induction of plant defences, and effector proteins are among the tools needed to do this, indeed candidate effectors are now being found in some endophytes. For instance, *S. indica* uses a battery of effectors to suppress host defences (Jacobs *et al.*, 2011; Rafiqi *et al.*, 2013), although the expression of many defence-related genes is induced (Schäfer *et al.*, 2009). Studies comparing plant responses to pathogenic and endophytic organisms do indicate that both are recognised in the same manner. However, they may differ in induction of the full defence response, because the endophytic organisms lack some unidentified pathogenic determinants (Wani *et al.*, 2015; Xu *et al.*, 2015). Thus, although the process of initial recognition is the same for both endophyte and pathogen, downstream regulatory mechanisms must be different. The question is therefore, what are these mechanisms? Are they determined by the endophyte or the plant, or more likely by crosstalk between the two (Kusari *et al.*, 2012)?

2.8 Specialised or Secondary Metabolites in Plant–Endophyte Interactions

It is well known that plants produce compounds, termed specialised (traditionally termed secondary) metabolites, collectively known as phytoalexins or phytoanticipins, to prevent or restrict the growth of fungi within the tissue (Ludwig-Müller, 2015; Rook, 2016). On the other hand, many pathogens produce phytotoxins to facilitate infection, i.e. metabolites toxic to plants. Many endophytes are also capable of inducing plant-based production of specialised metabolites upon colonisation. These metabolites have roles in competition between microbes and in some cases communication (Kusari *et al.*, 2012; Dupont *et al.*, 2015). The diversity of specialised metabolites produced in plant–microbe interactions is enormous and the range includes alkaloids, polyketides, terpenoids, phenylpropanoids, flavonoids, steroids, quinones, xanthones, benzopyranones, tetralones, cytochalasines and enniatines. Despite the inherent limitations in trying to study chemically diverse compounds simultaneously, the technologies of metabolomics for the study of the entire cellular diversity of these metabolites has improved substantially in recent years (Tenenboim and Brotman, 2016). A major role for these metabolites is to help one or the other partner in competitive interactions with other organisms they encounter, these can be: (1) other microorganisms they compete with; or (2) their hosts. Specific compounds play an important role in plant–endophyte communication and therefore in the establishment of a successful interaction (Kusari *et al.*, 2012). Specialised metabolic pathways of host plant and endophyte can interact in many ways (Kusari *et al.*, 2012). The metabolism of the host plant and/or endophyte can be influenced by the other partner (Ludwig-Müller, 2015): (1) the endophyte may induce host biosynthetic activity; (2) the host can induce endophyte biosynthetic activity; (3) the host and endophyte may share parts of a specific biosynthetic pathway and thus each may contribute partially to producing novel specialised metabolites; and (4) both the host and/or the endophyte may metabolise products made by the other partner.

Compared to other types of microbial interactions, our knowledge of the metabolites from endophyte interactions is both rudimentary and of great interest as a source of new metabolites with industrial potential, not least for the pharmaceutical industry (Kusari *et al.*, 2012). Many studies have demonstrated that pathogens adapted to a particular host are well capable of dealing with the phytoalexins and phytoanticipins produced in their host (Rook, 2016). This is apparently also true for endophytic fungi. For example, the majority of isolated oat root endophytes were resistant to the phytoanticipin avenicin, a saponin (i.e. terpenoid derivative) (Carter *et al.*, 1999). This raises an issue: one of the strategies for achieving transgenic disease resistance against pathogens is to move the biosynthetic apparatus for specific phytoalexins and phytoanticipins to plant species where the pathogens have not evolved to overcome them. This may have undesirable negative effects on the endophytic microbiome of the recipient since the natural endophytes adapted to a particular species would not be adapted to the phytoalexins and phytoanticipins originating from the gene donor, and this potential negative effect should be looked for when using this plant protection strategy.

It is also well known that many fungi produce specialised metabolites to give them a competitive advantage in interactions with other fungi or their hosts. Where these metabolites are measurably toxic for the host, they are termed phytotoxins. By definition, an organism behaving as an endophyte does not produce phytotoxins while exhibiting this lifestyle since this defines the necrotrophic phase of a fungal infection (Howlett, 2006). Endophytes may nevertheless produce secondary metabolites for their own benefit, e.g. to suppress the plant's defence system (which represents a conceptual overlap with hormones) or growth of competitors within the plant. In the case of the ryegrass endophyte *Epichlöe festucae*, a massive host response was induced at the level of induced gene expression, and much could be attributed to the production of specialised metabolites at the expense of primary metabolism

(Dupont *et al.*, 2015). The presence of several different endophyte species can lead to metabolite production by endophytes and/or host, because of competition (Kusari *et al.*, 2012). For example, the specialised metabolite production (specifically polyketide biosynthesis), differed in the endophyte *Alternaria tenuissima* depending on the presence of another endophytic fungus, *Nigrospora sphaerica*, in an *in vitro* system where the former inhibited growth of the latter (Chagas *et al.*, 2013). Sometimes, the metabolites shift the balance in the mutualistic interaction, turning the endophyte into a pathogen (Schulz and Boyle, 2005). Specialised metabolites produced by endophytes can be used by the host. For example, *Epichlöe* species produce ergot alkaloids which protect the plant against herbivores (Schardl and Phillips, 1997).

2.9 Selecting Endophytes as Potential BCAs

Many researchers use high-throughput *in vitro* methods to screen for potential biocontrol agents by isolating cultivatable organisms and test for their potential as BCAs. Typically, *in vitro* confrontation assays (dual culture) are performed to look for the ability of the potential BCA to inhibit growth of the target pathogen directly, i.e. direct antimicrobial activity (Zachow *et al.*, 2008; Furnkranz *et al.*, 2012; Gdanetz and Trail, 2017; Kosawang *et al.*, 2018) before moving on to *in planta* assays. The *in vitro* approach is often successful as evidenced by numerous reports of seemingly successful identification of new BCAs where this approach was employed. However, drawbacks of using this *in vitro* approach are: (1) they often do not correlate with disease control efficacy *in planta*; (2) the mechanism of induced resistance cannot be discovered in the absence of the plant. So, even if the *in vitro* approach is seemingly successful, it is not known how many potentially effective strains were discarded in the screening process. Despite these drawbacks, *in vitro* testing is often used for high-throughput screening of existing collections. In contrast to the *in vitro* approach, and in common with others (Schisler and Slininger, 1997; Comby *et al.*, 2016; Zhao *et al.*, 2017; Kernaghan *et al.*, 2017), our approach is ecological or niche-based: to look for novel fungal BCAs in the habitat where they can interact with pathogens (Figure 2.2). Specifically, the idea is to look for more healthy plants (i.e. less stressed) in these habitats to test the hypothesis that these plants manage better due to the presence of a biocontrol microorganism in their microbiome. In other words, our thesis is that endophytes isolated from the environment where they are intended for use are more likely to be adapted to these specific environmental conditions (e.g. water availability, temperature, UV radiation, competing organisms, etc.).

There are relatively few examples of beneficial endophytic fungi identified using niche-based screening strategies (Lugtenberg *et al.*, 2016; Kernaghan *et al.*, 2017; Zhao *et al.*, 2017). On the other hand, there are several examples of effective fungal

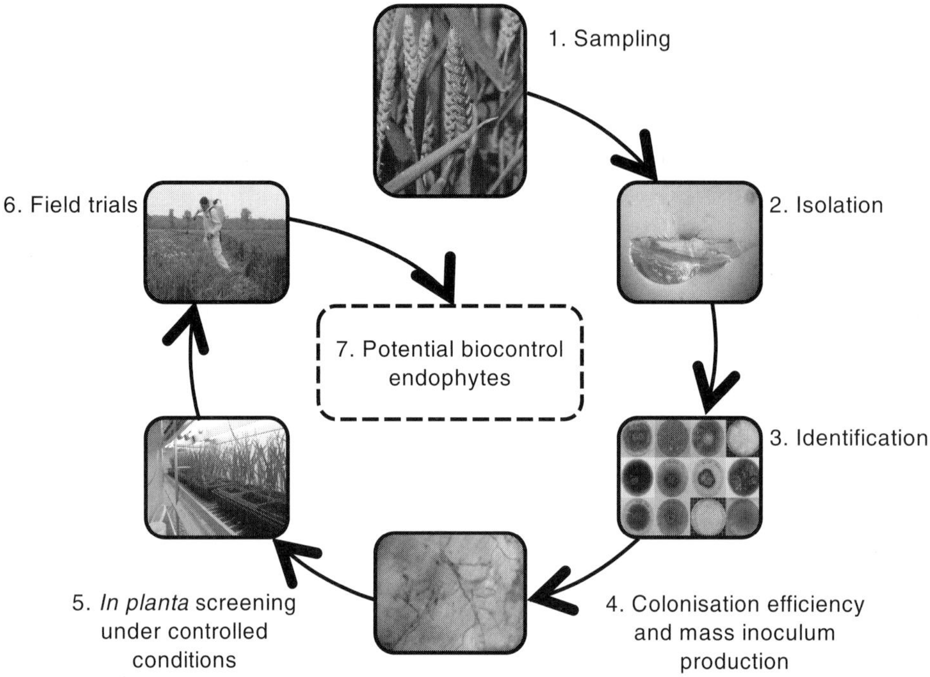

Figure 2.2 The process of discovery of biocontrol endophytes: 1. Sampling healthy plants in areas under disease pressure. 2. Isolation on artificial media from surfaced sterilised tissues (wheat glume). 3. Molecular identification of isolates using ribosomal genes. 4. Colonisation efficiency and mass production tests can reduce the number of potential isolates, as well as detect latent pathogens. 5. Screening of endophytes against diseases in small *in planta* assays provides a better observation of potential candidates. 6. Successful isolates can be tested in field conditions on larger-scale experimental units in order to assess field consistency. 7. Potential biocontrol endophytes can enter an industrial phase or mass production and marketing or continued biological studies about their mode of action and ecological significance. (A black and white version of this figure will appear in some formats. For the colour version, please refer to the plate section.)

BCAs from other microbiome components. An antagonistic isolate of *Cladosporium cladosporioides* was isolated from a sporulating colony of *Venturia inaequalis* and was shown to be as effective as fungicides for the control of apple scab in the field (Köhl *et al.*, 2014). The fungus *Clonostachys rosea* was originally isolated from *Fusarium*-infected barley roots (Knudsen *et al.*, 1995) and control of seed-borne and root-infecting diseases was demonstrated in the field (Jensen *et al.*, 2000). However, the exceptions prove the rule: *C. rosea* is also effective in controlling diseases in carrot (Jensen *et al.*, 2004; Koch *et al.*, 2010), Chinese cabbage (Møller *et al.*, 2003), oak (Knudsen *et al.*, 2004) and strawberry (Mamarabadi *et al.*, 2008), as well as foliar diseases in cereals (Jensen *et al.*, 2016a), demonstrating that it is competitive in habitats other than those it was isolated from. Another example is fungal root endophyte *S. indica*, which also exhibits a broad host range and acts as an effective BCA against several diseases (Franken, 2012; Card *et al.*, 2016).

Having isolated the organisms, a common approach is to design assays for biological control, which resemble the conditions where the potential BCAs will be used. This is done by screening the organisms *in planta* against pathogens under controlled conditions with an adequate number of replications since BCAs tend to show variable efficiency. However, for some pathosystems, large-scale *in planta* screening may not be feasible, for example due to the limited availability of material or timescale for bioassays (Kosawang *et al.*, 2018). Small-scale field experiments may be performed to assess the development of disease symptoms and optimise time points and application methodologies. Field trials at different locations with larger plots are necessary to determine effects on yield and pathogen infection. For some diseases, artificial inoculation coupled with environmental manipulation (e.g. irrigation) may be necessary since natural infection is unpredictable (e.g. for *Fusarium* head blight; Rojas *et al.*, 2018).

The information acquired will help in developing protocols for improving the reliable use of endophytes. It is also biologically interesting to record the endophytic lifestyle under controlled/sterile conditions, to confirm they are not latent plant pathogens of the test or other crop, and confirm that they are not potential human pathogens (see Section 2.11).

2.10 Exploiting Endophytes as BCAs

Once biocontrol has been demonstrated, it is very important to study the mechanisms used by the organisms to reduce disease, i.e. mode of action. Interactions between plants and microbes are complex and there is currently no way of predicting whether a particular BCA will have a positive or negative influence on the ability of a specific pathogen to cause disease (Busby *et al.*, 2016b) and therefore we need to test for effect. The mechanisms contributing to biological control include antibiosis, parasitism, competition and induced resistance (Hardoim *et al.*, 2015; Ludwig-Müller, 2015; Jensen *et al.*, 2016b; Card *et al.*, 2016). Can individual mechanisms contribute in concert in a particular three-way interaction among plant, pathogen and antagonist? It is believed that components of microbial communities (e.g. rhizosphere and endosphere) also stimulate plant growth and thereby can act to control disease by strengthening the plant (Card *et al.*, 2016). It is generally believed that the most important mechanisms employed by endophytic BCAs are antibiosis and induced resistance. We address the mechanisms employed by endophytic BCAs in more detail elsewhere (Latz *et al.*, 2018).

An important potential advantage for an established endophyte is that it is better protected from the external environment than microorganisms applied to the phyllosphere or rhizosphere (Schulz and Boyle, 2005). Since endophytes are inside the plant, we can predict that some can provide effective and robust means for controlling disease, assuming that any effect as a BCA depends on it being an endophyte.

What comes out must be put back to be of use! It may not be easy to introduce the endophyte into a new plant and considerable developmental effort should be expected for potential products both to ensure reliable inoculation and stable effect under various environmental conditions. The challenge is greatest for field use (especially for foliar applications) and rather less for controlled conditions (greenhouse). Seed treatment is very attractive since it requires no further action by the grower. This may also be achieved by introducing the endophytic BCA to the flowers during pollination. Genetic variation in the host is also an important factor affecting endophyte colonisation and is an important aspect which needs to be addressed (Busby *et al.*, 2016b; Kroll *et al.*, 2017). Once introduced to the host, some endophytes can be transmitted through the seed, e.g. *Epichlöe* spp. (Rodriguez *et al.*, 2009; Card *et al.*, 2016). Some BCAs can be used as seed treatments, and can provide sustained control against root and even foliar infections. For example the protection afforded by *S. indica* in barley suggests maintenance of effects for weeks (Waller *et al.*, 2008). Aerial parts – leaves and flowers – are more sensitive to variation in environmental conditions, in particular, humidity and UV light can pose problems.

2.11 Concerns for the Use of Endophytes

There is clearly an ecological risk in moving endophytes between continents and it is not possible to predict a problem in advance. Thus, effects on natural microbial communities may possess a risk of causing trophic imbalance. Botanists and horticulturalists have moved many plant species and plant products, presumably with their accompanying microbiomes, for centuries apparently without ill effect. Occasionally, a catastrophe occurs, and plant pathology text books name the worst cases where pathogens have spread from an important introduced plant species to a native species with devastating effect. The mechanisms behind this are often obscure (Woolhouse *et al.*, 2005; Giraud *et al.*, 2010). An intriguing story has emerged with the relatively new disease ash dieback caused by the ascomycete fungal pathogen *Hymenoscyphus fraxineus* in European ash (*Fraxinus excelsior*). This disease emerged in the 1990s in the Baltic region and has subsequently spread west (McKinney *et al.*, 2014). The pathogen was first identified some 10 years after the disease was discovered and it was subsequently found that this is a natural endophyte or perhaps a weak pathogen in ash leaves of the eastern Asian species

Fraxinus mandchurensis. Perhaps native endophytes can be found which could be used to combat this disease (Kosawang *et al.*, 2018; Lahiri *et al.*, 2019, Chapter 15)?

That an endophyte can find a new host and become a pathogen after intercontinental transfer represents one risk, and it is clear that intercontinental transfer should in general be avoided or at least evaluated carefully. Are there other risks? Potential BCAs, including endophytes, should be tested thoroughly to ensure they are not pathogenic or otherwise harmful to the environment. Not easy! They may be present naturally in low amounts, but what happens if they are present in large quantities after application for agronomic purposes? One criterion could be to ensure that the BCA cannot persist for long in the environment in the absence of the host. However, there is a balance here. Should the product be difficult to use and inefficient, arguably it would be less attractive to the market than a more persistent (aggressive BCA). Whereas a non-persistent product may be attractive to the supplier, who could sell more, this option is less attractive to the grower.

A prerequisite for an endophytic BCA is that it should not be pathogenic towards humans (or allergenic or produce mycotoxins), even the immunocompromised, as it is very unlikely that these would be approved as products during the registration process. An early identification can be useful and with the ease of molecular identification, known human (and plant!) pathogens can be dropped from further study (Alabouvette *et al.*, 2012). A simple screen for pathogenic potential is to compare the growth in culture of the isolated endophytes at ambient temperature, where they would be used as BCAs or stimulants, with growth at 37°C or thereabouts. Ideally, growth should be prolific at the lower temperature as this would be useful for product formulation and production, but not at the latter temperature (Köhl *et al.*, 2011).

Finally, there is a myth that something natural is intrinsically safer than something that is artificial. Like other fungi, a BCA may produce mycotoxins or other harmful metabolites (Alabouvette *et al.*, 2012). The grass endophytes *Epichlöe* spp. are a case in point, and considerable effort has been put into developing mycotoxin-free *Epichlöe* BCA products (Card *et al.*, 2016). Part of the development of a potential BCA for use should therefore include checking for the production of potentially harmful metabolites.

The legislation concerning the use of BCAs in the EU follows the Registration of Biological Control Agents (REBECA) policy (Anon, 2009; Ehlers, 2011), and the Organisation for Economic Co-operation and Development (OECD) has provided guidelines to help harmonise legislation (Alabouvette *et al.*, 2012; OEDC, 2012). Currently there is an illogical discrepancy in the legal frameworks for using beneficial microorganisms in agriculture in some (mostly OECD) countries: an organism can be sold as a biofertiliser or biostimulant without addressing BCA activity, but all BCAs are subject to an extensive registration process and this is hampering the development of products (OEDC, 2012; Villaverde *et al.*, 2014; Anon, 2016).

2.12 Perspectives and Further Research

It is becoming increasingly clear that biological control will have to play a larger role in the plant protection programmes of the future. However, we still have much to learn. BCAs are often trickier to use than chemicals or disease resistance, but they can offer control solutions or contribute where chemicals and disease resistance are unavailable. We see an increasing role for endophytes in the BCA armoury. There are several areas which need to be focused on in order to achieve this; for example, the biology of the microorganisms concerned, including the potential for developing crops producing useful metabolites. Another important area concerns the development of reliable delivery systems for endophytes to the crop. Examples of delivery can include the ability to infect seed via the flower, seed coating or spraying the crop. Finally, we need a better understanding of the effects of host genotype and role of environment on specific disease risks. Most of all there are many exciting biological questions to be answered about the recruitment of endophytes, and concerning the nature of the interactions of endophytic microorganisms with each other, with other microorganisms in the plant (pathogens and the classic symbionts, e.g. mycorrhiza) and with the physiology of their hosts.

Consortia are combinations of organisms which can grow together giving an improved effect against a specific disease or perhaps several different beneficial effects. Small consortia can be designed to combine different microorganisms with different targets; for example, combining *Clonostachys rosea* with *Metarhizium brunneum* or *Beauveria bassiana*, which target fungal pathogens and insect pests, respectively (Kapongo *et al.*, 2008; Keyser *et al.*, 2016;). Consortia offer both advantages and disadvantages.

Advantages:

1. They may be designed to provide a combination of different modes of action to give broad-spectrum effect against several pathogens.
2. Organisms with different environmental optima may be combined to secure an effect under different environmental conditions.

Disadvantages:

1. Companies prefer single strains due to difficulties in approval.
2. They may not always give an enhanced effect (Xu *et al.*, 2011).

With the predicted 20% human population growth over the next 30 years, the effects of urbanisation and climate changes, global agricultural productivity needs a boost. Simultaneously, this boost must be implemented under higher environmental standards and global sustainability goals. Plant pathogens, pests and abiotic stress are agricultural challenges where microbiomes and specifically endophytes have shown potential for contribution to the alleviation of crop loss and stimulation of crop yields. Our responsibility is to provide alternatives and to ensure that these achieve the potential by translating laboratory results into improved crop production globally.

Acknowledgements

We are grateful to Anna Kaja Høyer and John Larsen for critical comments. Our research receives funding from the European Union's Horizon 2020 research and innovation programme under the Marie Skłodowska-Curie grant agreements Nos. 674964 and 676480.

References

Abdelfattah, A., Wisniewski, M., Droby, S. and Schena, L. (2016). Spatial and compositional variation in the fungal communities of organic and conventionally grown apple fruit at the consumer point-of-purchase. *Horticulture Research*, **3**, 16047.

Aguiar-Pulido, V., Huang, W., Suarez-Ulloa, V. *et al.* (2016). Metagenomics, metatranscriptomics, and metabolomics approaches for microbiome analysis. *Evolutionary Bioinformatics Online*, **12**, 5–16.

Ahmed, A. A., McLellan, H., Aguilar, G. B. *et al.* (2016). Engineering barriers to infection by undermining pathogen effector function or by gaining effector recognition. In *Plant Pathogen Resistance Biotechnology*, ed. D. B. Collinge. New York and London: Wiley-Blackwell, pp. 23–50.

Alabouvette, C., Heilig, U. and Cordier, C. (2012). Microbial Control of Plant Diseases. In *Beneficial Microorganisms in Agriculture, Food and the Environment: Safety Assessment and Regulation*, ed. I. Sundh, A. Wilcks and M. Goettel. Oxfordshire, UK: CAB International, pp. 96–111.

Alonso-Ramírez, A., Poveda, J., Martín, I. *et al.* (2014). Salicylic acid prevents *Trichoderma harzianum* from entering the vascular system of roots. *Molecular Plant Pathology*, **15**, 823–831.

Anon. (2009). Regulation (EC) No 1107/2009 of the European Parliament and of the Council of 21 October 2009 concerning the placing of plant protection products on the market and repealing Council Directives 79/117/EEC and 91/414/EEC. *Official Journal of the European Union*, **52**, 1.

Anon. (2016). Guidance on active micro-organisms and biocidal products. *ECHA*, doi: 10.2823/82218.

Berger, S., El Chazli, Y., Babu, A. F. and Coste, A. T. (2017). Azole resistance in *Aspergillus fumigatus*: a consequence of antifungal use in agriculture? *Frontiers in Microbiology*, **8**, 1024.

Bulgarelli, D., Rott, M., Schlaeppi, K. *et al.* (2012). Revealing structure and assembly cues for Arabidopsis root-inhabiting bacterial microbiota. *Nature*, **488**, 91–95.

Busby, P. E., Peay, K. G. and Newcombe, G. (2016a). Common foliar fungi of *Populus trichocarpa* modify

Melampsora rust disease severity. *New Phytologist*, **209,** 1681–1692.

Busby, P. E., Ridout, M. and Newcombe, G. (2016b). Fungal endophytes: modifiers of plant disease. *Plant Molecular Biology*, **90**, 645–655.

Card, S., Johnson, L. E. B., Teasdale, S. and Caradus, J. (2016). Deciphering endophyte behaviour: the link between endophyte biology and efficacious biological control agents. *FEMS Microbiology Ecology*, **92**, fiw114.

Carter, J. P., Spink, J., Cannon, P. F., Daniels, M. J. and Osbourn, A. E. (1999). Isolation, characterization, and avenacin sensitivity of a diverse collection of cereal-root-colonizing fungi. *Applied and Environmental Microbiology*, **65**, 3364–3372.

Chagas, F. O., Dias, L. G. and Pupo, M. T. (2013). A mixed culture of endophytic fungi increases production of antifungal polyketides. *Journal of Chemical Ecology*, **39**, 1335–1342.

Coleman-Derr, D. and Tringe, S. G. (2014). Building the crops of tomorrow: advantages of symbiont-based approaches to improving abiotic stress tolerance. *Frontiers in Microbiology*, **5**, 283.

Collinge, D. B. (2018). Transgenic crops and beyond: how can biotechnology contribute to the sustainable control of plant diseases? *European Journal of Plant Pathology*, **152**, 977–986.

Collinge, D. B., Jørgensen, H. J. L., Lund, O. S. and Lyngkjær, M. F. (2010). Engineering pathogen resistance in crop plants – current trends and future prospects. *Annual Review of Phytopathology*, **48**, 269–291.

Collinge, D. B., Mullins, E., Jensen, B. and Jørgensen, H. J. L. (2016). The status and prospects for biotechnological approaches to attaining sustainable disease resistance. In *Plant Pathogen Resistance Biotechnology*, ed. D. B. Collinge. New York and London: Wiley-Blackwell, pp. 1–20.

Comby, M., Lacoste, S., Baillieul, F., Profizi, C. and Dupont, J. (2016). Spatial and temporal variation of cultivable communities of co-occurring endophytes and pathogens in wheat. *Frontiers in Microbiology*, **7**, 403.

de Jonge, R., Peter van Esse, H., Kombrink, A. *et al.* (2010). Conserved fungal LysM effector Ecp6 prevents chitin-triggered immunity in plants. *Science*, **329**, 953.

De Silva, D. D., Crous, P. W., Ades, P. K., Hyde, K. D. and Taylor, P. W. J. (2017). Life styles of *Colletotrichum* species and implications for plant biosecurity. *Fungal Biology Reviews*, **31**, 155–168.

De Vleesschauwer, D., Gheysen, G. and Hofte, M. (2013). Hormone defense networking in rice: tales from a different world. *Trends in Plant Science*, **18**, 555–565.

Diaz, P. L., Hennell, J. R. and Sucher, N. J. (2012). Genomic DNA extraction and barcoding of endophytic fungi. In *Plant DNA Fingerprinting and Barcoding: Methods and Protocols*, ed. N. J. Sucher, J. R. Hennell and M. C. Carles. Totowa, NJ: Humana Press, pp. 171–179.

Dupont, P.-Y., Eaton, C. J., Wargent, J. J. *et al.* (2015). Fungal endophyte infection of ryegrass reprograms host metabolism and alters development. *New Phytologist*, **208**, 1227–1240.

Eevers, N., Gielen, M., Sánchez-López, A. *et al.* (2015). Optimization of isolation and cultivation of bacterial endophytes through addition of plant extract to nutrient media. *Microbial Biotechnology*, **8**, 707–715.

Ehlers, R.-U. (2011). Regulation of biological control agents and the EU policy support action REBECA. In *Regulation of Biological Control Agents*, ed. R.-U. Ehlers. Dordrecht, The Netherlands: Springer, pp. 3–23.

Evangelisti, E., Rey, T. and Schornack, S. (2014). Cross-interference of plant development and plant–microbe interactions. *Current Opinion in Plant Biology*, **20**, 118–126.

Finkel, O. M., Castrillo, G., Herrera Paredes, S., Salas González, I. and Dangl, J. L. (2017). Understanding and exploiting plant beneficial microbes. *Current Opinion in Plant Biology*, **38**, 155–163.

Franken, P. (2012). The plant strengthening root endophyte *Piriformospora indica*: potential application and the biology behind. *Applied Microbiology and Biotechnology*, **96**, 1455–1464.

Fravel, D., Olivain, C. and Alabouvette, C. (2003). *Fusarium oxysporum* and its biocontrol. *New Phytologist*, **157**, 493–502.

Furnkranz, M., Lukesch, B., Muller, H. *et al.* (2012). Microbial diversity inside pumpkins: microhabitat-specific communities display a high antagonistic potential against phytopathogens. *Microbial Ecology*, **63**, 418–428.

Gdanetz, K. and Trail, F. (2017). The wheat microbiome under four management strategies, and potential for endophytes in disease protection. *Phytobiomes*, **1**, 158–168.

Giraud, T., Gladieux, P. and Gavrilets, S. (2010). Linking the emergence of fungal plant diseases with ecological speciation. *Trends in Ecology & Evolution*, **25**, 387–395.

Großkinsky, D. K., van der Graaff, E. E. and Roitsch, T. (2016). Regulation of abiotic and biotic stress responses by plant hormones. In *Plant Pathogen Resistance Biotechnology*, ed. D. B. Collinge. New York and London: Wiley-Blackwell, pp. 131–154.

Guimil, S., Chang, H. S., Zhu, T. *et al.* (2005). Comparative transcriptomics of rice reveals an ancient pattern of response to microbial colonization. *Proceedings of the National Academy of Sciences of the United States of America*, **102**, 8066–8070.

Gutjahr, C. (2014). Phytohormone signaling in arbuscular mycorhiza development. *Current Opinion in Plant Biology*, **20**, 26–34.

Hardoim, P. R., van Overbeek, L. S., Berg, G. *et al.* (2015). The hidden world within plants: ecological and evolutionary considerations for defining functioning of microbial endophytes. *Microbiology and Molecular Biology Reviews*, **79**, 293–320.

Hertz, M., Jensen, I. R., Jensen, L. Ø. *et al.* (2016). The fungal community changes over time in developing wheat heads. *International Journal of Food Microbiology*, **222**, 30–39.

Hilbert, M., Voll, L. M., Ding, Y. *et al.* (2012). Indole derivative production by the root endophyte *Piriformospora indica* is not required for growth promotion but for biotrophic colonization of barley roots. *New Phytologist*, **196**, 520–534.

Houterman, P. M., Cornelissen, B. J. C. and Rep, M. (2008). Suppression of plant resistance gene-based immunity by a fungal effector. *PLoS Pathogens*, **4**, e1000061.

Howlett, B. J. (2006). Secondary metabolite toxins and nutrition of plant pathogenic fungi. *Current Opinion in Plant Biology*, **9**, 371–375.

Ionescu, I. A., López-Ortega, G., Burow, M. *et al.* (2017). Transcriptome and metabolite changes during hydrogen cyanamide-induced floral bud break in sweet cherry. *Frontiers in Plant Science*, **8**, 1233.

Jacobs, S., Zechmann, B., Molitor, A. *et al.* (2011). Broad-spectrum suppression of innate immunity is required for colonization of *Arabidopsis* roots by the fungus *Piriformospora indica*. *Plant Physiology*, **156**, 726–740.

Jensen, B., Knudsen, I. M. B. and Jensen, D. F. (2000). Biological seed treatment of cereals with fresh and long-term stored formulations of *Clonostachys rosea*: biocontrol efficacy against *Fusarium culmorum*. *European Journal of Plant Pathology*, **106**, 233–242.

Jensen, B., Knudsen, I. M. B., Madsen, M. and Jensen, D. F. (2004). Biopriming of infected carrot seed with an antagonist, *Clonostachys rosea*, selected for control of seedborne *Alternaria* spp. *Phytopathology*, **94**, 551–560.

Jensen, B., Lübeck, P. S. and Jørgensen, H. J. L. (2016a). *Clonostachys rosea* reduces spot blotch in barley by inhibiting prepenetration growth and sporulation of *Bipolaris sorokiniana* without inducing resistance. *Pest Management Science*, **72**, 2231–2239.

Jensen, D. F., Karlsson, M., Sarrocco, S. and Vannacci, G. (2016b). Biological control using microorganisms as an alternative to disease resistance. In *Plant Pathogen Resistance Biotechnology*, ed. D. B. Collinge. New York and London: Wiley-Blackwell, pp. 341–363.

Kapongo, J. P., Shipp, L., Kevan, P. and Sutton, J. C. (2008). Co-vectoring of *Beauveria bassiana* and *Clonostachys rosea* by bumble bees (*Bombus impatiens*) for control of insect pests and suppression of grey mould in greenhouse tomato and sweet pepper. *Biological Control*, **46**, 508–514.

Kaul, S., Sharma, T. and Dhar, M. K. (2016). 'Omics' tools for better understanding the plant–endophyte interactions. *Frontiers in Plant Science*, **7**, 955.

Kernaghan, G., Mayerhofer, M. and Griffin, A. (2017). Fungal endophytes of wild and hybrid *Vitis* leaves and their potential for vineyard biocontrol. *Canadian Journal of Microbiology*, **63**, 583–595.

Keyser, C. A., Jensen, B. and Meyling, N. V. (2016). Dual effects of *Metarhizium* spp. and *Clonostachys rosea* against an insect and a seed-borne pathogen in wheat. *Pest Management Science*, **72**, 517–526.

Khan, A. L., Hamayun, M., Kang, S.-M. *et al.* (2012). Endophytic fungal association via gibberellins and indole acetic acid can improve plant growth under abiotic stress: an example of *Paecilomyces formosus* LHL10. *BMC Microbiology*, **12**, 3.

Khatabi, B., Molitor, A., Lindermayr, C. (2012). Ethylene supports colonization of plant roots by the mutualistic fungus *Piriformospora indica*. *PLoS One*, **7**, e35502.

Knudsen, I. M. B., Hockenhull, J. and Jensen, D. F. (1995). Biocontrol of seedling diseases of barley and wheat caused by *Fusarium culmorum* and *Bipolaris sorokiniana*: effects of selected fungal antagonists on growth and yield components. *Plant Pathology*, **44**, 467–477.

Knudsen, I. M. B., Thomsen, K. A., Jensen, B. and Poulsen, K. M. (2004). Effects of hot water treatment, biocontrol agents, disinfectants and a fungicide on storability of English oak acorns and control of the pathogen, *Ciboria batschiana*. *Forest Pathology*, **34**, 47–64.

Koch, E., Schmitt, A., Stephan, D. *et al.* (2010). Evaluation of non-chemical seed treatment methods for the control of *Alternaria dauci* and *A. radicina* on carrot seeds. *European Journal of Plant Pathology*, **127**, 99–112.

Köhl, J., Postma, J., Nicot, P., Ruocco, M. and Blum, B. (2011). Stepwise screening of microorganisms for commercial use in biological control of plant-pathogenic fungi and bacteria. *Biological Control*, **57**, 1–12.

Köhl, J., Scheer, C., Holb, I. J., Masny, S. and Molhoek, W. (2014). Toward an integrated use of biological control by *Cladosporium cladosporioides* H39 in apple scab (*Venturia inaequalis*) management. *Plant Disease*, **99**, 535–543.

Kojima, M., Kamada-Nobusada, T., Komatsu, H. *et al.* (2009). Highly sensitive and high-throughput analysis of plant hormones using MS-probe modification and liquid chromatography–tandem mass spectrometry: an application for hormone profiling in *Oryza sativa*. *Plant and Cell Physiology*, **50**, 1201–1214.

Kosawang, C., Amby, D. B., Bussaban, B. *et al.* (2018). Fungal communities associated with species of Fraxinus tolerant to ash dieback, and their potential for biological control. *Fungal Biology*, **122**, 2110–2120.

Kroll, S., Agler, M. T. and Kemen, E. (2017). Genomic dissection of host–microbe and microbe–microbe interactions for advanced plant breeding. *Current Opinion in Plant Biology*, **36**, 71–78.

Kurose, D., Furuya, N., Tsuchiya, K., Tsushima, S. and Evans, H. C. (2012). Endophytic fungi associated with *Fallopia japonica* (Polygonaceae) in Japan and their interactions with *Puccinia polygoni-amphibii* var. *tovariae*, a candidate for classical biological control. *Fungal Biology*, **116**, 785–791.

Kusari, S., Hertweck, C. and Spiteller, M. (2012). Chemical ecology of endophytic fungi: origins of secondary metabolites. *Chemistry & Biology*, **19**, 792–798.

Lahiri, A., Douglas, G. C., Murphy, B. R. and Hodkinson, T. R. (2019). *In vitro* methods for plant–microbe interaction and biocontrol studies in European ash (*Fraxinus excelsior* L.). In *Endophytes for a Growing World*, ed. T. R. Hodkinson, F. M. Doohan, M. J. Saunders and B. R. Murphy. Cambridge: Cambridge University Press, Chapter 15.

Latz, M. A. C., Jensen, B., Collinge, D. B. and Jørgensen, H. J. L. (2018). Endophytic fungi as biocontrol agents: elucidating mechanisms in disease suppression. *Plant Ecology and Diversity*, doi: 10.1080/17550874.2018.1534146.

Lo Presti, L., Lanver, D., Schweizer, G. *et al.* (2015). Fungal effectors and plant susceptibility. *Annual Review of Plant Biology*, **66**, 513–545.

Lofgren, L. A., LeBlanc, N. R., Certano, A. K. *et al.* (2018). *Fusarium graminearum*: pathogen or endophyte of North American grasses? *New Phytologist*, **217**, 1203–1212.

Louarn, S., Nawrocki, A., Thorup-Kristensen, K. *et al.* (2013). Proteomic changes and endophytic micromycota during storage of organically and conventionally grown carrots. *Postharvest Biology and Technology*, **76**, 26–33.

Lucas, J. A., Hawkins, N. J. and Fraaije, B. A. (2015). The evolution of fungicide resistance. *Advances in Applied Microbiology*, **90**, 29–92.

Ludwig-Müller, J. (2015). Plants and endophytes: equal partners in secondary metabolite production? *Biotechnology Letters*, **37**, 1325–1334.

Lugtenberg, B. J. J., Caradus, J. R. and Johnson, L. J. (2016). Fungal endophytes for sustainable crop production. *FEMS Microbiology Ecology*, **92**, fiw194.

Lundberg, D. S., Lebeis, S. L., Paredes, S. H. *et al.* (2012). Defining the core *Arabidopsis thaliana* root microbiome. *Nature*, **488**, 86–90.

Ma, K. W. and Ma, W. B. (2016). Phytohormone pathways as targets of pathogens to facilitate infection. *Plant Molecular Biology*, **91**, 713–725.

Ma, L. J., van der Does, H. C., Borkovich, K. A. *et al.* (2010). Comparative genomics reveals mobile pathogenicity chromosomes in Fusarium. *Nature*, **464**, 367–373.

Malinovsky, F. G., Fangel, J. U. and Willats, W. G. T. (2014). The role of the cell wall in plant immunity. *Frontiers in Plant Science*, **5**, 178.

Mamarabadi, M., Jensen, B., Jensen, D. F. and Lübeck, M. (2008). Real-time RT-PCR expression analysis of chitinase and endoglucanase genes in the three-way interaction between the biocontrol strain *Clonostachys rosea* IK726, *Botrytis cinerea* and strawberry. *FEMS Microbiology Letters*, **285**, 101–110.

McGrann, G. R. D., Stavrinides, A., Russell, J. *et al.* (2014). A trade off between *mlo* resistance to powdery mildew and increased susceptibility of barley to a newly important disease, Ramularia leaf spot. *Journal of Experimental Botany*, **65**, 1025–1037.

McGrann, G. R. D., Andongabo, A., Sjökvist, E. *et al.* (2016). The genome of the emerging barley pathogen *Ramularia collo-cygni*. *BMC Genomics*, **17**, 584.

McKinney, L. V., Nielsen, L. R., Collinge, D. B. *et al.* (2014). The ash dieback crisis; genetic variation in resistance can prove a long-term solution. *Plant Pathology*, **63**, 485–499.

Moissl-Eichinger, C., Pausan, M., Taffner, J. *et al.* (2018). Archaea are interactive components of complex microbiomes. *Trends in Microbiology*, **26**, 70–85.

Møller, K., Jensen, B., Andersen, H. P., Stryhn, H. and Hockenhull, J. (2003). Biocontrol of *Pythium tracheiphilum* in Chinese cabbage by *Clonostachys rosea* under field conditions. *Biocontrol Science and Technology*, **13**, 171–182.

Mukherjee, M., Mukherjee, P. K., Horwitz, B. A. *et al.* (2012). *Trichoderma*-plant-pathogen interactions: advances in genetics of biological control. *Indian Journal of Microbiology*, **52**, 522–529.

Müller, C. B. and Krauss, J. (2005). Symbiosis between grasses and asexual fungal endophytes. *Current Opinion in Plant Biology*, **8**, 450–456.

Murphy, B. R., Batke, S. P., Doohan, F. M. and Hodkinson, T. R. (2015). Media manipulations and the culture of beneficial fungal root endophytes. *International Journal of Biology*, **7**, 94–102.

Newsham, K. K. (2011). A meta-analysis of plant responses to dark septate root endophytes. *New Phytologist*, **190**, 783–793.

Nicolaisen, M., Justesen, A. F., Knorr, K., Wang, J. and Pinnschmidt, H. O. (2014). Fungal communities in wheat grain show significant co-existence patterns among species. *Fungal Ecology*, **11**, 145–153.

Nowara, D., Gay, A. P., Lacomme, C. *et al.* (2010). HIGS: Host-Induced Gene Silencing in the obligate biotrophic fungal pathogen *Blumeria graminis*. *The Plant Cell*, **22**, 3130–3141.

OEDC (2012). OECD guidance to the environmental safety evaluation of microbial biocontrol agents. *OECD Environment, Health and Safety Publications, Series on Pesticides and Biocides*, No. 67, Paris: OECD Publishing, pp. 63.

Peskan-Berghöfer, T., Vilches-Barro, A., Müller, T. M. *et al.* (2015). Sustained exposure to abscisic acid enhances the colonization potential of the mutualist fungus *Piriformospora indica* on *Arabidopsis thaliana* roots. *New Phytologist*, **208**, 873–886.

Pieterse, C. M. J., Van der Does, D., Zamioudis, C., Leon-Reyes, A. and van Wees, S. C. M. (2012). Hormonal modulation of plant immunity. *Annual Review of Cell and Developmental Biology*, **28**, 489–521.

Ploch, S. and Thines, M. (2011). Obligate biotrophic pathogens of the genus *Albugo* are widespread as asymptomatic endophytes in natural populations of Brassicaceae. *Molecular Ecology*, **20**, 3692–3699.

Rafiqi, M., Jelonek, L., Akum, N., Zhang, F. and Kogel, K.-H. (2013). Effector candidates in the secretome of *Piriformospora indica*, a ubiquitous plant-associated fungus. *Frontiers in Plant Science*, **4**, 228.

Rodriguez, R. J., White Jr, J. F., Arnold, A. E. and Redman, R. S. (2009). Fungal endophytes: diversity and functional roles. *New Phytologist*, **182**, 314–330.

Rojas, E. C., Jørgensen, H. J. L., Jensen, B. and Collinge, D. B. (2018). Fusarium diseases: biology and management perspectives. In *Integrated Disease Management of Wheat and Barley*, ed. R. P. Oliver. Cambridge, UK: Burleigh Dodds Science Publishing. doi: 10.19103/AS.2018.0039.02

Rook, F. (2016). Metabolic engineering of chemical defence pathways in plant disease control. In *Plant Pathogen Resistance Biotechnology*, ed. D. B. Collinge. New York and London: Wiley-Blackwell, pp. 71–89.

Rovenich, H., Boshoven, J. C. and Thomma, B. P. H. J. (2014). Filamentous pathogen effector functions: of pathogens, hosts and microbiomes. *Current Opinion in Plant Biology*, **20**, 96–103.

Sánchez-Vallet, A., Saleem-Batcha, R., Kombrink, A. *et al.* (2013). Fungal effector Ecp6 outcompetes host immune receptor for chitin binding through intrachain LysM dimerization. *eLife*, **2**, e00790.

Sánchez-Vallet, A., McDonald, M. C., Solomon, P. S. and McDonald, B. A. (2015). Is *Zymoseptoria tritici* a hemibiotroph? *Fungal Genetics and Biology*, **79**, 29–32.

Sapkota, R., Jørgensen, L. N. and Nicolaisen, M. (2017). Spatiotemporal variation and networks in the mycobiome of the wheat canopy. *Frontiers in Plant Science*, **8**, 1357.

Schäfer, P., Pfiffi, S., Voll, L. M. *et al.* (2009). Manipulation of plant innate immunity and gibberellin as factor of compatibility in the mutualistic association of barley roots with *Piriformospora indica*. *The Plant Journal*, **59**, 461–474.

Schardl, C. L. and Phillips, T. D. (1997). Protective grass endophytes: where are they from and where are they going? *Plant Disease*, **81**, 430–438.

Schisler, D. A. and Slininger, P. J. (1997). Microbial selection strategies that enhance the likelihood of developing commercial biological control products. *Journal of Industrial Microbiology and Biotechnology*, **19**, 172–179.

Schulz, B. and Boyle, C. (2005). The endophytic continuum. *Mycological Research*, **109**, 661–686.

Shetty, N. P., Kristensen, B. K., Newman, M. A. *et al.* (2003). Association of hydrogen peroxide with restriction of *Septoria tritici* in resistant wheat. *Physiological and Molecular Plant Pathology*, **62**, 333–346.

Stein, E., Molitor, A., Kogel, K. H. and Waller, F. (2008). Systemic resistance in *Arabidopsis* conferred by the mycorrhizal fungus *Piriformospora indica* requires Jasmonic Acid signaling and the cytoplasmic function of NPR1. *Plant and Cell Physiology*, **49**, 1747–1751.

Tenenboim, H. and Brotman, Y. (2016). Omic relief for the biotically stressed: metabolomics of plant biotic interactions. *Trends in Plant Science*, **21**, 781–791.

Tian, B.-Y., Cao, Y. and Zhang, K.-Q. (2015). Metagenomic insights into communities, functions of endophytes, and their associates with infection by root-knot nematode, *Meloidogyne incognita*, in tomato roots. *Scientific Reports*, **5**, 17087.

Toju, H., Tanabe, A. S., Yamamoto, S. and Sato, H. (2012). High-coverage ITS primers for the DNA-based identification of ascomycetes and basidiomycetes in environmental samples. *PLoS One*, **7**, e40863.

van den Burg, H. A., Harrison, S. J., Joosten, M. H. A. J., Vervoort, J. and de Wit, P. J. G. M. (2006). *Cladosporium fulvum* Avr4 protects fungal cell walls against hydrolysis by plant chitinases accumulating during infection. *Molecular Plant–Microbe Interactions*, **19**, 1420–1430.

Villaverde, J. J., Sevilla-Morán, B., Sandín-España, P., López-Goti, C. and Alonso-Prados, J. L. (2014). Biopesticides in the framework of the European Pesticide Regulation (EC) No. 1107/2009. *Pest Management Science*, **70**, 2–5.

Waller, F., Mukherjee, K., Deshmukh, S. D. *et al.* (2008). Systemic and local modulation of plant responses by *Piriformospora indica* and related Sebacinales species. *Journal of Plant Physiology*, **165**, 60–70.

Wani, Z. A., Ashraf, N., Mohiuddin, T. and Riyaz-Ul-Hassan, S. (2015). Plant-endophyte symbiosis, an ecological perspective. *Applied Microbiology and Biotechnology*, **99**, 2955–2965.

Waqas, M., Khan, A. L., Kamran, M. *et al.* (2012). Endophytic fungi produce gibberellins and indoleacetic acid and promotes host-plant growth during stress. *Molecules*, **17**, 10754–10773.

Waqas, M., Khan, A. L., Shahzad, R., Ullah, I., Khan, A. R. and Lee, I. J. (2015). Mutualistic fungal endophytes produce phytohormones and organic acids that promote japonica rice plant growth under prolonged heat stress. *Journal of Zhejiang University Science B*, **16**, 1011–1018.

Weiß, M., Waller, F., Zuccaro, A. and Selosse, M.-A. (2016). Sebacinales: one thousand and one interactions with land plants. *New Phytologist*, **211**, 20–40.

Woolhouse, M. E. J., Haydon, D. T. and Antia, R. (2005). Emerging pathogens: the epidemiology and evolution of species jumps. *Trends in Ecology & Evolution*, **20**, 238–244.

Xu X., Wang, C., Li, S. *et al.* (2015). Friend or foe: differential responses of rice to invasion by mutualistic or pathogenic fungi revealed by RNAseq and metabolite profiling. *Nature Reports*, **5**(13624), 1–14.

Xu, X. M., Jeffries, P., Pautasso, M. and Jeger, M. J. (2011). A numerical study of combined use of two biocontrol agents with different biocontrol mechanisms in controlling foliar pathogens. *Phytopathology*, **101**, 1032–1044.

Ye, W., Shen, C.-H., Lin, Y. *et al.* (2014). Growth promotion-related miRNAs in *Oncidium* orchid roots colonized by the endophytic fungus *Piriformospora indica*. *PLoS One*, **9**, e84920.

Zachow, C., Tilcher, R. and Berg, G. (2008). Sugar beet-associated bacterial and fungal communities show a high indigenous antagonistic potential against plant pathogens. *Microbial Ecology*, **55**, 119–129.

Zamioudis, C. and Pieterse, C. M. J. (2011). Modulation of host immunity by beneficial microbes. *Molecular Plant-Microbe Interactions*, **25**, 139–150.

Zeilinger, S., Gupta, V. K., Dahms, T. E. S. *et al.* (2016). Friends or foes? Emerging insights from fungal interactions with plants. *FEMS Microbiology Reviews*, **40**, 182–207.

Zhao, Y., Gao, Z., Tian, B. *et al.* (2017). Endosphere microbiome comparison between symptomatic and asymptomatic roots of *Brassica napus* infected with *Plasmodiophora brassicae*. *PLoS One*, **12**, e0185907.

Zuccaro, A., Basiewicz, M., Zurawska, M., Biedenkopf, D. and Kogel, K.-H. (2009). Karyotype analysis, genome organization, and stable genetic transformation of the root colonizing fungus *Piriformospora indica*. *Fungal Genetics and Biology*, **46**, 543–550.

Zuccaro, A., Lahrmann, U., Güldener, U. *et al.* (2011). Endophytic life strategies decoded by genome and transcriptome analyses of the mutualistic root symbiont *Piriformospora indica*. *PLoS Pathogens*, **7**, e1002290.

3

Application of Formulated Endophytic Entomopathogenic Fungi for Novel Plant Protection Strategies

VIVIEN KRELL, DESIRÉE JAKOBS-SCHOENWANDT AND ANANT V. PATEL

Abstract

In recent years, research into novel plant protection strategies with endophytic entomopathogenic fungi has increased markedly. However, current applications of these fungi are mostly not supported by targeted formulation strategies which should enhance fungal establishment in close proximity to plant tissue and promote endophytism. Further drawbacks are low stability of fungal propagules during drying and storage, difficult handling as well as high dosages and costs per hectare. Formulation has the potential to substantially improve all of these characteristics to come closer to implementation of these fungi in integrated pest management. This chapter reviews the currently available literature on application of formulated endophytic entomopathogenic fungi via sprays, seed treatments, and sowing of granules and beads. It further addresses future trends in formulation science to overcome current challenges of endophytic entomopathogenic fungi regarding consistent efficacy particularly under field conditions.

3.1 Introduction

It has been largely recognised that plants are holobionts commonly inhabited by endophytic microbial communities comprising bacteria, archaea, protozoa and fungi (Hardoim *et al.*, 2015). The term endophyte, coined by de Bary (1866), refers

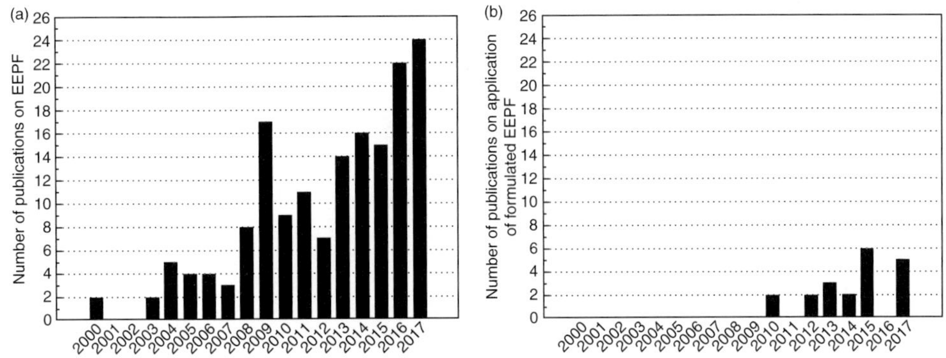

Figure 3.1 Number of publications on EEPF per year (Web of Science) searched on 6 December, 2017: (a) all items; (b) application techniques of EEPF including sprays, seed treatments and sowing of granules and beads.

to microorganisms living internally in plant tissues for at least part of their life cycle without causing apparent plant symptoms (Schulz and Boyle, 2005). Endophytes may hold multifaceted roles within plants, such as improving plant fitness by conferring resistance against abiotic and biotic stressors and promoting plant growth (Hardoim *et al.*, 2015).

Basically, fungal endophytes are categorised into two groups: clavicipitaceous and non-clavicipitaceous endophytes (Rodriguez *et al.*, 2009). Clavicipitaceous endophytes, such as fungi from the genus *Epichloë*, mostly reside within grasses and have been shown to decrease insect herbivory by production of fungal secondary metabolites, including alkaloid compounds (Clay, 1989; Johnson *et al.*, 2013). Non-clavicipitaceous endophytes are ubiquitous, less specialised and dominated by Ascomycota (Arnold and Lutzoni, 2007), which include most species of endophytic entomopathogenic fungi (EEPF) in the order Hypocreales (Vidal and Jaber, 2015).

Pioneering work on EEPF has been conducted by Bing and Lewis (1991). Since their finding that application of *Beauveria bassiana* (Balsamo-Crivelli) Vuillemin to corn and subsequent colonisation of plant tissue suppressed the European corn borer *Ostrinia nubilalis* (Huebner), more research into EEPF has followed. However, the number of publications on EEPF is still relatively low, with only 24 items in 2017 (Figure 3.1a) compared to 325 publications on entomopathogenic fungi in the same year. Undoubtedly, most studies on fungal endophytes have focused on species from the genus *Beauveria* found in various important crop plants (Posada and Vega, 2005; Gómez-Vidal *et al.*, 2006; Posada *et al.*, 2007; Akello *et al.*, 2008; Tefera and Vidal, 2009; Reay *et al.*, 2010; Parsa *et al.*, 2013). Other genera of EEPF include species from *Metarhizium*, *Lecanicillium*, *Acremonium*, *Isaria*, *Cladosporium* and *Clonostachys* (Vega, 2008; Moonjely *et al.*, 2016).

While entomopathogenic fungi have been used commercially in biocontrol strategies with direct modes of action against pests for decades (de Faria and Wraight, 2007), applications of EEPF for insect control and improved plant health have drawn interest only recently (Figure 3.1b). Although there is growing evidence of improved EEPF-mediated plant fitness (Ownley *et al.*, 2010; Barelli *et al.*, 2016), there are gaps in our understanding of multitrophic interactions between endophyte, plant and insect. Furthermore, current applications of fungi aiming at endophytism are mostly not supported by formulation strategies limiting efficacy particularly under field conditions (Vidal and Jaber, 2015).

Prior to formulation, EEPF have to be mass produced in a cost-effective fermentation process. In most commercialised products of entomopathogenic fungi, aerial conidia (AC) from *Beauveria* spp. and *Metarhizium* spp. are the active ingredient (de Faria and Wraight, 2007). Although AC are often tolerant to desiccation and storage, their mass production via solid-state fermentation is often non-economic, difficult to control and labour intensive (Ravensberg, 2011). With submerged cultivation techniques, fungal propagules can be produced in a short time to high yields with more process flexibility for production of specific morphologies (Lohse *et al.*, 2014). In liquid culture, *Beauveria* spp. and *Metarhizium* spp. generally form vegetative propagules, such as finely dispersed or aggregated mycelium (Adamek, 1965; Rombach *et al.*, 1988; Humphreys *et al.*, 1989; Andersch *et al.*, 1990; Magalhäes *et al.*, 1994; Leckie *et al.*, 2008) and blastospores (BS) (Bidochka *et al.*, 1987,; Thomas *et al.*, 1987; Hegedus *et al.*, 1990; Kleespies and Zimmermann, 1992; Issaly *et al.*, 2005). For some species, submerged conidia (SC) have been described (Thomas *et al.*, 1987; Hegedus *et al.*, 1992; Jenkins and Prior, 1993; Jenkins and Goettel, 1997; Kassa *et al.*, 2004; Holder *et al.*, 2007; Lohse *et al.*, 2014) and a few *Metarhizium* spp. may form microsclerotia (MS) in liquid media (Jackson and Jaronski, 2009b; Mascarin *et al.*, 2014). Although less sensitive to environmental stresses than BS (Leland *et al.*, 2005), neither SC nor MS have been commercialised yet. The first *Metarhizium*-based product containing mycelial aggregates 'BIO2010' made its appearance in the 1990s (Andersch *et al.*, 1990).

Apart from efficient mass production, other constraints to the development of EEPF as biocontrol agents have been low survival after drying and storing, poor handling and detrimental effects of environmental factors, as well as inconsistent and low efficacy. Application of formulated fungal propagules has the potential to overcome these problems as will be addressed in the following chapter.

3.2 Formulation and Application

To achieve stability and efficacy, fungi-based formulations have to ensure viability after drying, dormancy of fungal propagules during storage ideally at ambient temperature and support fungal activity continuously after application (Jaronski and Mascarin,

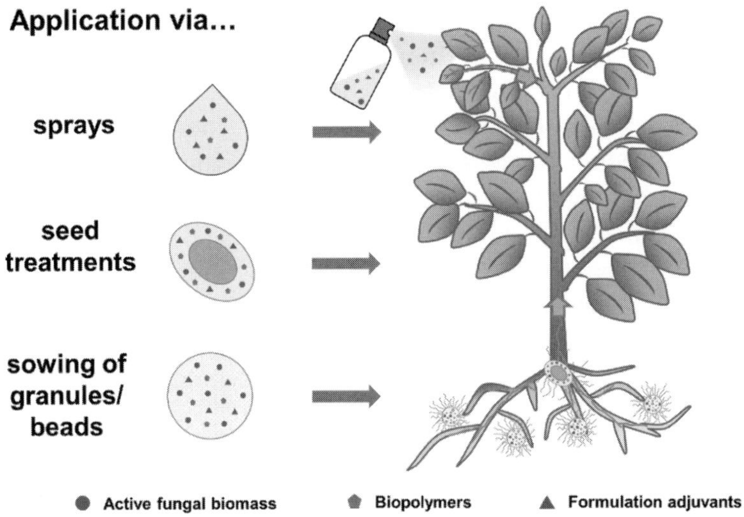

Figure 3.2 Overview of applications of formulated EEPF via sprays, seed treatments or sowing of granules and beads aiming at fungal colonisation of internal plant tissues. (A black and white version of this figure will appear in some formats. For the colour version, please refer to the plate section.)

2016). EEPF may be formulated as wettable powders and suspension concentrates for spray application or as granules and beads for soil treatment (Figure 3.2). All of these formulations may be utilised for seed treatments, such as seed coats or pills. The outcome of EEPF treatments may be more successful when the application is chosen depending on the plant tissue where the pest is most likely to occur.

Since the mode of action of EEPF has not been completely elucidated yet, their formulation remains a scientific challenge. Plant penetration by an entomopathogenic fungus was described for the first time by Wagner and Lewis (2000), who suggested the involvement of mechanical pressure and enzymatic degradation of plant cell walls. A formulation for EEPF would therefore need to deliver the fungus to the plant ensuring direct contact with plant surfaces and support establishment in plant tissues, e.g. by incorporation of penetration aids. In order to investigate endophytism and associated benefits, EEPF have to be reliably detected in plant tissues. The success may depend on the methodology and surface-sterilisation protocols as recently reviewed by McKinnon *et al.* (2017).

In the next sections, EEPF formulations will be discussed according to their application via sprays, seed treatments or sowing of granules and beads.

3.2.1 Application via Sprays

The most prevalent inoculation method for entomopathogenic fungi is application via the spraying of a spore suspension onto aboveground plant tissues for direct effects against the target herbivore. In 2007, more than 170 mycoinsecticides and mycoacaricides were commercially available. Most comprise emulsifiable suspensions or wettable powders with AC from *Beauveria* spp. and *Metarhizium* spp. and some with AC from *Isaria* spp. (de Faria and Wraight, 2007). They sometimes contain mycelium or a mixture of BS, SC and mycelium when propagules are produced by liquid fermentation. Generally, mycelium or derivatives thereof, such as MS, are not suitable for spraying due to their size. *Metarhizium*-based mycelial pellets may reach diameters of 1.0–1.5 mm (Andersch *et al.*, 1990). Though not as large, MS may occur in aggregates of 50–600 μm in diameter (Jackson and Schisler, 1995; Shearer and Jackson, 2006; Jackson and Jaronski, 2009b). Most water-based insecticide spraying is conducted with droplets of 50–150 μm to maximise coverage (Bateman and Chapple, 2001). For oil-based formulations, the optimal droplet size may be as low as 30 μm (Adams *et al.*, 1990).

Particularly for difficult-to-reach pests, because of tunnelling behaviour or their location at the bottom side of leaves, spraying of entomopathogenic fungi that become established as endophytes in leaves is attractive (Bing and Lewis, 1991; Greenfield *et al.*, 2016). Furthermore, once endophytic, entomopathogenic fungi are less affected by detrimental abiotic factors, such as solar radiation, temperature, rainfall and low humidity (Inglis *et al.*, 2001). Developing spray formulations is complex (Bejarano *et al.*, 2017) since several aspects have to be considered particularly for application of EEPF. First, coverage of leaves depends on the contact angle indicating the wettability of a surface (Yuan and Lee, 2013). Small contact angles correspond to high wettability, which can be achieved by incorporation of surfactants, such as tritons (Singh *et al.*, 1984). If coverage is high, the chance of endophytism may be increased due to the availability of larger plant surfaces for penetration. Second, retention of the active ingredient is crucial especially under field conditions with frequent rainfall or after application of large volumes when wash-off is a concern. By supplementation of adhesion aids, such as carboxymethyl cellulose (CMC), xanthan or low volatile oils, retention on leaves can be increased remarkably (Hazen, 2000). Liquid oil formulations have further advantages: they are excellent carriers for hydrophobic AC, can be atomised into small droplets of 50–100 μm and are easily dispensed in water in the presence of emulsifiers (Wraight *et al.*, 2001). Furthermore, they may serve as humectants providing the prolonged humidity required for spore germination and penetration of plant tissue. However, mineral oil formulations have to be applied with caution since they may have phytotoxic effects (Hodgkinson *et al.*, 2002).

Endophytism may be supported additionally by nutrients and some of them, such as molasses, may also provide protection from solar radiation important for UV sensitive fungal spores (Zimmermann, 1982, Lohse *et al.*, 2015). Further UV protectants incorporated into liquid formulations of entomopathogenic fungi include oils (Moore *et al.*, 1993), clay, brighteners (Inglis *et al.*, 1995) or chemical sunscreens, like oxybenzone (Shah *et al.*, 1998). By applying during the late afternoon or early evening, fungal establishment on the plant surface may be further supported since UV radiation is minimised and humidity levels are higher.

So far, most research groups have utilised simple water-based spray formulations with a surfactant required for hydrophobic AC. Gurulingappa *et al.* (2010) formulated *B. bassiana*, *L. lecanii* and *A. parasiticus* AC in water with 0.02% Triton X-100 and achieved endophytism in cotton, wheat, bean, tomato and pumpkin. Posada *et al.* (2007) successfully inoculated coffee plant leaves with *B. bassiana* AC suspended in 0.1% Triton X-100. Different concentrations of Tween 80 in water ranging from 0.01 to 0.1% were utilised for preparation of *Beauveria* spp. and *Metarhizium* spp. AC suspensions which resulted in endophytic establishment of fungi after application to leaves of artichoke (Guesmi-Jouini *et al.*, 2014), broad bean (Jaber and Enkerli, 2017), poppy (Quesada-Moraga *et al.*, 2006; Quesada-Moraga *et al.*, 2009), potato (Ríos-Moreno *et al.*, 2016), sorghum (Tefera and Vidal, 2009; Raya-Díaz *et al.*, 2017), alfalfa, tomato and melon (Resquín-Romero *et al.*, 2016).

While several studies have utilised simple AC formulations to investigate endophytism after EEPF spraying, no commercial formulations with BS as the active ingredient have been developed for endophytic applications so far. Lohse *et al.* (2015) looked into the effects of a more sophisticated spray formulation consisting of 0.1% Triton X-114, 1% sugar beet molasses, 1% titanium dioxide and a mixture of BS and SC from *B. bassiana* fermentation to support fungal endophytism. Non-ionic triton surfactants increase the contact area and may support transcuticular penetration (Boeger *et al.*, 2012). Sugar beet molasses mainly serve as a nutrient to enhance germination and plant penetration. Titanium dioxide was incorporated into the formulation to protect the fungus from detrimental UV radiation, a major drawback under field conditions (Burges, 2012). Application of this formulation to oilseed rape plant leaves resulted in endophytic establishment of *B. bassiana* determined after 14 days (Lohse *et al.*, 2015).

3.2.2 Application via Seed Treatments

A further technique to inoculate plants with EEPF is to coat seeds with a formulation of AC, BS, SC or mycelial structures able to colonise the seed, adjacent roots or the shoot during germination and plant growth. Seed treatments with chemical fungicides or pesticides are commonly conducted in conventional plant protection and the technique may be adapted to EEPF applications. Apart from the active ingredient, seed coating formulations consist of carrier materials, often polymeric and hygroscopic in nature, that may include nutrients or fillers (Pedrini *et al.*, 2017). Common carrier materials include CMC, gum arabic, sucrose solutions or vegetable oils (Bashan *et al.*, 2014). In some cases, additional pelleting of seeds may be beneficial to increase loading of seeds with fungal propagules. It is crucial that all formulation adjuvants are compatible with the active ingredient and do not inhibit but rather promote seed germination and plant growth. Furthermore, the formulation should be stable against abrasion and during drying and should provide protection from the suppressive effects of the surrounding soil, such as antagonistic microorganisms. For integrated pest management, the formulations should also tolerate or preferably act synergistically with other commonly applied seed treatments. Particularly for the delivery of EEPF, integration of additives supporting fungal growth and plant penetration may be beneficial.

Several reports on seed treatments with EEPF, commonly termed seed dipping, dressing or soaking, utilised simple formulations based on AC from *Beauveria* spp. and *Metarhizium* spp. in 0.01–0.1% Tween 80 in water. After sowing, thus applied fungi established endophytically in broad bean (Jaber and Enkerli, 2016), poppy (Quesada-Moraga *et al.*, 2009), sorghum (Tefera and Vidal, 2009; Raya-Díaz *et al.*, 2017), tomato and wheat (Sánchez-Rodríguez *et al.*, 2015).

Studies investigating more sophisticated formulations are limited and most of the developed seed coat/pill formulations are protected by biocontrol companies. However, some reports on seed coats for EEPF with endophytic establishment in plants can be found. Powell *et al.* (2009) coated tomato seeds with 2% CMC and *B. bassiana* AC. Similarly, Brownbridge *et al.* (2012) utilised 2% CMC or 0.2% xanthan to treat radiata pine seeds.

Since commercial seeds are commonly treated with fungicides, the compatibility of the seed treatment with the coated compound has to be ensured. Lohse *et al.* (2015) incorporated 2% gelatine and 1–4% autoclaved baker's yeast into a suspension of 2% CMC and a *B. bassiana* BS/SC mixture to coat oilseed rape plant seeds previously treated with the fungicides Thiram and Dimethomorph. Indeed, *B. bassiana* was able to tolerate the fungicides to a certain degree indicating that some EEPF-fungicide combinations may be feasible in the field.

3.2.3 Application via Sowing of Granules and Beads

For soil applications, the delivery of immobilised EEPF via granules or beads is promising. While granules are produced by agglomeration of particles (Shanmugam, 2015), beads are formed from droplets (Vemmer and Patel, 2013). Granules and beads may be applied during sowing under, above or alongside the seed with conventional sowing or fertiliser equipment, such as granulate spreaders, so that the fungus may be able to colonise the plant after having grown out of the carrier depot. For small particles <200 μm, application onto the foliage may be considered (Bing and Lewis, 1991). Basically, granules and beads offer freedom of choice for fungal morphologies to be incorporated but mycelial structures which include finely dispersed or aggregated mycelium as well as mycelial derivatives like MS seem favourable due to their robustness, shelf life and conidiation potential. The characteristics of the entrapped EEPF regarding improved handling, extended shelf life, protection from abiotic and biotic stress factors, as well as enhanced endophytism can be substantially improved by encapsulation systems (Vemmer and Patel, 2013; Krell *et al.*, 2017b). Furthermore, biomass contents may be reduced since the fungus is incorporated into a microbioreactor providing good conditions for proliferation (Przyklenk *et al.*, 2017). There is a wide range of encapsulation methods and matrix materials, such as alginate, starch and cellulose as well as formulation additives to be chosen from including peat, perlite, charcoal, diatomaceous earth, talc, clay, vermiculite, corn grits, inactivated baker's yeast, gelatine and synthetic polymers (Pereira and Roberts, 1990; Pereira and Roberts, 1991; Bernhard *et al.*, 1998; Marques *et al.*, 1999; Jackson and Jaronski, 2009a; Bashan *et al.*, 2014; Mayerhofer *et al.*, 2017; Przyklenk *et al.*, 2017). For endophytic applications, supplementation of plant cell wall degrading enzymes can be beneficial (Krell *et al.*, 2018).

For entomopathogenic fungi, there are numerous granular formulations and bead systems available and some of them have already been commercialised. However, of those, only a few have been utilised in endophytic studies. Bing and Lewis (1991) applied granules consisting of corn grits (30–40 mesh) sprayed with *B. bassiana* AC onto the foliage of corn plants and found the fungus established endophytically. Lohse *et al.* (2015) found that encapsulation of 1% of a *B. bassiana* BS/SC mixture in 2% Ca-alginate with 1% autoclaved baker's yeast and 2% gelatine significantly increased the radial mycelial growth from beads. Fungal endophytism was detected in the stem of oilseed rape plants when applied to sterile soil. Krell *et al.* (2017b) encapsulated 1.5% *M. brunneum* mycelium with 20% corn starch in 2% Ca-alginate with the aim of enhancing endophytism in tomato plants. Due to the protective and nutritive microenvironment, *M. brunneum* was able to grow out of the bead on both sterile and non-sterile soil and endophytism was found to be enhanced by encapsulation. A follow-up study investigated whether endophytism in potato plants could be improved by penetration aids. Applying the available knowledge on penetration mechanisms of EEPF, it was first investigated whether the addition of plant cell wall degrading enzymes, pectinase, hemicellulase and cellulose, to non-formulated mycelium would benefit *M. brunneum* endophytism. Interestingly, cellulase significantly improved endophytism compared to non-treated mycelium and mycelium treated with pectinase or hemicellulase (Krell *et al.*, unpublished). Supplementation of 1 Ug^{-1} cellulase to beads with 2% Ca-pectinamide, 20% corn starch, 2% autoclaved baker's yeast, 2% cellulose and 1.5% *M. brunneum* mycelium substantially improved spore formation on the bead surface and this correlated with enhanced endophytism by 61.2% (Krell *et al.*, 2018).

3.3 Conclusions and Prospects

In recent years, evidence of the potential of EEPF to control herbivorous pests and to improve plant health has accumulated but the mode of action remains mostly obscured. The current lack of formulation strategies supporting endophytism further limits consistency and efficacy particularly under field conditions. Applications via spray formulations, seed treatments, granules or beads are suitable delivery systems for EEPF allowing for endophytic establishment when in contact with plant surfaces. The range of available formulation techniques and materials offers great potential for improving the characteristics of EEPF-based product candidates. Endophytism may be improved by protecting the active ingredient from adverse environmental conditions and by promoting fungal growth at the place of application. Addition of plant cell wall degrading enzymes can support natural traits of EEPF to achieve entry through plant cell walls. Conditioning of EEPF for improved endophytism may be an additional strategy. By incorporating plant cell wall materials, such as cellulose, into fermentation medium, cellulase expression can be triggered allowing for enhanced establishment in plant tissue (Krell *et al.*, 2017a). Similarly, supplementation of cellulose into beads may induce cellulase activity. Encapsulation systems consisting entirely of plant cell wall materials, such as pectin, cellulose and hemicellulose, may also promote endophytism.

Undeniably, the potential of EEPF has not been fully exploited yet. By advancing in our understanding of the mode of action of EEPF and their ecology, we may apply this knowledge to develop targeted formulation strategies for future applications of EEPF in integrated pest management.

References

Adamek, A. (1965). Submerse cultivation of the fungus *Metarhizium anisopliae*. *Folia Microbiologica*, **4**, 255–257.

Adams, A. J., Chappie, A. and Hall, F. R. (1990). Droplet spectra for some agricultural fan nozzles, with respect to drift and biological efficiency. In *Pesticide Formulations and Application Systems*, ed. L. E. Bode, J. L. Hazen and D. G. Chasin. West Conshohocken, PA: ASTM International, pp. 156–169.

Akello, J., Dubois, T., Coyne, D. and Kyamanywa, S. (2008). Effect of endophytic *Beauveria bassiana* on populations of the banana weevil, *Cosmopolites sordidus*, and their damage in tissue-cultured banana plants. *Entomologia Experimentalis et Applicata*, **129**, 157–165.

Andersch, W., Hartwig, J., Reinecke, P. and Stenzel, K. (1990). Production of mycelial granules of the entomopathogenic fungus *Metarhizium anisopliae* for biological control of soil pests. In *5th International Colloquium on Invertebrate Pathology and Microbial Control: incorporating the 23rd Annual Meeting of the Society for Invertebrate Pathology: Proceedings and Abstracts.* Adelaide: Society for Invertebrate Pathology, pp. 2–5.

Arnold, A. E. and Lutzoni, F. (2007). Diversity and host range of foliar fungal endophytes: are tropical leaves biodiversity hotspots? *Ecology*, **88**, 541–549.

Barelli, L., Moonjely, S., Behie, S. W. and Bidochka, M. J. (2016). Fungi with multifunctional lifestyles: endophytic insect pathogenic fungi. *Plant Molecular Biology*, **90**, 657–664.

Bashan, Y., De-Bashan, L. E., Prabhu, S. and Hernandez, J.-P. (2014). Advances in plant growth-promoting bacterial inoculant technology: formulations and practical perspectives (1998–2013). *Plant and Soil*, **378**, 1–33.

Bateman, R. and Chapple, A. (2001). The spray application of mycopesticide formulations. In *Fungi as Biocontrol Agents: Progress Problems and Potential*, ed. T. M. Butt, C. Jackson and N. Magan. Wallingford: CABI Publishing, pp. 289–309.

Bejarano, A., Sauer, U. and Preininger, C. (2017). Design and development of a workflow for microbial spray formulations including decision criteria. *Applied Microbiology and Biotechnology*, **101**, 7335–7346.

Bernhard, K., Holloway, P. J. and Burges, H. D. (1998). A catalogue of formulation additives: Function, nomenclature, properties and suppliers. In *Formulation of Microbial Biopesticides: Beneficial Microorganisms, Nematodes and Seed Treatments*, ed. H. D. Burges. Dordrecht, The Netherlands: Kluwer Academic Publishers, pp. 333–365.

Bidochka, M. J., Pfeifer, T. A. and Khachatourians, G. G. (1987). Development of the entomopathogenic fungus *Beauveria bassiana* in liquid cultures. *Mycopathologia*, **99**, 77–83.

Bing, L. A. and Lewis, L. C. (1991). Suppression of *Ostrinia nubilalis* (Huebner) (Lepidoptera: Pyralidae) by endophytic *Beauveria bassiana* (Balsamo) Vuillemin. *Environmental Entomology*, **20**, 1207–1211.

Boeger, P., Wakabayashi, K. and Hirai, K., (2012). *Herbicide Classes in Development: Mode of Action, Targets, Genetic Engineering, Chemistry*. Heidelberg, Germany: Springer.

Brownbridge, M., Reay, S. D., Nelson, T. L. and Glare, T. R. (2012). Persistence of *Beauveria bassiana* (Ascomycota: Hypocreales) as an endophyte following inoculation of radiata pine seed and seedlings. *Biological Control*, **61**, 194–200.

Burges, H. D. (2012). Formulation of mycoinsecticides. In *Formulation of Microbial Biopesticides: Beneficial Microorganisms, Nematodes and Seed Treatments*, ed. H. D. Burges. Dordrecht, The Netherlands: Kluwer Academic Publishers, pp. 131–186.

Clay, K. (1989). Clavicipitaceous endophytes of grasses: their potential as biocontrol agents. *Mycological Research*, **92**, 1–12.

De Bary, A. (1866). *Morphologie und Physiologie der Pilze, Flechten und Myxomyceten*. Leipzig: W. Engelmann.

De Faria, M. R. and Wraight, S. P. (2007). Mycoinsecticides and mycoacaricides: a comprehensive list with worldwide coverage and international classification of formulation types. *Biological Control*, **43**, 237–256.

Gómez-Vidal, S., Lopez-Llorca, L., Jansson, H.-B. and Salinas, J. (2006). Endophytic colonization of date palm (*Phoenix dactylifera* L.) leaves by entomopathogenic fungi. *Micron*, **37**, 624–632.

Greenfield, M., Gómez-Jiménez, M. I., Ortiz, V., Vega, F. E., Kramer, M. and Parsa, S. (2016). *Beauveria bassiana* and *Metarhizium anisopliae* endophytically colonize cassava roots following soil drench inoculation. *Biological Control*, **95**, 40–48.

Guesmi-Jouini, J., Garrido-Jurado, I., López-Díaz, C., Halima-Kamel, M. B. and Quesada-Moraga, E. (2014). Establishment of fungal entomopathogens *Beauveria bassiana* and *Bionectria ochroleuca* (Ascomycota: Hypocreales) as endophytes on artichoke *Cynara scolymus*. *Journal of Invertebrate Pathology*, **119**, 1–4.

Gurulingappa, P., Sword, G. A., Murdoch, G. and McGee, P. A. (2010). Colonization of crop plants by fungal entomopathogens and their effects on two insect pests when in planta. *Biological Control*, **55**, 34–41.

Hardoim, P. R., Van Overbeek, L. S., Berg, G. *et al.* (2015). The hidden world within plants: ecological and evolutionary considerations for defining functioning of microbial endophytes. *Microbiology and Molecular Biology Reviews*, **79**, 293–320.

Hazen, J. L. (2000). Adjuvants – terminology, classification, and chemistry. *Weed Technology*, **14**, 773–784.

Hegedus, D. D., Bidochka, M. J. and Khachatourians, G. G. (1990). *Beauveria bassiana* submerged conidia production in a defined medium containing chitin, two hexosamines or glucose. *Applied Microbiology and Biotechnology*, **33**, 641–647.

Hegedus, D. D., Bidochka, M. J., Miranpuri, G. S. and Khachatourians, G. G. (1992). A comparison of the virulence, stability and cell-wall-surface characteristics of three spore types produced by the entomopathogenic fungus *Beauveria bassiana*. *Applied Microbiology and Biotechnology*, **36**, 785–789.

Hodgkinson, M. C., Johnson, D., Smith, G. *et al.* (2002). Causes of phytotoxicity induced by petroleum-derived spray oil. In *Spray Oils -Beyond 2000*, ed. G. A. C. Beattie, D. M. Watson, M. L. Stevens, D. J. Rae and R. N. Spooner-Hart. Sydney, Australia: University of Western Sydney, pp. 170–178.

Holder, D. J., Kirkland, B. H., Lewis, M. W. and Keyhani, N. O. (2007). Surface characteristics of the entomopathogenic fungus *Beauveria* (Cordyceps) *bassiana*. *Microbiology*, **153**, 3448–3457.

Humphreys, A. M., Matewele, P., Trinci, A. P. and Gillespie, A. T. (1989). Effects of water activity on morphology, growth and blastospore production of *Metarhizium anisopliae*, *Beauveria bassiana* and *Paecilomyces farinosus* in batch and fed-batch culture. *Mycological Research*, **92**, 257–264.

Inglis, G. D., Goettel, M. and Johnson, D. (1995). Influence of ultraviolet light protectants on persistence of the entomopathogenic fungus, *Beauveria bassiana*. *Biological Control*, **5**, 581–590.

Inglis, G. D., Goettel, M. S., Butt, T. M. and Strasser, H. (2001). Use of hyphomycetous fungi for managing insect pests. In *Fungi as Biocontrol Agents: Progress Problems and Potential*, ed. T. M. Butt, C. Jackson and N. Magan. Wallingford: CABI Publishing, pp. 23–70.

Issaly, N., Chauveau, H., Aglevor, F., Fargues, J. and Durand, A. (2005). Influence of nutrient, pH and dissolved oxygen on the production of *Metarhizium flavoviride* Mf189 blastospores in submerged batch culture. *Process Biochemistry*, **40**, 1425–1431.

Jaber, L. R. and Enkerli, J. (2016). Effect of seed treatment duration on growth and colonization of *Vicia faba* by endophytic *Beauveria bassiana* and *Metarhizium brunneum*. *Biological Control*, **103**, 187–195.

Jaber, L. R. and Enkerli, J. (2017). Fungal entomopathogens as endophytes: can they promote plant growth? *Biocontrol Science and Technology*, **27**, 28–41.

Jackson, M. A. and Jaronski, S. T. (2009a). Composition of entomopathogenic fungus and method of production and application for insect control. USA patent application 11/901,547. 19th March, 2009.

Jackson, M. A. and Jaronski, S. T. (2009b). Production of microsclerotia of the fungal entomopathogen *Metarhizium anisopliae* and their potential for use as a biocontrol agent for soil-inhabiting insects. *Mycological Research*, **113**, 842–850.

Jackson, M. A. and Schisler, D. A. (1995). Liquid culture production of microsclerotia of *Colletotrichum truncatum* for use as bioherbicidal propagules. *Mycological Research*, **99**, 879–884.

Jaronski, S. T. and Mascarin, G. M. (2016). Mass production of fungal entomopathogens. In *Microbial Control of Insect and Mite Pests: From Theory to Practice*, ed. L. A. Lacey. London: Elsevier Science, pp. 141–156.

Jenkins, N. E. and Goettel, M. S. (1997). Methods for mass-production of microbial control agents of grasshoppers and locusts. *Memoirs of the Entomological Society of Canada*, **129**, 37–48.

Jenkins, N. and Prior, C. (1993). Growth and formation of true conidia by *Metarhizium flavoviride* in a simple liquid medium. *Mycological Research*, **97**, 1489–1494.

Johnson, L. J., De Bonth, A. C., Briggs, L. R. *et al.* (2013). The exploitation of epichloae endophytes for agricultural benefit. *Fungal Diversity*, **60**, 171–188.

Kassa, A., Stephan, D., Vidal, S. and Zimmermann, G. (2004). Production and processing of *Metarhizium anisopliae* var. *acridum* submerged conidia for locust and grasshopper control. *Mycological Research*, **108**, 93–100.

Kleespies, R. and Zimmermann, G. (1992). Production of blastospores by three strains of *Metarhizium anisopliae* (Metch.) Sorokin in submerged culture. *Biocontrol Science and Technology*, **2**, 127–135.

Krell, V., Jakobs-Schoenwandt, D., Hettlage, L., Vidal, S. and Patel, A. V. (2017a). Exoenzymes improve penetration and colonization of potato plants by endophytic entomopathogenic *Metarhizium brunneum*. In *16th meeting of the IOBC-WPRS Working Group Microbial and Nematode Control of Invertebrate Pests*, ed. E. Tarasco, J. A. Jehle, M. Burjanadze, L. Ruiu, V. Půža, E. Quesada-Moraga, M. Lopez-Ferber and D. Stephan. Tbilisi, Georgia: IOBC-WPRS, 65–69.

Krell, V., Jakobs-Schoenwandt, D., Vidal, S. and Patel, A. V. (2017b). Encapsulation of *Metarhizium brunneum* enhances endophytism in tomato plants. *Biological Control*, **116**, 62–73.

Krell, V., Jakobs-Schoenwandt, D., Vidal, S. and Patel, A. V. (2018). Cellulase enhances endophytism of encapsulated *Metarhizium brunneum* in potato plants. *Fungal Biology*, **122**, 373–378.

Leckie, B. M., Ownley, B. H., Pereira, R. M. *et al.* (2008). Mycelia and spent fermentation broth of *Beauveria bassiana* incorporated into synthetic diets affect mortality, growth and development of larval *Helicoverpa zea* (Lepidoptera: Noctuidae). *Biocontrol Science and Technology*, **18**, 697–710.

Leland, J. E., Mullins, D. E., Vaughan, L. J. and Warren, H. L. (2005). Effects of media composition on submerged culture spores of the entomopathogenic fungus, *Metarhizium anisopliae* var.

acridum, part 1: comparison of cell wall characteristics and drying stability among three spore types. *Biocontrol Science and Technology*, **15**, 379–392.

Lohse, R., Jakobs-Schönwandt, D. and Patel, A. V. (2014). Screening of liquid media and fermentation of an endophytic *Beauveria bassiana* strain in a bioreactor. *AMB Express*, **4**, 47.

Lohse, R., Jakobs-Schönwandt, D., Vidal, S. and Patel, A. V. (2015). Evaluation of new fermentation and formulation strategies for a high endophytic establishment of *Beauveria bassiana* in oilseed rape plants. *Biological Control*, **88**, 26–36.

Magalhäes, B. P., Dias, J. M. C. D. S. and Ferreira, C. M. (1994). Mycelial production of *Metarhizium anisopliae* in liquid culture using different sources of carbon and nitrogen. *Revista de Microbiologia*, **25**, 181–187.

Marques, E. J., Alves, S. B. and Marques, I. M. (1999). Effects of the temperature and storage on formulations with mycelia of *Beauveria bassiana* (Bals.) Vuill and *Metarhizium anisopliae* (Metschn.) Sorok. *Brazilian Archives of Biology and Technology*, **42**, 153–160.

Mascarin, G. M., Kobori, N. N., De Jesus Vital, R. C., Jackson, M. A. and Quintela, E. D. (2014). Production of microsclerotia by Brazilian strains of *Metarhizium* spp. using submerged liquid culture fermentation. *World Journal of Microbiology and Biotechnology*, **30**, 1583–1590.

Mayerhofer, J., Eckard, S., Hartmann, M. *et al.* (2017). Assessing effects of the entomopathogenic fungus *Metarhizium brunneum* on soil microbial communities in *Agriotes* spp. biological pest control. *FEMS Microbiology Ecology*, **93**, fix117, 1–15.

McKinnon, A. C., Saari, S., Moran-Diez, M. E. *et al.* (2017). *Beauveria bassiana* as an endophyte: a critical review on associated methodology and biocontrol potential. *BioControl*, **62**, 1–17.

Moonjely, S., Barelli, L. and Bidochka, M. (2016). Insect pathogenic fungi as endophytes. *Advances in Genetics*, **94**, 107–135.

Moore, D., Bridge, P., Higgins, P., Bateman, R. and Prior, C. (1993). Ultra-violet radiation damage to *Metarhizium flavoviride* conidia and the protection given by vegetable and mineral oils and chemical sunscreens. *Annals of Applied Biology*, **122**, 605–616.

Ownley, B. H., Gwinn, K. D. and Vega, F. E. (2010). Endophytic fungal entomopathogens with activity against plant pathogens: ecology and evolution. *BioControl*, **55**, 113–128.

Parsa, S., Ortiz, V. and Vega, F. E. (2013). Establishing fungal entomopathogens as endophytes: towards endophytic biological control. *Journal of Visualized Experiments*, **74,** 50360.

Pedrini, S., Merritt, D. J., Stevens, J. and Dixon, K. (2017). Seed coating: science or marketing spin? *Trends in Plant Science*, **22**, 106–116.

Pereira, R. M. and Roberts, D. W. (1990). Dry mycelium preparations of entomopathogenic fungi, *Metarhizium anisopliae* and *Beauveria bassiana*. *Journal of Invertebrate Pathology*, **56**, 39–46.

Pereira, R. M. and Roberts, D. W. (1991). Alginate and cornstarch mycelial formulations of entomopathogenic fungi, *Beauveria bassiana* and *Metarhizium anisopliae*. *Journal of Economic Entomology*, **84**, 1657–1661.

Posada, F. and Vega, F. E. (2005). Establishment of the fungal entomopathogen *Beauveria bassiana*

(Ascomycota: Hypocreales) as an endophyte in cocoa seedlings (*Theobroma cacao*). *Mycologia*, **97**, 1195–1200.

Posada, F., Aime, M. C., Peterson, S. W., Rehner, S. A. and Vega, F. E. (2007). Inoculation of coffee plants with the fungal entomopathogen *Beauveria bassiana* (Ascomycota: Hypocreales). *Mycological research*, **111**, 748–757.

Powell, W. A., Klingeman, W. E., Ownley, B. H. and Gwinn, K. D. (2009). Evidence of endophytic *Beauveria bassiana* in seed-treated tomato plants acting as a systemic entomopathogen to larval *Helicoverpa zea* (Lepidoptera: Noctuidae). *Journal of Entomological Science*, **44**, 391–396.

Przyklenk, M., Vemmer, M., Hanitzsch, M. and Patel, A. (2017). A bioencapsulation and drying method increases shelf life and efficacy of *Metarhizium brunneum* conidia. *Journal of Microencapsulation*, **34**, 498–512.

Quesada-Moraga, E., Landa, B., Muñoz-Ledesma, J., Jiménez-Diáz, R. and Santiago-Alvarez, C. (2006). Endophytic colonisation of opium poppy, *Papaver somniferum*, by an entomopathogenic *Beauveria bassiana* strain. *Mycopathologia*, **161**, 323–329.

Quesada-Moraga, E., Munoz-Ledesma, F. and Santiago-Alvarez, C. (2009). Systemic protection of *Papaver somniferum* L. against *Iraella luteipes* (Hymenoptera: Cynipidae) by an endophytic strain of *Beauveria bassiana* (Ascomycota: Hypocreales). *Environmental Entomology*, **38**, 723–730.

Ravensberg, W. J. (2011). *A Roadmap to the Successful Development and Commercialization of Microbial Pest Control Products for Control of Arthropods*. Luxembourg: Springer Science & Business Media.

Raya-Díaz, S., Sánchez-Rodríguez, A. R., Segura-Fernández, J. M., Del Campillo, M. D. C. and Quesada-Moraga, E. (2017). Entomopathogenic fungi-based mechanisms for improved Fe nutrition in sorghum plants grown on calcareous substrates. *PloS One*, **12**, e0185903.

Reay, S., Brownbridge, M., Gicquel, B., Cummings, N. and Nelson, T. (2010). Isolation and characterization of endophytic *Beauveria* spp. (Ascomycota: Hypocreales) from *Pinus radiata* in New Zealand forests. *Biological Control*, **54**, 52–60.

Resquín-Romero, G., Garrido-Jurado, I., Delso, C., Ríos-Moreno, A. and Quesada-Moraga, E. (2016). Transient endophytic colonizations of plants improve the outcome of foliar applications of mycoinsecticides against chewing insects. *Journal of Invertebrate Pathology*, **136**, 23–31.

Ríos-Moreno, A., Garrido-Jurado, I., Resquín-Romero, G., Arroyo-Manzanares, N., Arce, L. and Quesada-Moraga, E. (2016). Destruxin A production by *Metarhizium brunneum* strains during transient endophytic colonisation of *Solanum tuberosum*. *Biocontrol Science and Technology*, **26**, 1574–1585.

Rodriguez, R., White Jr, J., Arnold, A. and Redman, A. R. A. (2009). Fungal endophytes: diversity and functional roles. *New Phytologist*, **182**, 314–330.

Rombach, M., Aguda, R. and Roberts, D. W. (1988). Production of *Beauveria bassiana* [Deuteromycotina: Hyphomycetes] in different liquid media and subsequent conidiation of dry mycelium. *Entomophaga*, **33**, 315–324.

Sánchez-Rodríguez, A. R., Del Campillo, M. C. and Quesada-Moraga, E. (2015). *Beauveria bassiana*: An entomopathogenic fungus alleviates Fe chlorosis symptoms in plants grown on calcareous substrates. *Scientia Horticulturae*, **197**, 193–202.

Schulz, B. and Boyle, C. (2005). The endophytic continuum. *Mycological Research*, **109**, 661–686.

Shah, P., Douro-Kpindou, O.-K., Sidibe, A., Daffe, C., Van Der Pauuw, H. and Lomer, C. (1998). Effects of the sunscreen oxybenzone on field efficacy and persistence of *Metarhizium flavoviride* conidia against *Kraussella amabile* (Orthoptera: Acrididae) in Mali, West Africa. *Biocontrol Science and Technology*, **8**, 357–364.

Shanmugam, S. (2015). Granulation techniques and technologies: recent progresses. *BioImpacts*, **5**, 55–63.

Shearer, J. F. and Jackson, M. A. (2006). Liquid culturing of microsclerotia of *Mycoleptodiscus terrestris*, a potential biological control agent for the management of hydrilla. *Biological Control*, **38**, 298–306.

Singh, M., Orsenigo, J. and Shah, D. (1984). Surface tension and contact angle of herbicide solutions affected by surfactants. *Journal of the American Oil Chemists' Society*, **61**, 596–600.

Tefera, T. and Vidal, S. (2009). Effect of inoculation method and plant growth medium on endophytic colonization of sorghum by the entomopathogenic fungus *Beauveria bassiana*. *BioControl*, **54**, 663–669.

Thomas, K., Khachatourians, G. and Ingledew, W. (1987). Production and properties of *Beauveria bassiana* conidia cultivated in submerged culture. *Canadian Journal of Microbiology*, **33**, 12–20.

Vega, F. E. (2008). Insect pathology and fungal endophytes. *Journal of Invertebrate Pathology*, **98**, 277–279.

Vemmer, M. and Patel, A. V. (2013). Review of encapsulation methods suitable for microbial biological control agents. *Biological Control*, **67**, 380–389.

Vidal, S. and Jaber, L. R. (2015). Entomopathogenic fungi as endophytes: plant–endophyte–herbivore interactions and prospects for use in biological control. *Current Science*, **109**, 46–54.

Wagner, B. L. and Lewis, L. C. (2000). Colonization of corn, *Zea mays*, by the entomopathogenic fungus *Beauveria bassiana*. *Applied and Environmental Microbiology*, **66**, 3468–3473.

Wraight, S. P., Jackson, M. A. and De Kock, S. L. (2001). Production, stabilization and formulation of fungal biocontrol agents. In *Fungi as Biocontrol Agents: Progress Problems and Potential*, ed. T. M. Butt, C. Jackson and N. Magan. Wallingford: CABI Publishing, pp. 253–288.

Yuan, Y. and Lee, T. R. (2013). Contact angle and wetting properties. In *Surface Science Techniques*, ed. B. Holst and G. Bracco. Heidelberg, Germany: Springer, pp. 3–34.

Zimmermann, G. (1982). Effect of high temperatures and artificial sunlight on the viability of conidia of *Metarhizium anisopliae*. *Journal of Invertebrate Pathology*, **40**, 36–40.

4

Crossing Frontiers: Endophytic Entomopathogenic Fungi for Biological Control of Plant Diseases

Lorena Barra-Bucarei, Andrés France and Paz Millas

Abstract

World population growth has generated a demand for increased healthy food production and the development of sustainable agricultural technologies to replace environmentally damaging farming practices, including the overuse of pesticides. Biological control using entomopathogenic fungi, as natural pathogens of arthropod pests, is an alternative method to meet this objective. These fungi have been traditionally studied to control insects, but recent studies have begun to examine their activity as plant endophytes to protect plants against phytopathogens and improve other aspects of crop production. This chapter reviews the importance of these entomopathogenic fungi as endophytes in the context of biological control. Our studies focus on determining the ability of native strains of entomopathogenic fungi for endophytic colonisation and their potential application for the control of diseases in tomato. The chapter also discusses aspects to consider for their development as commercial biopesticides and suggests ways to make this control method available to producers of agricultural crops.

4.1 Introduction

Entomopathogenic fungi are a unique and highly specialised group of microbes that colonise and kill insects. They have several desirable traits making them potential candidates for biopesticides (Lacey *et al.*, 2015). Some genera, such as *Beauveria* and *Metarhizium*, consist of species that are distributed worldwide, live in the soil and are insect pathogens. *Beauveria bassiana* (Balsamo-Crivelli) Vuillemin and *M. anisopliae* (Metchnikoff) Sorokin (Ascomycota: Hypocreales), have been reported as natural and facultative endophytes of many plant species, including some agricultural crops. Endophytic fungi are able to invade plant tissues and live inside the plant during all or part of their life cycle (Wilson, 1995). Facultative entomopathogenic fungal endophytes have been the focus of much research because of their potential in the development of commercial products aimed at pest and/or biological control of plant diseases. Some authors have studied a dual effect, that is the ability of these fungi to act as insect pathogens and be antagonistic to phytopathogens at the same time (Ownley *et al.*, 2004).

It is important to consider that there are several biotic (Vicari *et al.*, 2002) and abiotic (Bultman and Bell, 2003) factors affecting the establishment of entomopathogenic fungi as endophytes. Among the biotic factors, those associated with the plant include host species/variety, age, tissue type and endophytic microbiota (Parsa *et al.*, 2013; Posada *et al.*, 2010). Studies conducted by Vega *et al.* (2008) determined that the presence of endophytic fungi in coffee plants can affect the establishment of another endophytic fungus, *B. bassiana*. Other aspects that influence establishment are related to the fungus, such as the strain, inoculum concentration (Parsa *et al.*, 2013; Vidal and Lara, 2015), and association with other microorganisms such as bacteria (Fisher and Petrini, 1987). Among abiotic factors, the type of substrate/soil where the plants grow has been shown to be important (Lingg and Donaldson, 1981; Pereira *et al.*, 1993). So too have other environmental factors, such as temperature, radiation and humidity, to which the plants are exposed, as well as their nutritional status (Vidal and Lara, 2015). The mode of application also affects establishment and determines persistence as an endophyte (Quesada-Moraga *et al.*, 2014).

There is a need to increase the number of studies on entomopathogenic fungi and also explore new areas, like field adaptations, effectiveness against insect and microbial diseases, biological mechanisms and compatibility with other biological agents. Therefore, this chapter aims to review the importance of entomopathogenic fungi in the context of sustainable agriculture and crop production, review their use as endophytes for biological control of pests and diseases, and evaluate their potential in the development of commercial biocontrol products.

4.2 Importance of Biological Control in the Context of Sustainable Agricultural Production

Growing world demand for food has intensified agricultural production, resulting in adverse effects on the environment and people's health (Tilman *et al.*, 2002) including the indiscriminate use of pesticides. It is estimated that yield losses caused by pests (including invertebrates, pathogens and weeds) can vary from 27% to 42%; this range could increase from 48% to 83% if crops were not protected (Oerke and Dehne, 2004). A worldwide consensus has emerged about the necessity to develop new strategies for achieving sustainable agricultural production that does not compromise public health or the integrity of the environment. The use of chemical pesticides is creating environmental toxicity problems and expected control efficiency levels are often not met because the phytopathogen can develop different mechanisms of resistance to the chemical agents (Grille *et al.*, 2001; Leroux *et al.*, 2010). These chemicals also increase production costs. There is, therefore, a need to develop biological control as an alternative to reduce chemical inputs.

Biological control reduces pest populations using natural enemies and normally involves human intervention for its application (Hoffmann and Frodsham, 1993). Biological control agents are classified as predators, parasitoids and pathogens (pathogens are produced by microorganisms and cause diseases). Once identified, isolated and reproduced, biocontrol microorganisms can be applied in dilutions or released in a targeted manner on crops to reduce insect pests and diseases. These microorganisms can affect pests and diseases directly or indirectly and can reduce the damage below the economic threshold of the producer. Many microorganisms have been used as biopesticides because they offer a number of additional benefits beyond their intended use and known function (Glare *et al.*, 2012). Entomopathogenic fungi hold high potential in this group. More than a century ago, Louis Pasteur predicted the potential of entomopathogenic fungi for pest bioregulation (as parasites of insects harmful to plants). Since then, more than 700 fungal species have been identified as affecting insects (of different orders), and their use as biopesticides has increased over the last few decades (Shah *et al.*, 2009).

Mazón (2001) indicated that the most important entomopathogenic fungi with practical uses for pest control belong to *Metarhizium* and *Beauveria*. These genera are distributed worldwide, exhibit great metabolic and ecological versatility (Behie *et al.*, 2012; Lacey *et al.*, 2015), infect more than 200 insect species (Roberts and Hajek, 1992), have been extensively studied as biological control agents, and are commercially available (Lacey *et al.*, 2015). Entomopathogenic fungi parasitise insects and can cause their death. The traditional way to infect insects by entomopathogenic fungi is based on conidia inoculation of the insect cuticle. The germination tube penetrates the cuticle to the hemocele by mechanical and enzymatic action (Hajek and St. Leger, 1994). After invading the hemocele, the fungi have the ability to break through the host's cuticle once again and emerge outside where it can continue to develop saprophytically on the corpses, sporulating and converting them to new focal points of conidia dissemination (Meyling and Eilenberg, 2007). Insect pest control by entomopathogenic fungi is well known, but the same fungi have also been shown to protect the plant from damage caused by pathogenic microorganisms (Arnold *et al.*, 2003; Ownley, 2010). Recently, attention has been centred on developing *B. bassiana* and *M. anisopliae* as inundative biopesticides (De Faria and Wraight, 2007). Licensed commercial products exist whose formulations contain conidia and/or mycelia that allow their application in the field (De Faria and Wraight, 2007; Lacey *et al.*, 2015). Furthermore, the durability of biological control has long been assumed to be higher than that of chemical control (Holt and Hochberg, 1997).

4.3 Entomopathogenic Fungal Endophytes

Endophytes are defined as microorganisms (fungi and bacteria) that colonise the inside of the plant without causing any apparent damage during part or all of their life cycle (De Bary, 1884; Wilson, 1995), that is, they can live asymptomatically inside plant tissues (Petrini, 1991; Hyde and Soytong, 2008). The term endophyte originated in the nineteenth century and has evolved significantly since its first use, because associations with different plant parts have been taken into account and the types of association with their hosts explored further (Carroll, 1988). Endophytes are associated with most plant species and have been found naturally in all ecosystems studied so far (Arnol *et al.*, 2000). Fungal endophytes mainly belong to the taxonomic divisions Ascomycota, Basidiomycota and Zygomycota. They are common in nature and can produce a huge diversity of secondary metabolites. These metabolites have high commercial value and can have multiple applications in the fields of human health, biotechnology and agriculture (Strobel and Daisy, 2003; Wang *et al.*, 2013).

These fungi are widely studied in agriculture because of the beneficial characteristics that are conferred to their hosts (Aly *et al.*, 2010; Murphy *et al.*, 2015; 2018). Endophytes accomplish a variety of symbiotic and ecological functions (Rodríguez *et al.*, 2009) that are translated into the promotion of plant growth, inhibition of pathogenic organisms, removal of soil contaminants, as well as increased tolerance to extreme conditions such as temperature, water availability, and salinity (Quesada-Moraga *et al.*, 2009; Ownley *et al.*, 2010; Kauppinen *et al.*, 2016). The host plant benefits in numerous ways from the interaction with the endophyte in exchange for its carbon-based resources (Herre *et al.*, 2007).

The range of hosts of different endophyte species varies, going from highly specific single species to those with multiple host species (Stone *et al.*, 2004; Sánchez-Márquez *et al.*, 2011). Endophyte transmission can be vertical (from one generation to the next) or horizontal (from one individual to another) (Quezada-Moraga *et al.*, 2014; Majumder *et al.*, 2016). They can remain in plant tissues for long periods of their life cycle, protecting its host from the attack of pathogens and avoiding the external environmental changes that could threaten their survival and biocontrol effectiveness (Card *et al.*, 2016).

For many years, research has focused on clavicipitaceous endophytes associated with pasture grasses (Johnson and Caradus, 2019, Chapter 15). However, there is growing interest in non-clavicipitaceous endophytes (many of which are usually horizontally transmitted), associated with different species that are widespread in nature and dominated by the ascomycetes in which several entomopathogenic fungal genera are found (Vega *et al.*, 2008; Lopez and Sword, 2015).

Research studies have reported that entomopathogenic fungi can act as endophytes of a large variety of plants, including ornamental, agricultural, silvacultural and medicinal plants. Various species of the genera *Acremonium*, *Beauveria*, *Cladosporium*, *Clonostachys*, *Fusarium*, *Metarhizium*, *Paecilomyces*, *Trichoderma* and *Verticillium* (Gómez-Vidal *et al.*, 2006; Vega *et al.*, 2008; Powell *et al.*, 2009; Akello and Sikora, 2012; Akutse *et al.*, 2013) have been reported as natural or facultative endophytes of different plant species. Of these genera, *Beauveria* and *Metarhizium* are explored further here.

Beauveria bassiana is the species of entomopathogens most reported as an endophyte; it is able to colonise different plant species in specific plant organs, such as leaves, stems, flowers, and roots (Behie *et al.*, 2015; Krell *et al.*, 2017). Studies by Wagner and Lewis (2000) demonstrated that *B. bassiana* penetrates maize leaves through natural openings or the cuticle, which is facilitated by the mechanical force exerted by the infection structure (Bidochka and Khachatourians, 1991), the enzymatic dissolution of the cuticle, or a combination of both (Ferron, 1978; Kolattukudy, 1984); these mechanisms are similar to those used by the fungus to penetrate insects. Wagner and Lewis (2000) also showed that hyphae grow through the spaces between the parenchyma cells without haustorium formation. The authors found hyphae in the xylem vessels and concluded that *B. bassiana* can grow from the epidermis to the vascular tissues and achieve a systemic colonisation of the whole plant. Thus, this species has the ability to colonise the plant apoplast, parenchyma and vascular elements (Wagner and Lewis, 2000; Gomez-Vidal *et al.*, 2006; Quesada-Moraga *et al.*, 2006).

The persistence of *B. bassiana* in several plants has also been evaluated and it can remain inside the plant for several months (Brownbridge *et al.*, 2012; Quesada-Moraga *et al.*, 2014). Studies conducted by Akello *et al.* (2007) demonstrated that *B. bassiana* persisted for at least 4 months inside *Musa* spp. plants (banana). Publications have shown the potential of this fungus as an endophyte to decrease insect pest populations of banana (Akello *et al.*, 2008), maize (Bing and Lewis, 1991; Wagner and Lewis, 2000), and tomato (Powell *et al.*, 2009). On the other hand, species of *Metarhizium* mainly colonise the root system (Liao *et al.*, 2014; Barelli *et al.*, 2016; Greenfield *et al.*, 2016).

Studies of entomopathogenic fungal endophytes (EFEs) suggest that several of these fungi have never given up their role as plant symbionts, and that insect pathogenicity is a derived adaptation that allowed certain species to have access to a specialised source of nitrogen and other nutrients. The EFEs can substitute nutrients derived from insects with plant carbohydrates (Barelli *et al.*, 2016). For example, Behie *et al.* (2012) demonstrated that *Metarhizium robertsii* strains can translocate insect-derived nitrogen to plants. Furthermore, several studies have indicated that EFE applications do not negatively affect the yield of treated plants (Wagner and Lewis, 2000; Akello *et al.*, 2009; Tefera and Vidal, 2009), a desirable characteristic if they are to be used commercially.

EFEs can exhibit a systemic colonisation pattern in the whole plant (Quesada-Moraga *et al.*, 2006; Rodríguez *et al.*, 2009) or be located in a specific tissue/organ type such as leaves, flowers, roots, stems, bark and seeds (Tefera and Vidal, 2009; Behie *et al.*, 2015). Because of this property, *B. bassiana* and *M. anisopliae* have been artificially inoculated into plants using different techniques (Quesada-Moraga *et al.*, 2014) such as foliar spraying, stem injections, liquid applications to the soil, and seed treatment with conidial suspensions (Bing and Lewis, 1992; Quesada-Moraga *et al.*, 2006; Posada *et al.*, 2007; Terefa and Vidal, 2009).

The methodologies to study artificially inoculated EFEs usually include several steps. Reisolation of the plant endophyte requires rigorous surface sterilisation of the tissues. However, chemical product diffusion can affect the endophyte and result in underestimated colonisation rates (Ownley *et al.*, 2008; Lohse *et al.*, 2015). Furthermore, the techniques used, such as reisolation in different culture media, sometimes cannot differentiate between endophytic and epiphytic colonisations. For example, *B. bassiana* is horizontally transmitted and can simultaneously colonise external and internal tissues.

Morphological characters, taxonomic keys and molecular techniques are used to identify EFEs. In the first case, the fungus is identified when it is isolated from the tissues and grown in culture. For molecular techniques, the fungus can be identified when it is either inside the plant or grown in culture. The most widely used molecular techniques identify fungi by sequencing the nuclear ribosomal internal transcribed spacers 1 and 2 (nrITS1 and ITS2 regions) with universal primers (White, 1990). These primers work well for isolated fungi but exhibit difficulties when the fungus is inside the plant because the primers can amplify both plant and fungal DNA (Martin and Rygiewicz, 2005). Other regions that are used for identification of isolated EFE include: the intron sequences from β-tubulin (tubB), translation elongation factor 1-α (tefA) and γ-actin (actA) (Craven *et al.*, 2001), the largest and second largest subunits of ribosomal polymerase B (RPB1 and RPB2), and the nuclear intergenic region bloc (Rehner *et al.*, 2011). Intergenic specific markers (IGS) have also been designed to differentiate between species of the *M. anisopliae* complex (Kepler and Rehner, 2013). All these markers improve identification accuracy compared to that obtained with nrITS1 and ITS2 regions alone.

To detect the endophyte inside the plant it is possible to use specific primers for the target species (Garrido-Jurado *et al.*, 2016). It can be difficult to detect the endophyte inoculated by foliar spraying due to the probability of contamination with epiphytic progagules (Clayton *et al.*, 2017). Genome sequencing can also be applied and this allows the identification and further functional characterisation of the endophyte. For example, it can help in elucidating the role of the endophyte in plant secondary metabolite production (Venugopalan and Srivastava, 2015), to determinate its usefulness as a biocontrol agent in the agricultural industry or evaluate its potential as an aromatic compound degrader in environmental biotechnology (Kwak *et al.*, 2012). In addition, genome comparisons with related microorganisms can help predict the benefits the endophyte receives from the plant (Hardoim *et al.*, 2015).

4.4 Entomopathogenic Fungal Endophytes for Plant Disease Control

Entomopathogenic fungi have an ability to become endophytes in different plant species (Jaber and Ownley, 2017). Together with their insect control activity, this ability presents an opportunity for biological pest control. Several researchers suggest that EFEs could be able to simultaneously control pests and diseases (Kim *et al.*, 2007; Ownley, 2008). Among the EFE genera reported with phytopathogenic activity, *Beauveria*, *Lecanicillium* and *Metarhizium* are further described here. *Beauveria bassiana* has been most often reported as an antagonistic endophyte to plant diseases, caused not only by fungi but also bacteria and viruses. This could be due to the great diversity of bioactive and antimicrobial metabolites that it is able to produce (Warner and Lewis, 2000; Parine *et al.*, 2010) including destruxins, oosporein, beauvericin, bassianolide, bassianin, beauveriolide, bassiacridin, cordycepin and ciclosporin (Susuki *et al.*, 1977; Logrieco *et al.*, 1998; Quesada-Moraga and Vey, 2004; Wang and Wang, 2017). Studies conducted by Feng *et al.* (2015) determined that the *B. bassiana* genome exhibits at least 45 groups of different secondary metabolite biosynthesis genes.

Jaber (2015) showed that *Beauveria bassiana* applied by foliar spraying can colonise grapevine (*Vitis vinifera*) leaves to protect them from the pathogen *Plasmopara viticola*. All the strains evaluated in the study, independent of the endophyte colonisation percentage reached, had an antagonistic effect against the pathogen. Several mechanisms might explain its antagonistic ability, including competition for niche or resources, antibiosis, parasitism and induced systemic resistance (Jaber, 2015). This endophyte has also been shown to control levels of *Xanthomonas axonopodis* pv. *malvacearum* in *Gossypium hirsutum* roots. The authors attribute this bacterial control to induced systemic resistance (Ownley *et al.*, 2008). In other studies, *G. hirsutum* seeds were inoculated with different *B. bassiana* conidium concentrations in loamy soil and exhibited a significant biocontrol level of the pathogen *Rhizoctonia solani* (Griffin, 2007).

Naturalis®, a commercial product based on *B. bassiana*, has showed good endophytic colonisation ability and antagonistic action against zucchini yellow mosaic virus (ZYMV) in *Cucurbita pepo* (gourds, pumpkins, squashes). This is one of the few studies reporting the use of EFEs against plant diseases caused by viruses and could be attributable to the ability of *B. bassiana* to systemically colonise the plant and inhibit or interfere with the systemic movement of ZYMV from cell to cell (Jaber and Salem, 2014). This finding could open an important line of research given the few biological tools currently available for virus control.

Lecanicillum lecanii (= *Verticillium lecanni*) is another EFE that has been reported as an antagonist to plant diseases. Studies by Benhamou and Brodeur (2001) determined that it can colonise the root of *Cucumis sativus* (cucumber) and significantly reduce the incidence and severity of the oomycete pathogen *Pythium ultimum*. Hirano *et al.* (2008) later demonstrated that inoculating *C. sativus* cv. Hokushin plants with *L. lecanii* induces systemic resistance against the *P. ultimum*. Few papers report the use of *Metarhizium* spp. as an endophytic mycopathogen. However, the bean endophyte *M. robertsii* was shown to exhibit an antagonistic effect against *Fusarium* spp. in both *in vitro* and *in vivo* assays (Sasan and Bidochka, 2013).

The ability of EFEs to act antagonisticly to phytopathogens could be due to different mechanisms, however, knowledge in this field is still limited. They are attributed with direct mechanisms, such as competition and parasitism, and indirect mechanisms, such as antibiosis (production of primary and secondary metabolites, enzymes, or volatile compounds), resistance induction (Arnold *et al.*, 2003; Herre *et al.*, 2007) or a combination of the abovementioned mechanisms (Vega *et al.*, 2009; Ownley *et al.*, 2010), as outlined below:

1. Competition between the endophyte and pathogen inside the plant can be for space and/or food. The EFEs are able to successfully colonise different tissues and use available sources of food provided by the plant, which can reduce the probability of colonisation by pathogens (Ownley *et al.*, 2004). Early inoculation of the endophyte could be advantageous because, by being the first to colonise the plant, it depletes the plant's resources, thus leaving less available for the pathogen (especially carbon sources). The plant's development is not significantly affected by the endophyte and it can improve growth by several mechanisms related to nutrition or stress resistance (Behie and Bidochka, 2014).
2. Mycoparasitism is an antagonistic interaction between a fungal parasite and fungal host that usually involves the action of extracellular lytic enzymes that break down the host's cell walls (Sharma, 2011). Although there are several studies about the action of EFEs as mycoparasites, they only show results at the *in vitro* level (Fenice and Gooday, 2006) and little is known about their action in plants as endophytes.

3. Antibiosis is the direct action of antibiotics or toxic metabolites produced by microorganisms on another microorganism that is sensitive to them. Boucias and Pendland (1998) suggest that opportunistic microorganisms, such as phytopathogens, could be kept at bay by the antimicrobial substances produced by EFEs. For example, Beauvericin is a well-known mycotoxin produced by EFEs such as *B. bassiana*, *Paecilomyces fumosoroseus*, and *Fusarium* spp. (Bernardini *et al.*, 1975). This metabolite has high crop protection potential because it exhibits important antibacterial, antifungal, and antiviral activities (Xu *et al.*, 2010; Wang and Xu, 2012). Bassianolide is another metabolite with antibiotic activity that has been reported in *B. bassiana* and *L. lecanii* (Suzuki *et al.*, 1977). This metabolite plays a significant role in virulence of *B. bassiana* against insects (Xu *et al.*, 2009) and currently little is known about its toxicity in animals and plants. On the other hand, *Metarhizium* produce destruxins (Roberts and Leger, 2004) and they have been successfully evaluated to control the pathogens *Fusarium oxysporum* and *Cladosporium herbarum* (Ravindran *et al.*, 2014).
4. Systemic resistance can be an important antagonistic mechanism against phytopathogens and could be triggered in the plant by the presence of non-pathogenic fungi (Kavroulakis *et al.*, 2007) such as endophytes (Shoresh *et al.*, 2010). Plants that are endophytically colonised by fungi can generate a response more rapidly, which can be associated with its ability to produce secondary metabolites that act as elicitors of phytoalexin production (Gao *et al.*, 2010). In systemic resistance, the endophyte and the phytopathogen do not need to be in direct contact for the plant to react with a rapid expression of defence-related genes (Ownley *et al.*, 2010). There are two main types of induced resistance: systemic acquired resistance (SAR) and induced systemic resistance (ISR), which are differentiated on the basis of the nature of the elicitor and their regulatory pathways. SAR is induced by pathogen infection or chemical compounds, is dependent on salicylic acid and is associated with the accumulation of pathogenesis-related (PR) proteins. ISR, in contrast, is induced by other non-pathogenic microorganisms including endophytes and rhizobacteria. It is generally independent of salicylic acid but dependent on jasmonate and ethylene and is not associated with the accumulation of PR proteins (Pieterse *et al.*, 1998). However, some studies show that some rhizobacteria, including some of endophytic strains, that elicit ISR may be dependent on salicylic acid and independent of jasmonate or ethylene, in contrast to the model proposed by Pieterse *et al.* (1998) (Ryu *et al.*, 2003). Moreover, some strains of bacteria and fungal endophytes that induce ISR were associated with the expression of PR genes (Ryu *et al.*, 2003; Kavroulakis *et al.*, 2007).

The following section reports data on the endophytic colonisation ability in tomato of six native strains of *B. bassiana, M. anisopliae* and *M. robertsii* that form part of the Chilean Collection of Microbial Genetic Resources (CChRGM) and their antagonistic effects (*in vitro*) against *Botrytis cinerea, Sclerotinia sclerotiorum,* and *F. oxysporum.*

4.5 Materials and Methods

4.5.1 Fungal Isolates

Six native strains of *B. bassiana* and *M. anisopliae* were used in the assays (Table 4.1). These were isolated using the methodology described by France *et al.* (2000). Fungal conidia were sown on Petri dishes with 100% potato dextrose agar (PDA, Disfco™) and 150 mg L^{-1} chloramphenicol to prevent the growth of contaminating bacteria. The dishes were incubated in the dark at 25 ± 2°C for 10 d. Viability of the harvested conidia was determined according to the methodology described by Moore *et al.* (1995). They were suspended in test tubes with 10 ml sterile distilled water and 0.01% (v/v) Tween 80% (Difco™), and dilutions were performed until a concentration of 1×10^6 conidia ml^{-1} was reached. The concentration of conidia was determined with a Neubauer counting chamber (BOECO, Germany), and the inocula were used for all the experiments. The pathogens were used at concentrations of 1×10^6 conidia ml^{-1}.

4.5.2 Plant Inoculation and Endophytic Colonisation

Ten-day old samples of *Beauveria* and *Metarhizium* were removed from the surface of Petri dishes under sterile conditions, sown in 100 ml test tubes containing 15 ml PDA, and inoculated with a solution of 1×10^6 conidia ml^{-1} (Griffin, 2007). One test tube was left without fungal inocula as a control. Test tubes were incubated in the dark at 25 ± 2°C for 4 d. Once growth of the fungi was observed in each tube, 20 ml of substrate was added, which consisted of a mixture of perlite, peat, compost, and vermiculite (2:2:2:1) sterilised twice in an autoclave at 120°C and 115 psi for 1 h.

Tomato plants (var. Limachino; an INIA traditional cultivar) were disinfected according to a protocol adapted from Griffin (2007). Seeds were submerged in a 95% ethanol solution for 1 min and then in 1.5% NaOCl for 3 min; they were washed three times for 1 min in sterile distilled water. A control to demonstrate disinfection was made following the protocol of Ownley *et al.* (2008). Two seeds were planted for each test tube and, after the emergence, one plant per tube was left.

The tubes were incubated for 30 d in growth chambers at 25 ± 2°C with 12 h photoperiod and arranged in a randomised design with three replications. Plants were then removed from the tubes, washed with tap water and two cuts were made to separate roots, stems and leaves. The separate plant parts were subjected to disinfection with 70% ethanol for 2 min, 1.5% NaOCl for 5 min, and rinsed three times

Table 4.1 Fungal strains selected for use in colonisation and radial growth studies.

Bank code	Species	Collection location in Chile	Origin
RGM 673	*Metarhizium anisopliae*	Futaleufú, Los Lagos Region	Natural pasture soil
RGM 677	*Metarhizium anisopliae*	Puyehe, Los Lagos Region	Natural pasture soil
JFD1 23	*Metarhizium robertsii*	Robinson Crusoe, Valparaíso Region	Natural pasture soil
RGM 1237	*Beauveria bassiana*	Pozo Almonte, Tarapacá	Arable soil, *Citrullus lanatus* cultivated
JFD1 49	*Beauveria bassiana*	Robinson Crusoe, Valparaíso Region	Arable soil in national park
JFD2 1	*Beauveria bassiana*	Robinson Crusoe, Valparaíso Region	Native forest soil
RGM 2519	*Botrytis cinerea*	Colín, Maule Region	Tomato plant
RGM 1630	*Fusarium oxysporum*	San Ignacio, Biobío Region	Arable vegetable soil
RGM 1420	*Sclerotinia sclerotiurum*	Chillán, Biobío Region	Arable vegetable soil

for 1 min with sterile distilled water. The parts were left to dry on sterile absorbent paper. A disinfection control was included using the same procedure used for the seeds. Once sterilisation was done, five samples were cut from each of the plant part (roots, stem and leaves), resulting in a total of 15 parts per plant. Root and stem samples were 10 mm long, whereas leaves were cut in 6 mm discs. The cuttings were distributed in the Petri dishes with Noble agar medium and incubated in the dark at 25 ± 2°C for 30 d. The percentage of endophytic colonisation was evaluated.

4.5.3 *In Vitro* Antifungal Activity

Mycelium discs (3-d old, 5 mm diameter) from the pure *Beauveria* and *Metarhizium* cultures were placed 1.5 cm from the edge of a 9 cm Petri dish containing PDA medium (20 ml). Discs with each of pathogen were placed equidistantly opposite the endophyte and incubated in the dark at 25 ± 2°C for 10 d. Five replicate dishes per treatment were placed in a completely randomised design. At the same time, control dishes (from 3-d old *B. cinerea* culture) containing only pathogen discs were prepared. When the pathogen colonised the whole dish, the percentage of radial growth inhibition of the pathogen (PRGIP) was determined using equation (4.1). The radius of the colonies was measured (mm) in each of the treatments using a digital caliper.

$$\text{PRGIP} = [(R1 - R2)/R1] \times 100 \tag{4.1}$$

where R1 is the radius of the pathogen colony (mm) and R2 is the radius of the pathogen colony against the endophyte (mm).

4.6 Results and Discussion

Our study demonstrated that the six evaluated strains can be established as endophytes in tomato leaves, stems and roots, by inoculating plants in soil. Epiphytic colonisations could not be detected in the disinfection control samples, which demonstrates that all the observed mycelium growth came from internal tissues. Means of the colonisation percentages fluctuated between 0% and 100% in the leaves, and 0% and 40% in stems and roots (Figure 4.1). Furthermore, four strains demonstrated a localised pattern of colonisation in leaves, two in stems and two could colonise both leaves and roots. The colonisation of the different plant parts indicates that the EFEs move within the whole plant. Roots and stem colonisation was much lower than leaf colonisation. The reason for this difference could be explained by differences in microbial and physiological conditions in the different plant parts, according to Terefa and Vidal (2009). The EFEs can be adapted to particular conditions present in a given plant tissue/organ.

The native strains of endophytic *Beauveria* and *Metarhizium* demonstrated significant levels of radial growth inhibition of the pathogen. The pathogen (control) was able to totally colonise the dish (attained a 75 mm radius) at 7 d post-inoculation in the case of *B. cinerea*, 7 d for *F. oxysporum* and 5 d for *S. sclerotiorum*. The PRGIP fluctuated for *B. cinerea*, *F. oxysporum* and *S. sclerotiorum*: 26%–34%; 37%–47% and 26%–30%, respectively (Figure 4.2). These results might be explained by the ability of these fungi to produce metabolites with antimicrobial activities (Ownley *et al.*, 2010). When we observed the advancing hyphal zone under the microscope mycoparasitism mechanisms were not detected; this suggests that the evaluated strains have no ability to parasitise the phytopathogens evaluated.

Our results provide evidence of endophytic colonisation of tomato by native strains of *B. bassiana*, *M. anisopliae* and *M. robertsii*. These EFEs can colonise some part plants more efficiently than others, which might consequently influence the level of plant protection against different phytopathogens. All the evaluated strains showed some level of inhibition of pathogens and hence have potential as innovative management tools against plant diseases. Research is required to assess the antifungal effect of the EFEs on plants under laboratory and field conditions against different types of phytopathogens. Despite highly promising laboratory results, a number of challenges need also be addressed before the EFEs can be commercially exploited for crop protection as outlined below.

4.6.1 Commercial Potential of EFEs in Disease and Pest Biocontrol

The current use of chemical pesticides is not sustainable because of their negative impact on the environment and people's health (Levitan *et al.*, 1995; Kim *et al.*, 2017). Chemical pesticides are broad spectrum and affect non-target organisms, including predators and parasites as well as humans (Kumar, 2015). This has

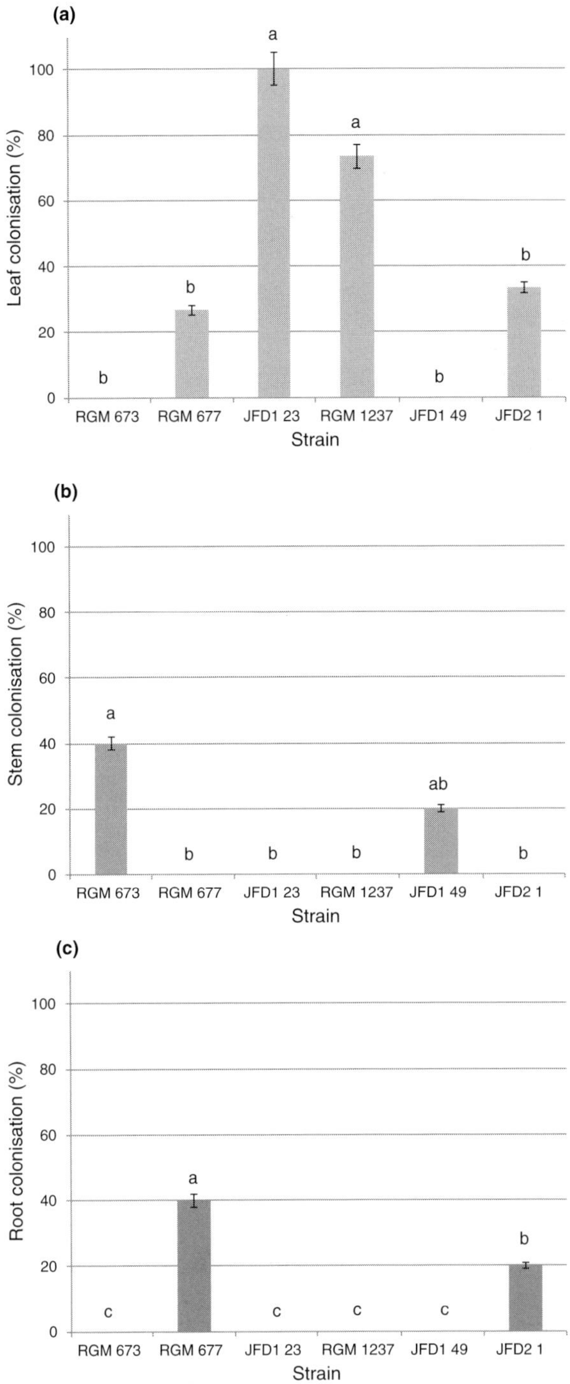

Figure 4.1 Percentage colonisation of tomato by *Beauveria bassiana*, *Metarhizium anisopliae* and *M. robertsii*: (a) leaves, (b) stems and (c) roots. Different letters over the bars indicate significant differences according to Fisher's least significant difference (LSD) test. Each bar represents the mean (± SE). Endophyte strain codes are given in Table 4.1.

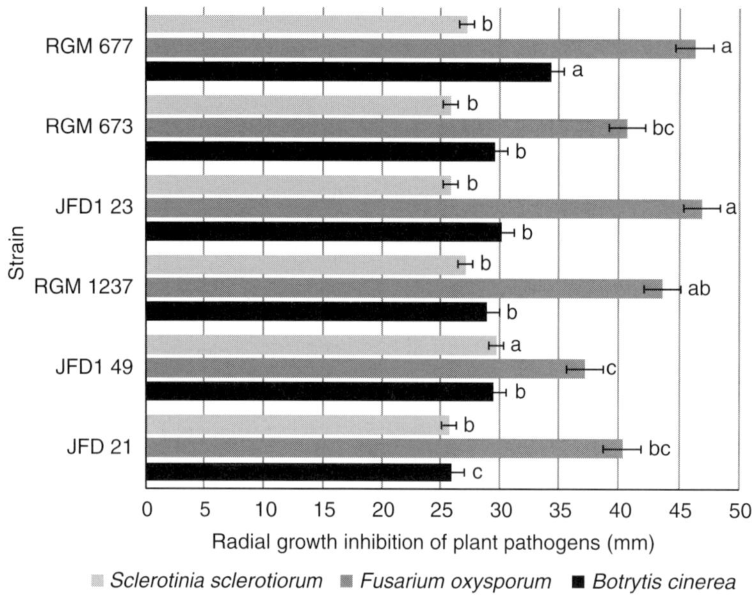

Figure 4.2 Radial growth inhibition of the plant pathogens *Botrytis cinerea, Fusarium oxysporum and Sclerotinia sclerotiorum* (%) by six native strains of endophytic *Beauveria bassiana, Metarhizium anisopliae* and *M. robertsii* in tomato. Different letters over the bars indicate significant differences according to Fisher's least significant difference (LSD) test. Each bar represents the mean (± SE). Endophyte codes are given in Table 4.1.

resulted in a more restricted use of agrochemicals in many countries. Biopesticides are a sustainable alternative and the incorporation of biopesticides in different agricultural systems has significantly increased over time. Biopesticides are highly specific affecting only the targeted pest or closely related pests, without harm to humans, other beneficial organisms or members of agricultural landscape ecosystems. Thus, many institutions worldwide have directed actions to increase research and development in this field, aimed at developing strategies for promoting their use.

Entomopathogenic fungi have been widely studied for their ability to control insect pests, and they are currently active ingredients of several commercial products (De Faria and Wraight, 2007). More than 170 products have been developed to control pests, which use approximately 12 fungal species (Lacey *et al.*, 2015). Soper and Ward (1981) suggest that a number of aspects associated with their biology need to be considered (including different types of mycelial propagules, conidiospore and blastospores; high pathogenicity to the target pest). Several aspects associated with their mass production also need to be considered (rapid growth, asexual reproduction, easy culture in different media, production of large amounts of conidia).

There is also a desire to develop the use of the fungi as plant endophytes instead of derived biopesticides. Their action as living endophytes could help address several of the commercial deficiencies they have exhibited as biopesticides, as they act of from inside the plant against multiple factors, including plant pathogens. However, there are several aspects that should be considered to assess the potential of EFEs for commercial development which are outlined below.

1. **Type of symbiosis, host specificity and multifunctionality.** The fact that EFEs are facultative plant colonisers gives them an advantage from the commercial point of view. This is because they could remain as saprophytes in the soil after their application and enter the plant when needed. Alternatively, they can be applied as a seed dressing and colonise the germinating seed (Murphy *et al.*, 2019, Chapter 18). Many of the published studies clearly show the low specificity of EFEs to the host plant. For example, *B. bassiana* has been reported as an endophyte in more than 30 plant species. This characteristic can be commercially advantageous because the endophyte could be used in a variety of crops. Furthermore, EFEs are known to perform multiple functions to protect plants against biotic (pests and diseases) and abiotic (such as high temperatures and water deficit) stresses, as well as promote growth (Quesada-Moraga *et al.*, 2009; Ownley *et al.*, 2010; Elena *et al.*, 2011). This characteristic provides an element of differentiation regarding the use of chemical products developed for plant protection.
2. **Dispersal, type of colonisation and persistence.** Entomopathogenic fungi are dispersed in the air and soil by living infected hosts that migrate and die in a site other than where they were infected (Hajek, 1997; Feng *et al.*, 2004). In addition, dispersal in localised plant parts can occur through the rain (Bruck and Lewis, 2002). Alternative crop inoculation methods are therefore required that do not rely on insects. EFEs can exhibit localised or systemic colonisation patterns. For example, *B. bassiana* strains applied to the roots have been reisolated from the leaves. Some species prefer certain plant parts, such as *Metarhrizium* spp., which prefers roots as an endophyte (Behie *et al.*, 2015), and *B. bassiana*, which prefers aerial plant parts. However, the differential preference in plant organs would not be a limiting factor for the commercial development of these fungi, because of the endophyte's ability to execute its antagonistic action through indirect mechanisms that act independently of the colonisation site (Jaber and Vidal, 2010).

   There are reports that *B. bassiana* persisted for more than 9 months inside *Pinus radiata* plants under controlled conditions (Brownbridge *et al.*, 2012). Mantzoukas *et al.* (2015) also reported that *B. bassiana, M. robertsii* and *Isaria fumosorosea* persisted 30 days after being inoculated by foliar pulverisation in *Sorghum bicolor* under field conditions. This level of persistence presents advantages of endophytes over chemical products that usually do not persist for over 10 days.

3. **Crop application.** Entomopathogenic fungi have been known to colonise plants after being inoculated by different methods, such as foliar spraying, root drenching and seed immersion in fungal solutions. This approach requires high conidial concentrations and an appropriate formulation, which allows the EFE to efficiently carry out its action. The efficiency of these applications is diminished because conidia normally spend a significant amount of time on the plant surface where they are exposed to suboptimal temperatures, UV radiation and low humidity that can affect their germination and viability (Fargues *et al.*, 1996; Vega *et al.*, 2012). Given this, EFEs could have an advantage in commercial development because it would be sufficient for only some of the conidia to grow and enter the plant to trigger a biocontrol action. Thus, low volumes of inocula could be used. Endophytic foliar spraying could be used to control diseases, such as powdery mildew, or EFEs could be applied to seeds before sowing (Taylor and Harman, 1990). In such situations their action starts at germination and could be used to protect the plants from seedling diseases such as damping off. Soil EFE applications would allow application of treatments to places where chemical products cannot reach. Furthermore, they could help manage saprophytes by competing with soil fungi such as *Sclerotinia* spp., *Phytophthoptora* spp. and *Rhizoctonia solani*. Moreover, endophytes could provide additional benefits to the plant by, for example, promoting growth.
4. **Compatibility with chemical treatments and the plant microbiome.** More studies are required to determine the different susceptibility levels of EFEs to chemical fungicides and insecticides. Combining these fungi with other products could produce higher control effectiveness levels (Gurulingappa *et al.*, 2011). Furthermore, little is known about the compatibility and competition with other endophytes existing in the plant. The plant microbiome could, for example, promote or hamper the establishment and persistence of the inoculated endophytes (Posada *et al.*, 2007; Jaber and Enkerli, 2016).
5. **Field studies, application and widespread use.** Many of the evaluations related to the ability of endophytic colonisation are carried out under controlled laboratory/growth conditions. Although this provides important scientific validity, it does not necessarily guarantee similar action under field conditions. The same occurs when evaluating the antagonistic effect against phytopathogens or pests because most studies have been conducted *in vitro*; very few studies involve plants and even fewer include plants under field conditions.

   Research is also required to optimise application and widespread use/adoption by the farming community. Fermentation and formulation of endophyte treatments need optimisation for commercial use to maximise endophytic colonisation in the different plant tissues, as well as increasing their persistence. The use of endophyte mixtures can be a strategy to increase the variety of action modes and expand the range of controlled pests and pathogens. Currently, there are few reports of compatible consortia involving endophytes and other biological agents to simultaneously control pests and disease, but this approach is likely to increase in the future (Karthiba *et al.*, 2010).

6. **Safety and registrations.** An advantage of entomopathogenic fungi over pesticides is that they are usually safe for humans. There is also no evidence that they have a negative impact on the environment (Vega *et al.*, 2008) and some, such as *B. bassiana*, have also been reported as natural endophytes of some cultivated and wild plants. Given that the same entomopathogenic fungi are already used in commercial products to control insects, the registration process as antagonistic endophytes of phytopathogens should be facilitated. This could be an opportunity for products, such as Naturalis® and BotaniGard®, which are being adapted for potential application as entomopathogenic and phytopathogenic endophytes (Powell *et al.*, 2009). Both products are widely commercialised in America and Europe. Naturalis® is used for biological control of the whitefly and has been tested as an endophyte to control *Tuta absoluta* and *Planococcus ficus* (Rondot and Reineke, 2016).

4.7 Conclusions

EFEs have high potential for the biological control of pests and diseases by acting from inside the plants instead of the outside. EFEs can also provide additional benefits to plants in terms of yield improvement and stress tolerance. There are social and environmental benefits associated with the use of these technologies compared to the application of synthetic chemical pesticides as they present fewer problems for human health and less pollution of soil and water. There have been very few research studies focused on the antagonistic activity of EFEs against plant-pathogenic microorganisms. We have demonstrated their ability *in vitro* to be antagonistic against different phytopathogens. This suggests great potential to control diseases in the whole plant. To take advantage of this potential, more research is required in which endophytes are used against plant pathogens, ideally under field conditions in order to increase the external validity of these assays. It is necessary to expand the number of applied studies aimed at increasing their compatibility and maximise their effectiveness when trying to incorporate them in different agricultural production systems.

The increase in the supply of bioinsecticides formulated on the basis of entomopathogenic fungi presents an opportunity to commercially develop EFEs as biofungicides. It should be possible to overcome registration hurdles associated with the use of these fungi because they include the same active ingredients, and already registered products, to those already commercialised in different markets.

Thus, EFEs represent an exciting development for more sustainable plant production by contributing to both pest and disease management in agriculture.

Acknowledgements

We are grateful to the organisers of the international conference in Trinity College Dublin 'SUSTAIN: Endophytes for a growing world' for recognising the value of our work and inviting us to contribute to the knowledge of this interesting discipline.

References

Aly, A. H., Debbab, A., Kjer, J. and Proksch, P. (2010). Fungal endophytes from higher plants: a prolific source of phytochemicals and other bioactive natural products. *Fungal Diversity*, **41**, 1–16.

Akello, J. and Sikora, R. (2012). Systemic acropedal influence of endophyte seed treatment on *Acyrthosiphon pisum* and *Aphis fabae* offspring development and reproductive fitness. *Biological Control*, **61**, 215–221.

Akello, J., Dubois, T., Coyne, D. and Hillnhutter, C. (2007). *Beauveria bassiana* as an endophyte in tissue-cultured banana plants: a novel way to combat the banana weevil *Cosmopolites sordidus*. In *III International Symposium on Banana: ISHS-ProMusa Symposium on Recent Advances in Banana Crop Protection for Sustainable Production and Improved Livelihoods*, ed. D. Jones and I. Van den Bergh. *Acta Horticulturae*, 828, 129–138.

Akello, J., Dubois, T., Coyne, D. and Kyamanywa, S. (2008). Endophytic *Beauveria bassiana* in banana (*Musa spp.*) reduces banana weevil (*Cosmopolites sordidus*) fitness and damage. *Crop Protection*, **27**, 1437–1441.

Akello, J., Dubois, T., Coyne, D. and Kyamanywa, S. (2009). The effects of *Beauveria bassiana* dose and exposure duration on colonization and growth of tissue cultured banana (*Musa sp.*) plants. *Biological Control*, **49**, 6–10.

Akutse, K. S., Maniania, N. K., Fiaboe, K. K. M., Van den Berg, J. and Ekesi, S. (2013). Endophytic colonization of *Vicia faba* and *Phaseolus vulgaris* (Fabaceae) by fungal pathogens and their effects on the life-history parameters of *Liriomyza huidobrensis* (Diptera: Agromyzidae). *Fungal Ecology*, **6**, 293–301.

Arnold, A., Maynard, Z., Gilbert, G., Coley, P. and Kursar, T. (2000). Are tropical endophytic fungi hyper diverse? *Ecology Letters*, **3**, 267–274.

Arnold, A., Mejía, L., Kyllo, D. *et al.* (2003). Fungal endophytes limit pathogen damage in a tropical tree. *Proceedings of the National Academy of Sciences of the United States of America*, **100**, 15649–15654.

Barelli, L., Moonjely, S., Behie, S. W. and Bidochka, M. J. (2016). Fungi with multifunctional lifestyles: endophytic insect pathogenic fungi. *Plant Molecular Biology*, **90**, 657–664.

Behie, S. W. and Bidochka, M. J. (2014). Ubiquity of insect-derived nitrogen transfer to plants by endophytic insect-pathogenic fungi: an additional branch of the soil nitrogen cycle. *Applied and Environmental Microbiology*, **80**, 1553–1560.

Behie, S. W., Zelisko, P. M. and Bidochka, M. J. (2012). Endophytic insect-parasitic fungi translocate nitrogen directly from insects to plants. *Science*, **336**, 1576–1577.

Behie, S. W., Jones, S. J. and Bidochka, M. J. (2015). Plant tissue localization of the endophytic insect pathogenic fungi *Metarhizium* and *Beauveria*. *Fungal Ecology*, **13**, 112–119.

Benhamou, N. and Brodeur, J. (2001). Pre-inoculation of Ri T-DNA transformed cucumber roots with the mycoparasite, *Verticillium lecanii*, induces host defense reactions against *Pythium ultimum* infection. *Physiological and Molecular Plant Pathology*, **58**, 133–146.

Bernardini, M., Carilli, A., Pacioni, G. and Santurbano, B. (1975). Isolation of beauvericin from *Paecilomyces fumoso-roseus*. *Phytochemistry*, **14**, 1865.

Bidochka, M. and Khachatourians, C. (1991). The implication of metabolic acids produced by *Beauveria bassiana* in pathogenesis of the migratory grasshopper, *Melanoplus sanguinipes*. *Journal of Invertebrate Pathology*, **58**, 106–117.

Bing, L. and Lewis, L. (1992). Endophytic *Beauveria bassiana* in corn, the influence of the plant growth stage and *Ostrimia nubilalis*. *Biocontrol Science and Technology*, **2**, 39–47.

Boucias, D. G. and Pendland, J. C. (1998). Entomopathogenic fungi: fungi imperfecti. In *Principles of Insect Pathology*, ed. D. Boucias and Jacquelyn C. Pendland. Boston, MA: Springer, pp. 321–364.

Brownbridge, M., Reay, S. D., Nelson, T. L. and Glare, T. R. (2012). Persistence of *Beauveria bassiana* (Ascomycota: Hypocreales) as an endophyte following inoculation of radiata pine seed and seedlings. *Biological Control*, **61**, 194–200.

Bruck, D. J. and Lewis, L. C. (2002). Rainfall and crop residue effects on soil dispersion and *Beauveria bassiana* spread to corn. *Applied Soil Ecology*, **20**, 183–190.

Bultman, T. L. and Bell, G. D. (2003). Interaction between fungal endophytes and environmental stressors influences plant resistance to insects. *Oikos*, **103**, 182–190.

Card, S., Johnson, L., Teasdale, S. and Caradus, J. (2016). Deciphering endophyte behaviour: the link between endophyte biology and efficacious biological control agents. *FEMS Microbiology Ecology*, **92**, fiw114.

Carroll, G. (1988). Fungal endophytes in stems and leaves: from latent pathogens to mutualistic symbiont. *Ecology*, **69**, 2–9.

Clayton, W., Eaton, C. J., Dupont, P.-Y. *et al.* (2017). Analysis of simple sequence repeat (SSR) structure and sequence within *Epichloë* endophyte genomes reveals impacts on gene structure and insights into ancestral hybridization events. *PLoS One* **12**, e0183748.

Craven, K. D., Hsiau, P. T. W., Leuchtmann, A., Hollin, W. and Schardl, C. L. (2001). Multigene phylogeny of Epichloë species, fungal symbionts of grasses. *Annals of the Missouri Botanical Garden*, **88**, 14–34.

De Bary, A. (1884). *Vergleichende Morphologie und Biologie der Pilze, Mycetozoen und Bacterien.* Leipzig: Wilhelm Engelmann.

De Faria, M. R. and Wraight, S. P. (2007). Mycoinsecticides and mycoacaricides: a comprehensive list with worldwide coverage and international classification of formulation types. *Biological Control*, **43**, 237–256.

Elena, G. J., Beatriz, P. J., Alejandro, P. and Lecuona, R. E. (2011). *Metarhizium anisopliae* (Metschnikoff) Sorokin promotes growth and has endophytic activity in tomato plants. *Advances in Biological Research*, **5**, 22–27.

Fargues, J., Goettel, M. S., Smits, N. *et al.* (1996). Variability in susceptibility to simulated sunlight of conidia among isolates of entomopathogenic Hyphomycetes. *Mycopathologia*, **135**, 171–181.

Feng, M. G., Chen, C. and Chen, B. (2004). Wide dispersal of aphid pathogenic Entomophthorales among aphids relies upon migratory alates. *Environmental Microbiology*, **6**, 510–516.

Feng, P., Shang, Y., Cen, K. and Wang, C. (2015). Fungal biosynthesis of the bibenzoquinone oosporein to evade insect immunity. *Proceedings of the National Academy of Sciences of the United States of America*, **112**, 11365–11370.

Fenice, M. and Gooday, G. W. (2006). Mycoparasitic actions against fungi and oomycetes by a strain (CCFEE 5003) of the fungus *Lecanicillium muscarium* isolated in Continental Antarctica. *Annals of Microbiology*, **56**, 1.

Ferron, P. (1978). Biological control of insect pests by entomogenous fungi. *Annual Review of Entomology*, **23**, 409–442.

Fisher, P. J. and Petrini, O. (1987). Location of fungal endophytes in tissues of *Suaeda fruticosa*: a preliminary study. *Transactions of the British Mycological Society*, **89**, 246–249.

France, I., Gerding. G., Gerding, M. and Sandoval, A. (2000). Patogenicidad de una colección de cepas nativas de *Metarhizium* spp. y *Beauveria* spp. en *Aegorhinus superciliosus*, *Asynonychus cervinus* y *Otiorhynchus sulcatus*. *Agricultura Técnica*, **60**, 205–215.

Gao, F. K., Dai, C. C. and Liu, X. Z. (2010). Mechanisms of fungal endophytes in plant protection against pathogens. *African Journal of Microbiology Research*, **4**, 1346–1351.

Garrido-Jurado, I., Landa, B. B. and Quesada-Moraga, E. (2016). Detection and quantification of the entomopathogenic fungal endophyte *Beauveria bassiana* in plants by nested and quantitative PCR. *Microbial-Based Biopesticides: Methods and Protocols*, pp. 161–166.

Glare, T., Caradus, J., Gelernter, W. *et al.* (2012). Have biopesticides come of age? *Trends in Biotechnology*, **30**, 250–258.

Greenfield, M., Gómez-Jiménez, M. I., Ortiz, V. *et al.* (2016). *Beauveria bassiana* and *Metarhizium anisopliae* endophytically colonize cassava roots following soil drench inoculation. *Biological Control*, **95**, 40–48.

Griffin, M. R. (2007). *Beauveria bassiana*, a cotton endophyte with biocontrol activity against seedling disease. PhD Dissertation. The University of Tennessee, Knoxville, TN.

Grille, G., Pascal, C., Franco, J. and Basso, C. (2001). Distribución espacial de *Trialeurodes vaporariorum* (Homoptera: Aleyrodidae) en plantas de tomate. *Boletín de Sanidad Vegetal. Plagas*, **27**, 475–488.

Gómez-Vidal, S., Lopez-Lorca L., Jansson, H. and Salinas, J. (2006). Endophytic colonization of date palm (*Phoenix dactylifera* L.) leaves by entomopathogenic fungi. *Micron*, **37**, 624–632.

Gurulingappa, P., McGee, P. A. and Sword, G. (2011). Endophytic *Lecanicillium lecanii and Beauveria bassiana* reduce the survival and fecundity of *Aphis gossypii* following contact with conidia and secondary metabolites. *Crop Protection*, **30**, 349–353.

Hajek, A. and St. Leger, F. (1994). Interactions between fungal pathogens and insect hosts. *Annual Review of Entomology*, **39**, 293–322.

Hajek, A. E. (1997). Ecology of terrestrial fungal entomopathogens. In *Advances in Microbial Ecology*, Vol. 15, ed. J. G. Jones. Boston, MA: Springer, pp. 193–249.

Hamil R., Higgens C., Boaz H. and Gorman M. (1969). The structure of beauvericin, a new depsipeptide antibiotic toxic to *Artemia salina*. *Tetrahedron Letters*, **49**, 4255–4258.

Hardoim, P. R., Van Overbeek, L. S., Berg, G. *et al.* (2015). The hidden world within plants: ecological and evolutionary considerations for defining functioning of microbial endophytes. *Microbiology and Molecular Biology Reviews*, **79**, 293–320.

Herre, E. A., Mejía, L. C., Kyllo, D. A. *et al.* (2007). Ecological implications of antipathogen effects of tropical fungal endophytes and mycorrhizae. *Ecology*, **88**, 550–558.

Hirano, E., Koike, M., Aiuchi, D. and Tani, M. (2008). Pre-inoculation of cucumber roots with *Verticillium lecanii* (*Lecanicillium muscarium*) induces resistance to powdery mildew. *Research Bulletin of Obihiro University*, **29**, 82–94.

Holt, R. D. and Hochberg, M. E. (1997). When is biological control evolutionarily stable (or is it)? *Ecology*, **78**, 1673–1683.

Hoffmann, M. and Frodsham, A. (1993). *Los Enemigos Naturales de las Plagas de Insectos Vegetales*. Ithaca, New York: Extensión Cooperativa de la Universidad de Cornell.

Hu, Q., Li, F. and Zhang, Y. (2016). Risks of mycotoxins from mycoinsecticides to humans. *BioMed Research International*, **2016**, ID 3194321, 1–13.

Hyde, K. D. and Soytong, K. (2008). The fungal endophyte dilemma. *Fungal Diversity*, **33**, e173.

Jaber, L. R. (2015). Grapevine leaf tissue colonization by the fungal entomopathogen *Beauveria bassiana s. l.* and its effect against downy mildew. *BioControl*, **60**, 103–112.

Jaber, L. R. and Enkerli, J. (2016). Effect of seed treatment duration on growth and colonization of *Vicia faba* by endophytic *Beauveria bassiana* and *Metarhizium brunneum*. *Biological Control*, **103**, 187–195.

Jaber, L. R. and Ownley, B. H. (2017). Can we use entomopathogenic fungi as endophytes for dual biological control of insect pests and plant pathogens? *Biological Control*, **116**, 35–45.

Jaber, L. R. and Salem, N. M. (2014). Endophytic colonisation of squash by the fungal entomopathogen *Beauveria bassiana* (Ascomycota: Hypocreales) for managing Zucchini yellow mosaic virus in cucurbits. *Biocontrol Science and Technology*, **24**, 1096–1109.

Jaber, L. R. and Vidal, S. (2010). Fungal endophyte negative effects on herbivory are enhanced on intact plants and maintained in a subsequent generation. *Ecological Entomology*, **35**, 25–36.

Karthiba, L., Saveetha, K., Suresh, S. *et al.* (2010). PGPR and entomopathogenic fungus bioformulation for the synchronous management of leaffolder pest and sheath blight disease of rice. *Pest Management Science*, **66**, 555–564.

Kauppinen M., Saikkonen K., Helander M., Pirttilä A. M. and Wäli P. R. (2016). *Epichloë* grass endophytes in sustainable agriculture. *Nature Plants*, **2**, 15224.

Kavroulakis, N., Ntougias, S., Zervakis, G. I. *et al.* (2007). Role of ethylene in the protection of tomato plants against soil-borne fungal pathogens conferred by an endophytic *Fusarium solani* strain. *Journal of Experimental Botany*, **58**, 3853–3864.

Kepler, R. M. and Rehner, S. A. (2013). Genome-assisted development of nuclear intergenic sequence markers for entomopathogenic fungi of the *Metarhizium anisopliae* species complex. *Molecular Ecology Resources*, **13**, 210–217.

Kim, J. J., Goettel, M. S. and Gillespie, D. R. (2007). Potential of *Lecanicillium* species for dual microbial control of aphids and the cucumber powdery mildew fungus, *Sphaerotheca fuliginea*. *Biological Control*, **40**, 327–332.

Kim, K. H., Kabir, E. and Jahan, S. A. (2017). Exposure to pesticides and the associated human health effects. *Science of The Total Environment*, **575**, 525–535.

Kolattukudy, P. (1984). Biochemistry and function of cutin and suberin. *Canadian Journal of Botany*, **62**, 2918–2933.

Kumar, V. (2015). A review on efficacy of biopesticides to control the agricultural insect's pest. *International Journal of Agricultural Science Research*, **4**, 168–179.

Krell, V., Jakobs-Schoenwandt, D., Vidal, S. and Patel, A. V. (2017). Encapsulation of *Metarhizium brunneum* enhances endophytism in tomato plants. *Biological Control*, **116**, 62–73.

Kwak, M. J., Song, J. Y., Kim, S. Y. *et al.* (2012). Complete genome sequence of the endophytic bacterium *Burkholderia* sp. strain KJ006. *Journal of Bacteriology*, **194**, 4432–4433.

Lacey, L. A., Grzywacz, D., Shapiro-Ilan, D. I. *et al.* (2015). Insect pathogens as biological control agents: back to the future. *Journal of Invertebrate Pathology*, **132**, 1–41.

Lahiri, A., Douglas, G. C., Murphy, B. R. and Hodkinson, T. R. (2019). *In vitro* methods for plant–microbe interaction and biocontrol studies in european ash (*Fraxinus excelsior* L.). In *Endophytes for a Growing World*, ed. T. R. Hodkinson, F. M. Doohan, M. J. Saunders and B. R. Murphy. Cambridge: Cambridge University Press, Chapter 15.

Leroux, P., Gredt, M., Leroch, M. and Walker, A. S. (2010). Exploring mechanisms of resistance to respiratory inhibitors in field strains of *Botrytis cinerea*, the causal agent of gray mold. *Applied and Environmental Microbiology*, **76**, 6615–6630.

Levitan, L., Merwin, I. and Kovach, J. (1995). Assessing the relative environmental impacts of agricultural pesticides: the quest for a holistic method. *Agriculture, Ecosystems & Environment*, **55**, 153–168.

Liao, X., O'Brien, T. R., Fang, W. and Leger, R. J. S. (2014). The plant beneficial effects of *Metarhizium* species correlate with their association with roots. *Applied Microbiology and Biotechnology*, **98**, 7089–7096.

Lingg A. and Donaldson M. (1981). Biotic and abiotic factors affecting stability of *B. bassiana* conidia in soil. *Journal of Invertebrate Pathology*, **38**, 191–200.

Logrieco A., Moretti A., Castella G. *et al.* (1998). Beauvericin production by *Fusarium* species. *Applied Environmental Microbiology*, **64**, 3084–3088.

Lohse, R., Jakobs-Schönwandt, D., Vidal, S. and Patel, A. V. (2015). Evaluation of new fermentation and formulation strategies for a high endophytic establishment of *Beauveria bassiana* in oilseed rape plants. *Biological Control*, **88**, 26–36.

Lopez, D. C. and Sword, G. A. (2015). The endophytic fungal entomopathogens *Beauveria bassiana* and *Purpureocillium lilacinum* enhance the growth of cultivated cotton (*Gossypium hirsutum*) and negatively affect survival of the cotton bollworm (*Helicoverpa zea*). *Biological Control*, **89**, 53–60.

Majumder, D., Kangjam, B., Devi, K. J. *et al.* (2016). Endophytes: an emerging tool for plant disease management. In *The Handbook of Microbial Bioresources*, ed. V. K. Gupta, G. D. Sharma, M. G. Tuohy and R. Gaur. Oxfordshire: CABI, pp. 179–192.

Mantzoukas, S., Chondrogiannis, C. and Grammatikopoulos, G. (2015). Effects of three endophytic entomopathogens on sweet sorghum and on the larvae of the stalk borer *Sesamia nonagrioides*. *Entomologia Experimentalis et Applicata*, **154**, 78–87.

Martin, K. J. and Rygiewicz, P. T. (2005). Fungal-specific PCR primers developed for analysis of the ITS region of environmental DNA extracts. *BMC Microbiology*, **5**, 28.

Mazón, A. (2001). Producción, uso y control de calidad de hongos entomopatógenos en Nicaragua. *Manejo Integrado de plagas, CATIE, Costa Rica*, **63**, 95–103.

Meyling, N. V. and Eilenberg, J. (2007). Ecology of the entomopathogenic fungi *Beauveria bassiana* and *Metarhizium anisopliae* in temperate agroecosystems: potential for conservation biological control. *Biological Control*, **43**, 145–155.

Moore D., Bateman R. P., Carey M. and Prior C. (1995). Long-term storage of *Metarhizium flavoviride* conidia in oil formulations for the control of locusts and grasshoppers. *Biocontrol Science and Technology*, **5**, 193–200.

Murphy, B. R., Martin Nieto, L., Doohan, F. M. and Hodkinson, T. R. (2015). Fungal endophytes enhance agronomically important traits in severely drought-stressed barley. *Journal of Agronomy and Crop Science*, **201**, 419–427.

Murphy, B. R., Doohan, F. M., Hodkinson, T. R. (2018). From concept to commerce: developing a successful fungal endophyte inoculant for agricultural crops. *Journal of Fungi*, **4**, 24.

Murphy, B. R., Doohan F. M and Hodkinson, T. R. (2019). Prospecting crop wild relatives for beneficial endophytes. In *Endophytes for a Growing World*, ed. T. R. Hodkinson, F. M. Doohan, M. J. Saunders and B. R. Murphy. Cambridge: Cambridge University Press, Chapter 18.

Oerke, E. C. and Dehne, H. W. (2004). Safeguarding production: losses in major crops and the role of crop protection. *Crop Protection*, **23**, 275–285.

Ownley, B. H., Pereira, R. M., Klingeman, W. E., Quigley, N. B. and Leckie, B. M. (2004). *Beauveria bassiana*, a dual purpose biocontrol organism, with activity against insect pests and plant pathogens. In *Emerging Concepts in Plant Health* Management, ed. R. T. Lartey and A. J. Caesar. Kerala, India: Research Signposts, pp. 255–269.

Ownley, B. H., Griffin, M. R., Klingeman, W. E. *et al.* (2008). *Beauveria bassiana*: endophytic colonization and plant disease control. *Journal of Invertebrate Pathology*, **98**, 267–270.

Ownley, B., Gwinn, K. and Vega, F. (2010). Endophytic fungal entomopathogens with activity against plant pathogens: ecology and evolution. *Biocontrol*, **55**, 113–128.

Parine, N. R., Devinder, K., Khan, P. A. A. and Varaprasad, B. (2010). Antifungal efficacy of secondary metabolites from entomopathogenic fungi *Beauveria bassiana*. *Journal of Pharmacy Research*, **3**, 855–856.

Parsa, S., Ortiz, V. and Vega, F. E. (2013). Establishing fungal entomopathogens as endophytes: towards endophytic biological control. *JoVE*, **74**, e50360.

Pereira, R., Stimac, J. and Alves, S. (1993). Soil antagonism affecting the dose- response of workers of the red imported fire ant *Solenopsis invicta*, to *Beauveria bassiana* conidia. *Journal of Invertebrate Pathology*, **61**, 156–161.

Petrini, O. (1991). Fungal endophytes of tree leaves. In *Microbial Ecology of Leaves*, ed. J. H. Andrews and S. S. Hirano. New York, NY: Springer, pp. 179–197.

Pieterse, C. M., Van Wees, S. C., Van Pelt, J. A. *et al.* (1998). A novel signaling pathway controlling induced systemic resistance in *Arabidopsis*. *The Plant Cell*, **10**, 1571–1580.

Posada, F., Aime, M. C., Peterson, S. W., Rehner, S. A. and Vega, F. E. (2007). Inoculation of coffee plants with the fungal entomopathogen *Beauveria bassiana* (Ascomycota: Hypocreales). *Mycological Research*, **111**, 748–757.

Posada, F. J., Chaves, F. C., Gianfagna, T. J., Pava-Ripoll, M. and Hebbar, P. (2010). Establishment of the fungal entomopathogen *Beauveria bassiana* as an endophyte in cocoa pods (*Theobroma cacao* L.). *Revista UDCA Actualidad & Divulgación Científica*, **13**, 71–78.

Powell, W. A., Klingeman, W. E., Ownley, B. H. and Gwinn, K. D. (2009). Evidence of endophytic *Beauveria bassiana* in seed-treated tomato plants acting as a systemic entomopathogen to larval *Helicoverpa zea* (Lepidoptera: Noctuidae). *Journal of Entomological Science*, **44**, 391–396.

Quesada-Moraga, E. and Vey, A. (2004). Bassiacridin, a protein toxic for locusts secreted by the entomopathogenic fungus *Beauveria bassiana*. *Mycological Research*, **108**, 441–452.

Quesada-Moraga, E., Ruiz-García, A. and Santiago-Alvarez, C. (2006). Laboratory evaluation of entomopathogenic fungi *Beauveria bassiana* and *Metarhizium anisopliae* against puparia and adults of *Ceratitis capitata* (Diptera: Tephritidae). *Journal of Economic Entomology*, **99**, 1955–1966.

Quesada-Moraga, E., Muñoz-Ledesma, F. J. and Santiago-Alvarez, C. (2009). Systemic protection of *Papaver somniferum* L. against *Iraella luteipes* (Hymenoptera: Cynipidae) by an endophytic strain of *Beauveria bassiana* (Ascomycota: Hypocreales). *Environmental Entomology*, **38**, 723–730.

Quesada-Moraga, E., López-Díaz, C. and Landa, B. B. (2014). The hidden habit of the entomopathogenic fungus *Beauveria bassiana*: first demonstration of vertical plant transmission. *PloS One*, **9**, e89278.

Ravindran, K., Chitra, S., Wilson, A. and Sivaramakrishnan, S. (2014). Evaluation of antifungal activity of *Metarhizium anisopliae* against plant phytopathogenic fungi. In *Microbial Diversity and Biotechnology in Food Security*, ed. R. N. Kharwar, R. S. Upadhyay, N. K. Dubey and R. Raghuwanshi. India: Springer, pp. 251–255.

Rehner, S. A., Minnis, A. M., Sung, G.-H., Luangsa-ard, J. J., Devotto, L. and

Humber, R. A. (2011). Phylogeny and systematics of the anamorphic, entomopathogenic genus *Beauveria*. *Mycologia*, **103**, 1055–1073.

Roberts, D. W. and Hajek, A. E. (1992). Entomopathogenic fungi as bioinsecticides. In *Frontiers in Industrial Mycology*, ed. G. Leatham. New York: Chapman and Hall, pp. 144–159.

Roberts, D. W. and Leger, R. J. S. (2004). *Metarhizium* spp., cosmopolitan insect-pathogenic fungi: mycological aspects. *Advances in Applied Microbiology*, **54**, 1–70.

Rodríguez, R., White Jr, J., Arnold, A. and Redman, R. (2009). Fungal endophytes: diversity and functional roles. *New Phytologist*, **182**, 314–330.

Rondot, Y. and Reineke, A. (2016). Endophytic *Beauveria bassiana* in grapevine *Vitis vinifera* (L.) reduces infestation with piercing-sucking insects. *Biological Control*, **116**, 82–89.

Ryu, C. M., Hu, C. H., Reddy, M. S. and Kloepper, J. W. (2003). Different signaling pathways of induced resistance by rhizobacteria in *Arabidopsis thaliana* against two pathovars of *Pseudomonas syringae*. *New Phytologist*, **160**, 413–420.

Sánchez Márquez, S., Bills, G. F., Herrero, N. and Zabalgogeazcoa, I. (2011). Non-systemic fungal endophytes of grasses. *Fungal Ecology*, **5**, 289–297.

Sasan, R. K. and Bidochka, M. J. (2012). The insect-pathogenic fungus *Metarhizium robertsii* (Clavicipitaceae) is also an endophyte that stimulates plant root development. *American Journal of Botany*, **99**, 101–107.

Sasan, R. K. and Bidochka, M. J. (2013). Antagonism of the endophytic insect pathogenic fungus *Metarhizium robertsii* against the bean plant pathogen *Fusarium solani* f. sp. *phaseoli*. *Canadian Journal of Plant Pathology*, **35**, 288–293.

Shah, F. A., Ansari, M. A., Watkins, J. *et al.* (2009). Influence of commercial fungicides on the germination, growth and virulence of four species of entomopathogenic fungi. *Biocontrol Science and Technology*, **19**, 743–753.

Sharma, P. (2011). Complexity of '*Trichoderma-fusarium*' interaction and manifestation of biological control. *Australian Journal of Crop Science*, **5**, 1027–1038.

Shoresh, M., Harman, G. E. and Mastouri, F. (2010). Induced systemic resistance and plant responses to fungal biocontrol agents. *Annual Review of Phytopathology*, **48**, 21–43.

Soper, R. S. and Ward, M. G. (1981). Production, formulation and application of fungi for insect control. In *Biological Control in Crop Production*, BARC Symposium No. 5, ed. G. C. Papavizas. Montclair, NJ: Allanheld, Osmun, pp. 161–180.

Stone, J., Polishook, J. and White, J. (2004). Endophytic fungi. In *Biodiversity of Fungi, Inventory and Monitoring Methods*, ed. G. M. Mueller, G. F. Bills and M. S. Foster. Amsterdam: Elsevier, pp. 241–270.

Strobel, G. and Daisy, B. (2003). Bioprospecting for microbial endophytes and their natural products. *Microbiology and Molecular Biology Reviews*, **67**, 491–502.

Susuki, A., Kanaoka, M., Isogai, A. *et al.* (1977). Bassianolide, a new insecticidal cyclodepsipeptide from *Beauveria bassiana* and *Verticillium lecanni*. *Tetrahedrom*, **25**, 2167–2170.

Tefera, T. and Vidal, S. (2009). Effect of inoculation method and plant growth medium on endophytic colonization of sorghum by the entomopathogenic fungus *Beauveria bassiana. Biocontrol,* **4**, 663–669.

Taylor, A. G. and Harman, G. E. (1990). Concepts and technologies of selected seed treatments. *Annual Review of Phytopathology*, **28**, 321–339.

Tilman, D., Cassman, K. G., Matson, P. A., Naylor, R. and Polasky, S. (2002). Agricultural sustainability and intensive production practices. *Nature*, **418**, 671–677.

Vega, F., Posada, F., Aime, M. *et al.* (2008). Entomopathogenic fungal endophytes. *Biological Control*, **46**, 72–82.

Vega, F. E., Goettel, M. S., Blackwell, M. *et al.* (2009). Fungal entomopathogens: new insights on their ecology. *Fungal Ecology*, **2**, 149–159.

Vega, F. E., Meyling, N. V., Luangsaard, J. J. and Blackwell, M. (2012). Fungal entomopathogens. *Insect Pathology*, **2**, 171–220.

Venugopalan, A. and Srivastava, S. (2015). Endophytes as *in vitro* production platforms of high value plant secondary metabolites. *Biotechnology Advances*, **33**, 873–887.

Vicari, M., Hatcher, P. and Ayres, P. (2002). Combined effect of foliar and mycorrhizal endophytes on an insect herbivore. *Ecology*, **83**, 2452–2464.

Vidal, S. and Jaber, L. R. (2015). Entomopathogenic fungi as endophytes: plant–endophyte–herbivore interactions and prospects for use in biological control. *Current Science*, **109**, 46–54.

Vining, L., Kelleher, W. and Schwarting, A. (1962). Oosporein production by a strain of *Beauveria bassiana* originally identified as *Amanita muscaria. Canadian Journal of Microbiology*, **8**, 931–933.

Wagner, B. L. and Lewis, L. C., (2000). Colonization of corn, *Zea mays*, by the entomopathogenic fungus *Beauveria bassiana. Applied and Environmental Microbiology* **66**, 3468–3473.

Wang, C. and Wang, S. (2017). Insect pathogenic fungi: genomics, molecular interactions, and genetic improvements. *Annual Review of Entomology*, **62**, 73–90.

Wang, Q. and Xu, L. (2012). Beauvericin, a bioactive compound produced by fungi: a short review. *Molecules*, **17**, 2367–2377.

Wang, X., Radwan, M. M., Taráwneh, A. H. *et al.* (2013). Antifungal activity against plant pathogens of metabolites from the endophytic fungus *Cladosporium cladosporioides. Journal of Agricultural and Food Chemistry*, **61**, 4551–4555.

White, T. J., Bruns, T. D., Lee, S. and Taylor, J. (1990). Amplification and direct sequencing of fungal ribosomal RNA genes for phylogenetics. In *PCR Protocols: A Guide to Methods and Applications*, ed. M. A. Innis, D. H. Gelfand, J. J. Sninsky and T. J. White. New York: Academic Press, pp. 315–322.

Wilson, D. (1995). Endophyte: the evolution of a term, and clarification of its use and definition. *Oikos*, **73**, 274–276.

Xu, Y., Orozco, R., Wijeratne, E. K. *et al.* (2009). Biosynthesis of the cyclooligomer depsipeptide bassianolide, an insecticidal virulence factor of *Beauveria bassiana. Fungal Genetics and Biology*, **46**, 353–364.

Xu, L., Wang, J., Zhao, J. *et al.* (2010). Beauvericin from the endophytic fungus, *Fusarium redolens*, isolated from *Dioscorea zingiberensis* and its antibacterial activity. *Natural Product Communications*, **5**, 811–814.

5

Emerging Methods for Biological Control of Barley Diseases Including the Role of Endophytes

ANNA K. HØYER, HANS J. L. JØRGENSEN, BIRGIT JENSEN, BRIAN R. MURPHY AND TREVOR R. HODKINSON

Abstract

Barley is an important crop worldwide with production largely used for animal feed and alcoholic beverages. Diseases are a major limiting factor to its production. These have, up until recently, been controlled by agrochemicals. However, legislation on the use of agrochemicals, especially within the European Union, is being tightened and there is growing interest in integrated pest management. This means that there is an increasing focus on controlling diseases using biological control. Living microorganisms that are applied as biological control agents (BCAs) to either soil, seed or leaves can have difficulty in persisting. Therefore, the focus of this review is on endophytes, which are microorganisms that live inside the plant without causing symptoms of disease and have the potential of staying protected as well as being beneficial to the plant and effective against multiple diseases. In this review, we discuss the different approaches for finding and testing beneficial endophytes and for determining the endophyte host range. Furthermore, we undertook a literature search to summarise previous studies that have investigated the use of endophytes as well as BCAs against barley diseases.

5.1 Important Diseases of Barley in Northern Europe

Barley is the fourth largest cereal crop in the world with a global production of more than 141 million tonnes (FAOSTAT, 2018). Barley is used primarily as animal feed (55–60% of total production), for seeds (5%) and to a much lesser extent for human consumption (2–3%). In addition, malting barley is used for the production of alcoholic beverages (30–40%) (Ullrich, 2011). In cereals, it is estimated that pre-harvest losses, due to disease, weeds and environmental stresses, can reach up to approximately 35% of the yield (Akar *et al.*, 2004). Losses due to disease are particularly damaging, and there are many important barley diseases worldwide. Some of the most important in Northern Europe are: powdery mildew (*Blumeria graminis* f. sp. *hordei*), leaf blotch or scald (*Rhynchosporium graminicola*), rusts such as brown rust (*Puccinia hordei*) and yellow rust (*P. striiformis* f. sp. *hordei*), net blotch (*Pyrenophora teres*), Ramularia leaf spot (*Ramularia collo-cygni*) and barley yellow dwarf (*BYDV*) (Oldach, pers. comm.; Walters *et al.*, 2012), which all cause symptoms on leaves (Mathre, 1982; Videira *et al.*, 2016). In addition, Fusarium head blight (*Fusarium* spp.) affects the malting quality and yield of malt when producing beer (Oliveira *et al.*, 2012; Nielsen *et al.*, 2014) with symptoms occurring in the heads (Mathre, 1982). It is expected that the use of some agrochemicals will be banned or restricted in the future (Oldach, pers. comm.), which means that the reliance on chemical inputs will need to be reduced. Furthermore, organic agriculture has increased by almost 20% a year globally (Nandwani and Nwosisi, 2016) and farmers would also benefit from an alternative non-synthetic solution.

A European Union regulation (Directive 2009/128/EC; European Parliament, 2009) has already been implemented, which is encouraging the sustainable use of pesticides and promoting integrated pest management (Department of Agriculture, 2013). Integrated pest management is defined by ENDURE (2008) as 'a sustainable approach to managing pests by combining biological, cultural and chemical tools in a way that minimises economic, environmental and health risks'. The use of biopesticides is promoted as an environmentally friendly alternative to synthetic pesticides. Within Europe, the term 'biopesticides' is often used to cover a range of products that can be used to protect crops from diseases, pests and weeds. The products can be divided into two subgroups that either rely on the use of (1) living organisms or (2) naturally occurring compounds, including extracts from plants and microorganisms as well as volatiles (Balog *et al.*, 2017). There are a limited number of registered biopesticide products in Europe and they are based on bacteria, fungi or viruses (Kabaluk *et al.*, 2010; Balog *et al.*, 2017). In most cases, the control agent is a different species to the plant pathogen. However, in some cases weak strains or non-pathogenic isolates of the same pathogen species are used (Punja, 1997; Kabaluk *et al.*, 2010). Within the scientific community, biological control has most often been defined as the use of living organisms for crop protection (Pal and Gardener, 2006) and here the term biological control is more precisely defined as the control of plant diseases by microorganisms. Biological control of plant diseases can work through one or a combination of four general mechanisms which are (1) parasitism, (2) antibiosis, (3) competition and (4) induction of host resistance

(Guetsky *et al.*, 2002; Alabouvette *et al.*, 2006). Integrated control of foliar barley diseases is reviewed by Walters *et al.* (2012). However, they do not include the use of biological control.

5.1.1 Endophytes Used for Biological Control of Plant Diseases

The interest in using microorganisms as biological control agents of plant diseases is increasing (Broadfoot, 2016), especially for diseases that are otherwise difficult to control (Walters, 2009). The plant microbiome consists of epiphytes and endophytes (Lindow and Brandl, 2003; Müller *et al.*, 2016). Epiphytes are the microorganisms that live on plant surfaces and they can be further divided into organisms that inhabit the rhizosphere, the phyllosphere (Müller *et al.*, 2016) and the spermosphere (Lindsey *et al.*, 2017). The term 'endophyte' was first used by Anton de Bary in 1884. He described an endophyte as a parasite living inside its host's organ (de Bary, 1884). The definition has since then been broadened and endophytes are generally defined as microorganisms living inside plants without causing symptoms of disease (Wilson, 1995). Some endophytes have been shown to provide plants with benefits such as drought tolerance (Naveed *et al.*, 2014), heat tolerance (Hubbard *et al.*, 2014), improved mineral nutrition (Murphy *et al.*, 2015a), salt-stress tolerance (Rodriguez *et al.*, 2008) and protection against disease (Maciá-Vicente *et al.*, 2009). While endophytes are not the only biocontrol approach to have received attention in recent years, they compare favourably to other microorganisms which may have difficulty persisting and/or remaining competent when they are applied to the leaves, the seeds or the soil (Walker *et al.*, 2002; Ting *et al.*, 2009; Buddrus-Schiemann *et al.*, 2010). Thus, the use of endophytes may keep the biological control agents (BCAs) protected within the plant (Eevers *et al.*, 2015) and provides the possibility of control against several stresses without losing efficacy over the growing season (Wilkinson *et al.*, 2000). In our research, focus is put on generalist endophytes, which can be transferred from crop wild relatives (CWRs) and promising results have been obtained in barley (A. K. Høyer, unpublished results). Thus, this review will emphasise the targeted search for plant protecting endophytes as well as previous studies of biocontrol in barley.

5.1.2 Endophyte Host Range and the Targeted Search for Beneficial Endophytes

All plants in natural habitats are believed to harbour endophytes (Aly *et al.*, 2011) and they can be tissue-type specific or systemic (Zabalgogeazcoa, 2008). The life cycles of the majority of endophytes are not completely understood, but it is clear that some endophytes do not remain exclusively within the plant throughout their whole life cycle, which means that they can potentially be latent pathogens (Comby *et al.*, 2016) or latent saprotrophs, or can represent early colonisation by rhizobia or mycorrhizal fungi (Porras-Alfaro and Bayman, 2011). The diversity of different taxonomic groups of endophytes that has been elucidated recently has been summarised in the meta-analysis by Hardoim *et al.* (2015). The most frequently reported sequences of prokaryotic endophytes were from Proteobacteria (54%), Actinobacteria (almost 20%) and Bacilli (15%), whereas eukaryotic sequences were mostly from Glomeromycota (40%, arbuscular mycorrhizal fungi), Ascomycota (almost 31%, with the subordinate class Dothideomycetes accounting for 15%) and Basidiomycota (20%, with Agaricomycetes accounting for 18%).

Many factors have been shown to influence the endophyte community composition and one of the important factors is host plant species (Nissinen *et al.*, 2012; Wearn *et al.*, 2012). Nissinen *et al.* (2012) showed that several bacterial genera were tightly associated with particular arcto-alpine plant species (*Oxyria digyna*, *Diapensia lapponica* and *Juncus trifidus*). In total, they identified 58 different bacterial genera. Of the major bacterial genera, five were exclusively associated with *J. trifidus* (*Acido Gp1*, *Arthrobacter*, *Knoellia*, *Paenibacillus*, *Paracoccus* and *Rhodanobacter*), four were specific to *O. digyna* (*Agreia*, *Ancylobacter*, *Rhizobium* and *Rhodococcus*) and one was exclusively associated with *D. lapponica* (*Pedobacter*). However, some groups of endophytes are generalist and are able to colonise plants of unrelated taxonomic identity. Interestingly, all three plant species were colonised by *Burkholderia*, *Mucilaginibacter*, *Nocardioides* and *Sphingomonas*. Wearn *et al.* (2012) explained that part of the fungal communities of grassland forbs (*Cirsium arvense*, *Plantago lanceolata* and *Rumex acetosa*) were host-plant specific. Thus, 48% of the fungal community belonging to *C. arvense* was generalist endophytes, with 58% and 72% generalists for *P. lanceolata* and *R. acetosa*, respectively (Wearn *et al.*, 2012). In grasses, generalist endophytes are, for instance, found in the groups of clavicipitaceous endophytes and dark septate endophytes (DSEs) (Clay, 1990; Jumpponen and Trappe, 1998; Mandyam *et al.*, 2010). Known generalists of clavicipitaceous endophytes in temperate grasses are *Epichloë coenophiala* and other *Epichloë* ssp., which infect grasses in the subfamily Pooideae (Hodkinson, 2018) and *Atkinsonella* spp. which infects *Danthonia* spp. and *Stipa* spp. (Clay, 1990). For generalists within the DSEs, Mandyam *et al.* (2010) showed that the roots of four C_4 grasses (*Andropogon gerardii*, *Sorghastrum nutans*, *Schizachyrium scoparium* and *Panicum virgatum*) normally had two DSEs in common, i.e. *Periconia macrospinosa* and *Microdochium* sp..

Several approaches have been explored to isolate potential endophytes that confer protection against diseases. Most studies have cultured endophytes from healthy looking plants that live in an environment that has a particular disease stress. In this case, it is hypothesised that the endophytes contribute to plant health and that they are able to relieve the stress (Araujo *et al.*, 2002). An alternative strategy has been suggested by Ellis (2017), who proposed looking for biocontrol agents in diseased tissue because organisms can persist in a pathogen-infected tissue and hence potentially act as control agents. Although this may appear as counter-intuitive,

Ellis (2017) gives an example of control of crown gall in stone fruit and, furthermore, Köhl *et al.* (2009) found antagonists supressing apple scab using this approach.

There are different approaches when it comes to both selecting target plant species as sources of beneficial endophytes and selecting a plant species to test the biological control properties. Most studies have isolated endophytes from a crop species and then tested the biological control effects in the original crop (Kirk and Deacon, 1987; Coombs *et al.*, 2004; Silva *et al.*, 2012). Some studies isolated endophytes from related taxonomic groups of a crop species and tested the effect in the crop (Maciá-Vicente *et al.*, 2008). CWRs are valuable resources in crop breeding programmes and have been used to transfer disease resistance (Zeng *et al.*, 2013; Brar and Hucl, 2017; Fedak *et al.*, 2017). Likewise, CWRs can be a unique source of potential biocontrol agents (Maciá-Vicente *et al.* 2008; Murphy *et al.* 2015b). CWRs will have their own microbiome and, although not adequately tested, could be expected to host some endophytes not ordinarily present in the crop species. Due to the close taxonomic affinity to the crop plants, they could also be expected to be more compatible to the target species than endophytes isolated from an unrelated species (Murphy *et al.*, 2018). In addition, some endophytes are isolated from an unrelated plant and then tested on crops. For example, *Serendipita indica* (formerly *Piriformospora indica*; Weiß *et al.*, 2016) is a basidiomycete endophyte that has been tested on many different crops, which are not closely related to the original host (Kumar *et al.*, 2009; Harrach *et al.*, 2013; Rabiey *et al.*, 2015; Wang *et al.*, 2015). *Serendipita indica* was isolated from the rhizosphere of two woody shrubs, *Prosopis juliflora* and *Zizyphus nummularia*, in desert soils of Rajasthan in India (Varma *et al.*, 2012). This fungus has been tested as a BCA in several crop species including wheat (*Triticum aestivum*; Serfling *et al.*, 2007; Rabiey *et al.*, 2015), barley (*Hordeum vulgare;* Harrach *et al.*, 2013), maize (*Zea mays*; Kumar *et al.*, 2009) and tomato (*Solanum lycopersicum;* Roylawar *et al.*, 2015; Wang *et al.*, 2015).

5.2 Previous Studies of Endophytes and Other BCAs Controlling Barley Diseases

An extensive literature survey, conducted here, revealed a total of 8 studies reporting the control of barley diseases by endophytes and 21 studies reporting control by other BCAs (Table 5.1). Only studies using living microorganisms were included in the review. Eight different fungal endophyte species were tested in the endophyte studies and *Serendipita indica* was tested in four of the investigations. In the studies reporting control by BCAs, several different organisms were used, with the majority using fungi. *Pseudomonas* spp. strains were widely used followed by *Trichoderma* spp. and *Clonostachys rosea*. Although not tested as endophytes in the investigations reviewed here, these commonly used organisms have often been widely isolated as

Table 5.1 Overview of the investigations on biological control of barley diseases.

The first part of the table lists studies using endophytes and the second part lists studies using other types of biological control agents. The investigations are organised first according to pathogen, second according to efficiency of disease control and third according to year of publication. The investigations are evaluated on a scale where '0' is given to reports where there was no disease control, '+' is given to reports were disease control efficiency of 0.01–33% are described, '+ +' are reports of 33–66% efficiency and '+ + +' are reports of 66–100% efficiency. The minus symbol indicates that the investigation did not look into disease control or the mechanism of control. Evalution of disease control and mechanism is reported according to the test system. Names of organisms are given according to Species Fungorum (www.speciesfungorum.org) and Catalogue of Life. Footnotes are listed at the end of the table.

Pathogen	**Endophyte**	**Endophyte origin**	**Test system**	**Disease control efficiency**	**Suggested mechanism of control**	**Author**
Bipolaris sorokiniana	*Chaetomium globosum*	Barley leaves	Dual culture	**+[1]/++[2,3]**	Antibiosis and competition[4]	Moya *et al.*, 2016
Pyrenophora teres					Competition and mycoparasitism[5]	
Bipolaris sorokiniana	*Serendipita indica*	Roots of woody shrubs	Detached-leaf-segment assay	**++[6]**	Systemic induction of resistance associated	Waller *et al.*, 2005
Blumeria graminis f. sp. *hordei*			Pot trials	**0[7]**	with elevated	
Fusarium culmorum					antioxidative capacity	
Blumeria graminis f. sp. *hordei*	*Dichotomopilus funicola*	Tomato roots	*In vitro* spore germination test	**++**	Antibiosis	Vilich *et al.*, 1998
	Chaetomium globosum	Laboratory strain[8]	Pot trials	**+/++[9]**	-	
Fusarium culmorum	*Serendipita indica*	Roots of woody shrubs	Pot trials with soil	**+**	Elevated antioxidative capacity	Harrach *et al.*, 2013

Fusarium graminearum	*Serendipita indica*	Roots of woody shrubs	Dual culture	**0**[10]	Not antibiosis	Deshmukh and Kogel, 2007
			Pot trials with soil	**+++**	Not induced resistance	
Fusarium graminearum	*Serendipita indica*	Roots of woody shrubs	Pot trial with soil	**0**[11]	Growth acceleration	Achatz *et al.*, 2010
Gaeumannomyces tritici	*Acremonium furcatum*	24 plant species	Dual plate bioassay	**+**	-	Maciá-Vicente *et al.*, 2008
	Dactylaria sp.		Tube with vermiculite	**+++**	-	
	Fusarium equiseti					
	Phoma herbarum					
Gaeumannomyces tritici	*Fusarium equiseti*	*Corynephorus canescens* and *Lygeum spartum*	Tube with vermiculite	**++**	-	Maciá-Vicente *et al.*, 2009
	Metacordyceps chlamydosporia	*Heterodera avenae* infected eggs and *Meloidogyne* sp.				
Pathogen	**BCA**	**BCA origin**	**Test system**	**Disease control efficiency**	**Suggested mechanism of control**	**Author**
Bipolaris sorokiniana	*Clonostachys rosea*	Barley roots	Pot trial with sand	**++/+++**[12]	-	Jensen *et al.*, 2002

(continued)

Table 5.1 (Cont.)

Pathogen	BCA	BCA origin	Test system	Disease control efficiency	Suggested mechanism of control	Author
Bipolaris sorokiniana	*Bipolaris maydis*	Maize	Pot trial with soil	**+**[13]/**++**[14]/ **+++**[15]	-	Jørgensen *et al.*, 1996
Blumeria graminis f. sp. hordei	*Parastagonospora nodorum*	Wheat				
Pyrenophora teres						
Rhynchosporium graminicola						
Bipolaris sorokiniana	*Clonostachys rosea*	Barley roots	Pot trial with soil	**0**[16]/**++**[17]/ **+++**[18]	Direct inhibition. Reduced conidial	Jensen *et al.*, 2016
Blumeria graminis f. sp. *hordei*					germination and	
Pyrenophora teres					appressorium formation[19]	
Rhynchosporium graminicola						
Bipolaris sorokiniana	*Pseudomonas chlororaphis*	Soil	Field experiments	**+**[20]/**++**[21]/ **+++**[22]	-	Johnsson *et al.*, 1998
Microdochium nivale						
Pyrenophora graminea						

Pyrenophora teres						
Ustilago hordei						
Ustilago nuda						
Bipolaris sorokiniana	*Acremonium sp.*	Barley seed	Pot trial with sand	**+/++**	-	Knudsen *et al.*, 1995
	Chaetomium sp.	Barley seed	Pot trial with soil	**+/++**	-	
	Clonostachys rosea	Organic soil	Field experiments	**+/++**	-	
	Fusarium roseum	Conventional soil				
	Humicola sp.	Conventional hay				
	Microdochium bolleyi	Organic hay				
	Trichoderma sp.	Roots				
Bipolaris sorokiniana	*Microdochium bolleyi*	Grassland	Cellulose filter paper rolls	**+**	Induced resistance	Liljeroth and Bryngelsson, 2002
			Pot trials with soil	**++**		
			Field experiment	**-**[23]		
Bipolaris sorokiniana	*Microdochium bolleyi*	Wheat coleoptile	Field experiments	**+**	-	Duczek, 1997
Blumeria graminis f. sp. *hordei*	*Bacillus thuringiensis* (14 isolates)	Faecal samples from mammalian species	Pot trial	**+/+++**	**-**	Choi *et al.*, 2007

(*continued*)

Table 5.1 (Cont.)

Pathogen	BCA	BCA origin	Test system	Disease control efficiency	Suggested mechanism of control	Author
Fusarium culmorum	*Clonostachys rosea*	Barley roots	Pot trial with sand	**++/+++**	-	Jensen *et al.*, 2000
			Field experiments	**++/+++**		
Fusarium culmorum	*Actinomycetes* (133 strains)	Saharan soil	Streak method	-[24]	-	Yekkour *et al.*, 2012
			Petri dishes with filter paper	**+/++/+++**	-	
Fusarium culmorum	*Acinetobacter* sp.	Cereal plant and soil	Dual culture	**+/++/+++**	Not antibiosis	Khan *et al.*, 2006
Fusarium graminearum	*Chryseobacterium* sp.		Pot trial with soil	**+/++/+++**	Induced resistance[25]	
Fusarium poae	*Pseudomonas fluorescens*					
	Pseudomonas frederiksbergensis					
	Pseudomonas sp.					
	unidentified isolate					
Fusarium culmorum	*Acinetobacter* sp.	Cereal plant and soil	Pot trial with soil	**+/++**[26]	-	Khan and Doohan, 2008a

Fusarium graminearum	*Chryseobacterium* sp.		Field experiment	**++**	-	
Fusarium poae	*Pseudomonas fluorescens*					
	Pseudomonas frederiksbergensis					
	Pseudomonas sp.					
Fusarium culmorum	*Pseudomonas fluorescens*	Barley rhizosphere soil	Pot trial with soil	**++**	-	Khan and Doohan, 2008b
			Field experiment	**++**	-	
Gaeumannomyces tritici	*Metacordyceps chlamydosporia*	Nematophagous fungi	Petri dish experiment	**++/+++**	-	Monfort *et al.*, 2005
	Metapochonia rubescens		Tube with vermiculite	**+**	-	
	Lecanicillium lecanii					
Pyrenophora graminea	*Streptomyces* ssp. (5 strains)	-	Pot trial with soil	**+/++/+++**	-	Koch *et al.*, 2006
Pyrenophora teres	*Trichoderma* spp. (5 strains)	-				
Pyrenophora teres	Actinomycete	Soil and straw	*In vitro* straw assay	**+++**	-	Mostafa, 1993
	Albifimbria verrucaria		Tube with cotton	**++/+++**	-	

(*continued*)

Table 5.1 (Cont.)

Pathogen	BCA	BCA origin	Test system	Disease control efficiency	Suggested mechanism of control	Author
	Trichoderma viride		Pot trial[27]	+++	-	
	Trichoderma pseudokoningii					
	Trichoderma sp.					
Pyrenophora teres	400 bacterial isolates	Roots of wild and	Pot trial with soil	+/++/+++	-	Hökeberg *et al.*, 1997
	Pseudomonas isolates	cultivated plants	Field experiment	+++	-	
Pyrenophora teres	*Micromonospora*	Barley straw	*In vitro* straw assay	+/++/+++	-	Ali-Hoimoud *et al.*, 1993
	Trichoderma koningii					
	Trichoderma pseudokoningii					
	Trichoderma viride					
	Five unidentified fungi					
Pyrenophora teres	Nine fungal antagonists[28]	Soil and roots of barley sprouts	Pot trial with sand	+/++/+++	-	Abrahamsen, 1992
Pyrenophora teres	*Bipolaris maydis*	Maize	Pot trial with soil	+/+++	Induced resistance	Jørgensen *et al.*, 1998
	Parastagonospora nodorum	Wheat				

Pseudomonas syringae	*Pantoea agglomerans*[29]	Barley seeds	Pot trial[30]	**++/+++**	-[31]	Braun-Kiewnick *et al.*, 2000
			Field experiments	**+/++/+++**		

1 For *Bipolaris sorokiniana*
2 For *Pyrenophora teres*
3 Some treatments gave reductions of 0–33% / others gave reductions of 33–66% / while others gave reductions of 66–100%
4 For *Bipolaris sorokiniana*
5 For *Pyrenophora teres*
6 For *Blumeria graminis* f. sp. *hordei*
7 Methods of evaluation was insufficient
8 Perhaps it originated from twigs of *Quercus* sp., however it is unclear
9 For *Chaetomium globosum*
10 No inhibition zone was formed
11 Methods of evaluation were insufficient
12 Dependent on storage conditions
13 For *Bipolaris sorokiniana* and *Blumeria graminis* f. sp. *hordei.* Results are cultivar dependent
14 For *Bipolaris sorokiniana*, *Pyrenophora teres* and *Rhynchosporium graminicola.* Results are cultivar dependent
15 For *Rhynchosporium graminicola.* Results are cultivar dependent
16 For *Blumeria graminis* f. sp. *hordei*
17 For *Bipolaris sorokiniana* and *Pyrenophora teres*
18 For *Rhynchosporium graminicola*
19 For *Bipolaris sorokiniana*
20 Against *Ustilago nuda*
21 Against *Bipolaris sorokiniana* and *Microdochium nivale*
22 Against *Pyrenophora graminea*, *Pyrenophora teres*, *Ustilago hordei* and *Ustilago nuda*
23 No disease symptoms occurred before ear emergence
24 No control treatment – difficult to estimate the disease reduction. Antagonists were tested against several *Fusarium* ssp., *Aspergillus* ssp. and *Penicillium* ssp. at this step
25 Results are from wheat not barley
26 Dependent on timing of BCA application
27 Substrate not mentioned
28 Identity not mentioned
29 Several strains
30 Substrate not mentioned
31 Investigated but mentioned in a different paper

endophytes (Evans *et al.*, 2003; Høyer *et al.*, 2016; Mercado-Blanco *et al.*, 2016). All the studies collectively focus on a limited number of pathogens and they include the leaf pathogens *Bipolaris sorokiniana, Blumeria graminis* f. sp. *hordei, Pseudomonas syringae, Pyrenophora graminea, Pyrenophora teres, Rhynchosporium graminicola, Ustilago hordei, U. nuda,* as well as the soil pathogens *Bipolaris sorokiniana, Fusarium culmorum, F. graminearum, F. poae, Gaeumannomyces tritici* and *Microdochium nivale.*

The rationale behind the selection of host plant species as a source of biological control agents has often not been described sufficiently well in the studies. The endophytes were sourced from barley (Moya *et al.*, 2016), grasses including *Ammophila arenaria* ssp. *australis, Corynephorus canescens* and *Lygeum spartum* (Maciá-Vicente *et al.*, 2008, 2009) and unrelated plant species (Vilich *et al.*, 1998; Waller *et al.*, 2005; Maciá-Vicente *et al.*, 2008; Deshmukh and Kogel, 2007; Achatz *et al.*, 2010; Harrach *et al.*, 2013). Among the studies using other BCAs, one study did not describe where the BCA originated from (Koch *et al.*, 2006). Seven out of 21 studies used antagonists that originated from barley (Abrahamsen, 1992; Ali-Hoimoud *et al.*, 1993; Knudsen *et al.*, 1995; Braun-Kiewnick *et al.*, 2000; Jensen *et al.*, 2000, 2002, 2016), five studies used BCAs from other cereals or what was termed 'straw' (Mostafa, 1993; Knudsen *et al.*, 1995; Duczek, 1997; Jørgensen *et al.*, 1996, 1998) and two studies used antagonists originating from unspecified grasses (Hökeberg *et al.*, 1997; Liljeroth and Bryngelsson, 2002). One study used BCAs from unrelated plant species (Hökeberg *et al.*, 1997) and seven used samples from soil (Mostafa, 1993; Knudsen *et al.*, 1995; Johnsson *et al.*, 1998; Khan *et al.*, 2006; Khan and Doohan, 2008a, 2008b; Yekkour *et al.*, 2012). Two studies used nematophagous fungi (Monfort *et al.*, 2005; Maciá-Vicente *et al.*, 2009), one used fungi isolated from insects (Maciá-Vicente *et al.*, 2009) and one study found their control agents in mammalian faeces (Choi *et al.*, 2007).

Four different reasons for working with a specific endophyte species were given (Table 5.1) including getting good results from a preliminary *in vitro* study (Maciá-Vicente *et al.*, 2008), the fact that the endophyte belonged to a genus which is known for the production of secondary metabolites (Vilich *et al.*, 1998), previous success with the organism within the research group (Maciá-Vicente *et al.*, 2009; Achatz *et al.*, 2010) or a literature review (Deshmukh and Kogel, 2007; Achatz *et al.*, 2010; Harrach *et al.*, 2013; Moya *et al.*, 2016). The choice of BCA species in the other studies was often explained by the fact that the organisms had been used successfully in previous studies by the same authors or research group (Knudsen *et al.*, 1995; Jensen *et al.*, 2000, 2002, Koch *et al.*, 2006; Khan and Doohan, 2008b; Jensen *et al.*, 2016). However, a short literature review of the abilities of the BCAs in different crops or against specific diseases was more common (Mostafa, 1993; Duczek, 1997; Braun-Kiewnick *et al.*, 2000; Liljeroth and Bryngelsson, 2002; Monfort *et al.*, 2005; Choi *et al.*, 2007; Khan *et al.*, 2006; Khan and Doohan, 2008a; Yekkour *et al.*, 2012).

Much rarer reasoning was that the organisms were screened *in planta* in the actual study (Hökeberg *et al.*, 1997; Johnsson *et al.*, 1998), or were non-pathogens of barley (Jørgensen *et al.*, 1996, 1998). Sometimes no reason was given (Abrahamsen, 1992; Ali-Hoimoud *et al.*, 1993).

5.2.1 Experimental Test Systems

There is increasing financial expense in progressing from *in vitro* studies, to pot trials, to field experiments. There is, however, also an increase in the value of the knowledge produced, at least if the aim is to reduce disease pressure in the field. *In vitro* studies are, in general, controversial because there is often not a good correlation between *in vitro* results and results obtained from more complex growth systems (Renwick *et al.*, 1991; Fravel, 1988; Khan *et al.*, 2006; Deshmukh and Kogel, 2007).

In two of the biocontrol investigations (Table 5.1), long-term field trials were used (Duczek, 1997; Johnsson *et al.*, 1998). However, the most common experimental test system is pot trials (Abrahamsen, 1992; Jørgensen *et al.*, 1996, 1998; Jensen *et al.*, 2002; Koch *et al.*, 2006; Choi *et al.*, 2007; Achatz *et al.*, 2010; Harrach *et al.*, 2013; Jensen *et al.*, 2016), combined with *in vitro* testing (Mostafa, 1993; Vilich *et al.*, 1998; Khan *et al.*, 2006; Waller *et al.*, 2005; Deshmukh and Kogel, 2007) or followed by field experiments (Knudsen *et al.*, 1995; Hökeberg *et al.*, 1997; Jensen *et al.*, 2000; Braun-Kiewnick *et al.*, 2000; Liljeroth and Bryngelsson, 2002; Khan and Doohan, 2008a, b). Different substrates have been used in pot trials including vermiculite (Ali-Hoimoud *et al.*, 1993), sand (Jensen *et al.*, 2002) and soil (Jørgensen *et al.*, 1996). The more complex the pot trial system, the better it represents field conditions; thus it is preferable to use soil instead of vermiculite. However, when disease symptoms are evaluated on roots, it can ease the work flow not to use soil. At the less complex end of the spectrum, Yekkour *et al.* (2012) used Petri dishes with filter paper and four studies used tube assays with either vermiculite (Monfort *et al.*, 2005; Maciá-Vicente *et al.*, 2008, 2009) or cotton (Mostafa, 1993). Two studies used *in vitro* experiments of BCA and pathogen only (Ali-Hoimoud *et al.*, 1993; Moya *et al.*, 2016). Ali-Hoimoud *et al.* (1993) used a cut straw assay because they were interested in biocontrol of the survival structures of *Pyrenophora teres* on crop residues. The study by Moya *et al.* (2016) performed a 'classical' dual culture test using only one type of medium. This could be considered controversial because various studies have shown that type of media and water potential within the medium will influence growth rates, production of secondary metabolites and hyphal interactions between antagonist and pathogen (Whipps, 1987; Whipps and Magan, 1987).

Six of the eight endophyte studies checked whether their control agent could colonise barley as an endophyte (Vilich *et al.*, 1998; Waller *et al.*, 2005; Deshmukh and Kogel, 2007; Maciá-Vicente *et al.*, 2008, 2009 Achatz *et al.*, 2010). If the reduction in disease symptoms is linked to the lifestyle of the microorganism as an endophyte then it is relevant to show that the endophyte colonises the plant in question as an

Table 5.2 Summary of the best biological control results from the most complex systems obtained against five of the most commonly studied pathogens in barley.

The origin of the biological control organism is also listed. Names of organisms are given according to Species Fungorum (www.speciesfungorum.org).

Pathogen	Disease control (%)	Test system	Origin of BCA or endophyte	Author
Bipolaris sorokiniana	43	Field experiments	Soil	Knudsen *et al.*, 1995
Blumeria graminis f.sp. *hordei*	70	Pot experiment	Mammalian faeces	Choi *et al.*, 2007
Fusarium culmorum	73	Field experiment	Barley roots	Jensen *et al.*, 2000
Gaeumannomyces tritici	88	Tube with vermiculite	Endophyte of *Ammophila arenaria* ssp. *australis* (Poaceae)	Maciá-Vicente *et al.*, 2008
Pyrenophora teres	98	Field experiments	Roots of wild and cultivated plants	Hökeberg *et al.*, 1997

endophyte, especially, but not exclusively, if the endophyte has been sourced from a different species than the crop.

5.2.2 Biological Control Efficiency

Biological control efficiency varies among experiments with the best results for the most complex test systems summarised in Table 5.2. It is clear that barley diseases can be controlled using BCAs as well as endophytes in pot and in field trials. Beneficial microorganisms have been discovered from many and varying places and the best ones have originated from barley itself, marram grass (*Ammophila arenaria* ssp. *australis*), wild and cultivated plants and mammalian faeces (Table 5.2).

The experimental test system will influence the reported outcomes. The fewer the variables in the experiments, the easier it will be to obtain efficient biocontrol results. In the investigations, which tested biological control agents first in pot trials and later in the field, there was a tendency for the efficiency of the control agents to be 4–35% lower in the field (Knudsen *et al.*, 1995; Hökeberg *et al.*, 1997; Braun-Kiewnick *et al.*, 2000; Jensen *et al.*, 2000; Khan and Doohan, 2008a, 2008b). As an exception, Hökeberg *et al.* (1997) reported a specific *Pseudomonas* strain (MA 342), which controlled disease slightly better in the field (98%) compared to the pot trial (75%). If the treatment works in a pot experiment it will have a higher likelihood of success in the field than if the BCA was identified *in vitro*.

5.2.3 Biocontrol Mechanisms Used Against Barley Diseases

Ten studies (Table 5.1) have investigated the mechanisms behind the biological control, but rigorous evaluations are rare. In many cases, the potential involvement of all the possible mechanisms in biological control (antibiosis, competition, parasitism and induced resistance) have not been studied or even been possible to study simply because an appropriate experimental setup has not been applied. For example, to show that induced resistance is involved in biological control, requires plant experiments to be performed and defence responses to be studied.

Two of the studies used *Chaetomium* spp. endophytes as BCAs and only mechanisms inferred from *in vitro* assays were reported (Vilich *et al.*, 1998; Moya *et al.*, 2016). Both studies showed that *Chaetomium* spp. worked through antibiosis against leaf pathogens of barley *in vitro*. Vilich *et al.* (1998) concluded that their fungal isolate reduced spore germination of barley powdery mildew by antibiosis. They spread conidia of *Blumeria graminis* f. sp. *hordei* on malt extract agar plates that contained a filtrate of the BCA. However, they did not outline their control treatment, which makes it difficult to evaluate their findings and, furthermore, since the pathogen is an obligate biotroph, their *in vitro* setup may yield different results from a more realistic situation using barley leaves. In subsequent pot experiments, a BCA spore suspension was applied to the seeds and the pathogen was inoculated onto the leaves. It is, however, difficult to make firm conclusions on the mechanism *in planta* from the *in vitro* study. Thus, it is not known whether compounds of the endophyte reached the leaves, which would be a prerequisite for concluding that metabolites produced by the BCA was responsible for any disease reducing effect. Moya *et al.* (2016) performed a dual culture test where they placed a plug of the *Chaetomium* antagonist on a PDA plate and 3 days later placed a plug of either *Bipolaris sorokiniana* or *Pyrenophora teres* at a distance of 4 cm away from the first plug. The control treatment was the pathogen alone, which is perhaps not the optimal control as it may be argued that a proper control would have been a pure agar plug placed on a plate and a pathogen plug added 3 days later to exclude any effect of the agar. The conclusion was that the *Chaetomium globosum* isolates worked through antibiosis and competition against *Bipolaris sorokiniana* and through competition and mycoparasitism against *Pyrenophora teres.* These conclusions are all based on evaluations using a microscope and, unfortunately, these observations stand alone. Thus, it is unknown whether the endophytes had a similar behaviour *in planta* or whether they would be able to induce resistance against the pathogen.

Four studies investigated the mechanisms of control exerted by the endophyte *Serendipita indica* (Waller *et al.*, 2005; Deshmukh and Kogel, 2007; Achatz *et al.*, 2010; Harrach *et al.*, 2013). All studies used pathogens from the genus *Fusarium* and Waller *et al.* (2005) also included *Blumeria graminis* f. sp. *hordei* and *Bipolaris sorokiniana,* all in separate experiments. Harrach *et al.* (2013) and Waller *et al.* (2005) both concluded that elevated antioxidative capacity was the mechanism for disease control, whereas Achatz *et al.* (2010) suggested that the endophyte used plant growth promotion to avoid disease and Deshmukh and Kogel (2007) concluded that pathogenesis-related (PR) proteins were not involved in protection. The main aim of the investigation by Achatz *et al.* (2010) was to show that *S. indica* relieved plants from nutrient stress and *Fusarium* sp. was used as an additional biotic stress. They showed that plants with and without *Fusarium* infection had equivalent grain yields. As grain yield is not a reliable measure of biological control and disease symptoms were not evaluated, it is difficult to discern if the pathogen was established and one must therefore be cautious in interpreting the results. Harrach *et al.* (2013) used *S. indica* against *F. culmorum* in a pot trial. No direct symptom scoring was made, but they used the shoot/root biomass as a proxy for disease scoring and they did quantify pathogen biomass as an indication of disease pressure. Antioxidant status of the roots was examined through ascorbate and glutathione levels as well as antioxidant enzyme activity. It was concluded that *S. indica* altered the antioxidant status of the cells so that they could detoxify excess reactive oxygen species (ROS) produced by the pathogen. However, in the literature used to indicate how the pathogen is affected by ROS, the authors only show 'plausible' correlations between *F. culmorum* and ROS production in *Arabidopsis* floral tissue. So, it is not entirely clear whether these responses can explain reductions in disease in barley.

The study of Waller *et al.* (2005) also used shoot/root biomass as an indicator of biological control for *F. culmorum* and the data for *B. sorokiniana* are not shown. It is suggested that the mechanism cannot be antibiosis because this was ruled out in a study in axenic culture, but data are not shown. Furthermore, it is not clear how plant inoculation with *S. indica* took place and, therefore, it is difficult to evaluate the relevance of the *in vitro* study. Antioxidant capacity was also studied when inoculating roots with and without *S. indica*. Since the pathogen was not present in these experiments, it is difficult to make conclusions about the mechanisms of control. Waller *et al.* (2005) also examined the control of *Blumeria graminis* f. sp. *hordei* and used a disease index to show reduction in disease symptoms in a detached-leaf assay, but again the antagonist delivery system is not clear. This time, systemic resistance was suggested.

Deshmukh and Kogel (2007) also ruled out antibiosis based on dual culture tests, although the nature of these experiments was not fully described. Perhaps it is too early to rule out antibiosis when there have been no additional tests of whether the endophyte can produce antagonistic compounds within the plant. The authors found that PR-protein genes were expressed at lower levels when *S. indica* was present with *F. culmorum* compared to plants inoculated with *F. culmorum* alone. They, therefore, concluded that PR proteins were not involved in the protection induced by the endophyte.

Khan *et al.* (2006) also examined the biocontrol of *Fusarium* spp., but they used bacteria as their control agents. Antibiosis was excluded as a potential mechanism using *in vitro* inhibition zone studies on one type of medium. Again, it is perhaps premature to completely rule out antibiosis because of the absence of an inhibition zone when it is not clear what the BCA produces *in planta*. Their subsequent *in planta* study was only conducted on wheat and the results indicate that induced resistance is the mechanism involved. They examined the expression of a PR-gene (class III peroxidase), which is known to be upregulated in wheat in response to *Fusarium* infection. In this experiment, they worked with *Pseudomonas fluorescens* (MKB 156) and *Pseudomonas* sp. (MKB 158) and they were only able to show induced resistance for one of the strains (MKB 158).

The three last studies (Table 5.1) concern control of *Bipolaris sorokiniana* (Liljeroth and Bryngelsson, 2002; Jensen *et al.*, 2016) and/or *Pyrenophora teres* (Jørgensen *et al.*, 1998; Jensen *et al.*, 2016). Furthermore, Jensen *et al.* (2016) also included *Blumeria graminis* f. sp. *hordei* and *Rhynchosporium graminicola*. All three investigations used different BCAs. Thus, Jørgensen *et al.* (1998) used two non-barley pathogens to control diseases, whereas Liljeroth and Bryngelsson (2002) used *Microdochium bolleyi* as a BCA and Jensen *et al.* (2016) used *Clonostachys rosea*. Jørgensen *et al.* (1998) found that induced resistance was probably the main mechanism involved in the local protection exerted by the two non-barley pathogens. They showed that appressoria-formation was reduced and that papillae formation was increased. In the study by Jensen *et al.* (2016), *C. rosea* was able to control *Bipolaris sorokiniana*, *P. teres* and *R. graminicola*, but mechanisms of control were only evaluated for *B. sorokiniana*. It was concluded that the inhibition was direct and therefore probably involved mycoparasitism, competition and/or antibiosis. This was based on the fact that germination of pathogen conidia and inhibition of appressorial formation was observed. Induced resistance was ruled out because expression of three PR-protein genes was not increased in plants treated with antagonist and pathogen compared to the control, and furthermore, there was no increase in defence responses when evaluated under the microscope. In the study by Liljeroth and Bryngelsson (2002), *B. sorokiniana* was suggested to be controlled by induced resistance and this was shown by the elevated expression of PR-protein genes. However, results for pathogen and antagonist treated plants were performed separately and there was no treatment with both pathogen and antagonist together so it is not clear whether the antagonist can upregulate PR-protein genes in the presence of the pathogen.

5.3 Conclusion

There is a need to address the heavy reliance on agrochemicals in barley production and improve the environmental sustainability of the industry. Integrated pest management is encouraged within the European Union and biological control can be incorporated into this approach. It is evident from the literature review conducted here that living microorganisms can control barley diseases in controlled laboratory experiments and, more importantly, under field conditions. It is furthermore clear that some of the most problematic diseases of barley in Northern Europe, including rusts, Ramularia leaf spot and barley yellow dwarf have not been challenged using biological control in agricultural systems. There is a trend in legislation for restricting the use of certain agrochemicals and organic agriculture is increasing globally each year which means that there will be a huge demand for non-chemical control methods for these diseases in the future.

Another finding is that there is no particular, specific niche from where to isolate biocontrol antagonists. It appears that it is possible to find antagonists in many types of environments. However, the majority of studies sourced their control agents from barley plants, other cereals or wild grasses and some of the best results were also obtained with BCAs obtained from such hosts. Endophytes also showed good results and they were sourced from barley leaves, wild grasses and from unrelated plant species. There is very little known about the host range of endophytes and it seems theoretically more likely that reliable results will be obtained when looking for endophytes from the crop of interest or its wild relatives because the chance of successful establishment within the plant is increased. Also, work with endophytes is recommended because endophytes can be protected within the plant and also have a biocontrol potential for multiple diseases.

Very few investigations have examined the mechanisms behind the biological control reported in barley, and within these, rigorous investigations were found to be infrequent. There is a need for the biocontrol research community to agree on standards in order to conclusively demonstrate biological control and determine the mechanisms involved. Appropriate disease symptoms must be evaluated and it is essential to choose relevant control treatments. Furthermore, gene expression studies or other studies to quantify defence responses in plants need to include treatments with pathogen and antagonist present together to compare with treatments with pathogen alone and quantify defence responses with a documented effect against the pathogen in question. When using endophytes to control diseases, it is also important to show that the endophyte can establish within the plant.

The trend in biological control research is to isolate control agents that can reduce symptoms from more than one disease or combine control agents in synergistic consortia. Such BCAs should have different modes of action. Ideally, a control agent should also be found which controls the pathogens in such a way that the pathogen does not evolve quickly to overcome the mechanism. Induced resistance is one such example because it generally elicits multiple defence reactions in the plant and thereby becomes difficult to overcome. Conversely, antibiosis might not be the best approach for biocontrol in barley and other crops because the pathogen population might develop tolerance to the active compound, as they are known to do with agrochemicals.

Acknowledgements

This project has received funding from the European Union's Horizon 2020 research and innovation programme under the Marie Skłodowska-Curie grant agreement No. 674964.

References

Abrahamsen, S. (1992). Methods for testing fungal antagonists against *Drechslera teres* in barley seeds. *IOBC-WPRS Bulletin*, **15**, 27–29.

Achatz, B., von Rüden, S., Andrade, D. *et al.* (2010). Root colonization by *Piriformospora indica* enhances grain yield in barley under diverse nutrient regimes by accelerating plant development. *Plant and Soil*, **333**, 59–70.

Akar, T., Avci, M. and Dusunceli, F. (2004). *Barley: Post-Harvest Operations.* Rome: Food and Agricultre Organization of the United Nations.

Alabouvette, C., Olivain, C. and Steinberg, C. (2006). Biological control of plant diseases: the European situation. *European Journal of Plant Pathology*, **114**, 329–341.

Ali-Hoimoud, D.-E., Mostafa, M., Barrault, G. and Albertini, L. (1993). Evaluation of organisms antagonistic to the sclerorotid organs of *Drechslera teres*, the causal agent of barley net blotch. *Plant Disease*, **77**, 1251–1255.

Aly, A. H., Debbab, A. and Proksch, P. (2011). Fungal endophytes: unique plant inhabitants with great promises. *Applied Microbiology and Biotechnology*, **90**, 1829–1845.

Araujo, W. L., Marcon, J., Maccheroni, W. *et al.* (2002). Diversity of endophytic bacterial populations and their interaction with *Xylella fastidiosa* in citrus plants. *Applied and Environmental Microbiology*, **68**, 4906–4914.

Balog, A., Hartel, T., Loxdale, H. D. and Wilson, K. (2017). Differences in the progress of the biopesticide revolution between the EU and other major crop-growing regions. *Pest Management Science*, **73**, 2203–2208.

de Bary, A. (1884). *Vergleichende Morphologie und Biologie der Pilze, Mycetozoen, und Bacterien.* Leipzig, Germany: Verlag von Wilhelm Engelmann.

Brar, G. S. and Hucl, P. J. (2017). Cultivar description 00Ar134-1, a spring wheat line of intergeneric origin. *Canadian Journal of Plant Science*, **97**, 153–156.

Braun-Kiewnick, A., Jacobsen, B. J. and Sands, D. C. (2000). Biological control of *Pseudomonas syringae* pv. *syringae*, the causal agent of basal kernel blight of barley, by antagonistic *Pantoea agglomerans*. *Phytopathology*, **90**, 368–375.

Broadfoot, M. (2016). Microbes added to seeds could boost crop production. *Scientific American*. www.scientificamerican.com/article/microbes-added-to-seeds-could-boost-crop-production/

Buddrus-Schiemann, K., Schmid, M., Schreiner, K. *et al.* (2010). Root colonization by *Pseudomonas* sp. DSMZ 13134 and impact on the indigenous rhizosphere bacterial community of barley. *Microbial Ecology*, **60**, 381–393.

Choi, G. J., Kim, J. C., Jang, K. S. and Lee, D. H. (2007). Antifungal activities of *Bacillus thuringiensis* isolates on barley and cucumber powdery mildews. *Journal of Microbiology and Biotechnology*, **17**, 2071–2075.

Clay, K. (1990). Fungal endophytes of grasses. *Annual Review of Ecology and Systematics*, **21**, 275–297.

Comby, M., Lacoste, S., Baillieul, F., Profizi, C. and Dupont, J. (2016). Spatial and temporal variation of cultivable communities of co-occurring endophytes and pathogens in wheat. *Frontiers in Microbiology*, **7**, 403.

Coombs, J. T., Michelsen, P. P. and Franco, C. M. M. (2004). Evaluation of endophytic Actinobacteria as antagonists of *Gaeumannomyces graminis* var. *tritici* in wheat. *Biological Control*, **29**, 359–366.

Department of Agriculture (2013). *National Action Plan for the Sustainable Use of Pesticides*, Vol. II. Dublin: Pesticide Registration & Control Division, Department of Agriculture, Food and the Marine.

Deshmukh, S. D. and Kogel, K.-H. (2007). *Piriformospora indica* protects barley root rot caused by *Fusarium graminearum*. *Journal of Plant Diseases and Protection*, **114**, 1861–3829.

Duczek, L. J. (1997). Biological control of common root rot in barley by *Idriella bolleyi*. *Canadian Journal of Plant Pathology*, **194**, 402–405.

Eevers, N., Gielen, M., Sánchez-López, A. *et al.* (2015). Optimization of isolation and cultivation of bacterial endophytes through addition of plant extract to nutrient media. *Microbial Biotechnology*, **8**, 707–715.

Ellis, J. G. (2017). Can plant microbiome studies lead to effective biocontrol of plant diseases? *Molecular Plant–Microbe Interactions*, **30**, 190–193.

ENDURE (2008). *ENDURE IPM Training Guide, Chapter Arguments*. www.endure-network.eu/

European Parliament (2009). Directive 2009/128/EC of the European Parliament and the Council of 21 October 2009 Establishing a Framework for Community Action to Achieve the Sustainable Use of Pesticides. October, Vol. **309**.

Evans, H. C., Holmes, K. A. and Thomas, S. E. (2003). Endophytes and mycoparasites associated with an indigenous forest tree, *Theobroma gileri*, in Ecuador and a preliminary assessment of their potential as biocontrol agents of cocoa diseases. *Mycological Progress*, **2**, 149–160.

FAOSTAT (2018). *Statistical Databases*. Rome: Food and Agriculture Organization of the United Nations. www.fao.org/faostat/en/#home

Fedak, G., Cao, W., Wolfe, D., Chi, D. and Xue, A. (2017). Molecular characterization of Fusarium resistance from *Elymus repens* introgressed into bread wheat. *Cytology and Genetics*, **51**, 130–133.

Fravel, D. R. (1988). Role of antibiosis in the biocontrol of plant diseases. *Annual Review of Phytopathology*, **26**, 75–91.

Guetsky, R., Shtienberg, D., Elad, Y., Fischer, E. and Dinoor, A. (2002). Improving biological control by combining biocontrol agents each with several mechanisms of disease suppression. *Phytopathology*, **92**, 976–985.

Hardoim, P. R., van Overbeek, L. S., Berg, G. *et al.* (2015). The hidden world within plants: ecological and evolutionary considerations for defining functioning of microbial endophytes. *Microbiology and Molecular Biology Reviews*, **79**, 293–320.

Harrach, B. D., Baltruschat, H., Barna, B., Fodor, J. and Kogel, K.-H. (2013). The mutualistic fungus *Piriformospora indica* protects barley roots from a loss of antioxidant capacity caused by the necrotrophic pathogen *Fusarium culmorum*. *Molecular Plant–Microbe Interactions*, **26**, 599–605.

Hodkinson, T. R. (2018). Evolution and taxonomy of the grasses (Poaceae): a model family for the study of species-rich groups. *Annual Plant Reviews Online*, doi: 10.1002/9781119312994.apr0622.

Hökeberg, M., Gerhardson, B. and Johnsson, L. (1997). Biological control of cereal seed-borne diseases by seed bacterization with greenhouse-selected bacteria. *European Journal of Plant Pathology*, **103**, 25–33.

Høyer, A. K., Jørgensen, H. J. L., Amby, D. B. and Jensen, B. (2016). Endophytic colonisation of tomato plants by the biological control agent *Clonostachys rosea*. *IOBC-WPRS Bulletin*, **115**, 101–106.

Hubbard, M., Germida, J. J. and Vujanovic, V. (2014). Fungal endophytes enhance wheat heat and drought tolerance in terms of grain yield and second-generation seed viability. *Journal of Applied Microbiology*, **116**, 109–122.

Jensen, B., Knudsen, I. M. B. and Jensen, D. F. (2000). Biological seed treatment of cereals with fresh and long-term stored formulations of *Clonostachys rosea*: biocontrol efficacy against *Fusarium culmorum*. *European Journal of Plant Pathology*, **106**, 233–242.

Jensen, B., Knudsen, I. M. B. and Funck Jensen, D. (2002). Survival of conidia of *Clonostachys rosea* on stored barley seeds and their biocontrol efficacy against seed-borne *Bipolaris sorokiniana*. *Biocontrol Science and Technology*, **124**, 427–441.

Jensen, B., Lübeck, P. S. and Jørgensen, H. J. (2016). *Clonostachys rosea* reduces spot blotch in barley by inhibiting prepenetration growth and sporulation of *Bipolaris sorokiniana* without inducing resistance. *Pest Management Science*, **72**, 2231–2239.

Johnsson, L., Hökeberg, M. and Gerhardson, B. (1998). Performance of the *Pseudomonas chlororaphis* biocontrol agent MA 342 against cereal seed-borne diseases in field experiments. *European Journal of Plant Pathology*, **104**, 701–711.

Jørgensen, H. J. L., Andresen, H. and Smedegaard-Petersen, V. (1996). Control of *Drechslera teres* and other barley pathogens by preinoculation with *Bipolaris maydis* and *Septoria nodorum*. *Phytopathology*, **86**, 602–607.

Jørgensen, H. J. L., Lübeck, P. S. *et al.* (1998). Mechanisms of induced resistance in barley against *Drechslera teres*. *Phytopathology*, **88**, 698–707.

Jumpponen, A. and Trappe, J. M. (1998). Dark septate endophytes: a review of facultative biotrophic root-colonizing fungi. *New Phytologist*, **140**, 295–310.

Kabaluk, J. T., Svircev, M. S., Goettel, M. S. and Woo, S. G., eds. (2010). *The Use and Regulation of Microbial Pesticides in Representative Jurisdictions Worldwide*. IOBC Global, www.IOBC-Global.org.

Khan, M. R. and Doohan, F. M. (2008a). Bacterium-mediated control of Fusarium head blight disease of wheat and barley and associated mycotoxin contamination of grain. *Biological Control*, **48**, 42–47.

Khan, M. R. and Doohan, F. M. (2008b). Comparison of the efficacy of chitosan with that of a fluorescent *pseudomonad* for the control of Fusarium head blight disease of cereals and associated mycotoxin contamination of grain. *Biological Control*, **48**, 48–54.

Khan, M. R., Fischer, S., Egan, D. and Doohan, F. M. (2006). Biological control of Fusarium seedling blight disease of wheat and barley. *Phytopathology*, **96**, 386–394.

Kirk, J. J. and Deacon, J. W. (1987). Control of the take-all fungus by *Microdochium bolleyi*, and interactions involving *M. bolleyi*, *Phialophora graminicola* and *Periconia macrospinosa* on cereal roots. *Plant and Soil*, **98**, 231–237.

Knudsen, I. M. B., Hockenhull, J. and Jensen, D. F. (1995). Biocontrol of seedling diseases of barley and wheat caused by *Fusarium culmorum* and *Bipolaris sorokiniana*: effects of selected fungal antagonists on growth and yield components. *Plant Pathology*, **44**, 467–477.

Knudsen, I. M. B., Hockenhull, J., Funck Jensen, D. *et al.* (1997). Selection of biological control agents for controlling soil and seed-borne diseases in the field. *European Journal of Plant Pathology*, **103**, 775–784.

Koch, E., Weil, B., Wächter, R., Wohlleben, S. *et al.* (2006). Evaluation of selected microbial strains and commercial alternative products as seed treatments for the control of *Tilletia tritici*, *Fusarium culmorum*, *Drechslera graminea* and *D. teres*. *Journal of Plant Diseases and Protection*, **113**, 150–158.

Köhl, J. J., Molhoek, W. W. M. L. *et al.* (2009). Selection and orchard testing of antagonists suppressing conidial production by the apple scab pathogen *Venturia inaequalis*. *European Journal of Plant Pathology*, **123**, 401–414.

Kumar, M., Yadav, V., Tuteja, N. and Johri, A. K. (2009). Antioxidant enzyme activities in maize plants colonized with *Piriformospora indica*. *Microbiology*, **155**(3), 780–790.

Liljeroth, E. and Bryngelsson, T. (2002). Seed treatment of Barley with *Idriella bolleyi* causes systemically enhanced defence against root and leaf infection by *Bipolaris sorokiniana*. *Biocontrol Science and Technology*, **12**, 235–249.

Lindow, S. E. and Brandl, M. T. (2003). Microbiology of the phyllosphere. *Applied and Environmental Microbiology*, **69**, 1875–1883.

Lindsey III, B. E., Rivero, L., Calhoun, C. S., Grotewold, E. and Brkljacic, J. (2017). Standardized method for high-throughput sterilization of *Arabidopsis* seeds video link. *Journal of Visualized Experiments*, **128**, 56587.

Maciá-Vicente, J. G., Jansson, H.-B., Mendgen, K. and Lopez-Llorca, L. V. (2008). Colonization of barley roots by endophytic fungi and their reduction of take-all caused by *Gaeumannomyces graminis* var. *tritici*. *Canadian Journal of Microbiology*, **54**, 600–609.

Maciá-Vicente, J. G., Rosso, L. C., Ciancio, A. *et al.* (2009). Colonisation of barley roots by endophytic *Fusarium equiseti* and *Pochonia chlamydosporia*: effects on plant growth and disease. *Annals of Applied Biology*, **155**, 391–401.

Mandyam, K., Loughin, T. and Jumpponen, A. (2010). Isolation and morphological and metabolic characterization of common endophytes in annually burned tallgrass prairie. *Mycologia*, **102**, 813–821.

Mathre, D. E., ed. (1982). *Compendium of Barley Diseases*. St Paul, MN: APS Press.

Mercado-Blanco, J., Alos, E., Rey, M. D. and Prieto, P. (2016). *Pseudomonas fluorescens* PICF7 displays an endophytic lifestyle in cultivated cereals and enhances yield in barley. *FEMS Microbiology Ecology*, **92**, 1–13.

Monfort, E., Lopez-Llorca, L. V., Jansson, H.-B. *et al.* (2005). Colonisation of seminal roots of wheat and barley by egg-parasitic nematophagous fungi and their effects on *Gaeumannomyces graminis* var. *tritici* and development of root rot. *Soil Biology and Biochemistry*, **37**, 1229–1235.

Mostafa, M.-M. (1993). Biological control of *Drechslera teres*: ability of antagonists to reduce conidia formation, coleoptile infection and leaf infection in barley (*Hordeum vulgare*). *Cryptogamie Mycologie*, **14**, 287–295.

Moya, P., Pedemonte, D., Amengual, S., Franco, M. E. E. and Sisterna, M. N. (2016). Antagonism and modes of action of *Chaetomium globosum* species group, potential biocontrol agent of barley foliar diseases. *The Bulletin of the Botanical Society of Argentina*, **51**, 569–578.

Müller, D. B., Vogel, C., Bai, Y. and Vorholt, J. A. (2016). The plant microbiota: systems-level insights and perspectives. *Annual Review of Genetics*, **50**, 211–234.

Murphy, B. R., Doohan, F. M. and Hodkinson, T. R. (2015a). Fungal root endophytes of a wild barley species increase yield in a nutrient-stressed barley cultivar. *Symbiosis*, **65**, 1–7.

Murphy, B. R., Doohan, F. M. and Hodkinson, T. R. (2015b). Persistent fungal root endophytes isolated from a wild barley species suppress seed-borne infections in a barley cultivar. *BioControl*, **60**, 281–292.

Murphy, B., Doohan, F. M. and Hodkinson, T. R. (2018). From concept to commerce: developing a successful fungal endophyte inoculant for agricultural crops. *Journal of Fungi*, **4**, 24.

Nandwani, D. and Nwosisi, S. (2016). Global trends in organic agriculture. In *Organic Farming for Sustainable Agriculture*, ed. D. Nandwani. Berlin: Springer International Publishing, pp. 1–35.

Naveed, M., Mitter, B., Reichenauer, T. G., Wieczorek, K. and Sessitsch, A. (2014). Increased drought stress resilience of maize through endophytic colonization by *Burkholderia phytofirmans* PsJN and *Enterobacter* sp. FD17. *Environmental and Experimental Botany*, **97**, 30–39.

Nielsen, L. K., Cook, D. J., Edwards, S. G. and Ray, R. V. (2014). The prevalence and impact of Fusarium head blight pathogens and mycotoxins on malting barley quality in UK. *International Journal of Food Microbiology*, **179**, 38–49.

Nissinen, R. M., Männistö, M. K. and van Elsas, J. D. (2012). Endophytic bacterial communities in three arctic plants from low arctic fell tundra are cold-adapted and host-plant specific. *FEMS Microbiology Ecology*, **82**, 510–522.

Oliveira, P., Mauch, A., Jacob, F. and Arendt, E. K. (2012). Impact of *Fusarium culmorum*-infected barley malt grains on brewing and beer quality. *American Society of Brewing Chemists*, **70**, 186–194.

Pal, K. K. and Gardener, B. M. (2006). Biological control of plant pathogens. *The Plant Health Instructor*, doi: 10.1094/PHI-A-2006-1117-02.

Porras-Alfaro, A. and Bayman, P. (2011). Hidden fungi, emergent properties: endophytes and microbiomes. *Annual Review of Phytopathology*, **49**, 291–315.

Punja, Z. K. (1997). Comparative efficacy of bacteria, fungi, and yeasts as biological control agents for diseases of vegetable crops. *Canadian Journal of Plant Pathology*, **19**, 315–323.

Rabiey, M., Ullah, I. and Shaw, M. W. (2015). The endophytic fungus *Piriformospora indica* protects wheat from Fusarium crown rot disease in simulated UK autumn conditions. *Plant Pathology*, **64**, 1029–1040.

Renwick., A., Campbell, R. and Coe, S. (1991). Assessment of *in vivo* screening systems for potential biocontrol agents of *Gaeumannomyces graminis*. *Plant Pathology*, **40**, 524–532.

Rodriguez, R. J., Henson, J., Van Volkenburgh, E. *et al.* (2008). Stress tolerance in plants via habitat-adapted symbiosis. *The ISME Journal*, **2**, 404–416.

Roylawar, P., Panda, S. and Kamble, A. (2015). Comparative analysis of BABA and *Piriformospora indica* mediated priming of defence-related genes in tomato against early blight. *Physiological and Molecular Plant Pathology*, **91**, 88–95.

Serfling, A., Wirsel, S. G. R., Lind, V. and Deising, H. B. (2007). Performance of the biocontrol fungus *Piriformospora indica* on wheat under greenhouse and field conditions. *Phytopathology*, **97**, 523–531.

Silva, H. S. A., Tozzi, J. P. L., Terrasan, C. R. F. and Bettiol, W. (2012). Endophytic microorganisms from coffee tissues as plant growth promoters and biocontrol agents of coffee leaf rust. *Biological Control*, **63**, 62–67.

Ting, A. S. Y., Fang, M. T. and Tee, C. S. (2009). Assessment on the effect of formulative materials on the viability and efficacy of *Serratia marcescens*: a biocontrol agent against *Fusarium oxysporum* f. sp. *cubense* race 4. *American Journal of Agricultural and Biological Sciences*, **4**, 283–288.

Ullrich, S. E. (2011). Significance, adaptation, production, and trade of barley. In *Barley: Production Improvement and Uses*, ed. S. E. Ullrich. Ames, IA: Wiley-Blackwell, pp. 3–13.

Varma, A., Bakshi, M., Lou, B., Hartmann, A. and Oelmueller, R. (2012). *Piriformospora indica*: a novel plant growth-promoting mycorrhizal fungus. *Agricultural Research*, **1**, 117–131.

Videira, S. I. R., Groenewald, J. Z., Braun, U., Shin, H. D. and Crous, P. W. (2016). All that glitters is not *Ramularia*. *Studies in Mycology*, **83**, 49–163.

Vilich, V., Dolfen, M. and Sikora, R. A. (1998). *Chaetomium* spp. colonization of barley following seed treatment and its effect on plant growth and *Erysiphe graminis* f. sp. *hordei* disease severity. *Journal of Plant Diseases and Protection*, **105**, 130–139.

Walker, R., Rossall, S. and Asher, M. J. C. (2002). Colonization of the developing rhizosphere of sugar beet seedlings by potential biocontrol agents applied as seed treatments. *Journal of Applied Microbiology*, **92**, 228–237.

Waller, F., Achatz, B., Baltruschat, H. *et al.* (2005). The endophytic fungus *Piriformospora indica* reprograms barley to salt-stress tolerance, disease resistance, and higher yield. *Proceedings of the National Academy of Sciences of the United States of America*, **102**, 13386–13391.

Walters, D. (2009). Controlling plant disease using biological and environmentally friendly approaches: making it work in practice. In *Disease Control in Crops:*

Biological and Environmentally Friendly Approaches, ed. D. Walters. Oxford, UK: Wiley-Blackwell, pp. 257–261.

Walters, D. R., Avrova, A., Bingham, I. J. *et al.* (2012). Control of foliar diseases in barley: towards an integrated approach. *European Journal of Plant Pathology*, **133**, 33–73.

Wang, H., Zheng, J., Ren, X. *et al.* (2015). Effects of *Piriformospora indica* on the growth, fruit quality and interaction with tomato yellow leaf curl virus in tomato cultivars susceptible and resistant to TYCLV. *Plant Growth Regulation*, **76**, 303–313.

Wearn, J. A., Sutton, B. C., Morley, N. J. and Gange, A. C. (2012). Species and organ specificity of fungal endophytes in herbaceous grassland plants. *Journal of Ecology*, **100**, 1085–1092.

Weiß, M., Waller, F., Zuccaro, A. and Selosse, M.-A. (2016). Sebacinales: one thousand and one interactions with land plants. *New Phytologist*, **211**, 20–40.

Whipps, J. M. (1987). Effect of media on growth and interactions between a range of soil-borne glasshouse pathogens and antagonistic fungi. *New Phytologist*, **107**, 127–142.

Whipps, J. M. and Magan, N. (1987). Effects of nutrient status and water potential of media on fungal growth and antagonist-pathogen interactions. *EPPO Bulletin*, **17**, 581–591.

Wilkinson, H. H., Siegel, M. R., Blankenship, J. D. *et al.* (2000). Contribution of fungal loline alkaloids to protection from aphids in a grass–endophyte mutualism. *Molecular Plant–Microbe Interactions*, **13**, 1027–1033.

Wilson, D. (1995). Endophyte: the evolution of a term, and clarification of its use and definition. *Oikos*, **73**, 274–276.

Yekkour, A., Sabaou, N., Zitouni, A., Errakhi, R., Mathieu, F. and Lebrihi, A. (2012). Characterization and antagonistic properties of *Streptomyces* strains isolated from Saharan soils, and evaluation of their ability to control seedling blight of barley caused by *Fusarium culmorum*. *Letters in Applied Microbiology*, **55**, 427–435.

Zabalgogeazcoa, I. (2008). Fungal endophytes and their interaction with plant pathogens. *Spanish Journal Of Agricultural Research*, **6**, 138–146.

Zeng, J., Cao, W., Hucl, P. *et al.* (2013). Molecular cytogenetic analysis of wheat *Elymus repens* introgression lines with resistance to Fusarium head blight. *Genome*, **56**, 75–82.

6

Phosphate Nutrition in Root–Fungus Interactions

WAEL YAKTI, DIANA R. ANDRADE-LINARES, BERNARD NGWENE, MICHAEL BITTERLICH, GÁBOR M. KOVÁCS AND PHILIPP FRANKEN

Abstract

Phosphate belongs to the major mineral nutrient category in plants and is a non-renewable resource. Many natural soils are phosphate deficient, and phosphate fixation into insoluble mineral complexes limits plant growth by decreasing root uptake. Different strategies have appeared during the evolution of land plants to cope with this situation, one of which is to interact with various microbes (bacteria and fungi) located in the plant rhizosphere. This chapter will focus on three major groups of fungi that colonise the roots of most land plants: arbuscular mycorrhizal fungi (Glomeromycotina), fungi from the order Sebacinales (Basidiomycota) and the diverse form-group of dark septate endophytes (Ascomycota). Three major mechanisms of fungal contribution to plant nutrition will be discussed. First, fungi are able to solubilise phosphate from inorganic sources that are not available to plants. Second, fungi can set free mineral nutrients from organic compounds/sources. Third, fungi are able to transport phosphate along their hyphae towards the plant, thereby bridging phosphate depletion zones around the roots. In this chapter, we summarise published knowledge on this topic and present some new non-published data to complete our current model.

6.1 Introduction

Phosphorus (P) is the second most important macronutrient after nitrogen. P accounts for approximately 0.2% of total plant dry weight and plays a significant role in major cellular processes, including photosynthesis, nucleic acid synthesis, respiration, energy storage and transfer, redox reactions, signalling, metabolism and many other vital functions (Schachtman *et al.*, 1998; Hawkesford *et al.*, 2012). Even though P is highly abundant in most agricultural soils, most P is chemically fixed, and only a small proportion is accessible to plants.

P is present in soils in several organic and inorganic forms (Shen *et al.*, 2011). The inorganic P pool includes phosphate (Pi) minerals and secondary precipitates formed with calcium (Ca), sulfur (S), aluminium (Al) and iron (Fe), in addition to orthophosphate anions (HPO_4^{2-} and $H_2PO_4^-$). The organic P pool in soils contains compounds such as sugar phosphates, phosphoproteins, mononucleotides and inositol phosphates (Turner *et al.*, 2002). Among the different forms of P occurring in the soil, plants can only take up orthophosphate forms in soil solution (HPO_4^{2-} and $H_2PO_4^-$) (Hinsinger, 2000), which represent a very small fraction of the total P in soil. Due to the pH dependence of the dissociation of H ions, $H_2PO_4^-$ is the form that is most associated with acidic soils, while secondary HPO_4^{2-} is more prevalent when the soil is basic. In extreme acidic and basic soils, however, Pi solubility is decreased, and H_3PO_4 or PO_4^{-3} is formed, respectively (Brady, 1990). Thus, pH is an influential factor for P availability in soils.

Pi deficiency is considered a main growth-limiting factor in many horticultural and agricultural production systems (Miller *et al.*, 2010). Therefore, Pi needs to be supplied through fertilisers, increasing production costs and imposing ecological concerns and economic problems (Tilman *et al.*, 2001). The use of P fertilisers has increased five-fold in the last 50 years (Chen and Graedel, 2016). Nevertheless, supplied Pi can be rapidly fixed upon application due to the high reactivity of orthophosphates with cations such as Ca^{2+} in alkaline soils or Al^{3+} and Fe^{3+} in acid soils, forming secondary precipitates that are inaccessible to plants. Furthermore, a significant amount of soil P accumulates as microbial biomass. Microbes compete with plants for Pi and present a considerable immobilised P pool. However, this pool is highly dynamic and therefore potentially available for plants (Richardson and Simpson, 2011).

Applied Pi fertilisers may not exceed 30% efficiency (Sharma *et al.*, 2013). Additionally, the production of Pi fertilisers is mostly based on rock phosphate sources, which are non-renewable, finite and becoming more expensive to extract (Cordell *et al.*, 2009). The production of P fertilisers is also energy demanding, and reserves are unevenly distributed, thus being subject to geopolitical risk (Cordell and White, 2015).

A wide range of strategies have appeared during the evolution of land plants to cope with P deficiency (reviewed in Ramaekers *et al.* (2010)). Plants alter root growth to enhance soil exploration, increase their uptake surface, and prioritise the formation of roots that are more efficient in Pi uptake (Lynch *et al.*, 2005). In general, root-to-shoot ratios are increased (Hermans *et al.*, 2006); the formation of lateral roots over primary roots is preferred (Hodge, 2009); and more root hairs develop, resulting in increased Pi depletion zones foraging below surface soil that is rich in plant-available P (Lynch, 2007). In addition to root morphology, plants exhibit biochemical adaptations including the production of extracellular phosphatases, protons and organic anions, which could improve Pi availability and, thus, its uptake (Richardson, 2001; Richardson *et al.*, 2009).

In addition to morphological and biochemical alterations, plants form beneficial associations with a wide spectrum of microorganisms (mainly bacteria and fungi). These associations are significant for plant nutrition in general and for P acquisition in particular (Lambers *et al.*, 2008; Behie and Bidochka, 2014). In this chapter, we will focus on three main groups of root-associated fungi: the arbuscular mycorrhizal fungi (AMF), fungi of the order Sebacinales (with a special focus on the model fungus *Serendipita indica*), and the less investigated group, dark septate endophytes (DSEs).

6.2 Arbuscular Mycorrhizal Fungi (Subphylum: Glomeromycotina)

Most terrestrial plants establish arbuscular mycorrhizal (AM) symbioses, which are characterised by the provision of mineral nutrients to plants (Marschner and Dell, 1994; Wang *et al.*, 2017a). In exchange, plant-synthesised hexoses (Helber *et al.*, 2011; van der Heijden *et al.*, 2015) and lipids (Keymer *et al.*, 2017; Rich *et al.*, 2017) are transferred to fungi. AMF colonise the roots of more than 80% of plant species and proliferate beyond the ambit of roots (Smith and Read, 2008). In this section, we summarise important aspects of the contribution of AMF to plant Pi nutrition, which are undoubtedly of great economic and ecological significance.

Plant associations with members of Glomeromycotina (Spatafora *et al.*, 2016) are widespread, and such interactions are believed to be 400–460 million years old (Redecker and Raab, 2006), suggesting that these ancient symbioses have been pivotal for plant adaptation and persistence in terrestrial environments (Kenrick and Crane, 1997). AMF develop coenocytic hyphal networks that have hundreds of cytoplasm-sharing nuclei and produce multinuclear spores that are rich in storage lipids and protected by chitin-containing cell walls (Smith and Read, 2008). AMF are obligate biotrophic microbes, so they cannot be cultured without a plant partner. This feature results in strong limitations to their production and broad application for plant production (Bago and Cano, 2005).

Even though the responses of plants to AMF colonisation are diverse and many factors influence their effects (Tawaraya, 2003), the interspecific relationship can result in positive as well as negative effects on host plants (Johnson *et al.*, 1997). While AMF may have negative or neutral effects on plant growth (Graham and

Abbott, 2000; Li *et al.*, 2006) and P uptake (Smith *et al.*, 2003; Smith *et al.*, 2004), these fungi are known for their beneficial effects on plant P nutrition when P is sparingly available in soils. Improvement in P uptake is considered the primary benefit and is the best-studied outcome of these symbioses, with approximately 20% of all published studies on AMF dealing with P uptake (Rillig and Mummey, 2006).

AMF expedite plant P uptake through the delivery of Pi from distant soil patches that are inaccessible by roots, thus increasing the Pi depletion zone (George *et al.*, 1992). Experiments in which compartment systems are utilised provide clear evidence of direct transfer of P through hyphae towards the plant (George *et al.*, 1992; Kikuchi *et al.*, 2016). In these studies, the access of roots to soil compartments was controlled with a small-diameter mesh that allowed only fungal hyphae to pass through them and transfer nutrients, and P from these compartments could still be detected in plant tissues.

Plants can take up P via two main pathways, involving different cell types and Pi transporters (Bucher, 2007). The direct, AM-independent pathway of P uptake from soil solution occurs in roots and root hairs and allows Pi uptake from the rhizosphere around the roots. High-affinity Pi transporter (PT) genes are highly expressed in the root apex and root hairs, with less expression being observed in more mature root parts (Gordon-Weeks *et al.*, 2003). The expression of these genes is reduced when P is highly available and when plants are colonised by AMF (Liu *et al.*, 1998).

In the AM pathway, AMF form extraradical hyphal networks that actively take up Pi from soil areas beyond root depletion zones via a trans-membrane, high-affinity phosphate transporter. The first gene encoding such a transporter was identified in the AMF *Glomus versiforme* (Harrison and van Buuren, 1995). Maldonado-Mendoza *et al.*, (2001) identified a homologous PT-encoding gene in the model AMF *Glomus intraradices* (= *Rhizoglomus irregulare*) and showed that accumulation of its RNA is induced at low Pi levels. Phosphate is then translocated to root internal hyphae in the form of polyphosphate (Cox *et al.*, 1980; Kikuchi *et al.*, 2016) and released from specialised hyphae that form tree-like structures called arbuscules (Smith *et al.*, 2011). These fungal structures develop within plant cells and create interfaces for nutritional exchange between the fungus and the plant, involving the membranes of both partners with an apoplastic space between them (Smith and Read, 2008). After the release of Pi from the arbuscules, it is taken up by the plant cell via AM-specific phosphate transporters that belong to the Pht1 family of plant Pi transporters (Rausch *et al.*, 2001; Harrison *et al.*, 2002). These AM-specific plant P transporters are crucial not only for the plant but also for the fungus, as knock-down of the genes encoding these transporters reduces AMF root colonisation (Javot *et al.*, 2007).

AM and non-AM plants likely consume the same forms of Pi, including P adsorbed to soil minerals (Frossard *et al.*, 2011). Moreover, it appears that AMF are capable of hydrolysing Pi from organic sources (Joner and Johansen, 2000) and delivering it to the plant (Koide and Kabir, 2000), but its quantitative contribution to plant nutrition is thought to be small (Joner and Johansen, 2000). In red clover, organic P is delivered to the plant in significant amounts via AMF hyphae, but the rate is lower than that estimated for P from inorganic sources (Wolfe *et al.*, 2008). This transfer is due to the activity of phytases, rather than acidic phosphatases (Wang *et al.*, 2017b). In general, it is not clear whether AMF are able to solubilise P from organic and inorganic sources, or if this activity is accomplished by bacteria attached to the hyphae (Taktek *et al.*, 2017). The addition of rock phosphate to monoxenic cultures without bacteria indicated that AMF are not able to solubilise rock phosphate (Antunes *et al.*, 2007). However, acidic phosphatase activity was observed in the AMF *Rhizophagus clarus* (= *Rhizoglomus clarum*) in a monoxenic culture without the presence of such bacteria (Sato *et al.*, 2015). Therefore, P solubilisation by AMF is a topic of controversy, and its significance for plant nutrition is not clear.

Generally, the extraradical mycelium (ERM) harbours different bacteria due to its large, nutrient-rich surface (Qin *et al.*, 2016). Bacterial communities both attached and non-attached to the hyphae were characterised, and bacterial families such as *Oxalobacteraceae* were shown to occur on AMF hyphae (Scheublin *et al.*, 2010). AMF colonisation is also linked to the presence of the genus *Pseudomonas* in the rhizosphere (Andrade *et al.*, 1997). Hyphae-colonising bacteria have been demonstrated to improve hyphal growth and enhance AMF root colonisation (Frey-Klett *et al.*, 2007). In contrast, negative effects of bacteria on AMF growth are also observed (Li *et al.*, 2008; Larsen *et al.*, 2009). In a recent study, more than 370 bacterial strains were isolated from the spores of *Rhizophagus irregularis* (= *Rhizoglomus irregulare*), among which many of the isolates were able to solubilise P from phytate and mineral sources (Battini *et al.*, 2016). Such bacteria were further found to be able to improve AMF growth, facilitate P uptake by AMF hyphae from soil compartments and support the growth of host plants (Battini *et al.*, 2017).

Based on the assumption that P solubilisation does not play a major role in the mycorrhizal P uptake pathway, it is probable that the translocation of Pi across depletion zones is the most important factor allowing plant P nutrition to be supported to a great extent by the symbiont. It has recently been shown that P transport through the extraradical hyphae is not an active process but instead seems to be driven by water flow towards the plant (Kikuchi *et al.*, 2016). This water flow is driven by the opening of stomata in the leaves, which in turn determines the CO_2 assimilation rate, pointing to a common regulatory mechanism of symbiotic P nutrition and photosynthesis via the water status of the plant (Bitterlich and Franken, 2016).

The importance of the ERM for mycorrhizal P nutrition and the connection to water relations are demonstrated by a different line of inquiry. Fungal hyphae have a much smaller diameter than plant roots (Staddon *et al.*, 2003), allowing colonisation of smaller soil pores and increased exploration and depletion of the soil volume (Drew *et al.*, 2003). This might be a key mechanism by which the fungi support plant P nutrition. In addition, AMF hyphae constitute a network that spreads plant-derived carbon in the soil (Miller and Jastrow, 2000). Thus, AMF contribute to the formation and stabilisation of soil structure (i.e. aggregation) by enmeshment, entanglement, exudation and turnover (Rillig and Mummey, 2006). Aggregate stabilisation or *de novo* formation changes the hydraulic properties of the pore space in which solute transport occurs, and AMF have been shown to affect soil water retention (Auge *et al.*, 2001; Daynes *et al.*, 2013) and unsaturated hydraulic conductivity (Bitterlich *et al.*, 2018a; Bitterlich *et al.*, 2018b). Changes in hydraulic soil properties, pore geometry or tortuosity caused by AMF may also indirectly contribute to plant P nutrition, which is mainly restricted by diffusion resistance. Improvements in hydraulic conductivity upon AMF substrate colonisation (Bitterlich *et al.*, 2018a; Bitterlich *et al.*, 2018b) may be indicative of alleviation of the flow resistance of water and solutes and/or maintenance of P diffusion pathways in AMF substrates. Diffusion pathways for P in soils increase with desiccation (Gahoonia *et al.*, 1994), and hyphae may bridge air-filled pores or restore pore connectivity via particle reorientation (Rillig and Mummey, 2006) or smoothing of surface profiles. Theoretically, this could either alleviate diffusion resistance for P in the soil, or hyphae could provide an uptake network that is less dependent on the degree of soil moisture content, because they bridge air-filled porosity. If this is true, AMF may comparatively improve soil P mobility under drought conditions, which could have high ecological significance. However, data from the quantification of AMF effects on soil hydraulic properties are scarce, and distinction between intrahyphal P translocation and P translocation allowed by changes in pore space properties is lacking and methodologically demanding. However, such data would be required to elucidate the physiological relevance of indirect AMF effects on plant P nutrition mediated by soil hydraulic properties.

6.3 Sebacinales Fungi (Phylum: Basidiomycota)

6.3.1 The Model *Serendipita indica*

The lineage of 'Sebacinales' diverged 100–200 million years ago (Tedersoo *et al.*, 2014) and is considered to belong to the mushroom-forming basidiomycetes (Agaricomycotina) (Hibbett *et al.*, 2007). These fungi colonise herbaceous angiosperms, bryophytes (e.g. mosses), pteridophytes (e.g. ferns) and important crops, including the non-mycorrhizal plants of the Brassicaceae (cabbage family) (Selosse *et al.*, 2002; Weiß *et al.*, 2011).

Fungi of the Sebacinales order are suggested to form two distinct phylogenetic groups (Weiß *et al.*, 2004). The Sebacinaceae 'Sebacinales Group A' (Oberwinkler *et al.*, 2014; Weiß *et al.*, 2016) include basidiome-forming fungi, many of which form different types of mycorrhizal interactions with trees (ectomycorrhiza), orchids (orchid mycorrhiza) and plants of the order Ericales (ericoid mycorrhiza). Phosphate nutrition of plants mediated by these fungi can play a role in such symbioses but will not be discussed in this chapter. The clade previously known as 'Sebacinales Group B' now referred to as the Serendipitaceae (Weiß *et al.*, 2016) currently contains a single genus and a small number of described species. However, it is expected to include more species in the future because sequences apparently belonging to this groups have been found in roots in many different ecosystems (Weiß *et al.*, 2011), but cultures of these species appear to be difficult to obtain.

The most-studied serendipitaceous fungus is *Serendipita indica* (previously called *Piriformospora indica*), which was isolated from a spore of the AMF *Funneliformis mosseae* (formerly *Glomus mosseae*) in the Indian Thar desert (Verma *et al.*, 1998). *Serendipita indica* shows plant growth-promoting effects for a wide range of hosts (Varma *et al.*, 1999), including non-mycorrhizal plants such as *Arabidopsis thaliana* (Peskan-Berghofer *et al.*, 2004). Because of the ability to transform *S. indica* and its available genome data (Zuccaro *et al.*, 2009; Zuccaro *et al.*, 2011), this species represents a model for the study of plant–fungus interactions. Moreover, due to its positive effects on plant health and development (Oelmuller *et al.*, 2009; Franken, 2012) and the ability to grow this fungus easily in axenic culture, it is a potential candidate for application in sustainable plant production systems.

Concerning plant nutrition, it became apparent early on that *S. indica* interacts with plants in a different way than mycorrhizal fungi. If plants are fertilised with sufficient amounts of phosphate, they often show much lower colonisation by AMF (Breuillin *et al.*, 2010). This is not the case for *S. indica* (Varma *et al.*, 1999). Many studies have investigated the impact of *S. indica* on plant nutrition, and a contribution of the fungus to P nutrition has been shown in several cases. Growth promotion by *S. indica* was observed to be accompanied by a higher ^{32}P content per cotyledon in *A. thaliana* seedlings grown on ^{32}P-labelled orthophosphate medium (Shahollari *et al.*, 2005). Yadav *et al.*, (2010) showed that knocking out the phosphate transporter gene of *S. indica* (*PiPT*) reduces the P content and biomass of colonised maize plants. In other studies in which the P uptake of barley was measured in comparison with an AMF, no contribution of *S. indica* to P nutrition could be found (Achatz *et al.*, 2010). In the following section, we discuss the P solubilisation and P translocation activities of *S. indica*.

6.3.2 *Serendipita indica* Improves Plant P Uptake from Rock Phosphate

To study the ability of *S. indica* to support the P nutrition of the host plant from sources unavailable for plant usage, the P-solubilising activity of the fungus was tested (Ngwene *et al.*, 2016). The results showed that the fungus is more efficient in utilising inorganic P than organic P. *Serendipita indica* can solubilise P from $Ca_3(PO_4)_2$ and rock phosphate, probably via the production of organic acids rather

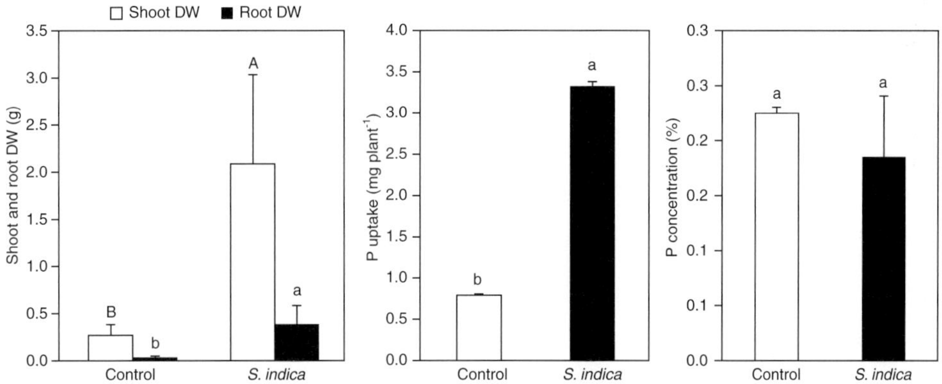

Figure 6.1 Influence of *Serendipita indica* inoculation on the growth of tomato plants. Plant growth (shoot and root dry weight) and P concentrations, as well as P uptake in inoculated (*S. indica*) and non-inoculated (control) tomato plants are shown. Plants were grown for 8 weeks in substrate fertilised with 180 mg P (rock phosphate) kg^{-1} dry substrate. The mean values ± standard deviations are shown. Mean values labelled with the same letter (lower or upper case) are not significantly different ($P = 0.05$; $n = 4$).

than through enzymatic (phosphatase) activity. Even though corresponding phosphatase genes were found to be expressed, no enzymatic activity was detected (Ngwene *et al.*, 2016). The ability of *S. indica* to support the P nutrition of plants from rock phosphate was tested in an experiment described in the following summary.

Surface-sterilised seeds of tomato (*Solanum lycopersicum* cv. Hildares) were germinated on sterile sand, then transferred to a loamy sand luvisol that was optimally fertilised with all nutrients other than P according to Müller *et al.*, (2013); 180 mg P kg^{-1} dry soil was added in the form of rock phosphate, and plants were either inoculated with *S. indica* or mock-inoculated and cultivated according to Fakhro *et al.* (2010). The plants were harvested after 8 weeks, and both their fresh and dry biomass were measured, and P concentrations were analysed as described by Boldt *et al.* (2011).

Tomato plants inoculated with *S. indica* exhibited much higher dry weights (DWs) of the shoots and roots than mock-inoculated control plants (Figure 6.1). This was accompanied by a significant increase in the rate of P uptake, although the P concentration did not differ between *S. indica*-inoculated and control plants (Figure 6.1). Two mechanisms may be involved in improved P nutrition: (1) Pi is solubilised from rock phosphate by *S. indica* and made available in the rhizosphere for plant uptake through the root pathway, or (2) Pi is solubilised from rock phosphate by *S. indica* (or by other microbes closely attached to the hyphae) and transferred to the plant through the fungal hyphae pathway, as observed in the AM symbiosis described above. Whether P is translocated in significant amounts through the hyphae of *S. indica* will be discussed in the next section.

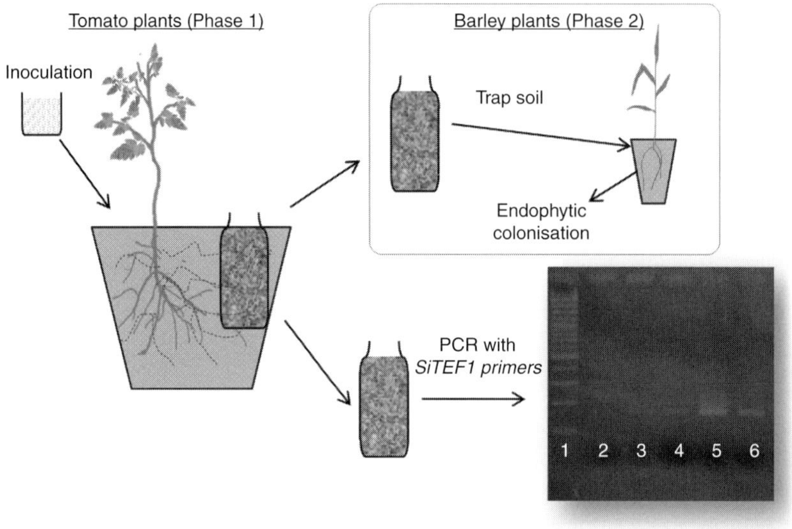

Figure 6.2 Experimental setup for the second experiment. Inoculated (*S. indica*, *R. irregulare* or dual inoculation) and mock-inoculated tomato plants were grown in C-loess substrate for 8 weeks in a greenhouse. Fungal compartments filled with a 1:1 mixture of fine soil (particle size < 40 μm) and glass beads were inserted in the planting substrate and supplied with 50 (low P, LP) or 200 (high P, HP) mg P kg^{-1} substrate. To confirm the extraradical hyphal growth of *S. indica*, compartment substrates from *S. indica*-inoculated plants were used either to inoculate barley trap plants or for molecular detection of *S. indica* in the substrate compartment via polymerase chain reaction (PCR) with an *S. indica*-specific primer pair. The results of gel electrophoresis are shown (lane 1: DNA ladder; lane 2: negative control; lanes 3 and 4: DNA extracted from compartment substrates; lanes 5 and 6: DNA extracted from inoculated roots).

6.3.3 P Transfer to the Plant Through the Extraradical Mycelium of *S. indica*

The question of whether there might be any direct transfer of P from the fungus to the plant via fungal hyphae could not be answered based on the previous approach and was further investigated in a second pot experiment using the same plant cultivar and growth conditions. In this second experiment, the main pots received 50 mg P (KH_2PO_4) kg^{-1} dry substrate instead of rock phosphate. Additional P (soluble Pi) was applied in fungal compartments inserted into the planting substrate (Figure 6.2).

Two categories of compartments that contained soluble Pi were introduced. The first category was used for DNA extraction and detection of ERM and was fertilised with 50 mg P kg^{-1} dry substrate (low P, LP). The second category was used for the inoculation of barley trap plants and contained 200 mg P kg^{-1} (high P, HP). The sides and bottom of the compartments were covered with meshes that allowed the ingrowth of fungal ERM but not roots. This experiment consisted of five inoculation treatments: *S. indica* and *S. indica* mock treatments, *R. irregulare* and *R. irregulare* mock treatments, and a dual inoculation (*S. indica* + *R. irregulare*) treatment. Inoculation with *S. indica* was performed similarly to the first experiment. In the *R. irregulare* treatments, a 10% (w/w) commercial inoculum of the fungus (INOQ GmbH; Schnega, Germany) was mixed with the pot substrate, and an autoclaved inoculum served as the mock treatment. After 8 weeks of cultivation, plant fresh and dry weights, as well as P concentrations, were assessed similarly to the first experiment. For the estimation of AMF colonisation, roots were stained with 0.05% trypan blue in lactic acid, and the AMF colonisation rate was quantified via a modified gridline intersection method (Tennant, 1975). For each sample, at least 200 intersections were analysed. *Serendipita indica* colonisation was examined microscopically following the method of Bütehorn *et al.* (2000). Compartment soils were used for the measurement of ERM (for *R. irregulare* treatments) and detection of *S. indica* via DNA extraction and polymerase chain reaction (PCR) (for *S. indica* treatments), using specific primers for the *Pitef1* gene, encoding the translation elongation factor EF-1alpha (Bütehorn *et al.*, 2000). Additionally, compartment soil from pots containing *S. indica*-inoculated plants was used to inoculate trap barley plants to confirm the growth of the fungus (Figure 6.2).

The results showed that *S. indica* was able to grow out of the roots into the soil. Barley plants were colonised by the fungus (data not shown), and PCR resulted in amplification of the fungal gene (*Pitef1*) from extracts obtained from the roots (as a positive control) as well as the compartment substrates (Figure 6.2). The DW of tomato shoots was influenced by inoculation, but the impact of the AMF was only marginal, and the phosphate levels in the compartments did not influence shoot biomass (Table 6.1). Inoculation with the fungi also had a significant effect on the shoot P concentration, but the difference between *S. indica*-colonised plants and the corresponding controls was not significant. *R. irregulare* clearly increased the shoot P concentration when applied alone (Table 6.1). This AMF-based increase was abolished by the addition of *S. indica* at the high phosphate level. P uptake was similarly increased by both fungi, and a synergistic effect was evident, but only at the low phosphate level (Figure 6.3). When plants were supplied with a higher concentration of phosphate in the compartment, a reduction in plant P uptake was observed.

Inoculation with *R. irregulare* both served as a positive control and allowed the interaction between *S. indica* and an AMF to be analysed. Plant roots were adequately colonised, but there was a reduction in AMF colonisation when a high level of P was applied to the compartment in the dual inoculation treatment (Table 6.1). The colonisation of *S. indica* was slightly reduced in this treatment as well. This negative effect was also observed for ERM, DW and plant P uptake in the same treatment. Whether the reduced AMF (and also *S. indica*) colonisation rate was a direct effect of the high level of localised P (Olsson *et al.*, 2002) or a result of competition for colonisation sites (Cano and Bago, 2005) is not clear.

Table 6.1 Shoot dry weight (DW), root DW, shoot P concentration (mg g^{-1}), fungal colonisation rate (%) and fungal extraradical mycelium DW at harvest.

Inoculated (*S. indica*, *R. irregulare* (AMF) or dual inoculation) and mock-inoculated tomato plants were grown for 8 weeks in a compartment system (Figure 6.2), where the compartments were supplied with 50 (low P, LP) or 200 (high P, HP) mg P kg-1 substrate. The mean values ± standard deviations are shown. Mean values for each parameter labelled with the same letter are not significantly different ($P = 0.05$; $n = 4$). The results of two-way ANOVA showing the effects of inoculation and the phosphate concentrations in compartments, as well as their interactions, on the listed parameters are presented below. The effects were considered significant when the *P*-values were below 0.05 (printed in bold).

		Shoot DW (g plant^{-1})	**Root DW (g plant^{-1})**	**Shoot P conc. (mg g^{-1})**	**S. indica colonisation (%)**	**AM fungal colonisation (%)**	**DW (mg) of mycelium**
S. indica control	LP	13.35 ± 1.3	3.34 ± 0.66	0.964 ± 0.02a	No colonisation	No colonisation	--
	HP	13.37 ± 0.2	3.13 ± 0.44	1.001 ± 0.00a			--
AMF control	LP	12.19 ± 1.2	2.88 ± 0.53	1.036 ± 0.05a	No colonisation	No colonisation	--
	HP	12.08 ± 0.5	3.48 ± 0.53	1.025 ± 0.05a			--
S. indica	LP	14.58 ± 0.8	3.24± 0.50	1.050 ± 0.04a	19 ± 12.4c	Not applicable	--
	HP	14.08 ± 0.5	2.81 ± 0.25	1.047 ± 0.05a	55 ± 10.6b		--
AMF	LP	12.60 ± 1.1	2.89 ± 0.16	1.188 ± 0.06bc	Not applicable	83.09 ± 2.08b	151.5 ± 51.8b
	HP	12.92 ± 0.6	2.89 ± 0.29	1.288 ± 0.09a		87.50 ± 5.59b	167.9 ± 77.1b
S. indica + AMF	LP	15.01 ± 0.4	2.98 ± 0.63	1.226 ± 0.06b	44 ± 9.7^{b}	88.04 ± 6.33b	190.8 ± 69.2b
	HP	14.78 ± 0.5	3.18 ± 0.34	1.068 ± 0.06ac	34 ± 12.4bc	64.74 ± 12.44c	57.9 ± 28.8c
Inoculation (Inoc)		**<0.001**	0.712	**<0.001**	0.963	**<0.001**	0.259
P level		0.697	0.725	0.679	0.086	0.096	0.075
Inoc*P level		0.882	0.242	**<0.001**	**0.002**	**0.006**	**0.028**

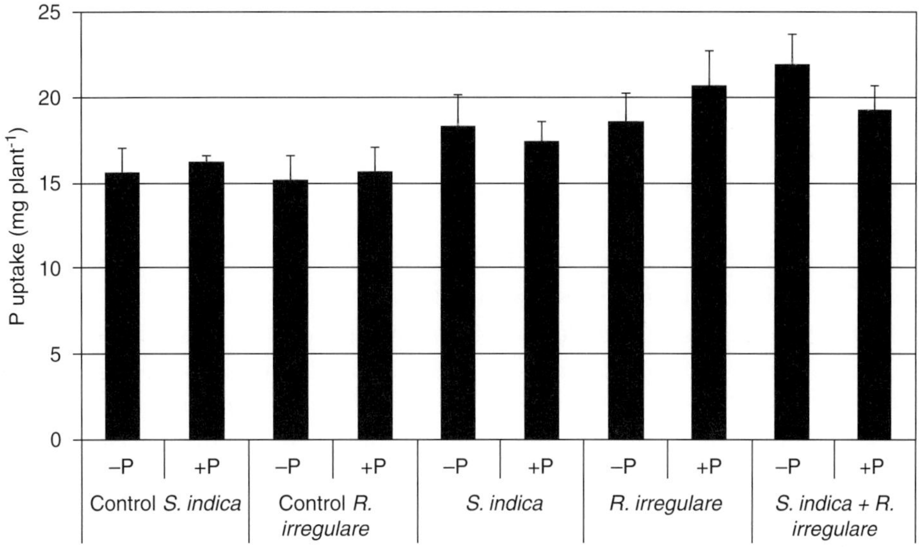

Figure 6.3 Comparison of the influences of *S. indica* and mycorrhizal fungi on P uptake. Inoculated (*S. indica, R. irregulare* or dual inoculation) and mock-inoculated tomato plants were grown for 8 weeks in a compartment system (Figure 6.2), where the compartments were supplied with 50 (low P, LP) or 200 (high P, HP) mg P kg^{-1} substrate. The mean values and standard deviations are shown. Two-way analysis of variannce (ANOVA) showed that P uptake was significantly increased by both fungi, and an interaction between inoculation and P levels was found ($P = 0.05$; $n = 4$).

Inoculation with either of the fungi increased plant P uptake, which could result from higher P availability for root uptake or from direct transfer by the fungus. In the case of AMF, direct transfer has been previously demonstrated and was confirmed in this study. The higher P uptake rates were more due to increased P concentrations than to increased DWs (Table 6.1 and Figure 6.3). In the case of *S. indica*, the opposite was true: the fungus increased shoot growth but not the P concentration. These findings indicated that *R. irregulare* improved P nutrition by translocating P from the compartments to the plant, while *S. indica* (at least) did not translocate significant amounts of P from the compartments. These different functions might be because the hyphae of *S. indica* are much thinner than those of AMF and are septate, which would hinder Pi translocation via mechanisms discussed above for AMF.

6.4 Dark Septate Endophytes (Phylum: Ascomycota)

DSEs belong to various orders of the phylum Ascomycota (Sieber and Grünig, 2013). They are very common root colonisers characterised by dark, melanised hyphae and the formation of structures known as microsclerotia inside root cells (Jumpponen and Trappe, 1998). These fungi colonise roots simultaneously with mycorrhizal fungi, but additionally colonise non-mycorrhizal plant species (Mandyam and Jumpponen, 2005). Even though DSEs are widely distributed and have a very broad host range, their ecological function is not well understood (Kivlin *et al.*, 2013). Because the DSE 'form taxon' includes genetically distant species that do not represent a monophyletic group, diversity of their functional traits and effects on plants is expected.

In addition to their association with plants, DSEs are successful saprotrophs and can be grown in axenic culture. Knapp and Kovács (2016) showed that DSE species exhibit a wide range of enzymatic activities, allowing them to degrade a broad spectrum of substrates. Genomic analyses of DSEs also showed that saprobic capacity might play crucial role in their endophytic lifestyle (Knapp *et al.*, 2018).

DSEs can have different effects on plant performance. Positive effects of these fungi on plant nutrition, especially on N uptake from organic sources, have been observed in experimental systems under controlled conditions, and the beneficial effects are intriguingly not necessarily accompanied by high fungal colonisation rates (Newsham, 2011). In extreme environments, DSEs are also assumed to improve the P nutrition of plants (Mandyam and Jumpponen, 2005; Almario *et al.*, 2017). An early study showed that inoculation with DSEs occurring on sedges improves plant biomass and P content (Haselwandter and Read, 1982). In other studies, improved plant growth and increased P and N levels have been observed (Jumpponen *et al.*, 1998; Newsham, 2000). The ability of DSEs to solubilise inorganic phosphate, including Ca and Al phosphates, has been demonstrated. These abilities are linked to a pH decrease in their growth media (Spagnoletti *et al.*, 2017) and result in better *in vitro* plant growth (Barrow and Osuna, 2002).

The ability of DSEs to efficiently utilise organic phosphate in pure culture was demonstrated by our group for two species (*Periconia macrospinosa* DSE2036 and *Cadophora* sp. DSE1049) whose complete genomes have recently been published (Knapp *et al.*, 2018), but P transfer through the DSE ERM was too low to be detected, if it occurred at all, and there was no increase in P uptake in plants fertilised with organic P and inoculated with DSEs (Yakti *et al.*, unpublished).

We also tested the ability of the two strains to use and solubilise different forms of inorganic phosphorus. The two fungi were grown on modified National Botanical Research Institute phosphate growth media (Nautiyal, 1999) supplied with rock phosphate (from Algeria, Cameroon, Morocco, Mexico or Mali), hydroxyl apatite or $Ca_3(PO_4)_2$ as the P source. Fungal plugs were transferred to the centre of Petri plates, and cultures were grown for 10 days at 24°C. Both DSEs grew on all P sources, and the fungi produced a halo around the colony in the case of Malian and Cameroonian rock phosphate as well as $Ca_3(PO_4)_2$ (Figure 6.4a). Such a halo is an indication of a drop in the pH of the medium, most likely due to the production of organic acids by the fungus as a mechanism of P solubilisation. We also

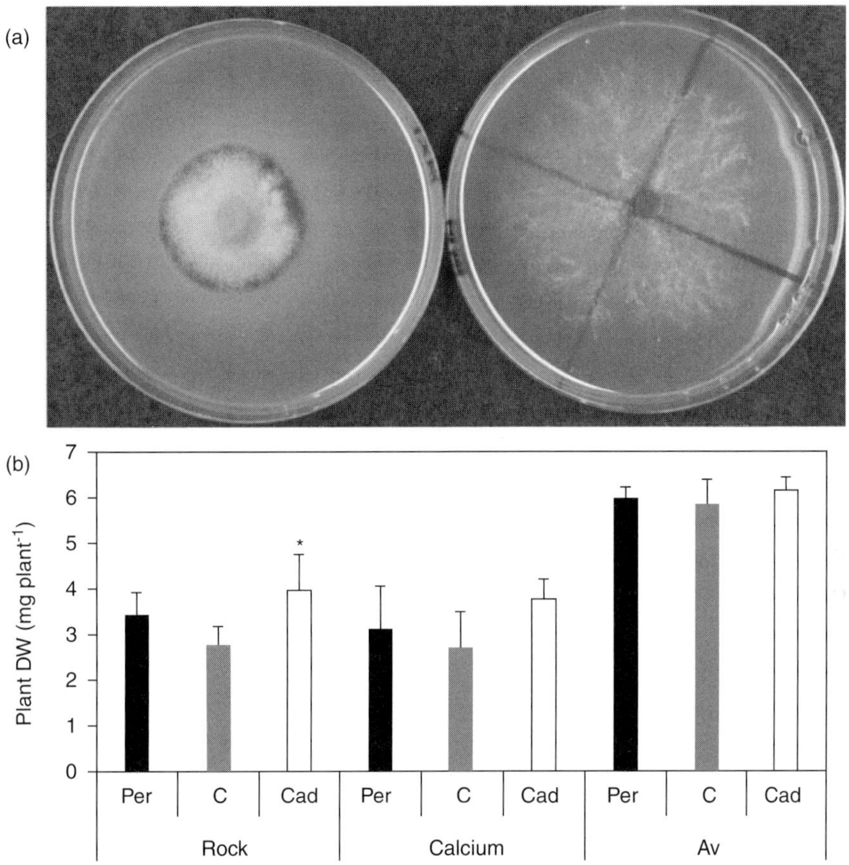

Figure 6.4 Growth of *Periconia macrospinosa* and *Cadophora* sp. on modified National Botanical Research Institute's phosphate (NBRIP) medium containing Malian rock phosphate, and their effects on seedling growth. (a) Growth of *Periconia macrospinosa* (right) and *Cadophora* sp. (left) on modified NBRIP medium containing Malian rock phosphate. White halos around colonies indicate phosphate solubilisation activity. (b) Effects of *P. macrospinosa* (Per) and *Cadophora* sp. (Cad) on seedlings grown on media containing rock phosphate (Rock), $Ca_3(PO_4)_2$ (Calcium) or KH_2PO_4 (Av). Plants were grown for 3 weeks, and their dry weights were measured. The Kruskal–Wallis test ($P = 0.05$, $n = 3$–4) showed significant effects of the two factors (DSE and inoculation) on plant dry weight. The Kruskal–Wallis median test ($P = 0.05$, $n = 3$-4) showed that *Cadophora* sp. significantly increased plant growth in rock phosphate medium.

investigated whether the P solubilisation activity of DSEs would benefit the *in vitro* growth of tomato plants when supplied with only insoluble forms of inorganic P. In this assay, a factorial experiment was carried out. Tomato seeds (cv. Hildares) were surface sterilised, germinated, and transferred to three different modified PNM media (Johnson *et al.*, 2011) in plant tissue culture bottles. P was then added as

$Ca_3(PO_4)_2$, Cameroonian rock phosphate or as KH_2PO_4, and the pH was adjusted to 6. The media were inoculated upon seedling transfer with one agar plug of *Periconia macrospinosa* or *Cadophora* sp. grown on PDA, or with sterile PDA as a control. The plants were grown for 3 weeks in a greenhouse where growth promotion could notably be observed. *Cadophora* sp., but not *P. macrospinosa*, increased the growth of tomato seedlings (Figure 6.4b) when the media contained rock phosphate or $Ca_3(PO_4)_2$. This effect was absent when the media were supplied with plant-available Pi (KH_2PO_4). Our observations indicated that *Cadophora* sp. is most likely able to support plant nutrition by making P available from rock phosphate and $Ca_3(PO_4)_2$. Further research on the *in vivo* plant growth effects of *Cadophora* sp. is still required to further assess its potential to support P nutrition from insoluble P sources, in addition to understanding the mechanism (P solubilisation vs. direct transport) underlying this effect.

6.5 Outlook and Concluding Remarks

The negative impacts of P deficiency and unavailability for plant production systems and the environmental risks of excess Pi fertiliser application mean that better, sustainable, eco-friendly alternatives are needed. The use of fungal inoculants could reduce the excessive use of hazardous chemical fertilisers by promoting the efficient use of available soil nutrients and providing significant positive impacts on plant P nutrition in production systems. Current knowledge of plant–fungal endophyte interactions is highly promising, and further research is required to better elucidate and understand the factors influencing plant-promoting effects. Exploration, screening and characterisation of additional endophytic fungal species that may exhibit more promising plant-supporting characteristics are crucial for field application. The three major groups discussed in this chapter interact with plants differently, and their abilities to access different forms of P vary. DSEs are able to free Pi from inorganic and organic sources, but their contributions to plant nutrition appear to be small. The representative Sebacinales species *S. indica* can solubilise only inorganic P sources but improves P uptake of the plant from such sources to a great extent. Solubilised Pi appears to then be taken up by plant roots. Finally, AMF are mainly involved in transporting P across depletion zones towards the roots. These characteristics indicate distinct and possibly complementary functions that play key roles, particularly in plant P nutrition and generally in ecosystem functioning. We therefore emphasise the importance of investigating intermicrobial interactions in the rhizosphere and their impacts on plant nutrition, which could be valuable information for promoting fungal performance and, consequently, for promoting plant access to nutrients.

Acknowledgements

This research was partially funded by the Ministries of Consumer Protection, Food and Agriculture of the Federal Republic of Germany, the Ministry of Science, Research and Culture of the State of Brandenburg, the Thuringian Ministry of Infrastructure and Agriculture, Germany, and the Hungarian National Research, Development and Innovation Office (NKFIH, /OTKA/ K109102). Wael Yakti was supported by ERASMUS MUNDUS grant and a short trip scientific mission in the frame of COST Action (FA1103).

References

Achatz, B., von Ruden, S., Andrade-Linares, D. R. *et al.* (2010). Root colonization by *Piriformospora indica* enhances grain yield in barley under diverse nutrient regimes by accelerating plant development. *Plant and Soil*, **333**, 59–70.

Almario, J., Jeena, G., Wunder, J. *et al.* (2017). Root-associated fungal microbiota of nonmycorrhizal *Arabis alpina* and its contribution to plant phosphorus nutrition. *Proceedings of the National Academy of Sciences of the United States of America*, **114**, E9403–E9412.

Andrade G., Mihara, K. L., Linderman, R. G. and Bethlenfalvay, G. J. (1997). Bacteria from rhizosphere and hyphosphere soils of different arbuscular-mycorrhizal fungi. *Plant and Soil*, **192**, 71–79.

Antunes, P. M., Schneider, K., Hillis, D. and Klironomos, J. N. (2007). Can the arbuscular mycorrhizal fungus *Glomus intraradices* actively mobilize P from rock phosphates? *Pedobiologia*, 51, 281–286.

Auge, R. M., Stodola, A. J. W., Tims, J. E. and Saxton, A. M. (2001). Moisture retention properties of a mycorrhizal soil. *Plant and Soil*, **230**, 87–97.

Bago, B. and Cano, C. (2005). *In Vitro Culture of Mycorrhizas*. Berlin: Springer.

Barrow, J. R. and Osuna, P. (2002). Phosphorus solubilization and uptake by dark septate fungi in fourwing saltbush, *Atriplex canescens* (Pursh) Nutt. *Journal of Arid Environments*, **51**, 449–459.

Battini, F., Cristani, C., Giovannetti, M. and Agnolucci, M. (2016). Multifunctionality and diversity of culturable bacterial communities strictly associated with spores of the plant beneficial symbiont *Rhizophagus intraradices*. *Microbiol Research*, **183**, 68–79.

Battini, F., Grønlund, M., Agnolucci, M., Giovannetti, M. and Jakobsen, I. (2017). Facilitation of phosphorus uptake in maize plants by mycorrhizosphere bacteria. *Scientific Reports*, **7**, 4686.

Behie, S. W. and Bidochka, M. J. (2014). Nutrient transfer in plant–fungal symbioses. *Trends in Plant Science*, **19**, 734–740.

Bitterlich, M. and Franken, P. (2016). Connecting polyphosphate translocation and hyphal water transport points to a key of mycorrhizal functioning. *New Phytologist*, **211**, 1147–1149.

Bitterlich, M., Franken, P. and Graefe, J. (2018a). Arbuscular mycorrhiza improves substrate hydraulic conductivity in the plant available moisture range under root growth exclusion. *Frontiers in Plant Science*, **9**, 301.

Bitterlich, M., Sandmann, M. and Graefe, J. (2018b). Arbuscular mycorrhiza alleviates restrictions to substrate water flow and delays transpiration limitation to stronger drought in tomato. *Frontiers in Plant Science*, **9**.

Boldt, K., Pörs, Y., Haupt, B. *et al.* (2011). Photochemical processes, carbon assimilation and RNA accumulation of sucrose transporter genes in tomato arbuscular mycorrhiza. *Journal of Plant Physiology*, **168**, 1256–1263.

Brady, N. C. (1990). *The Nature and Properties of the Soil.* New York, NY: Macmillan Publishing Co.

Breuillin, F., Schramm, J., Hajirezaei, M. *et al.* (2010). Phosphate systemically inhibits development of arbuscular mycorrhiza in *Petunia hybrida* and represses genes involved in mycorrhizal functioning. *Plant Journal*, **64**, 1002–1017.

Bucher, M. (2007). Functional biology of plant phosphate uptake at root and mycorrhiza interfaces. *New Phytologist*, **173**, 11–26.

Bütehorn, B., Rhody, D. and Franken, P. (2000). Isolation and characterisation of *Pitef1* encoding the translation elongation factor EF-1a of the root endophyte *Piriformospora indica*. *Plant Biology*, **49**, 687–692.

Cano, C. and Bago, A. (2005). Competition and substrate colonization strategies of three polyxenically grown arbuscular mycorrhizal fungi. *Mycologia*, **97**, 1201–1214.

Chen, M. and Graedel, T. E. (2016). A half-century of global phosphorus flows, stocks, production, consumption, recycling, and environmental impacts. *Global Environmental Change*, **36**, 139–152.

Cordell, D. and White, S. (2015). Tracking phosphorus security: indicators of phosphorus vulnerability in the global food system. *Food Security*, **7**, 337–350.

Cordell, D., Drangert, J. O. and White, S. (2009). The story of phosphorus: global food security and food for thought. *Global Environmental Change*, **19**, 292–305.

Cox, G., Moran, K. J., Sanders, F., Nockolds, C. and Tinker, P. B. (1980). Translocation and transfer of nutrients in vesicular-arbuscular mycorrhizas. III. Polyphosphate granules and phosphorus translocation. *New Phytologist*, **84**, 649–659.

Daynes, C. N., Field, D. J., Saleeba, J. A., Cole, M. A. and McGee, P. A. (2013). Development and stabilisation of soil structure via interactions between organic matter, arbuscular mycorrhizal fungi and plant roots. *Soil Biology and Biochemistry*, **57**, 683–694.

Drew, E. A., Murray, R. S., Smith, S. E. and Jakobsen, I. (2003). Beyond the rhizosphere: growth and function of arbuscular mycorrhizal external hyphae in sands of varying pore sizes. *Plant and Soil*, **251**, 105–114.

Fakhro, A., Andrade-Linares, D. R., von Bargen, S. *et al.* (2010). Impact of *Piriformospora indica* on tomato growth and on interaction with fungal and viral pathogens. *Mycorrhiza*, **20**, 191–200.

Franken, P. (2012). The plant strengthening root endophyte *Piriformospora indica*: potential application and the biology behind. *Applied Microbiology and Biotechnology*, **96**, 1455–1464.

Frey-Klett, P., Garbaye, J. and Tarkka, M. (2007). The mycorrhiza helper bacteria revisited. *New Phytologist*, **176**, 22–36.

Frossard, E., Achat, D. L., Bernasconi, S. M. *et al.* (2011). *Phosphorus in Action: Biological Processes in Soil Phosphorus Cycling*. Berlin: Springer.

Gahoonia, T. S., Raza, S. and Nielsen, N. E. (1994). Phosphorus depletion in the rhizosphere as influenced by soil moisture. *Plant and Soil*, **159**, 213–218.

George, E., Häussler, K.-U., Vetterlein, D., Gorgus, E. and Marschner, H. (1992). Water and nutrient translocation by hyphae of *Glomus mosseae*. *Canadian Journal of Botany*, **70**, 2130–2137.

Gordon-Weeks, R., Tong, Y., Davies, T. G. E. and Leggewie, G. (2003). Restricted spatial expression of a high-affinity phosphate transporter in potato roots. *Journal of Cell Science*, **116**, 3135–3144.

Graham, J. H. and Abbott, L. K. (2000). Wheat responses to aggressive and non-aggressive arbuscular mycorrhizal fungi. *Plant and Soil*, **220**, 207–218.

Harrison, M. J. and van Buuren, M. L. (1995). A phosphate transporter from the mycorrhizal fungus *Glomus versiforme*. *Nature*, **378**, 626–629.

Harrison, M. J., Dewbre, G. R. and Liu, J. Y. (2002). A phosphate transporter from *Medicago truncatula* involved in the acquisiton of phosphate released by arbuscular mycorrhizal fungi. *Plant Cell*, **14**, 2413–2429.

Haselwandter, K. and Read, D. J. (1982). The significance of a root-fungus association in two *Carex* species of high-alpine plant communities. *Oecologia*, **53**, 352–354.

Hawkesford, M., Horst, W., Kichey, T. *et al.* (2012). *Marschner's Mineral Nutrition of Higher Plants*, 3rd edn. San Diego, CA: Academic Press.

Helber, N., Wippel, K., Sauer, N. *et al.* (2011). A versatile monosaccharide transporter that operates in the arbuscular mycorrhizal fungus *Glomus* sp is crucial for the symbiotic relationship with plants. *Plant Cell*, **23**, 3812–23.

Hermans, C., Hammond, J. P., White, P. J. and Verbruggen, N. (2006). How do plants respond to nutrient shortage by biomass allocation? *Trends in Plant Science*, **11**, 610–617.

Hibbett, D. S., Binder, M., Bischoff, J. F. *et al.* (2007). A higher-level phylogenetic classification of the Fungi. *Mycological Research*, **111**, 509–47.

Hinsinger, P. (2000). *Trace Elements in the Rhizosphere*. Boca Raton, FL: CRC Press.

Hodge, A. (2009). Root decisions. *Plant, Cell & Environment*, **32**, 628–40.

Javot, H., Penmetsa, R. V., Terzaghi, N., Cook, D. R. and Harrison, M. J. (2007). A *Medicago truncatula* phosphate transporter indispensable for the arbuscular mycorrhizal symbiosis. *Proceedings of the National Academy of Sciences of the United States of America*, **104**, 1720–1725.

Johnson, J. M., Sherameti, I., Ludwig, A. *et al.* (2011). Protocols for *Arabidopsis thaliana* and *Piriformospora indica* co-cultivation: a model system to study plant beneficial traits. *Journal of Endocytobiosis and Cell Research*, **21**, 101–113.

Johnson, N. C., Graham, J. H. and Smith, F. A. (1997). Functioning of mycorrhizal associations along the mutualism–parasitism continuum. *New Phytologist*, **135**, 575–585.

Joner, E. J. and Johansen, A. (2000). Phosphatase activity of external hyphae of two arbuscular mycorrhizal fungi. *Mycological Research*, **104**, 81–86.

Jumpponen, A. and Trappe, J. M. (1998). Dark septate endophytes: a review of facultative biotrophic root-colonizing fungi. *New Phytologist*, **140**, 295–310.

Jumpponen, A., Mattson, K. G. and Trappe, J. M. (1998). Mycorrhizal functioning of *Phialocephala fortinii* with *Pinus contorta* on glacier forefront soil: interactions with soil nitrogen and organic matter. *Mycorrhiza*, **7**, 261–265.

Kenrick, P. and Crane, P. R. (1997). The origin and early evolution of plants on land. *Nature*, **389**, 33.

Keymer, A., Pimprikar, P., Wewer, V. *et al.* (2017). Lipid transfer from plants to arbuscular mycorrhiza fungi. *Elife*, **6**, e29107.

Kikuchi, Y., Hijikata, N., Ohtomo, R. *et al.* (2016). Aquaporin-mediated long-distance polyphosphate translocation directed towards the host in arbuscular mycorrhizal symbiosis: application of virus-induced gene silencing. *New Phytologist*, **211**, 1202–1208.

Kivlin, S. N., Emery, S. M. and Rudgers, J. A. (2013). Fungal symbionts alter plant responses to global change. *American Journal of Botany*, **100**, 1445–1457.

Knapp, D. G. and Kovács, G. M. (2016). Interspecific metabolic diversity of root-colonizing endophytic fungi revealed by enzyme activity tests. *FEMS Microbiology Ecology*, **92**, fiw190.

Knapp, D. G., Németh, J. B., Barry, K. *et al.* (2018). Comparative genomics provides insights into the lifestyle and reveals functional heterogeneity of dark septate endophytic fungi. *Scientific Reports*, **8**, 6321.

Koide, R. T. and Kabir, Z. (2000). Extraradical hyphae of the mycorrhizal fungus *Glomus intraradices* can hydrolyse organic phosphate. *New Phytologist*, **148**, 511–517.

Lambers, H., Raven, J. A., Shaver, G. R. and Smith, S. E. (2008). Plant nutrient-acquisition strategies change with soil age. *Trends in Ecology & Evolution*, **23**, 95–103.

Larsen, J., Cornejo, P. and Barea, J. M. (2009). Interactions between the arbuscular mycorrhizal fungus *Glomus intraradices* and the plant-growt-promoting rhizobacteria *Paenibacillus polymyxa* and *P. macerans* in the mycorrhizosphere of *Cucumis sativus*. *Soil Biology and Biochemistry*, **41**, 286–292.

Li, B., Ravnskov, S., Xie, G. and Larsen, J. (2008). Differential effects of *Paenibacillus spp.* on cucumber mycorrhizas. *Mycological Progress*, **7**, 277–284.

Li, H., Smith, S. E., Holloway, R. E., Zhu, Y. and Smith, F. A. (2006). Arbuscular mycorrhizal fungi contribute to phosphorus uptake by wheat grown in a phosphorus-fixing soil even in the absence of positive growth responses. *New Phytologist*, **172**, 536–543.

Liu, H., Trieu, A. T., Blayloock, L. A. and Harrison, M. J. (1998). Cloning and characterization of two phosphate transporters from *Medicago truncatula*: root regulation in response to phosphate and to colonization by arbuscular mycorrhizal fungi. *Molecular Plant–Microbe Interactions*, **11**, 14–22.

Lynch, J. P. (2007). Roots of the second Green Revolution. *Australian Journal of Botany*, **55**, 493–512.

Lynch, J. P. and Ho, M. D. (2005). Rhizoeconomics: carbon costs of phosphorus acquisition. *Plant and Soil*, **269**, 45–56.

Maldonado-Mendoza, I. E., Dewbre, G. R. and Harrison, M. J. (2001). A phosphate transporter gene from the extra-radical mycelium of an arbuscular mycorrhizal fungus *Glomus intraradices* is regulated in response to phosphate in the environment. *Molecular Plant–Microbe Interactions*, **14**, 1140–1148.

Mandyam, K. and Jumpponen, A. (2005). Seeking the elusive function of the root-colonising dark septate endophytic fungi. *Studies in Mycology*, 173–189.

Marschner, H. and Dell, B. (1994). Nutrient uptake in mycorrhizal symbiosis. *Plant and Soil*, **159**, 89–102.

Miller, R. M. and Jastrow, J. D. (2000). Mycorrhizal fungi influence soil structure. In *Arbuscular Mycorrhizas: Physiology and Function*, ed. Y. Kapulnik and D. D. Douds. Dordrecht, The Netherlands: Springer, pp. 3–18.

Miller, S. H., Browne, P., Prigent-Combaret, C. *et al.* (2010). Biochemical and genomic comparison of inorganic phosphate solubilization in *Pseudomonas* species. *Environmental Microbiology Reports*, **2**, 403–11.

Müller, A., George, E. and Gabriel-Neumann, E. (2013). The symbiotic recapture of nitrogen from dead mycorrhizal and non-mycorrhizal roots of tomato plants. *Plant and Soil*, **364**, 341–355.

Nautiyal, C. S. (1999). An efficient microbiological growth medium for screening phosphate solubilizing microorganisms. *FEMS Microbiology Letters*, **170**, 265–270.

Newsham, K. K. (2000). *Phialophora graminicola*, a dark septate fungus, is a beneficial associate of the grass *Vulpia ciliata* ssp. *ambigua*. *New Phytologist*, **144**, 517–524.

Newsham, K. K. (2011). A meta-analysis of plant responses to dark septate root endophytes. *New Phytologist*, **190**, 783–793.

Ngwene, B., Boukail, S., Söllner, L., Franken, P. and Andrade-Linares, D. R. (2016). Phosphate utilization by the fungal root endophyte *Piriformospora indica*. *Plant and Soil*, **405**, 231–241.

Oberwinkler, F., Riess, K., Bauer, R. and Garnica, S. (2014). Morphology and molecules: the Sebacinales, a case study. *Mycological Progress*, **13**, 445–470.

Oelmuller, R., Sherameti, I., Tripathi, S. and Varma, A. (2009). *Piriformospora indica*, a cultivable root endophyte with multiple biotechnological applications. *Symbiosis*, **49**, 1–17.

Olsson, P. A., van Aarle, I. M., Allaway, W. G., Ashford, A. E. and Rouhier, H. (2002). Phosphorus effects on metabolic processes in monoxenic arbuscular mycorrhiza cultures. *Plant Physiology*, **130**, 1162–1171.

Peskan-Berghofer, T., Shahollari, B., Giong, P. H. *et al.* (2004). Association of *Piriformospora indica* with *Arabidopsis thaliana* roots represents a novel system to study beneficial plant-microbe interactions and involves early plant protein modifications in the endoplasmic reticulum and at the plasma membrane. *Physiologia Plantarum*, **122**, 465–477.

Qin, H., Brookes, P. C. and Xu, J. (2016). Arbuscular mycorrhizal fungal hyphae alter soil bacterial community and enhance polychlorinated biphenyls dissipation. *Frontiers in Microbiology*, **7**, 939.

Ramaekers, L., Remans, R., Rao, I. M., Blair, M. W. and Vanderleyden, J. (2010). Strategies for improving phosphorus acquisition efficiency of crop plants. *Field Crops Research*, **117**, 169–176.

Rausch, C., Daram, P., Brunner, S. *et al.* (2001). A phosphate transporter expressed in arbuscule-containing cells in potato. *Nature*, **414**, 462–466.

Redecker, D. and Raab, P. (2006). Phylogeny of the Glomeromycota (arbuscular mycorrhizal fungi): recent developments and new gene markers. *Mycologia*, **98**, 885–895.

Rich, M. K., Nouri, E., Courty, P.-E. and Reinhardt, D. (2017). Diet of arbuscular mycorrhizal fungi: bread and butter? *Trends in Plant Science*, **22**, 652–660.

Richardson, A. E. (2001). Prospects for using soil microorganisms to improve the acquisition of phosphorus by plants. *Australian Journal of Plant Physiology*, **28**, 896–906.

Richardson, A. E. and Simpson, R. J. (2011). Soil microorganisms mediating phosphorus availability. *Plant Physiology*, **156**.

Richardson, A. E., Hocking, P. J., Simpson, R. J. and George, T. S. (2009). Plant mechanisms to optimise access to soil phosphorus. *Crop and Pasture Science*, **60**, 124–143.

Rillig, M. C. and Mummey, D. L. (2006). Mycorrhizas and soil structure. *New Phytologist*, **171**, 41–53.

Sato, T., Ezawa, T., Cheng, W. G. and Tawaraya, K. (2015). Release of acid phosphatase from extraradical hyphae of arbuscular mycorrhizal fungus *Rhizophagus clarus*. *Soil Science and Plant Nutrition*, **61**, 269–274.

Schachtman, D. P., Reid, R. J. and Ayling, S. M. (1998). Phosphorus uptake by plants: from soil to cell. *Plant Physiology*, **116**, 447–453.

Scheublin, T. R., Sanders, I. R., Keel, C. and van der Meer, J. R. (2010). Characterisation of microbial communities colonising the hyphal surfaces of arbuscular mycorrhizal fungi. *The ISME Journal*, **4**, 752.

Selosse, M. A., Bauer, R. and Moyersoen, B. (2002). Basal hymenomycetes belonging to the Sebacinaceae are ectomycorrhizal on temperate deciduous trees. *New Phytologist*, **155**, 183–195.

Shahollari, B., Varma, A. and Oelmuller, R. (2005). Expression of a receptor kinase in *Arabidopsis* roots is stimulated by the basidiomycete *Piriformospora indica* and the protein accumulates in Triton X-100 insoluble plasma membrane microdomains. *Journal of Plant Physiology*, **162**, 945–958.

Sharma, S. B., Sayyed, R. Z., Trivedi, M. H. and Gobi, T. A. (2013). Phosphate solubilizing microbes: sustainable approach for managing phosphorus deficiency in agricultural soils. *SpringerPlus*, **2**, 587.

Shen, J., Yuan, L., Zhang, J. *et al.* (2011). Phosphorus dynamics: from soil to plant. *Plant Physiology*, **156**, 997–1005.

Sieber, T. N. and Grünig, C. R. (2013). Fungal root endophytes. In *Plant Roots: The Hidden Half*, ed. Y. Wasel, A. Eshel and U. Kafkafi. New York: Marcel Dekker, pp. 1–49.

Smith, S. E. and Read, D. (2008). *Mycorrhizal Symbiosis*, 3rd edn. London: Academic Press.

Smith, S. E., Smith, F. A. and Jakobsen, I. (2003). Mycorrhizal fungi can dominate phosphate supply to plants irrespective of growth responses. *Plant Physiology*, **133**, 16–20.

Smith, S. E., Smith, F. A. and Jakobsen, I. (2004). Functional diversity in arbuscular mycorrhizal (AM) symbioses: the contribution of the mycorrhizal P uptake pathway is not correlated with mycorrhizal responses in growth or total P uptake. *New Phytologist*, **162**, 511–524.

Smith, S. E., Jakobsen, I., Grønlund, M. and Smith, F. A. (2011). Roles of arbuscular mycorrhizas in plant phosphorus nutrition. *Plant Physiology*, **156**, 1050–1057.

Spagnoletti, F. N., Tobar, N. E., Fernández Di Pardo, A., Chiocchio, V. M. and Lavado, R. S. (2017). Dark septate endophytes present different potential to solubilize calcium, iron and aluminum phosphates. *Applied Soil Ecology*, **111**, 25–32.

Spatafora, J. W., Chang, Y., Benny, G. L. *et al.* (2016). A phylum-level phylogenetic classification of zygomycete fungi based on genome-scale data. *Mycologia*, **108**, 1028–1046.

Staddon, P. L., Ramsey, C. B., Ostle, N., Ineson, P. and Fitter, A. H. (2003). Rapid turnover of hyphae of mycorrhizal fungi determined by AMS microanalysis of 14C. *Science*, **300**, 1138–40.

Taktek, S., St-Arnaud, M., Piché, Y., Fortin, J. A. and Antoun, H. (2017). Igneous phosphate rock solubilization by biofilm-forming mycorrhizobacteria and hyphobacteria associated with *Rhizoglomus irregulare* DAOM 197198. *Mycorrhiza*, **27**, 13–22.

Tawaraya, K. (2003). Arbuscular mycorrhizal dependency of different plant species and cultivars. *Soil Science and Plant Nutrition*, **49**, 655–668.

Tedersoo, L., Bahram, M., Ryberg, M. *et al.* (2014). Global biogeography of the ectomycorrhizal /sebacina lineage (Fungi, Sebacinales) as revealed from comparative phylogenetic analyses. *Molecular Ecology*, **23**, 4168–4183.

Tennant, D. (1975). A test of a modified line intersect method of estimating root length. *Journal of Ecology*, **63**, 995–1001.

Tilman, D., Fargione, J., Wolff, B. *et al.* (2001). Forecasting agriculturally driven global environmental change. *Science*, **292**, 281–284.

Turner, B. L., Papházy, M. J., Haygarth, P. M. and McKelvie, I. D. (2002). Inositol phosphates in the environment. *Philosophical Transactions of the Royal Society B: Biological Sciences*, **357**, 449–469.

van der Heijden, M. G., Martin, F. M., Selosse, M. A. and Sanders, I. R. (2015). Mycorrhizal ecology and evolution: the past, the present, and the future. *New Phytologist*, **205**, 1406–23.

Varma, A., Verma, S., Sudah, Sahay, N. and Franken, P. (1999). *Piriformospora indica*, a cultivable plant growth-promoting root endophyte. *Applied and Environmental Microbiology*, **65**, 2741–2744.

Verma, S., Varma, A., Rexer, K.-H. *et al.* (1998). *Piriformospora indica*, gen. nov. sp. nov., a new root-colonizing fungus. *Mycologia*, **90**, 896–903.

Wang, W., Shi, J., Xie, Q. *et al.* (2017a). Nutrient exchange and regulation in arbuscular mycorrhizal symbiosis. *Molecular Plant*, **10**, 1147–1158.

Wang, X. X., Hoffland, E., Feng, G. and Kuyper, T. W. (2017b). Phosphate uptake from phytate due to hyphae-mediated phytase activity by arbuscular mycorrhizal maize. *Frontiers in Plant Science*, **8**, 684.

Weiß, M., Selosse, M. A., Rexer, K. H., Urban, A. and Oberwinkler, F. (2004). Sebacinales: a hitherto overlooked cosm of heterobasidiomycetes with a broad mycorrhizal potential. *Mycological Research*, **108**, 1003–1010.

Weiß, M., Sykorova, Z., Garnica, S. *et al.* (2011). Sebacinales everywhere: previously overlooked ubiquitous fungal endophytes. *PLoS One*, **6**, e16793.

Weiß, M., Waller, F., Zuccaro, A. and Selosse, M.-A. (2016). Sebacinales: one thousand and one interactions with land plants. *New Phytologist*, **211**, 20–40.

Wolfe, M. S., Baresel, J. P., Desclaux, D. *et al.* (2008). Developments in breeding cereals for organic agriculture. *Euphytica*, **163**, 323.

Yadav, V., Kumar, M., Deep, D. K. *et al.* (2010). A phosphate transporter from the root endophytic fungus *Piriformospora indica* plays a role in phosphate transport to the host plant. *Journal of Biological Chemistry*, **285**, 26532–26544.

Zuccaro, A., Basiewicz, M., Zurawska, M., Biedenkopf, D. and Kogel, K. H. (2009). Karyotype analysis, genome organization, and stable genetic transformation of the root colonizing fungus *Piriformospora indica*. *Fungal Genetics and Biology*, **46**, 543–550.

Zuccaro, A., Lahrmann, U., Guldener, U. *et al.* (2011). Endophytic life strategies decoded by genome and transcriptome analyses of the mutualistic root symbiont *Piriformospora indica*. *PLoS Pathogens*, **7**, e1002290.

7

From Darkness to Light: Emergence of the Mysterious Dark Septate Endophytes in Plant Growth Promotion and Stress Alleviation

CHARLOTTE BERTHELOT, MICHEL CHALOT, CORINNE LEYVAL AND DAMIEN BLAUDEZ

Abstract

Dark septate endophytes (DSEs) are ascomycetous fungi whose structure is characterised by dark melanised hyphae and microsclerotia located in plant roots. Associations with DSEs are commonly found in various biomes and plant taxa. Although DSEs are commonly recorded, the effects of their colonisation on plant growth and fitness are unclear. This chapter summarises the state of knowledge about DSEs from the literature and personal data. The effects of DSEs on plant growth range from parasitism to mutualism. They can promote plant growth by improving nutrition (e.g. solubilisation of minerals, degradation of complex carbon compounds), producing secondary metabolites (e.g. phytohormones, volatile organic compounds) and protecting against phytopathogens. More particularly, the high tolerance of DSEs to abiotic stress and their relatively high abundance in trace element-contaminated and other stressful habitats suggest that they may have an important function for host survival under these conditions. Finally, this chapter outlines why additional research is required in the emerging field of plant–DSE interactions to address future challenges.

7.1 Introduction

Plant roots establish intimate associations with a variety of microorganisms such as bacteria, mycorrhizal fungi and endophytic fungi. The term 'endophyte' is used to describe microorganisms living in plant tissues without causing any apparent harm or eliciting a negative response from the host (Schulz and Boyle, 2005). The plant–fungal endophyte relationship is thought to differ from mycorrhizal symbiosis in that the following three key features are lacking: (1) a cellular interface where specialised structures (e.g. arbuscules) develop, (2) synchronised development between the plant and the fungal associate, and (3) significant benefits for both partners (Brundrett, 2006). Root endophytic associations involve a diverse array of fungi, the best known of which are dark septate endophytes (DSEs). DSEs are a polyphyletic assemblage of Ascomycetes mainly characterised by brown to blackish mycelium (Figure 7.1a). The accumulation of high concentrations of melanin in hyphae is specific to this group and probably reflects a common response to different environmental pressures (Butler and Day, 1998). Due to their singular morphology, they were initially described as 'mycelium radicis atrovirens' (MRA) or 'pseudomycorrhizal' fungi (Melin, 1921). MRA were often found to co-exist with mycorrhizal fungi, and the colonisation of 135 species of angiosperms by dark pigmented fungi was documented as early as 1924 (Peyronel, 1924). Thereafter, these fungi were introduced as fungi with sterile and dark septate hyphae that occur in the roots of various plants (Read and Haselwandter, 1981).

Few DSEs sporulate in culture, where they remain in the form of sterile mycelium, requiring the use of molecular techniques for taxonomic identification (Grünig *et al.*, 2011). Such analyses highlighted that DSEs are phylogenetically diverse and belong to different orders such as Helotiales, Xylariales, Pleosporales, Sordariales, Hypocreales and Chaetosphaeriales (Grünig *et al.*, 2011) (Figure 7.2). Most studies on these groups have focused on several *Cadophora*, *Exophiala*, *Leptodontidium* and *Phialocephala* species (Jumpponen and Trappe, 1998a; Grünig *et al.*, 2011). The *Gaeumannomyces–Phialophora* complex and the *Phialocephala fortinii–Acephala applanata* species complex are also among the best-studied major subgroups of DSEs (Grünig *et al.*, 2011).

DSEs form specific structures in roots including hyphal networks (Figure 7.1b,c) and microsclerotia that are thought to be resting structures or propagules (Ruotsalainen *et al.*, 2007; Peterson *et al.*, 2008) (Figure 7.1c,d). Moreover, DSE colonisation shows similarities with mycorrhizal colonisation, like hyphae narrowing when crossing plant cell walls, accumulation of polyphosphates (Yu *et al.*, 2001), or the development of a mantle and Hartig net as seen in ectomycorrhiza (Peterson *et al.*, 2008).

DSEs are widespread and have been recorded in more than 600 plant species spanning 100 families (Jumpponen and Trappe, 1998a). They colonise

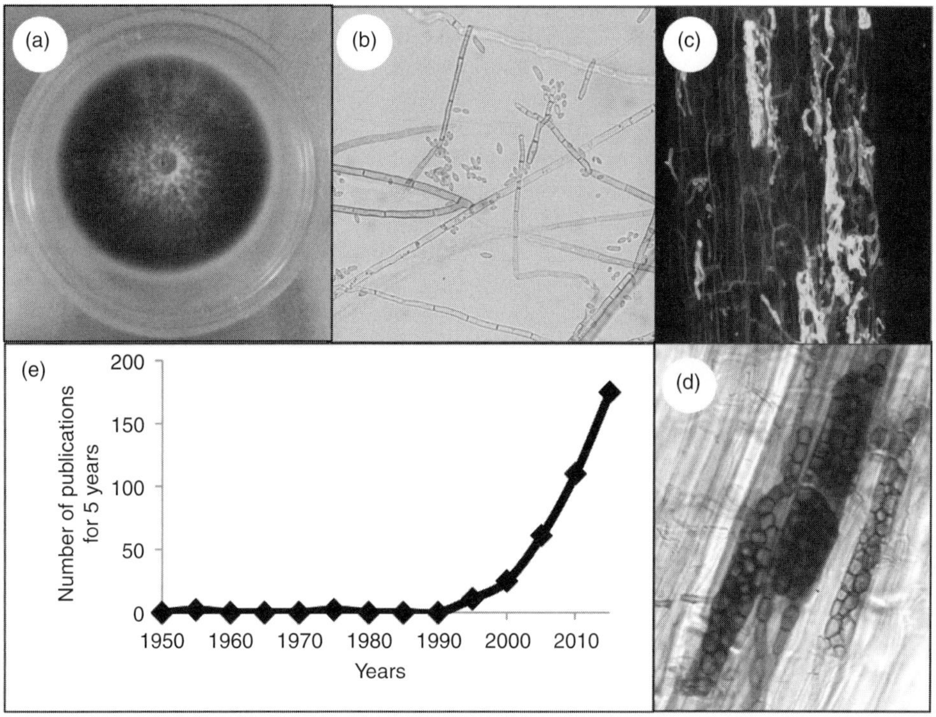

Figure 7.1 Typical characteristics of dark septate endophytes (DSEs) and emergence of DSE-related studies in the literature. (a) Two-week old colony of *Cadophora malorum* CBS 100584 on malt agar medium. (b) Septate hyphae and conidiophores of *Cladosporium cladosporioides* CBS101367. (c) Septate hyphae and microsclerotia of *Phialophora mustea* Pr27 in poplar roots; the fungal structures (green) were labelled with WGA-AF88. (d) Microsclerotia of *Leptodontidium* sp. Me10 in ryegrass (*Lolium*) roots. (e) Increase in the number of DSE-related publications since 1950. The survey was performed using Web of Science tools (http://apps.webofknowledge.com) using the following keywords: 'MRA'; 'DSE'; 'Dark septate endophytes'; and 'Mycelium Radicis Atrovirens'. (A black and white version of this figure will appear in some formats. For the colour version, please refer to the plate section.)

both herbaceous and woody plants, and more particularly the roots of metal hyperaccumulators, halophytes, or marine macrophytes (Mandyam and Jumpponen, 2005). They occur in diverse habitats/biomes such as arid, arctic, boreal, alpine, temperate and tropical (Jumpponen and Trappe, 1998a; Kovàcs and Szigetvari, 2002; Mandyam and Jumpponen, 2005; Knapp *et al.*, 2012).

Although DSEs are commonly found worldwide, they long remained understudied in comparison with other beneficial fungi such as mycorrhizal fungi (Figure 7.1e). Yet, the elucidation of their ecological and physiological roles has been considered to be of primary interest in the past decade because of their abundance

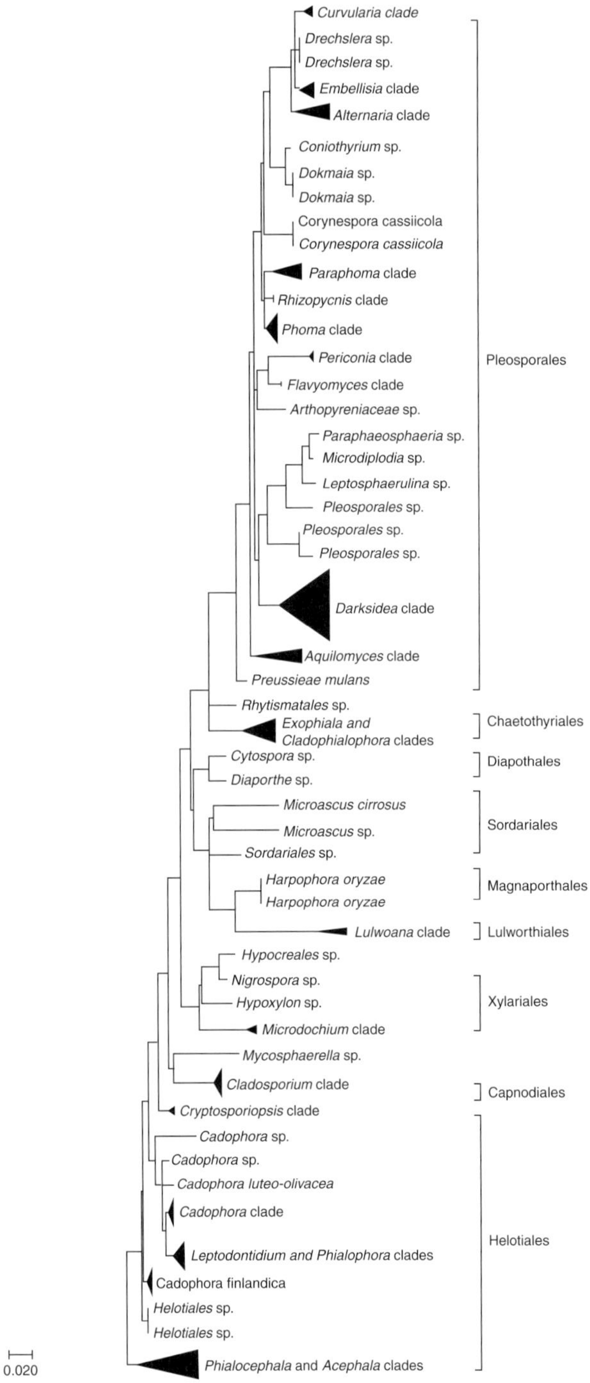

Figure 7.2 Minimum-evolution phylogenetic tree showing the position of DSEs based on nrITS sequences. Two hundred and forty internal transcribed spacer (ITS) sequences (ITS1-5.8S-ITS2) from DSEs were downloaded from National Center for Biotechnology Information (NCBI) GenBank. Sequence alignment was carried out using Clustal X. The minimum-evolution tree was drawn with the Kimura 2-parameter model and gamma for rate heterogeneity using MEGA 6 (Tamura *et al.*, 2013). Bootstrap analyses were performed with 1000 replications.

and widespread distribution in all environments. However, almost a century after their discovery, little is still known about the role of these mysterious and rather elusive fungal symbionts.

This chapter presents a historical overview of DSE research and summarises current understanding of these fungi. The review is based on the literature and on personal data with special reference to some well-studied DSE fungal lineages. We focus on their beneficial physiological interactions with plant hosts, which may expand our views on how plants accommodate fungal endophytes to adapt to adverse conditions and improve nutrient acquisition. We also discuss the use of these fungi in sustainable agriculture.

7.2 The DSE Paradigm: Mutualists Versus Parasites?

The 'mutualism–parasitism continuum' is a paradigm established to explain why plant–microorganism associations may vary from mutualism to parasitism (Mandyam and Jumpponen, 2015). According to this paradigm, plant–DSE associations produce host responses ranging from mutualistic interactions, in which DSEs are beneficial to the host plant, to parasitic interactions that lead to deleterious consequences for the host plant.

Generalisations about the exact nature/function of DSEs are nevertheless problematic owing to their taxonomic diversity and the limited number of available studies (Addy *et al.*, 2005; Alberton *et al.*, 2009; Newsham, 2011; Mayerhofer *et al.*, 2013). Alberton *et al.* (2009) suggested that host growth responses to colonisation by DSEs were variable. Meta-analyses by Newsham (2011) and Mayerhofer *et al.* (2013) underlined that both the experimental design and the environmental context had an impact on the outcome of the interaction. For instance, comparisons between greenhouse and *in vitro* experiments showed inconsistent results: when *Pinus contorta* was inoculated with *Phialocephala fortinii*, mesocosm experiments and *in vitro* culture assays differed in responses in terms of biomass and foliar nutrient concentrations (Jumpponen and Trappe, 1998b). Similarly, using diverse experimental systems such as bi-compartimentalised multi-well plates or Petri dishes, sterilised perlite or soil/loam mesocosms (Figure 7.3a–e), we noticed different host responses of a given plant species (eucalypt, ryegrass, poplar or alfalfa) to inoculation by *Leptodontidium* sp. Pr30. Newsham (2011) concluded that the outcome of DSE inoculation was strongly dependent on the nutritional status (i.e. the form of nitrogen supplied during the experiment). Moreover, positive or negative effects of DSEs may also depend on the developmental stage of the plant (Andrade-Linares *et al.*, 2011). For instance, tomato plants inoculated with DSE strains belonging to Pleosporales accelerated flowering and produced significantly higher fruit biomass at the beginning of the harvest. This enhancement of fruit production, however,

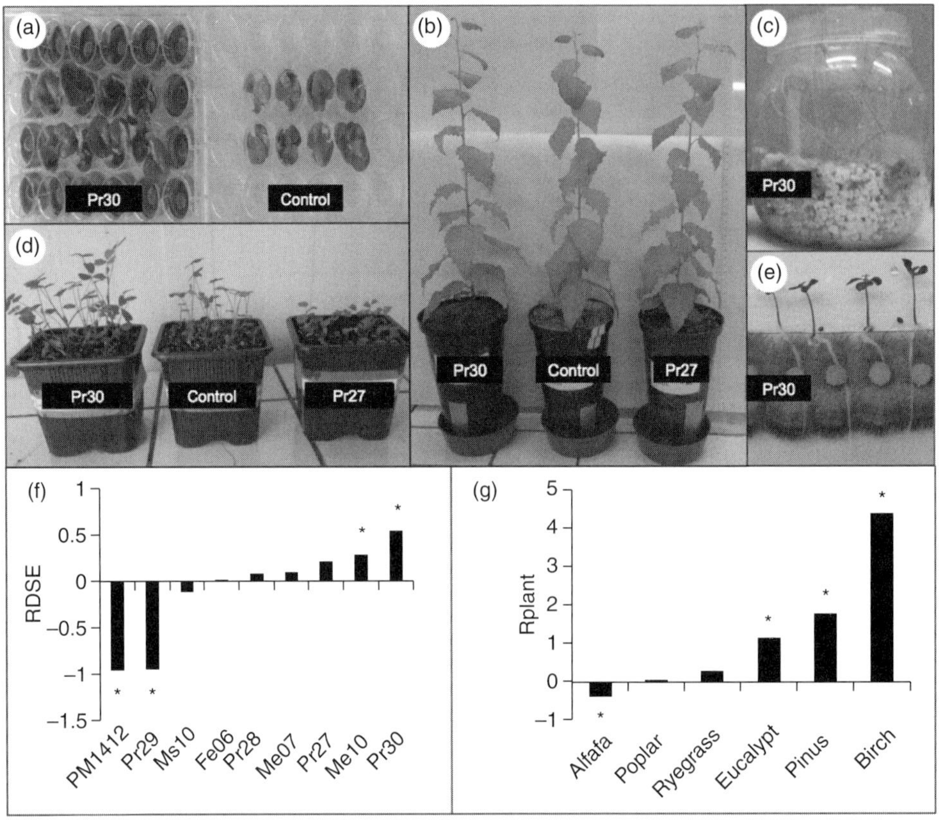

Figure 7.3 Different growth systems used to study DSE-plant interactions and the mutualism- parasitism paradigm. (a) 24-well culture plates used to test the growth-promoting effect of volatile organic compounds from *Leptodontidium* sp. Pr30 on tobacco in the absence of physical contact. (b) Axenic perlite system used to test the effect of *Leptodontidium* sp. Pr30 on ryegrass growth. (c, d) Contrasting effects of *Leptodontidium* sp. Pr30 and *Phialophora mustea* Pr27 on alfalfa (c) and poplar (d) grown in a sterilised trace element-polluted soil. (e) Bi-compartmentalisation plates used to co-culture eucalypt seedlings and *Phialophora mustea* Pr29. (f) Growth responses of ryegrass to colonisation by nine DSE strains. PMI_412, Pr30, Me07, Me10 belong to *Leptodontidium* sp., Pr27 and Pr29 belong to *Phialophora mustea*, Ms10 belongs to *Meliniomyces* sp., Fe06 belongs to *Cadophora* sp., and Pr28 belongs to *Phialocephala fortinii*. (g) Growth responses of six plant species to inoculation with *Phialophora mustea* Pr27. Response (RDSE or Rplant) to fungal inoculation indicates the ratio between the growth value of 'the inoculated minus the control treatments' to that of the control. Values above the *x*-axis indicate a positive response and values below the axis a negative response. Asterisks indicate significant differences between the control and inoculated plants (Kruskal-Wallis, $P < 0.05$). (A black and white version of this figure will appear in some formats. For the colour version, please refer to the plate section.)

disappeared in later harvests (Andrade-Linares *et al.*, 2011). In conclusion, a change in the host physiology or in the environment may trigger pathogenicity or neutrality in an endophyte that was previously considered as mutualist, and *vice versa* (Peterson *et al.*, 2008; Berthelot *et al.*, 2017).

However, several studies highlight the fact that the outcome of the plant–DSE interaction is thought to be predominantly influenced by the interacting plant and fungal species (Tellenbach and Sieber, 2012; Knapp *et al.*, 2012; Mayerhofer *et al.*, 2013). We found that ryegrass (*Lolium perenne*) growth was promoted by *Leptodontidium* sp. Me10 and Pr30, but inhibited by *Leptodontidium* sp. PMI 412 and *Phialophora mustea* Pr29 (Figure 7.3f) (Berthelot *et al.*, 2016). Conversely, Figure 7.3g depicts the differential effects of *Phialophora mustea* Pr27 on different plant hosts. We indeed noticed a range of effects from parasitism (alfalfa), neutralism (ryegrass) to mutualism (birch). Other observations are included in a meta-analysis by Mayerhofer *et al.* (2013). They showed that the effect of DSEs on plant growth promotion was greater in herbs and graminoids than in shrubs and trees. However, further studies are needed to confirm this finding. For instance, a comparison of the plant–DSE relationship with a large set of strains isolated from either herbs or trees and cross-inoculated with both herbs and trees would be of high interest. More particularly, the influence of the host genotype on the response to DSEs could also be studied at the ecotype level. Mandyam *et al.* (2013) indeed reported that the outcome of the relationship between three cultivars of *Arabidopsis thaliana* and 30 isolates of *Microdochium bolleyi* differed among plant cultivars and DSE isolates.

Part of the difficulty in distinguishing endophytes from pathogens is that many endophytes are closely related to pathogens. In many cases morphological comparisons do not provide sufficient resolution, but the use of multiple gene/loci sequence analyses allowed to differentiate related endophytes and pathogens (Porras-Alfaro and Bayman, 2011). More recently, the relationship between plant and DSEs was partially elucidated *via* transcriptomic analyses and comparative genomics. To exemplify the first point, Reininger and Schlegel (2016) analysed the transcriptome of *Phialocephala subalpina* during colonisation of its host plant *Picea abies*. They observed 2549 genes differentially regulated in the mycelium of *P. subalpina* in the presence of *P. abies* roots as compared to pure culture mycelium. Differentially expressed genes were found predominantly in the functional categories 'metabolism,' 'transport' and 'cell rescue, defence and virulence' (Reininger and Schlegel, 2016). Similarly, Mandyam and Jumpponen (2014) used microarrays to investigate the response of *Arabidopsis thaliana* colonised by *Microdochium* sp. Two thousand and one genes were differentially regulated in comparison with the non-inoculated control. Fungal colonisation altered cell wall modification, auxin and jasmonic acid biosynthesis, and induced defence and stress responses (Mandyam and Jumpponen, 2014). Importantly, the expression of many genes with putative roles in nutrient transfer was also altered in both partners during the DSE/plant interaction (Mandyam and Jumpponen, 2014; Reininger and Schlegel, 2016). Some of these functional categories are discussed below.

Fungal genome sequencing projects are blooming. The genomes of several DSEs have been fully sequenced (genome.jgi.doe.gov/programs/fungi/1000fungalgenomes.jsf) such as *Phialocephala subalpina* and *Microdochium bolleyi* (David *et al.*, 2016; Reininger and Schlegel, 2016). The comparative analyses of these genomes to those of saprophytic, mycorrhizal and pathogenic fungi should soon shed light on their detailed lifestyle. Interestingly, Xu *et al.* (2014) revealed the likely evolution of the DSE fungus *Harpophora oryzae* from a pathogenic ancestor *Magnaporthe oryzae*. Transcriptomic and genomic analyses reported that the genome of *H. oryzae* was 8% larger than the genome of *M. oryzae*. The gain of a high number of transposable elements in *H. oryzae*, and the loss of 73% of genes putatively involved in appressorium-mediated colonisation of leaves by the pathogen *M. oryzae* might explain the larger genome of *H. oryzae*. Taken together, these observations suggest a strikingly different evolution of the *H. oryzae* genome in comparison with the *M. oryzae* genome (Xu *et al.*, 2014). During the interactions of the two fungi with rice, the expression of 11.7% of the genes showed opposing trends, suggesting differences in gene regulation in the presence of the same plant host. Moreover, infection patterns and the related host defence responses, signal transduction and nutritional preferences exhibited remarkable differentiation between the two fungi (Xu *et al.*, 2014).

7.3 Plant Growth Promotion by DSEs

Plants acquire essential inorganic nutrients from the environment either directly via their roots or indirectly through symbiosis with microorganisms. Unfortunately, the contribution of DSEs to nutrient uptake and translocation to plant cells has not been fully determined. A few ultra-structural studies showed that DSE structures in root cells lacked a perifungal membrane and interfacial matrix material, so that they could not be regarded as specialised interfaces for nutrient transfer between fungal and plant cells (Peterson *et al.*, 2008). Positive effects of DSEs on plant biomass in the absence of detectable hyphae in roots could be considered as suspicious at first sight. Moreover, no relationship has been reported between plant growth and/or shoot nutrient accumulation and the intensity of root colonisation by DSEs (Ruotsalainen and Kytöviita, 2004; Berthelot *et al.*, 2017). In the absence of specialised interfaces for nutrient transfer, several hypotheses have been proposed to explain such PGP effects, including mineralisation of organic compounds or solubilisation of insoluble minerals by DSEs in the rhizosphere (Mandyam and Jumpponen, 2005; Newsham, 2011).

Several studies suggest that DSEs produce a variety of extracellular enzymes that break down complex forms of carbon, nitrogen and phosphorus commonly present in soils making them available to plants. Enzymatic activities related to the catabolism of complex carbon sources have been reported for two Pleosporales species (Knapp and Kovacs, 2016), whereas *in vitro* growth assays were performed on a set of Helotiales such as *Cadophora* sp., *Phialocephala fortinii* or *Phialophora* sp. and showed that these DSE were able to degrade both parietal- and storage-plant-derived compounds, such as starch, cellulose and xylose (Caldwell *et al.*, 2000; Mandyam and Jumpponen, 2005; Berthelot *et al.*, 2016; Surono and Narisawa, 2017). Interestingly, using api-ZYM and BioLog FF approaches, Knapp and Kovács (2016) investigated the functional diversity of 15 DSEs belonging to Helotiales, Pleosporales, Xylariales, Eurotiales and Hypocreales and their ability to use various organic compounds. Results highlighted striking differences among DSE species, but all substrates (*c.* 90) were utilised by the DSEs. The broad spectrum of utilised substrates (e.g. amino and organic acids, nucleotides, simple sugars, carbon polymers) illustrates that DSEs might play a role, not only in nutrient mobilisation and plant uptake, but also in general plant performance and ecosystem functioning.

Conversely, the performance of DSEs in the dissolution of insoluble nutrients such as phosphates, zinc or potassium is much less documented. *In vitro*, DSE strains belonging to Helotiales and Pleosporales could solubilise insoluble zinc, calcium and aluminium phosphates (Spagnoletti *et al.*, 2017). Similarly, our results (Figure 7.4a,b) show that *Leptodontidium* sp. and *Phialophora mustea* strains could solubilise various insoluble phosphorus or zinc compounds, and that the intensity of solubilisation is not genus- but strain dependent (Figure 7.4a). This solubilisation could be explained by the secretion of alkaline and acid phosphatases by DSEs (Della Monica *et al.*, 2015; Knapp and Kovács, 2016).

The ability of DSEs to solubilise and mineralise nutrients could improve plant growth and nutrient accumulation. Interrelations between PGP and the use of organic resources by DSEs have often been recorded (Usuki and Narisawa, 2007; Upson *et al.*, 2009; Newsham, 2011). For example, DSEs improved nitrogen uptake by plants, especially when this element was available in organic forms such as leucine, phenylalanine, valine and glycine (Usuki and Narisawa, 2007; Upson *et al.*, 2009; Mahmoud and Narisawa, 2013). Furthermore, fungi such as *Cadophora* sp. could stimulate *Cucumis melo* growth also in the presence of inorganic nitrogen (Khastini *et al.*, 2014). However, similarly to growth stimulation, this effect was strain and plant dependent for nitrogen accumulation in *Rhododendron* sp. associated with *P. fortinii* (Vohník *et al.*, 2005). Other studies have also reported an increase in phosphorus and microelement concentrations in plants such as *Carex* (Haselwandter and Read, 1982) or *Pinus contorta* (Jumpponen *et al.*, 1998). Additionally, our results (Figure 7.4c) show that alfalfa nutrition was improved after inoculation by *Leptodontidium* sp. Me07 and *Phialophora mustea* Pr27, whereas

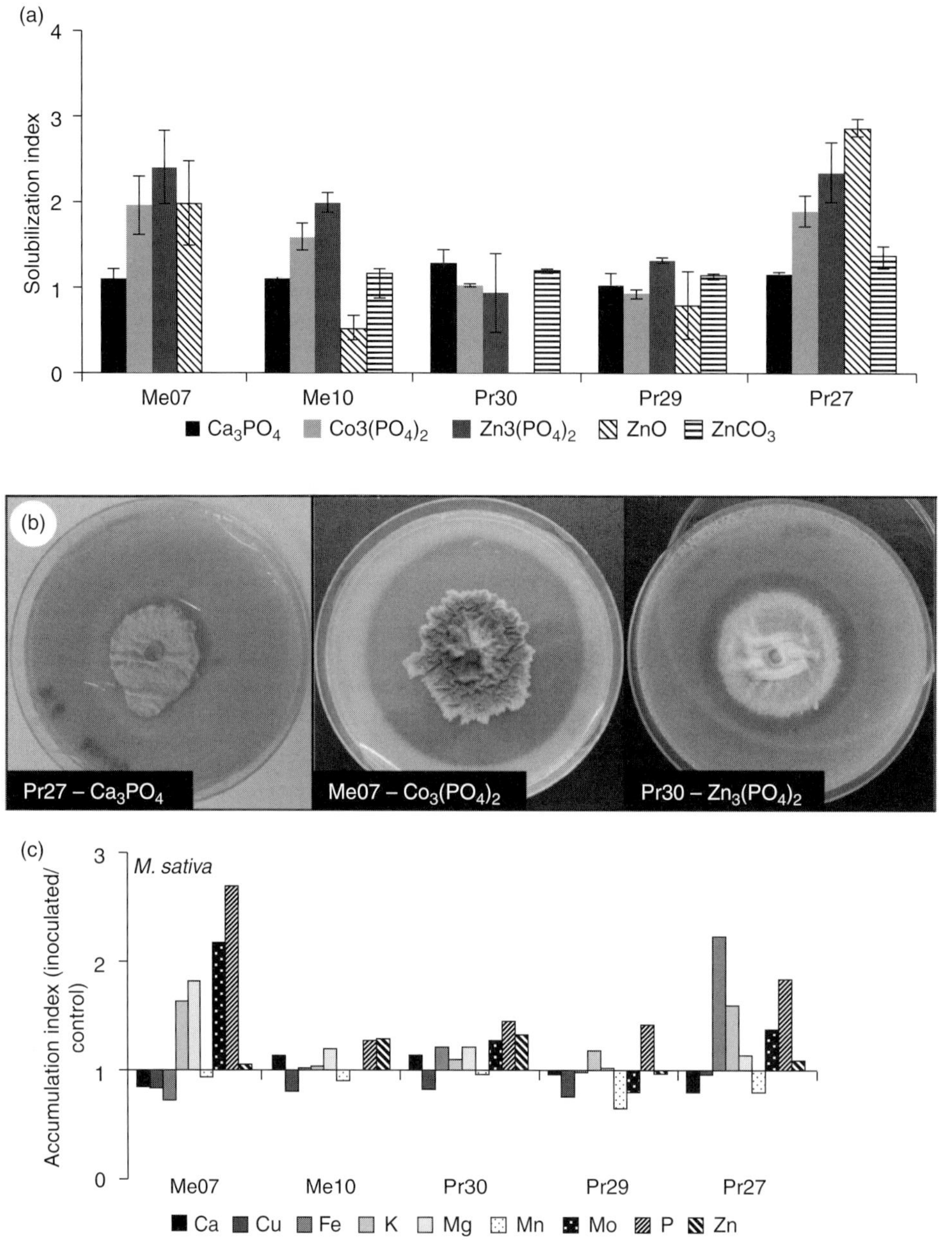

Figure 7.4 Solubilisation of minerals by dark septate endophytes (DSEs) and effect of inoculation on nutrient acquisition by *Medicago sativa*. (a) Solubilisation of Ca_3PO_4, $Co_3(PO_4)_2$, $Zn_3(PO_4)_2$, ZnO and $ZnCO_3$ by five DSE strains. The solubilisation index was calculated as the ratio 'diameter of the solubilization halo' over the 'diameter of the colony'. (b) Solubilisation of Ca_3PO_4, $CO_3(PO_4)_2$, and $Zn_3(PO_4)_2$ by *Phialophora mustea* Pr27, *Leptodontidium* sp. Me07, and *Leptodontidium* sp. Pr30, respectively. (c) Effect of five DSE strains on mineral accumulation in *Medicago sativa* shoots. Seedlings were inoculated with DSEs and grown for two weeks in a loam substrate. Elemental concentrations in the shoots were evaluated by ICP-AES. Strains Me07, Me10, Pr30 belong to *Leptodontidium*, while Pr27 and Pr29 belong to *Phialophora*. The five strains were previously characterised in Berthelot *et al.* (2016).

this effect was much less pronounced with the other three DSE strains tested. Again, this result highlights a strong strain-dependent effect.

DSEs may also produce bioactive substances to communicate or to improve growth of their host plants. Although this topic has been extensively studied for mycorrhizal and endophytic fungi, only a handful of studies have focused on DSEs. Among them, a few secondary metabolites have been isolated and characterised according to their ecological role and their potential application in phytopromotion (Khan *et al.*, 2011). DSEs could promote plant growth directly by releasing soluble compounds. For instance, *Chaetomium globosum* and *Exophiala* sp. respectively produced auxin (IAA) and gibberellin (GA) in liquid media (Khan *et al.*, 2011). Secretion of phytohormones by hyphae could contribute to increase root and shoot development and root surface absorption areas, resulting in enhanced water and nutrient uptake, and could promote root colonisation by other fungi (Hilbert *et al.*, 2012; Sukumar *et al.*, 2013). However, DSEs could also promote the biosynthesis of secondary metabolites by the host plant itself. For instance, colonisation of *Cucumis sativus* by *Exophiala* sp. resulted in increased amounts of salicylic acid, IAA and GA in shoots, while plant genes related to jasmonic acid and ethylene signals were upregulated in *Arabidopsis* sp. roots after inoculation by *Microdochium* sp. (Khan *et al.*, 2011; Mandyam and Jumpponen, 2014). Additionally, *Larix decidua* root colonisation by the DSEs *Cryptosporiopsis* sp. and *Phialophora* sp. resulted in increased concentrations of proanthocyanidins (Schulz *et al.*, 2002).

DSEs could also promote plant growth by emitting volatile organic compounds (VOCs). VOCs are low molecular weight compounds that can easily diffuse in the soil and the atmosphere. Their functions and exact mechanisms of action are poorly understood (Kramer and Abraham, 2012; Lemfack *et al.*, 2014), but several authors suggest a major role of VOCs emitted by rhizospheric and endophytic fungi in plant growth promotion (Hung *et al.*, 2013; Naznin *et al.*, 2013; Paul and Park, 2013). We showed that DSEs could promote tobacco growth in the absence of a physical contact (Berthelot *et al.*, 2016). The emitted VOCs were identified as 1,3-cyclopentadiene 1,3bis and n-methyl butanol (Berthelot *et al.*, 2016) (Figure 7.3a). These compounds were previously identified among the major VOCs released by different fungi and they are likely to play key roles in plant growth promotion (Strobel *et al.*, 2001; Naznin *et al.*, 2013; Paul and Park, 2013).

7.4 Impact of DSEs on Abiotic Stress Alleviation

Plants growing in natural ecosystems are exposed to adverse abiotic conditions such as high temperature, drought, salinity and pollution. Such stressful situations could be alleviated by DSEs. For example, DSE induced plant tolerance to drought and salt stress has been demonstrated by several authors (Mandyam and Jumpponen, 2005; Bois *et al.*, 2006; Zhang *et al.*, 2017). Bois *et al.* (2006) reported a high capacity of *Phialocephala* sp. to tolerate high concentrations of NaCl *in vitro*, whereas Zhang *et al.* (2017) observed that *Exophiala* sp. GM35 increased biomass and photosynthesis efficiency of *Sorghum bicolor* under drought conditions. Furthermore, a growing number of studies have focused on trace element (TE) stress alleviation by DSEs as explored below.

The hypothesis that DSEs had a high constitutive tolerance to TE emerged in the late 2000s (Vrålstad *et al.*, 2002). First, pure fungal culture studies revealed that DSEs were highly tolerant to several metal species (Zhang *et al.*, 2008; Ban *et al.*, 2012; Likar and Regvar, 2013; Lacercat-Didier *et al.*, 2016; Berthelot *et al.*, 2017), including strains isolated from non-contaminated sites. However, it is generally believed that pure culture TE tolerance tests should be interpreted with caution. Second, several field studies suggested that DSEs could be less impacted by TE than endomycorrhizal fungi, which share the same root zone. For instance, Ruotsalainen *et al.* (2007) and Deram *et al.* (2011) reported that DSE colonisation of herbaceous species growing at a highly polluted site was not affected, whereas mycorrhizal colonisation drastically decreased. Similarly, the levels of DSE colonisation of *Salix caprea* growing on a TE-polluted site even showed a positive correlation with higher soil metal contents, whereas endomycorrhizal colonisation was reduced (Likar and Regvar, 2009). Finally, several authors also reported a relatively high abundance of DSEs in TE-polluted soils even when these soils were highly contaminated (Zhang *et al.*, 2008; Op De Beeck *et al.*, 2015; Xu *et al.*, 2015).

These observations naturally prompted different research groups to study the effects of DSEs on the growth and fitness of host plants under TE-stress conditions. The role of DSEs in the alleviation of TE toxicity in plants was suggested in a large number of publications (Table 7.1). Out of a total of 42 strain/soil/plant combinations tested in these studies, 61% of them resulted in higher TE tolerance of the inoculated plant, 41% revealed either plant growth promotion and/or an improvement of plant mineral nutrition, and only 22% did not report any positive effects. Furthermore, we showed that the benefits of DSE inoculation to poplar and birch were soil-dependent (Berthelot *et al.*, 2017). In a low TE-contaminated soil, most of the DSE strains had a positive impact on plant growth, whereas few benefits were observed in a medium-polluted soil. In conclusion, plant–DSE interactions under TE stress may vary from mutualism to parasitism depending on the level of TE contamination, the DSE strain, and the plant species. Even if the mutualism–parasitism paradigm also occurs under TE-stress conditions, it is important to note that the mutualistic effect seems to dominate in a majority of cases.

Mechanistically speaking, several hypotheses have been proposed to explain the positive effects observed on host plants under TE stress. De Maria *et al.* (2011) showed that although *Cadophora finlandica* did not improve *Salix caprea* growth in a cadmium (Cd)/zinc (Zn)-polluted soil, it indirectly enhanced its mineral nutrition. DSEs could also mitigate metal toxicity in plants by indirectly increasing chlorophyll

Table 7.1 Studies reporting DSE and trace element stress experiments.

Trace elements	DSE	Plant species	Type of study	References
Cd, Pb, Zn	*Leohumicola* sp. *Leptodontidium* sp.	*Salix caprea*	PS	Likar and Regvar, 2013
Cd, Pb, Zn	*Exophiala pisciphila*	*Zea mays*	SS	Li *et al.*, 2011
Cd, Pb, Zn	*Cadophora finlandica*	*Salix caprea* *Salix smithiana*	PS	Utmazian *et al.*, 2007
Cd, Zn, Pb	*Leptodontidium* sp. *Cadophora* sp. *Phialophora mustea*	*Populus tremula x alba* *Betula pendula*	PS	Berthelot *et al.*, 2017
Cd, Zn	*Cadophora finlandica*	*Salix caprea*	PS	De Maria *et al.*, 2011
Cd	*Exophiala pisciphila*	*Zea mays*	SS	Wang *et al.*, 2016
Cd	*Exophiala pisciphila*	*Zea mays*	SS	Zhan *et al.*, 2017
Pb	*Gaeumannomyces cylindrosporus*	*Zea mays*	SS	Ban *et al.*, 2017
Cd	*Exophiala pisciphila*	*Zea mays*	SS	He *et al.*, 2017
Cd, Zn, Pb	*Cadophora* sp.	*Lolium perenne*	PS	Berthelot *et al.*, 2018

Note: SS refers to soils artificially spiked with TE; PS refers to soils contaminated by anthropogenic activities that were collected *in situ*.

concentrations and transpiration rates (Wang *et al.*, 2016; Likar and Regvar, 2013; Ban *et al.*, 2017). Additionally, they could also modulate TE accumulation in shoots. However, the effects of DSEs on TE accumulation by plants are diverse and may be perceived as an array of responses from decreasing (Li *et al.*, 2011; Likar and Regvar, 2013), not affecting (Utmazian *et al.*, 2007) or even increasing metal uptake (De Maria *et al.*, 2011), depending on different plant–DSE combinations and the TE species involved (Berthelot *et al.*, 2017). However the number of studies on this particular topic is limited so that it is premature to build a robust conclusion. Therefore this topic resembles the topic of mycorrhizal plant responses to TE in its infancy, when clear conclusions needed time to emerge and will benefit from meta-analyses.

The molecular mechanisms that underpin this apparent detoxification have received little attention. Wang *et al.* (2016) evidenced that inoculation of maize seedlings with *Exophiala pisciphila* decreased the expression of a ZIP (Zrt-Irt-like protein) transporter gene putatively involved in Cd supply to the cells. In line with this result, the increased expression of an MTP (metal transport protein) transporter gene and of other genes involved in vacuolar sequestration suggest induced removal of Cd from the cytosol. Moreover, plant inoculation with DSEs could also modify Cd speciation in roots, evidenced by a decrease of the available fraction of this element while the concentration of Cd-oxalate complexes increased (Wang *et al.*, 2016). In parallel, antioxidant enzymes such as superoxide dismutase and catalase were more active, and chelating compounds (e.g. glutathione) were produced in greater amounts, decreasing the oxidative stress generated by Cd (Wang *et al.*, 2016; Zhan *et al.*, 2017). Finally, the elevated TE tolerance of DSEs might be related to the characteristically high melanin content of their cell walls (Zhan *et al.*, 2016). The binding of metal ions to melanin functional groups (e.g. carboxyl, phenolic, amine) could indeed lower TE toxicity (Ban *et al.*, 2012; Zhan *et al.*, 2016). However, further studies are needed to experimentally demonstrate the role of melanin in the high TE tolerance of DSEs.

7.5 Effects of DSEs on Biotic Stress

Biological control of plant diseases using antagonistic microorganisms offers a highly effective, economical and environment-friendly alternative to the use of chemical compounds (Compant *et al.*, 2016). As previously reported with endophytes and mycorrhizal fungi, DSEs could support the host plant by protecting it from biotic stress such as phytopathogenic fungi (Table 7.2). For example, isolates of *Cadophora* sp. suppressed the wilting effect of *Fusarium* sp. in *Cucumis melo* plants *in vitro* (Khastini *et al.*, 2014). Furthermore, when *Heterochomium chaetospira* was used as a pre-inoculum, it suppressed the incidence of clubroot and *Verticillium* yellows when plants were post-inoculated with the causal agents of these diseases (Narisawa *et al.*, 2000). Nevertheless, these biocontrol effects could be attributed to different factors including plant and pathogen species, as well as the plant developmental stage. The disease index of the pathogenic fungus *Verticillium dahliae* was reduced in six week old tomato plants colonised by the DSE *Leptodontidium orchidicola*, whereas no effect was noted in 24-wk-old plants (Andrade-Linares *et al.*, 2011). In the same study, DSE inoculation was efficient only with low amounts of phytopathogens.

Pathogen control can be accomplished through: (1) antagonism, including competition, parasitism or production of secondary metabolites; (2) induction of host defences; or (3) stimulation of host growth and fitness. DSEs are suspected to protect plants by rapidly colonising roots and thereby exhausting the limited amounts of available substrates so that little will remain available for pathogens to grow. This competition for nutrients could be indirectly achieved through the release

Table 7.2 Studies reporting DSE and biocontrol experiments.

Pathogen	DSE	Plant species	Type of study	References
Heterobasidion parviporum	*Phialocephala fortinii*	–	*In vitro*	Terhonen *et al.*, 2014
Heterobasidion parviporum *Phytophtora pini* *Botrytis cinerea*	*Cryptosporiopsis* sp. *Phialocephala sphareoides*	*Picea abies*	*In vitro* *In planta*	Terhonen *et al.*, 2016
Verticillium dahliae	*Leptodontidium orchidicola*	*Solanum lycopersicum*	*In planta*	Andrade-Linares *et al.*, 2011
Magnaporthe oryzae	*Harpophora oryzae*	*Oryza sativa*	*In planta*	Su et *al.*, 2013
Phytophthora citricola	*Phialocephala fortinii* *Phialocephala subalpina*	–	*In vitro*	Tellenbach *et al.*, 2013
Fusarium oxysporum	*Cadophora* sp.	*Cucumis melo*	*In planta*	Khastini *et al.*, 2014
Phytophthora citricola *Pythium undulatum*	*Phialocephala subalpina*	*Picea abies*	*In planta*	Tellenbach and Sieber, 2012
Verticillium dahliae	*Heteroconium chaetospira* *Phialocoephala fortinii*	*Solanum melongena*	*In planta*	Narisawa *et al.*, 2002
Plasiodophora brassicacea *Verticillium sp.*	*Heterochomium chaetospira*	*Brassica campestris*	*In planta*	Narisawa *et al.*, 2000

of secondary metabolites such as siderophores. Siderophores are low molecular weight compounds produced by microorganisms to chelate available TE, especially iron. For instance, the secretion of hydroxamate by *Phialocephala fortinii* might be a strategy to impede phytopathogenic access to iron (Bartholdy *et al.*, 2001). Additionally, DSEs could also produce toxic metabolites inhibiting the growth of the pathogen or inducing morphological alterations of hyphae (Figure 7.5). This mechanism is called antibiosis. The secretion of echinocandin and cryptosporiopsin by *P. fortinii* affected the growth of *Heterobasidion parviporum*, while the growth of *Phytophthora citricola* was decreased by sclerin emitted by *Phialocephala europaea*

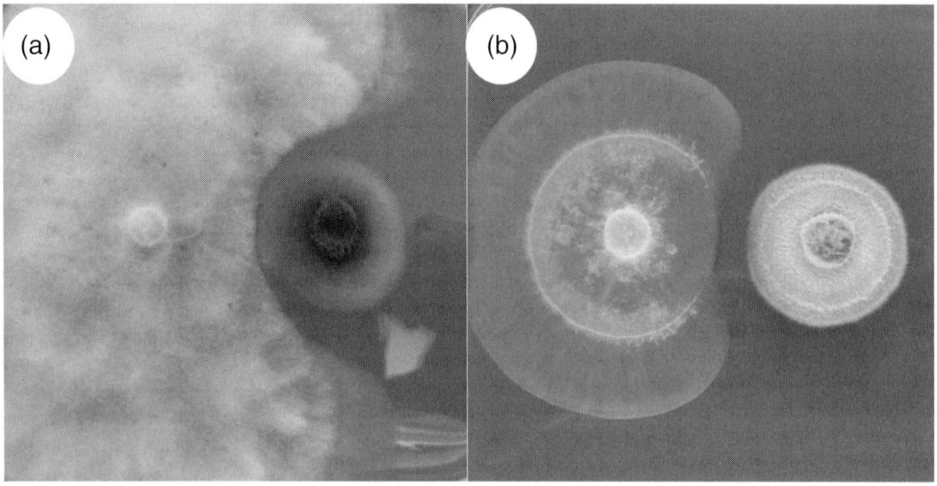

Figure 7.5 Dual inoculations with DSEs and phytopathogenic strains.

The pictures show the co-cultures of (a) *Heterobasidion annosum* / *Cadophora* sp. Fe06, (b) *Phytophthora citricola* / *Phialophora mustea* Pr27 after two weeks of growth at 24°C on malt extract agar medium.

(Tellenbach *et al.*, 2013; Terhonen *et al.*, 2016). DSE isolates do not all produce metabolites against pathogens, but could still protect plants through induced systemic resistance or systemic acquired resistance. In this line, Su *et al.* (2013) evidenced that colonisation of *Oryza sativa* by *Harpophora oryzae* increased the activity of superoxide dismutase, catalase, peroxidase and gluthatione reductase, all involved in plant protection against pathogens. Finally, competition for space could also explain DSE-mediated biocontrol effects. For instance, colonisation of rice roots by *Harpophora oryzae* involved a physical barrier against invasion by *Magnaporthe oryzae* (Su *et al.*, 2013), while the formation of an external mantle by *Cadophora* sp. protected *Cucumis melo* against *Fusarium* sp. (Khastini *et al.*, 2014). Furthermore, plants produce lignins and other cell wall deposits to limit the growth of endophytes (Harman *et al.*, 2004). As a result, the cell wall is re-reinforced after endophytic colonisation, so that cells are less infested by pathogens. The different mechanisms described above might overlap, and a given DSE strain may induce several effects. However, the underlying mechanisms of the beneficial activity of DSEs are not yet completely deciphered.

7.6 Conclusions

Both fundamental and applied investigations on DSEs are still in their infancy. However, thousands of DSEs potentially remain to be discovered, investigated, and analysed. The exponential development of metagenomics should help researchers to identify putative new species and better understand their ecological distribution and functions in different plant species and ecosystems. DSEs could improve growth and alleviate stress in several economically important plant species. While the use of arbuscular mycorrhizal fungi may be hampered by their obligate biotrophism, many DSEs might be easily isolated and maintained in pure cultures facilitating their mass production and application. The fact that fungal root endophytes are relatively easy to culture and have potential positive effects on host plants allows for attractive speculations on their use as biofertilisers or as biocontrol agents for sustainable agriculture. However, future research is needed (1) to better characterise the plant/DSE paradigm/relationship, (2) to develop efficient and cost-effective DSE inocula, and (3) to characterise the ecological impact of inoculum inputs in agricultural crops. Finally, the ubiquity of DSEs in all environments suggests that their metabolism facilitates adaptation to all kinds of lifestyles. This observation prompts us to think that DSEs could be a fertile ground for drug discovery and production.

Acknowledgements

This work was financially supported by the Lorraine Region and the European Regional Development Fund (ERDF) under the LORVER project, by the Agence Nationale de la Recherche (ANR) under the Blanc International (ANR-10-INTB-1703-01-BIOFILTREE) programme, and by the Agence de l'Environnement et de la Maitrise de l'Energie (ADEME) under the PROLIPHYT (1172C0053) project. We thank G. Bonito (Michigan State University, USA) for donating the *Leptodontidium* sp. PMI 412 strain.

References

Addy, H. D., Piercey, M. M. and Currah, R. S. (2005). Microfungal endophytes in roots. *Canadian Journal of Botany-Revue Canadienne de Botanique*, **83**, 1–13.

Alberton, O., Kuyper, T. W. and Summerbell, R. C. (2009). Dark septate root endophytic fungi increase growth of Scots pine seedlings under elevated CO_2 through enhanced nitrogen use efficiency. *Plant and Soil*, **328**, 459–470.

Andrade-Linares, D. R., Grosh, R., Restrepo, S. *et al.* (2011). Effects of dark septate endophytes on tomato plant performance. *Mycorrhiza*, **21**, 413–422.

Ban, Y., Tang, M., Chen, H. *et al.* (2012). The response of dark septate endophytes (DSE) to heavy metals in pure culture. *PLoS One*, **7**, e47968.

Ban, Y., Xu, Z., Yang, Y. *et al.* (2017). Effect of dark septate endophytic fungus *Gaeumannomyces cylindrosporus* on plant growth, photosynthesis and Pb tolerance of maize (*Zea mays* L.). *Pedosphere*, **27**, 283–292.

Bartholdy, B., Berreck, M. and Haselwandter, K. (2001). Hydroxamate siderophore synthesis by *Phialocephala fortinii*, a typical dark septate fungal root endophyte. *Biometals*, **14**, 33–42.

Berthelot, C., Leyval, C., Foulon, J. *et al.* (2016). Plant growth promotion, metabolite production and metal tolerance of dark septate endophytes isolated from metal-polluted poplar phytomanagement sites. *FEMS Microbiology Ecology*, **92**, fiw144.

Berthelot, C., Blaudez, D. and Leyval, C. (2017). Differential growth promotion of poplar and birch inoculated with three dark septate endophytes in two trace element-contaminated soils. *International Journal of Phytoremediation*, **19**,1118–1125.

Berthelot, C., Blaudez, D., Beguiristain T. *et al.* (2018). Co-inoculation of *Lolium perenne* with *Funneliformis mosseae* and the dark septate endophyte *Cadophora* sp. in a trace element-polluted soil. *Mycorrhiza*, **28,** 301–314.

Bois, G., Bertrand, A., Piché, Y. *et al.* (2006). Growth, compatible solute and salt accumulation of five mycorrhizal fungal species grown over a range of NaCl concentrations. *Mycorrhiza*, **16**, 99–109.

Brundrett, M. C. (2006). Understanding the roles of multifunctional mycorrhizal and endophytic fungi. In *Microbial Root Endophytes*, ed. B. Schulz, C. Boyle and T. N. Sieber. Berlin: Springer, pp. 281–298.

Butler, M. and Day, A. (1998). Fungal melanins: a review. *Canadian Journal of Microbiology*, **44**, 1115–1136.

Caldwell, B. A., Jumpponen, A. and Trappe, J. M. (2000). Utilization of major detrital substrates by dark-septate, root endophytes. *Mycologia*, **92**, 230–232.

Compant, S., Saikkonen, K., Mitter, B. *et al.* (2016). Editorial special issue: soil, plants and endophytes. *Plant and Soil*, **405, 1**

David, A. S., Haridas, S., LaButti, K. *et al.* (2016). Draft genome sequence of *Microdochium bolleyi*, a dark septate fungal endophyte of beach grass. *Genome Announcements*, **4**, e00270-16.

Della Monica, I. F., Saparrat, M. C. N., Godeas, A. M. *et al.* (2015). The co-existence between DSE and AMF symbionts affects plant P pools through P mineralization and solubilization processes. *Fungal Ecology*, **17**, 10–17.

De Maria, S. Rivelli, A. R., Kuffner, M. *et al.* (2011). Interactions between accumulation of trace elements and macronutrients in *Salix caprea* after inoculation with rhizosphere microorganisms. *Chemosphere*, **84**, 1256–1261.

Deram, A., Languereau, F. and Haluwyn, C. Van. (2011). Mycorrhizal and endophytic fungal colonization in *Arrhenatherum elatius* L. roots according to the soil contamination in heavy metals. *Soil and Sediment Contamination*, **20**, 114–127.

Grünig, C. R., Queloz, V. and Sieber, T. N. (2011). Structure of diversity in dark septate endophytes: from species to genes. In *Endophytes of forest trees*, ed. A. M. Pirtilla and A. C. Franck. London: Springer, pp. 3–30.

Harman, G. E., Howell, C., Viterbo, A. *et al.* (2004). *Trichoderma* species-opportunistic, avirulent plant symbionts. *Nature Reviews Microbiology*, **2**, 43–56.

Haselwandter, K. and Read, D. J. (1982). The significance of a root-fungus association in two *Carex* species of high-alpine plant communities. *Oecologia*, **53**, 352–354.

He, Y., Yang, Z., Li, M. *et al.* (2017). Effects of a dark septate endophyte (DSE) on growth, cadmium content, and physiology in maize under cadmium stress. *Environmental Science and Pollution Research International*, **24**. doi: 10.1007/s11356-017-9459-6.

Hilbert, M., Voll, L. M., Ding, L. *et al.* (2012). Indole derivative production by the root endophyte *Piriformospora indica* is not required for growth promotion but for biotrophic colonization of barley roots. *New Phytologist*, **196**, 520–534.

Hung, R., Lee, S. and Bennett, J. W. (2013). *Arabidopsis thaliana* as a model system for testing the effect of Trichoderma volatile organic compounds. *Fungal Ecology*, **6**, 19–26.

Jumpponen, A. and Trappe, J. M. (1998a). Dark septate endophytes: a review with special reference to facultative biotrophic root-colonizing fungi. *New Phytologist*, **140**, 295–310.

Jumpponen, A. and Trappe, J. M. (1998b). Performance of *Pinus contorta* inoculated with two strains of root endophytic fungus, *Phialocephala fortinii*: effects of synthesis system and glucose concentration. *Canadian Journal of Botany*, **76**, 1205–1213.

Jumpponen, A., Trappe, J. M. and Mattson, K. (1998). Mycorrhizal functioning of *Phialocephala fortinii* with *Pinus contorta* on glacier forefront soil: interactions with soil nitrogen and organic matter. *Mycorrhiza*, **7**, 261–265.

Khan, A.L., Hamayun, M., Ahmad, N. *et al.* (2011). *Exophiala* sp. LHL08 reprograms *Cucumis sativus* to higher growth under abiotic stresses. *Physiologia Plantarum*, **143**, 329–43.

Khastini, R. O., Ogawara, T., Sato, Y. *et al.* (2014). Control of *Fusarium* wilt in melon by the fungal endophyte, *Cadophora* sp. *European Journal of Plant Pathology*, **139**, 333–342.

Knapp, D. G. and Kovács, G. M. (2016). Interspecific metabolic diversity of root colonizind fungi reveals by enzyme activity tests. *FEMS Microbiology Ecology*, **92**, p.fiw190.

Knapp, D. G., Pintye, A. and Kovács, G. M. (2012). The dark side is not fastidious: dark septate endophytic fungi of native and invasive plants of semiarid sandy areas. *PLoS One*, **7**, p.e32570.

Kovàcs, G. M. and Szigetvari, C. (2002). Mycorrhizae and other root-associated fungal structures of the plants of a sandy grassland on the Great Hungarian Plain. *Phyton – Annales Rei Botanicae*, **42**, 211–223.

Kramer, R. and Abraham, W. R. (2012). Volatile sesquiterpenes from fungi: what are they good for? *Phytochemistry Reviews*, **11**, 15–37.

Lacercat-Didier, L., Berthelot, C., Foulon, J. *et al.* (2016). New mutualistic fungal endophytes isolated from poplar roots display high metal tolerance. *Mycorrhiza*, **26**, 1–15.

Lemfack, M. C., Nickel, J., Dunkel, M. *et al.* (2014). MVOC: a database of microbial volatiles. *Nucleic Acids Research*, **42**, 744–748.

Li, T., Liu, M. J., Zhang, X. T. *et al.* (2011). Improved tolerance of maize (*Zea mays* L.) to heavy metals by colonization of a dark septate endophyte (DSE) *Exophiala pisciphila. The Science of the Total Environment*, **409**, 1069–1074.

Likar, M. and Regvar, M. (2009). Application of temporal temperature gradient gel electrophoresis for characterisation of fungal endophyte communities of *Salix caprea* L. in a heavy metal polluted soil. *The Science of the Total Environment*, **407**, 6179–6187.

Likar, M. and Regvar, M. (2013). Isolates of dark septate endophytes reduce metal uptake and improve physiology of *Salix caprea* L. *Plant and Soil*, **370**, 593–604.

Mahmoud, R. S. and Narisawa, K. (2013). A new fungal endophyte, *Scolecobasidium humicola*, promotes tomato growth under organic nitrogen conditions. *PLoS One*, **8**, e78746.

Mandyam, K. G. and Jumpponen, A. (2005). Seeking the elusive function of the root-colonising dark septate endophytic fungi. *Studies in Mycology*, **53**, 173–189.

Mandyam, K. G. and Jumpponen, A. (2014). Unraveling the dark septate endophyte functions: insights from the *Arabidospis* model. In *Advances in Endophytic Research*, ed. V. C. Verma and A. C. Gange. India: Springer, pp. 115–141.

Mandyam, K. G. and Jumpponen, A. (2015). Mutualism–parasitism paradigm synthesized from results of root-endophyte models. *Frontiers in Microbiology*, **5**, 1–13.

Mandyam, K. G., Roe, J. and Jumpponen, A. (2013). *Arabidopsis thaliana* model system reveals a continuum of responses to root endophyte colonization. *Fungal Biology*, **117**, 250–260.

Mayerhofer, M. S., Kernaghan, G. and Harper, K. A. (2013). The effects of fungal root endophytes on plant growth: a meta-analysis. *Mycorrhiza*, **23**, 119–128.

Melin, E. (1921). Über die mykorrhizapilze von *Pinus sylvestris* L. und *Picea abies* (L.) Karst. (vorläufige mitteilungen). *Svensk Botanisk Tidskrift*, **15**, 192–203.

Narisawa, K., Ohki, K. T. and Hashiba, T. (2000). Suppression of clubroot formation in Chinese cabbage by the root endophytic fungus, *Heteroconium chaetospira*. *Plant Pathology*, **49**, 141–146.

Narisawa, K., Kawamata, H., Currah, S. H. *et al.* (2002). Suppression of *Verticillium* wilt in eggplant by some fungal root endophytes. *European Journal of Plant Pathology*, **108**, 103–109.

Naznin, H. A., Kimura, N., Miyasawa, M. *et al.* (2013). Analysis of volatile organic compounds emitted by plant growth-promoting fungus *Phoma* sp. GS8-3 for growth promotion effects on tobacco. *Microbes and Environments*, **28,** 42–49.

Newsham, K. K. (2011). A meta-analysis of plant responses to dark septate root endophytes. *New Phytologist*, **190**, 783–793.

Op De Beeck, M., Lievens, B., Busschaert, P. *et al.* (2015). Impact of metal pollution on fungal diversity and community structures. *Environmental Microbiology*, **17**, 2035–2047.

Paul, D. and Park, K. (2013). Identification of volatiles produced by *Cladosporium cladosporioides* CL-1, a fungal biocontrol agent that promotes plant growth. *Sensors (Basel, Switzerland)*, **13**, 13969–13977.

Peterson, R. L., Wagg, C. and Pautler, M. (2008). Associations between microfungal endophytes and roots: do structural features indicate function? *Botany*, **86**, 445–456.

Peyronel, B. (1924). Prime ricerche sulla micorizae endotrofiche e sulla microflora radicola normalle della fanerograme. *Revista Biologia*, **6**, 463–485.

Porras-Alfaro, A. and Bayman, P. (2011). Hidden fungi, emergent properties: endophytes and microbiomes. *Annual Review of Phytopathology*, **49**, 291–315.

Read, D. J. and Haselwandter, K. (1981). Observations on the mycorrhizal status of some alpine plant communities. *The New Phytologist*, **88**, 341–352.

Reininger, V. and Schlegel, M. (2016). Analysis of the *Phialocephala subalpina* transcriptome during colonization of its host plant *Picea abies*. *PLoS One*, **11**, 1–16.

Ruotsalainen, A. L. and Kytöviita, M. M. (2004). Mycorrhiza does not alter low temperature impact on *Gnaphalium norvegicum*. *Oecologia*, **140**, 226–233.

Ruotsalainen, A. L., Markkola, A. and Kozlov, M. V. (2007). Root fungal colonisation in *Deschampsia flexuosa*: Effects of pollution and neighbouring trees. *Environmental Pollution*, **147**, 723–728.

Schulz, B. and Boyle, C. (2005). The endophytic continuum. *Mycological Research*, **109**, 661–686.

Schulz, B., Boyle, C., Draeger, S. *et al.* (2002). Endophytic fungi: a source of novel biologically active secondary metabolites. *Mycological Research*, **106**, 996–1004.

Spagnoletti, F. N., Tobar, N., di Pardo, F. *et al.* (2017). Dark septate endophytes present different potential to solubilize calcium, iron and aluminum phosphates. *Applied Soil Ecology*, **111**, 25–32.

Strobel, G. A., Dirkse, E., Sears, J. *et al.* (2001). Volatile antimicrobials from *Muscodor albus*, a novel endophytic fungus. *Microbiology*, **147**, 2943–2950.

Su, Z. Z., Mao, L.-J., Li, N. *et al.* (2013). Evidence for biotrophic lifestyle and biocontrol potential of dark septate endophyte *Harpophora oryzae* to rice blast disease. *PLoS One*, **8**, e61332.

Sukumar, P., Legué V., Vayssière A. *et al.* (2013). Involvement of auxin pathways in modulating root architecture during beneficial plant-microorganism interactions. *Plant, Cell and Environment*, **36**, 909–919.

Surono and Narisawa, K. (2017). The dark septate endophytic fungus *Phialocephala fortinii* is a potential decomposer of soil organic compounds and a promoter of *Asparagus officinalis* growth. *Fungal Ecology*, **28**, 1–10.

Tamura, K., Stecher, D., Perterson, D. (2013). MEGA6: Molecular Evolutionary Genetics Analysis version 6.0. *Molecular Biology and Evolution*, **30**, 2725–2729.

Tellenbach, C. and Sieber, T. N. (2012). Do colonization by dark septate endophytes and elevated temperature affect pathogenicity of oomycetes? *FEMS Microbiology Ecology*, **82**, 157–168.

Tellenbach, C., Sumarah, M. W., Grunig, C. R. *et al.* (2013). Inhibition of *Phytophthora* species by secondary metabolites produced by the dark septate endophyte *Phialocephala europaea*. *Fungal Ecology*, **6**, 12–18.

Terhonen, E., Kerio, S., Sun, H. *et al.* (2014). Endophytic fungi of Norway spruce roots in boreal pristine mire, drained peatland and mineral soil and their inhibitory effect on *Heterobasidion parviporum* in vitro. *Fungal Ecology*, **9**, 17–26.

Terhonen, E., Sipari, N. and Asiegbu, F. O. (2016). Inhibition of phytopathogens by fungal root endophytes of Norway spruce. *Biological Control*, **99**, 53–63.

Upson, R., Read, D. J. and Newsham, K. K. (2009). Nitrogen form influences the response of *Deschampsia antarctica* to dark septate root endophytes. *Mycorrhiza*, **20**, 1–11.

Usuki, F. and Narisawa, K. (2007). A mutualistic symbiosis between a dark septate endophytic fungus, *Heteroconium chaetospira*, and a nonmycorrhizal plant, Chinese cabbage. *Mycologia*, **99**, 175–184.

Utmazian, M. N. D. S., Sweiger, P., Sommer, P. *et al.* (2007). Influence of *Cadophora finlandica* and other microbial treatments on cadmium and zinc uptake in willows grown on polluted soil. *Plant Soil and Environment*, **53**, 158–166.

Vohník, M., Albrechtová, J. and Vosátka, M. (2005). The inoculation with *Oidodendron maius* and *Phialocephala fortinii* alters phosphorus and nitrogen uptake, foliar C:N ratio and root biomass distribution in *Rhododendron* cv. Azurro. *Symbiosis*, **40**, 87–96.

Vrålstad, T., Myrhe, E. and Schumasher, T. (2002). Molecular diversity and phylogenetic affinities of symbiotic root-associated ascomycetes of the Helotiales in burnt and metal polluted habitats. *New Phytologist*, **155**, 131–148.

Wang, J., Li, T., Liu, G.-Y. *et al.* (2016). Unraveling the role of dark septate endophyte (DSE) colonizing maize (*Zea mays*) under cadmium stress: physiological, cytological and genic aspects. *Nature Scientific Reports*, **6**, 1–12.

Xu, R., Li, T., Cui, H. *et al.* (2015). Diversity and characterization of Cd-tolerant dark septate endophytes (DSEs) associated with the roots of Nepal alder (*Alnus nepalensis*) in a metal mine tailing of southwest China. *Applied Soil Ecology*, **93**, 11–18.

Xu, X. H., Su, Z.-Z., Wang, C. *et al.* (2014). The rice endophyte *Harpophora oryzae* genome reveals evolution from a pathogen to a mutualistic endophyte. *Scientific Reports*, **4**, 1–9.

Yu, T., Nassuth, A. and Peterson, R. L. (2001). Characterization of the interaction between the dark septate fungus *Phialocephala fortinii* and *Asparagus officinalis* roots. *Canadian Journal of Microbiology*, **47**, 741–753.

Zhan, F., Li, T., Cui, H. *et al.* (2016). Effects of tricyclazole on cadmium tolerance and accumulation characteristics of a dark septate endophyte (DSE), *Exophiala pisciphila. Bulletin of Environmental Contamination and Toxicology*, **96**, 235–241.

Zhan, F., Li, T., Cui, H. *et al.* (2017). Effects of a dark septate endophyte (DSE) on the glutathione metabolism in maize plants under cadmium. *Journal of Plant Interactions*, **12**, 421–428.

Zhang, Q., Gong, M., Yuan, J. *et al.* (2017). Dark septate endophyte improves drought tolerance in sorghum. *International Journal of Agriculture and Biology*, **19**, 53–60.

Zhang, Y., Zhang, Y., Liu, M. *et al.* (2008). Dark septate endophyte (DSE) fungi isolated from metal polluted soils: their taxonomic position, tolerance, and accumulation of heavy metals in vitro. *Journal of Microbiology*, **46**, 624–632.

PART III

Diversity and Community Ecology of Endophytes

8

Microbispora Dominate Diversity of Endophytic Actinobacteria from Australian Rice Plants

Fitri Widiantini and Christopher Franco

Abstract

A study to detect the diversity of endophytic Actinobacteria from Australian rice was conducted using culture-dependent and culture-independent methods. Rice samples were collected from the rice growing area near Yanco, New South Wales, Australia. Isolation of the endophytic Actinobacteria was done over two consecutive growing seasons. The results demonstrated that most isolates were obtained from plants 10 weeks and older, and only a few were found in younger plants. *Microbispora* spp. were the most commonly isolated endophytic Actinobacteria (94%) with *Streptomyces* spp. and other genera present at lower numbers (6%). The culture-dependent method findings were confirmed by T-RFLP profile analysis. Restriction digests using *HhaI* and *RsaI* also showed an abundance of terminal restriction fragments (TRFs) profiles related to the genus *Microbispora*. Furthermore, other biological properties of the endophytic Actinobacteria isolates were also determined. Four isolates, *Saccharothrix* OSH21, *Saccharopolyspora* OSR26, *Streptomyces* OSR46 and *Microbispora* OSR61, were found to suppress the growth of the pathogenic fungus *Rhizoctonia solani*. Moreover, these isolates might be able to promote plant growth by producing indole acetic acid or to solubilise phosphate making this nutrient available for plant uptake.

8.1 Introduction

Endophytic Actinobacteria are filamentous bacteria which reside within healthy plant tissues without causing any negative effect on the host plant. They have been isolated from a wide variety of plants, from cultivated plants, such as wheat (Coombs and Franco, 2003a), banana (Cao *et al.*, 2005), tomato (Tan *et al.*, 2006), rice (Tian *et al.*, 2004), maize (Araujo *et al.*, 2000) and cucumber (Shimizu *et al.*, 2009), to uncultivated native plants, such as native Australian trees (Kaewkla and Franco, 2013) and a Malaysian ethnomedicinal plant (*Portulaca oleracea)* (Sarmin *et al.*, 2013). Studies of endophytic Actinobacteria have revealed their benefits towards the plant host. Not only are they able to act as biocontrol agents (Coombs *et al.*, 2003), but can also promote plant growth and nitrogen fixation (Berg, 2009).

Many studies have demonstrated the potential of endophytic Actinobacteria as biocontrol agents. A recent study by Shimizu *et al.* (2009) showed that *Streptomyces* spp. isolated from cucumber had the potential ability to control cucumber anthracnose. In Australia, several endophytic Actinobacteria isolated from healthy wheat roots showed potential as biological control agents against several pathogenic fungi of wheat including *Gaeumannomyces graminis* var. *tritici* (Ggt) and *Rhizoctonia* (Barnett *et al.*, 2017). The isolates showed relatively stable activity when they were tested both *in vitro* as well as in the field (Barnett *et al.*, 2017; Franco *et al.*, 2007). The ability to act as biocontrol agents is likely related to Actinobacteria being prolific secondary metabolite producers. However, they have also been implicated in inducing systemic resistance and priming plant defence systems to microbial pathogens (Conn *et al.*, 2008). Thus, the use of endophytic Actinobacteria as biocontrol agents may provide a sustainable avenue for controlling fungal root pathogens.

The most common endophytic Actinobacteria isolated from rice plants (*Oryza sativa*) reported to date are *Streptomyces* (Tian *et al.*, 2004), though there are reports of *Microbispora* and *Kineococcus* spp. from Thai jasmine rice. From glutinous rice plants, other researchers found *Actinophytocola oryzae*, a new genus of the Pseudonocardiaceae family (Indananda *et al.*, 2010) and also *Actinoallomurus oryzae*, a new species of the *Actinoallomurus* genus (Indananda *et al.*, 2011). These recent findings demonstrated that rice plants are a great source for isolating novel Actinobacteria.

It is well known that the conventional isolation by culture-dependent methods is only able to recover a small percentage of the total bacterial communities. These culture-dependent methods are also affected by various factors including the type of isolation media, the size of the inoculum as well as the incubation time (Davis *et al.*, 2005). Knowing the actual diversity of endophytic Actinobacteria will help to plan strategies for isolating the desired isolates. Therefore, it is necessary to have a culture-independent method to study the variation of endophytic actinobacterial diversity in the rice samples. The method chosen was terminal restriction fragment length polymorphism (T-RFLP) that has been used to reveal microbial diversity in different environments (Marsh, 2005).

T-RFLP was used to detect diversity of endophytic Actinobacteria in rice stems and roots in China. The study demonstrated the same *Streptomyces* isolates had been found in both rice stems and roots using both culture-dependent and -independent methods (Tian *et al.*, 2007). In our laboratory, T-RFLP was employed to determine that the type of soil has a major influence on the type of endophytic Actinobacteria found inside wheat roots (Conn and Franco, 2004a). Furthermore, T-RFLP was also used to compare the effect of natural endophytic Actinobacteria, which were previously isolated from healthy roots of wheat (Coombs and Franco, 2003a), and commercially available microbial inoculants, on the diversity of the indigenous microbial populations of the wheat plant endo-rhizosphere (Conn and Franco, 2004b). The study has shown us that the addition of non-indigenous microbes reduced the balance of rhizosphere microbial diversity of the wheat plants. This is in contrast with the addition of endophytic Actinobacteria; though the proportion of colonisation by the endophyte had increased by 5–10% of the total population, the diversity of the indigenous microorganisms in the endo- and ecto-rhizosphere remained the same.

Australia has been growing rice for 80 years, with the main cultivation area located along the Murray and Murrumbidgee river irrigation system of New South Wales. Australia has had the highest rice yields for irrigated rice production at over 10 t ha^{-1}; while the average irrigated rice production is 5 ton ha^{-1}. Farmers will grow other crops such as wheat and lucerne to take advantage of the irrigated soil. Therefore, rice does not grow in these soils for some years. This unique management system has enabled Australian rice to be free from disease (Cother *et al.*, 2004). Diversity of rice root-associated bacteria across 10 rice cultivars was studied using PCR-based denaturing gradient gel electrophoresis (DGGE). The study indicated that the population of bacteria inhabiting the rice roots might be more affected by the soil type than plant genotype. Thus, soil management indirectly influences actinobacterial occupant capacity (Hardoim *et al.*, 2011).

The present study was carried out to isolate endophytic Actinobacteria from rice growing in Australia and to determine the endophytic actinobacterial diversity using both culture-dependent and -independent methods. The isolations were conducted in two consecutive growing seasons, 2009 and 2010; while, the T-RFLP was conducted to determine the endophytic Actinobacteria diversity in rice from the latter growing season only. In addition, primary screening for detection of potential inhibition by the endophyte was also done against several pathogenic fungi.

8.2 Material and Methods

8.2.1 Isolation and Characterisation of Endophytic Actinobacteria from Rice

Oryza sativa cv. Langi samples were collected from rice paddy fields near the Yanco Agricultural Institute, New South Wales, over two growing seasons: 2009 and 2010. The samples were collected when the rice was 7-, 10-, 13- and 16-weeks old. Only 10- and 16-week-old rice plants were collected in the 2010 growing season. Fifteen healthy rice plants were selected randomly, carefully dug out from the paddy fields and packed along with soil attached to the roots. The plants were packed using paper bags and sent to the laboratory. The samples were arrived 6–7 days after they were harvested.

Plant samples were washed thoroughly using running water to free the plants from soil. They were sectioned according to their organ, such as roots, stems, leaf sheaths and leaf blades. They were then air dried overnight at room temperature. The plant tissues were subjected to surface-sterilisation steps as described by Coombs and Franco (2003a). These included sonication of the plants tissues in 124 mM phosphate buffer solution for 1 min, and a series of washes in absolute ethanol and 3% sodium hypochlorite. The samples were rinsed in sterile water and air dried before being cut into 1 cm fragments. The fragments were plated onto isolation media and incubated at two temperatures (27°C and 37°C) for up to 8 weeks. The isolation media used were tryptic soy agar (TSA), modified humic acid vitamin B (HV) agar (1 g humic acid, 0.25 g Na_2HPO_4, 0.85 g KCl, 0.025 g $MgSO_4.7H_2O$, 0.05 g $FeSO_4.7H_2O$, 0.01 g $CaCO_3$, and 18 g agar per 1 l, reverse osmosis treated (RO) water, 1 ml filter-sterilised vitamin B stock solution added after autoclaving; Hayakawa and Nonomura, 1987), mannitol soy (MS) agar (20 g mannitol, 20 g soy flour, and 18 g agar per 1 l RO water) and tap water yeast extract (TWYE) agar (0.25 g yeast extract, 0.5 g K_2HPO_4 and 18 g agar in 1 l tap water). The isolation media were checked weekly for any colonies of endophytic Actinobacteria which were transferred to half-strength potato dextrose agar (HPDA). The isolation plates were kept for up to 8 weeks for weekly observation.

Validation of Surface-Sterilisation Method

The effectiveness of the surface-sterilisation method was validated in two ways: (1) The surface-sterilised fragments were rolled onto TSA plates and incubated at 27°C and 37°C. The plates were checked for the presence of any colonies weekly for up to 8 weeks. (2) Two hundred microlitres of the sterile water used for the final rinse was inoculated onto TSA medium and incubated as in the previous experiment. The surface-sterilisation method was considered to be successful when there was no growth on the TSA.

8.2.2 Characterisation of Endophytic Actinobacteria

The Actinobacteria isolates were tentatively grouped based on their morphological similarity on three media (MS, ISP2 and ISP3 agar). After 7 days of incubation at

27°C, they were observed for growth rate, spore production, the colour of aerial mycelium, reverse colony colour, soluble pigment and spore mass colour with reference to the Methuen Handbook of Colour (Kornerup and Wanscher, 1978).

DNA Extraction and 16S rRNA Gene Sequencing

The isolates selected for identification were grown on cellophane on HPDA for 7 days. The genomic DNA was extracted following the method of Conn and Franco (2004a). Nearly full sequences of the 16S rRNA gene were obtained by combining the first part of the 16S rRNA gene amplified with 27f (5′-AGAGTTTGATCMTGGCTCAG-3′) and 765r (5′-CTGTTTGCTCCCCACGCTTTC-3′) primers and the second part of the 16S rRNA gene were amplified using 704f (5′-GTAGCGGTGAAATGCGTAGA-3′) and 1492r (5′-TACGGGTACCTTGTTACGACTT-3′) primers (Coombs and Franco, 2003a). Each PCR mixture contained 2 μl of DNA sample, 5 μl of thermoPol buffer, 5 μmol of each primer, 10 mM of each dNTP (dATP, dGTP, dCTP and dTTP), 5 units of *Taq* DNA polymerase (BioLabs) and injection water to make up a final volume of 50 μl. The PCR programme used was an initial denaturation step at 94°C for 2 min; followed by 40 cycles of a denaturation step at 94°C for 1 min; an annealing step at 45°C for 1 min; and an elongation step at 72°C for 2 min; the last cycle represented an elongation step at 72°C for 10 min. The PCR product was purified using UltraClean™ PCR Clean-Up, DNA Purification Kit (MoBio Laboratories, Inc). The sample was sequenced by the Flinders Sequencing Facility at Flinders Medical Centre, Adelaide, South Australia. The nucleotide sequences were assembled to obtain the full 16S rRNA gene sequence which was compared with nucleotide sequences from the National Center for Biotechnology Information (NCBI) database using BLAST (Basic Local Alignment Search Tool). Emboss needle global pairwise alignment accessed from www.ebi.ac.uk/Tools/emboss/align/ was used to calculate the level of sequence similarity to the type cultures.

Detection of Endophytic Actinobacteria Diversity by Terminal Restriction Fragment Length Polymorphism (T-RFLP) Analysis

The 16S rRNA gene for T-RFLP analysis was amplified using Actinobacteria-specific/biased primers 6′-FAM (carboxyl-fluoroscein)-labelled S-C-Act-0235-a-S-20 (5′-CGCGGCCTATCAGCTTGTTG-3′) and S-C-Act-0878-a-A-19 (5′-CCGTACTCCCCAGGCGGGG-3′) (Stach *et al.*, 2003) prepared by GeneWorks, Adelaide, Australia. The PCR mixture (50 μl) contained 1x PCR buffer, 200 μM of dNTPs (Roche), 0.4 μM of each primers, 1 unit of DNA *Taq* polymerase (MoBio Labs) and 20–30 ng of DNA sample. The reactions were performed in an Axygene-Maxygene thermal cycler by using an initial denaturation step at 95°C for 5 min, following 25 cycles of 94°C for 1 min, 53°C for 1 min, 72°C for 1 min, and a final extension step of 72°C for 10 min (Jenkins *et al.*, 2009).

The PCR products were then purified using UltraClean™ PCR Clean-Up, DNA Purification Kit (MoBio Laboratories, Inc.) following the manufacturer's instructions. The purified DNA was checked for its purity by running a 1% agarose gel and visualised by staining with GelRed (Biotium). Approximately 200 ng of the labelled amplified gene were digested with restriction enzymes *HhaI* and *RsaI*, prepared following the manufacturer's instructions (BioLabs, Inc., New England) in separate reactions at 37°C for 16–18 h to achieve complete digestion. The enzyme was inactivated by incubating the samples at 65°C for 20 min and stored at -20°C prior to further analysis. For each sample, replicate T-RFLP profiles were obtained. Each of these replicates was generated from four replicates of the PCR product to avoid bias and ensure that sufficient amplicon was amplified for further analyses.

Analysis of TRF Profiles

The T-RFLP profiles were analysed using Peak Scanner v1.0 software (Applied Biosystem) to identify the TRFs after being run on an Applied Biosystems Genetic Analyzer. A TRF was not included in the further analysis if it was not present in both replicates or its size was < 60 bp and > 600 bp (Jenkins *et al.*, 2009). The profiles produced in each replicate were aligned by clustering together TRF profiles which were different by less than 0.5 bp and the average size of those TRF was determined. When fragment profiles within samples were clustered together, their area values were also added (Dunbar *et al.*, 2001). The relative abundance of each peak was calculated against the total peak area of the TRFs present to give the relative composition of each of the detected TRFs in the sample community (Jenkins *et al.*, 2009).

The identification was conducted using the T-RFLP analysis PAT+ (Phylogenetic Assignment Tool) accessed from a web-based software Microbial Community Analysis (MiCA) 3 available online at mica.ibest.uidaho.edu which is linked to a 16S bacterial rRNA gene database, RDP. The window size bin was set at ± 2 bps (Shyu *et al.*, 2007).

8.2.3 Antifungal Activity and Plant Growth Promotion Factors

All the isolates were tested for antifungal activity against three anastomoses groups of the pathogenic fungus *Rhizoctonia solani*, namely AG1, AG3 and AG8. The fungi were grown on PDA (potato dextrose agar, Oxoid). For the assay, a 6 mm diameter plug of actively grown fungus was placed in the centre of a corn meal agar (CMA, Oxoid) plate pre-inoculated with Actinobacteria placed on one side of the plate, 7 days earlier. The plates were incubated in the dark at 27°C for 7 days or until the fungal growth on the control plates (*Rhizoctonia* on CMA without Actinobacteria) had covered the whole plate. The width of the inhibition zone of the pathogen growth from the edge of the actinobacterial growth was measured and evaluated as follows: (+++) ≥ 20 mm, (++) 5–20 mm, (+) ≤ 5 mm and (-) no activity (Burch and Sarathchandra, 2006; Taechowisan *et al.*, 2003).

Isolates were also tested for their chitinase activity and for siderophore production. The chitinase activity assay was conducted by spot inoculation of the isolates onto a medium containing 0.2% colloidal chitin, 0.05% KCl, 0.1% K_2HPO_4, 0.05% $MgSO_4$. $7H_2O$, 0.001% $FeSO_4$, 0.05% yeast extract, 2% agar and RO water, adjusted to pH 7. Plates were then incubated for 3–14 days at 30°C and chitinase activity was detected by the presence of a clear zone surrounding the colonies (Kawase *et al.*, 2004). Siderophore production was done following the method of Alexander and Zuberer (1991). The presence of orange halos surrounding the isolates indicated its ability to produce siderophores.

The selected isolates were tested further for plant growth promotion activity by detecting indole acetic acid (IAA) production and phosphate solubilisation. IAA production was detected following Glickmann and Dessaux (1995) and Khamna *et al.* (2009). Furthermore, the isolates were tested for phosphate solubilisation activity following the method of Beneduzi *et al.* (2008).

8.3 Results and Discussion

8.3.1 Isolation and Characterisation of Endophytic Actinobacteria from Rice

A total of 264 actinobacterial strains were isolated from all isolation media used (Figure 8.1). More isolates were obtained from plates incubated at 37°C. Nutrient-poor media (TWYE and modified HV) gave similar numbers of Actinobacteria compared to the richer media (TSA and MS; Figure 8.1a). The frequency of isolation of endophytes varied between the plant age as well as the plant tissue. More isolates were found from plants 10 weeks and older and only 2% of the total collection was isolated from seven week old rice plants. The number of isolates found from 10, 13 and 16-week-old rice plants were 30%, 29% and 39% of the total isolates, respectively (Figure 8.1b). In general, the number of isolates found from the leaf sheath was higher when compared to the number of isolates found in the leaf blade (Figure 8.1b and 8.1c). The leaf sheath serves as a temporary storage organ for starch before being transported into grain (Watanabe *et al.*, 1997), therefore, potentially attracting microorganisms due to the presence of an available carbon source. This was also noted by Prakamhang *et al.* (2009) for the endophytic diazotrophic bacterial community.

Morphological characterisation resulted in 20 different groups. Most of them were non-*Streptomyces* isolates and only 5.7% were *Streptomyces*-like isolates (Table 8.1). Group representatives from the morphological study were chosen and identified further by partial 16S rRNA gene sequencing. The result demonstrated that the majority of endophytic Actinobacteria isolated from rice belonged to the genus *Microbispora* constituting 92.8% of the isolates. A similar result was obtained from a second round of isolations in the 2010 growing season (Figure 8.2), where 96% of the total endophytic Actinobacteria isolates were identified as *Microbispora* (Table 8.2). However,

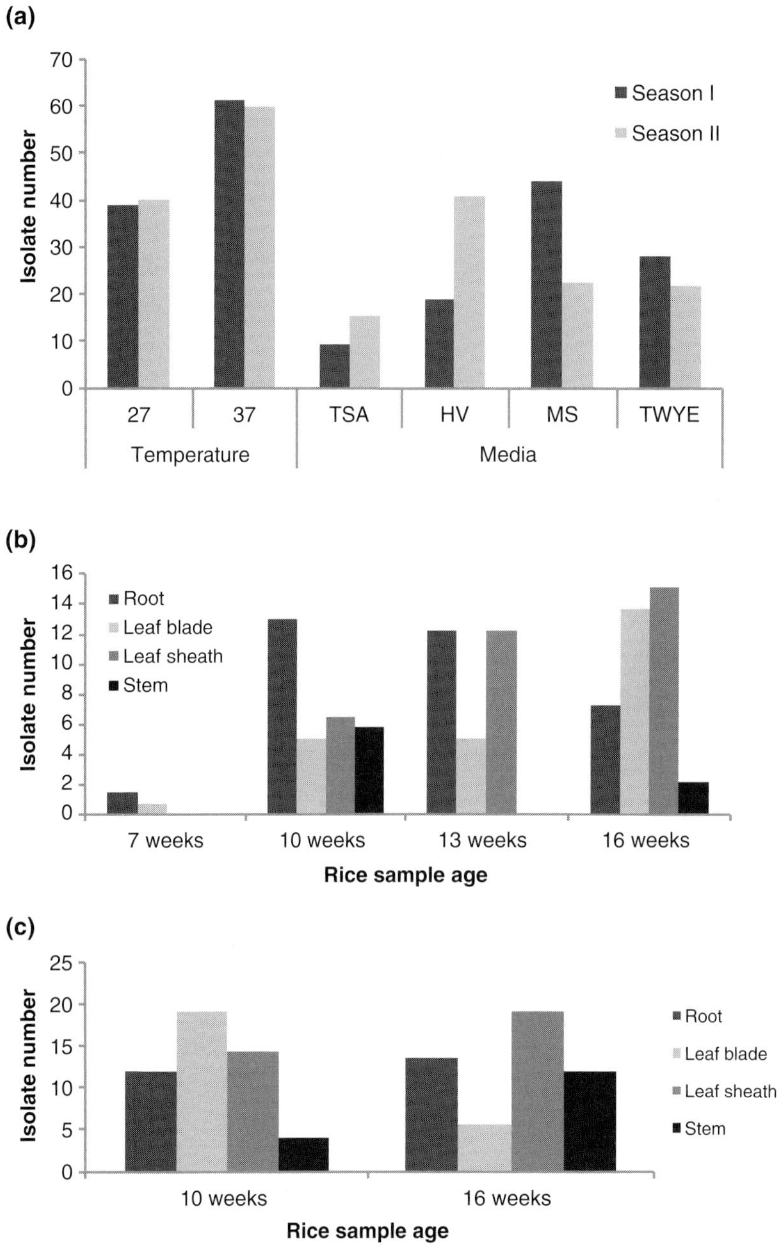

Figure 8.1 The number of endophytic Actinobacteria (a) isolated from rice plants found at different temperatures of incubation and different agar media (TSA, tryptic soy agar; HV, modified humic acid vitamin B; mannitol soy, MS; tap water yeast extract, TWYE); and isolated from different rice tissues at different ages obtained from (b) growing season I in 2009 and (c) growing season II in 2010.

Table 8.1 Percentage of *Streptomyces* and non-*Streptomyces*-like isolates found in different rice plant tissues growing in 2009 and 2010.

Rice plant tissue	Season I (2009) (%)		Season II (2010) (%)		Total
	non-Streptomyces	Streptomyces	non-Streptomyces	Streptomyces	
Root	15.53	1.89	11.74	0.38	29.54
Leaf blade	12.12	0.76	11.74	0	24.62
Leaf sheath	17.04	1.14	15.53	0.38	34.09
Stem	4.17	0	6.44	1.14	11.75
Total	**48.86**	**3.79**	**45.45**	**1.90**	**100**

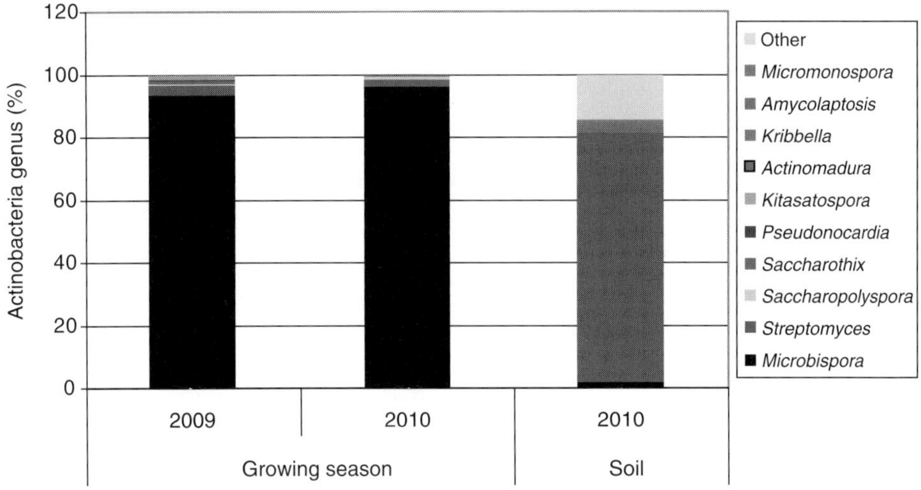

Figure 8.2 Distribution of endophytic Actinobacteria isolated from rice plants in growing season 2009 and 2010 and from soil. (A black and white version of this figure will appear in some formats. For the colour version, please refer to the plate section.)

as expected, Actinobacteria isolated from rice paddy soil were more diverse. The most predominant species at 79% of the isolates were identified as *Streptomyces* spp. (data not presented). This is contrast with previous studies by Tian *et al.* (2004, 2007) who found that *Streptomyces* was the most common actinobacterial genus isolated from rice stems and roots in rice plants growing in China. On the contrary, Kizuka *et al.* (1998) demonstrated that *Microbispora* sp. were more dominant at 23% among the actinobacterial isolates from a variety of fresh healthy leaves in Japan. Furthermore, Araujo *et al.* (2000) found that approximately 62% of the total isolated

Table 8.2 Results of 16S rRNA gene sequencing of the representative groups of endophytic Actinobacteria isolated from rice plants and their morphology characteristics on ISP2.

Isolate representative	Group	Aerial mycelium on ISP2	Result of 16S rRNA gene sequence comparison	
			Closest type strain	Similarity (%)
OST5	1	Pastel red	*M. karnatakensis* (AY445647)	99.6
OSR27	2	Reddish brown	*M. karnatakensis* (AY445647)	99.4
OSL20	3	Dark brown	*M. diastatica* (U48990)	99.0
OSH2	4	Dark brown	*M. chromogenes* (U48989)	99.0
OSL19	5	Dark brown	*M. karnatakensis* (AY445647)	98.9
OST9	6	Yellowish brown	*M. karnatakensis* (AY445647)	98.6
OSR7	7	Brown	*M. karnatakensis* (AY445647)	99.6
OSR24	8	Light brown	*M. karnatakensis* (AY445647)	99.5
OSL2	9	Pastel red	*M. karnatakensis* (AY445647)	99.2
OSL28	10	Yellowish white	*M. chromogenes* (U48989)	99.3
OSL6	11	Cream	*Streptomyces badius* (AY999783)	99.7
OSL32	12	Yellow	*Pseudonocardia adelaidensis* (FJ805427)	98.9
OSH7	13	Brown	*M. karnatakensis* (AY445647)	99.4
OSH19	14	Pale red	*M. karnatakensis* (AY445647), *M. chromogenes* (U48989)	98.1
OST1	15	Pale red	*M. karnatakensis* (AY445647)	99.2
OSR14	16	Lilac grey	*Kitasatospora phosalacinea* (AB184596)	98.8
OSR26	17	Greyish yellow	*Saccharopolyspora shandongensis* (EF104116)	99.3
OSH21	18	Greyish yellow	*Saccharothrix texasensis* (NR024966)	98.8
OST14	19	Greyish yellow	*Actinoallumurus spadix* (AB364581)	99.8
OSR46	20	Yellowish brown	*Streptomyces capoamus* (AB045877)	98.8

Table 8.2 (*Cont.*)

Isolate representative	Group	Aerial mycelium on ISP2	Result of 16S rRNA gene sequence comparison	
			Closest type strain	**Similarity (%)**
OSH4	21	Yellowish white	*Streptomyces heliomycini* (AB184712)	98.7
OSH11	22	Brownish orange	*M. parva* (U48985)	98.9
OSL55	23	Greyish red	*M. karnatakensis* (AY445647)	98.3
OST28	24	Brownish orange	*Streptomyces niveorubacter* (DQ445796)	98.0
OST29	25	Yellowish orange	*Actinomadura glauciflava* (AB184612)	99.1
OSR62	26	Greyish yellow	*Streptomyces cellostaticus* (AY999742)	98.6
OSR69	27	Dark brown	*M. karnatakensis* (AY445647)	99.5

endophytic actinobacterial isolates from maize (*Zea mays*) in Brazil, belonged to the genus *Microbispora*. Lee *et al.* (2008) demonstrated that the majority of endophytic Actinobacteria isolated from Chinese cabbage grown in a rice paddy field belonged to *Microbispora* (67%) followed by *Streptomyces* and *Micromonospora* at 12% and 11%, respectively. A recent study by Kampapongsa and Kaewkla (2016) using media similar to ours identified at least 40% of endophytic actinobacterial isolates as *Microbispora* spp. from rice plants grown in Thailand.

These findings indicate that rice as the host plant preferentially selects *Microbispora* spp. to inhabit its plant tissues. The genus has 19 type species including a number that grow at higher temperatures of 37 to 45°C (Franco, 2015). Rice is a summer crop which would favour strains that are able to grow at higher temperatures such as the *Microbispora*. Endophytic Actinobacteria and their host plants have mutualistic interactions from which each party gains benefits. The host plant provides nutrients for the microbes colonising the plant tissues as well as protection. In return, the endophytic Actinobacteria are able to enhance the plant fitness against environment stresses by evolving biocontrol properties for plant disease control and stimulants for plant growth promotion (Hasegawa *et al.*, 2006). Furthermore, the Australian rice management system might have cultivated an environment for the establishment of specific endophyte populations. Australian rice farmers implement plant rotation, thus the same soil would not be planted with rice up to 3–4 years later (Boerma, 1974). It is possible that rice plants are able to secrete compounds which affect the microenvironment of the soil. These compounds would vary between plant species (Hartmann *et al.*, 2009). A study of endophytic actinobacterial communities in rice conducted in China demonstrated that more diverse endophytic species live in poor-growing seedlings and susceptible rice varieties compared to the resistant varieties (Tian *et al.*, 2004). A study across 10 rice cultivars demonstrated that the population of endophytic Actinobacteria in rice roots were more affected by the type of soil rather than the plant genotype (Hardoim *et al.*, 2011).

8.3.2 Detection of Endophytic Actinobacteria Diversity by T-RFLP

Detection of endophytic Actinobacteria populations using T-RFLP solely aimed to confirm the culture-dependent results. The population of endophytic Actinobacteria as shown by culture-dependent methods demonstrated the dominance of the genus *Microbispora*. The analysis of the TRFs of the 16S rRNA gene from rice plant communities demonstrated relatively different actinobacterial populations between plants of different ages and also within the plant tissues from the same aged samples, as shown in Figure 8.3.

In general, both rice ages shared almost similar TRFs profiles, though the relative abundance was different. The value of the relative abundance represents the level of richness of certain peak fragments in the communities. Endophytic Actinobacteria populations in cultivated rice were found to have the greatest diversity in rice plants at 10 weeks compared to 16 weeks old rice. In general, TRF profiles varied more in stems compared to the other rice tissues.

The terminal fragments generated from both restriction enzymes were subjected to analysis using MiCA3. Not all of the TRFs were from Actinobacteria. The genera of Actinobacteria which were able to generate a particular size from both restriction cuts was compared manually. From this PAT+ result, *Microbispora* spp. gave a TRF size of 135 bp and 230 bp after *HhaI* and *RsaI* digestion, respectively (Figure 8.4). The *Streptomyces* genus comprises over 500 valid species (Lanoot *et al.*, 2004), thus the TRFs profiles given by this genus were varied and present in almost all the TRF profiles generated. However, *Streptomyces* generally gave a fragment size of 144–147 bp and 249 bp after *HhaI* and *RsaI* digestion, respectively.

In order to confirm that the TRFs profile results were from *Microbispora*, a pure culture of *Microbispora* isolate OST5, was used for confirmation. OST5 is a member of *Microbispora* group 1 which was present at a high percentage of the total isolates found from all the plant samples (data shown elsewhere). A fragment of 135 bp size was observed when the pure culture was digested with *HhaI*. Furthermore, when it was digested with *RsaI*, a fragment size of 230 was also present (data not presented). The result confirmed the presence of *Microbispora* spp. with these two TRFs thus indicating that the *Microbispora* genus is present at great abundance in the rice

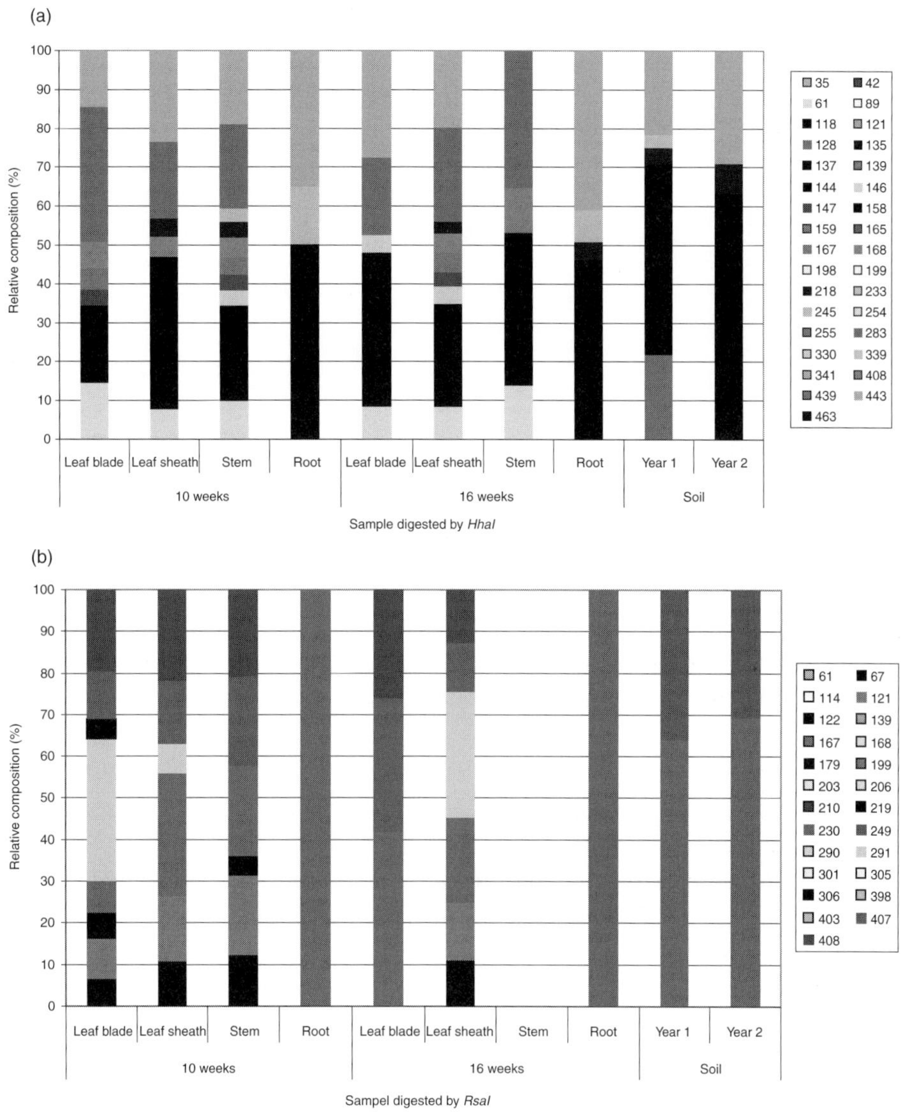

Figure 8.3 Terminal restriction fragments of actinobacterial communities from parts of 10 and 16 week old rice plants and soils after digestion with restriction enzymes: (a) *HhaI* and (b) *RsaI*. (A black and white version of this figure will appear in some formats. For the colour version, please refer to the plate section.)

samples (Figure 8.4). This finding is in agreement with the culture-dependent method with plants from the same growing season II in 2010 which demonstrated that 96% of the total endophytic Actinobacteria isolates belong to the genus of *Microbispora*.

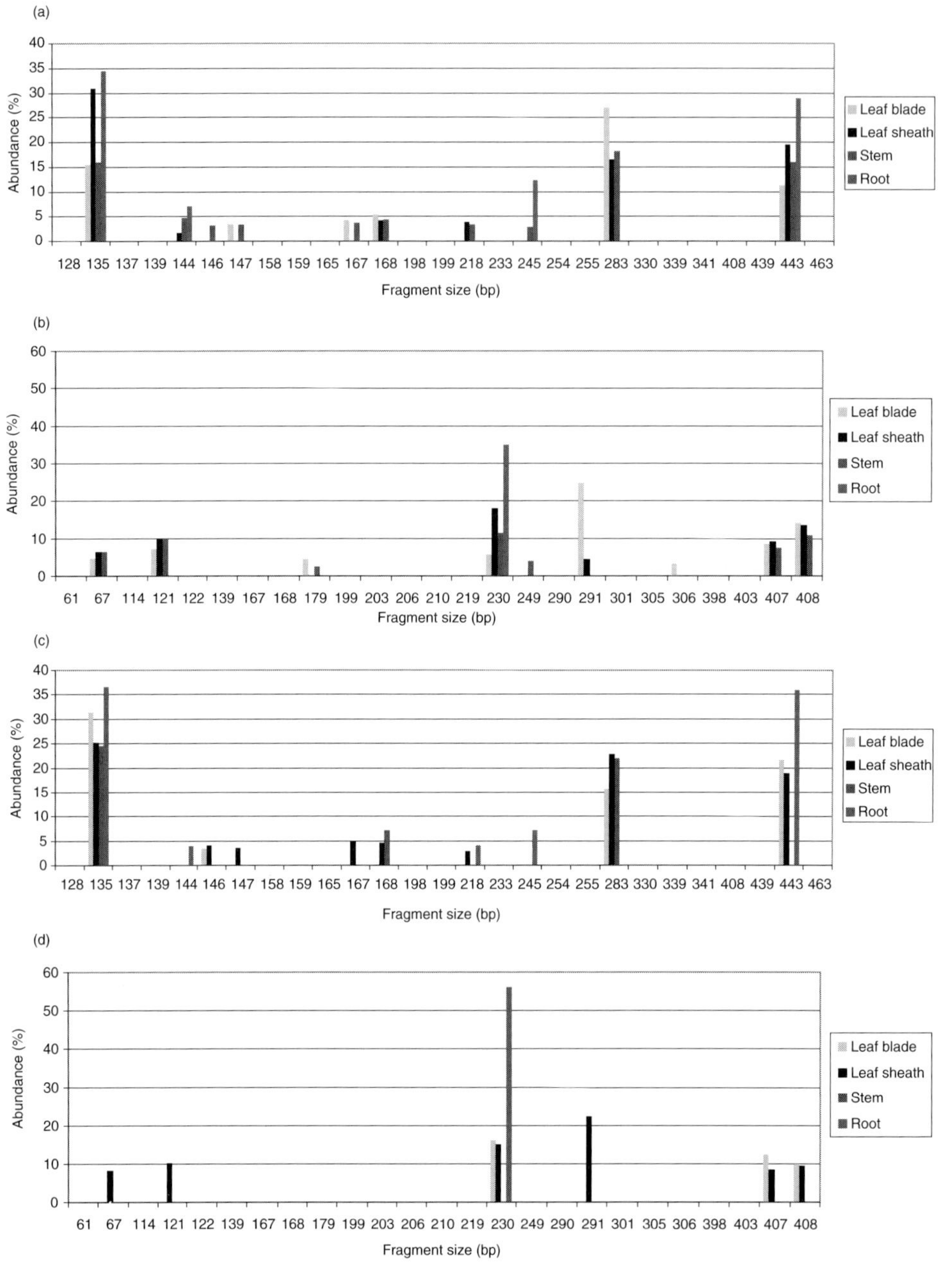

Figure 8.4 Terminal restriction fragments of actinobacterial communities from 10-week-old rice plants after digestion with restriction enzymes (a) *HhaI*, (b) *RsaI* and 16-week-old rice plants after digestion with restriction enzymes (c) *HhaI* and (d) *RsaI*. (A black and white version of this figure will appear in some formats. For the colour version, please refer to the plate section.)

Table 8.3 The antifungal activity of endophytic Actinobacteria isolated from rice plants.

Antifungal activity	**Number of isolates shown activity against test pathogen (%)**		
	***R. solani* AG1**	***R. solani* AG3**	***R. solani* AG8**
+++*	16 (6.1%)	14 (5.3%)	9 (3.4%)
++	9 (3.4%)	21 (8.0%)	27 (10.2%)
+	26 (9.8%)	40 (15.1%)	45 (17.1%)
-	213 (80.7%)	189 (71.6%)	183 (69.3%)

* +++ inhibition zone is ≥20 mm, ++ inhibition zone is 5–20 mm, + inhibition zone ≤5 mm, - not active.

(a) (b)

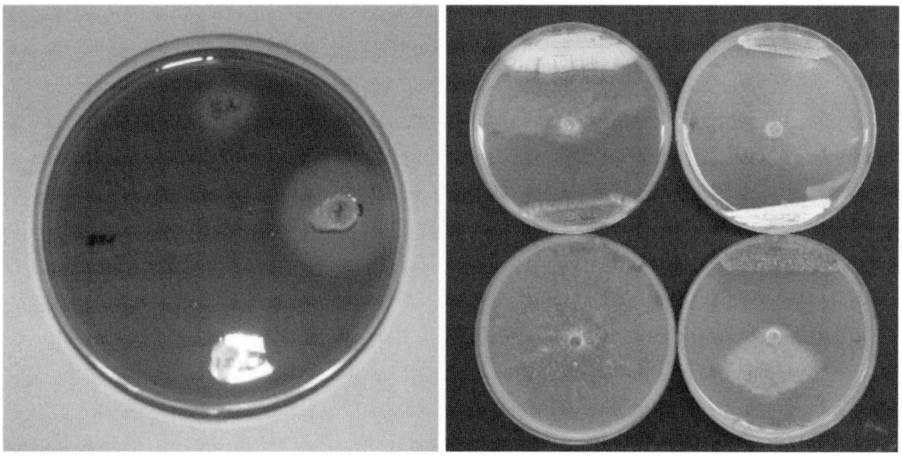

Figure 8.5 (a) Siderophore production by endophytic Actinobacteria detected by the presence of a halo surrounding the isolates and (b) antifungal activity of several endophytic Actinobacteria against *R. solani* AG1.

8.3.3 Detection of Potential Biological Control Agents

Approximately 43% of the total rice endophytic Actinobacteria population were able to inhibit at least one of the pathogens: *R. solani* AG1, AG3 and AG8 (Table 8.3). Of the 264 endophytic Actinobacteria isolates, 17 isolates were chosen for further *in vitro* antagonistic activity assays as they demonstrated moderate to strong inhibition activity against all the three fungal pathogens tested (Figure 8.5). They were also subjected to *in vitro* assays for plant growth promotion traits. Among 17 tested isolates, 7 isolates (OSH4, OSH21, OSR20, OSR26, OSR46, OSL66 and OSR61) demonstrated the ability to produce chitinase and siderophores (Table 8.4; Figure 8.5). Furthermore, only four - OSH21, OSR26, OSR46 and OSR61 - were detected as having either IAA production activity or phosphate solubilisation, both traits of benefit to plants.

Table 8.4 Biocontrol properties and plant growth promotion activities of selected rice endophytic Actinobacteria isolates.

Isolates	Genus	Antifungal activity[a]	Biocontrol properties[b]		Plant growth promotion[b]	
			Chitinase activity	Siderophore production	IAA production	Phosphate solubilisation
OST14	*Actinoallomurus*	+++	+	–	–	–
OSR14	*Kitasatospora*	+	–	+	–	+
OSL15	*Microbispora*	+++	+	–	+	–
OSL22	*Microbispora*	+++	–	+	–	–
OST9	*Microbispora*	+++	+	–	–	–
OSL42	*Microbispora*	+++	–	+	–	–
OSL43	*Microbispora*	+++	–	–	–	–
OSL65	*Microbispora*	++	–	+	–	–
OSL66	*Microbispora*	++	+	+	–	–
OSR61	*Microbispora*	+	+	+	+	–
OSR26	*Saccharopolyspora*	+++	+	+	+	–
OSH21	*Saccharothrix*	+++	+	+	+	–
OSH4	*Streptomyces*	+++	+	+	–	–
OSR20	*Streptomyces*	++	+	+	–	–
OSR46	*Streptomyces*	+	+	+	–	+
OST28	*Streptomyces*	+++	+	–	+	+
OSR62	*Streptomyces*	+	+	–	+	+

[a] Tested against *R. solani* AG1 (+) weak activity, producing clearing zone of <5 mm, (++) moderate activity, producing clearing zone of 5–20 mm, and (+++) strong activity, producing clearing zone of >20 mm observed 5 days after Actinobacteria inoculation.
[b] + produced, – not produced

8.4 Conclusions

The diversity of endophytic Actinobacteria from rice plants growing in Australia have been studied using both culture-dependent and culture-independent methods. An unexpected finding was the high percentage of isolates obtained from the culture-dependent method belong to the genus *Microbispora* from two consecutive growing seasons with only a few *Streptomyces* isolates present. Furthermore, the culture-independent method confirmed that *Microbispora* spp. were indeed present in great abundance as TRF profiles related to *Microbispora* were found in all of the tested samples. This molecular approach also revealed a more diverse population of endophytic Actinobacteria found in 10 week old plants as compared to the 16 week old plants. It showed that the rice growth stage and type of plant tissue affected the composition of the endophytic actinobacterial population.

In terms of the potential plant-growth-promoting activity, this study demonstrated that four endophytic actinobacterial isolates: *Saccharothrix* OSH21, *Saccharopolyspora* OSR26, *Streptomyces* OSR46 and *Microbispora* OSR61, demonstrated potential beneficial plant-growth-promoting traits. They were able to inhibit the growth of pathogenic fungi, produce chitinase and siderophores as well as solubilising phosphate and releasing the phytohormone, IAA. Further evaluation *in planta* needs to be done to assess these isolates.

References

Alexander, D. B. and Zuberer, D. A. (1991). Use of chrome azurol S reagents to evaluate siderophore production by rhizosphere bacteria. *Biology and Fertility of Soils*, **12**, 39–45.

Araujo, J. M., da Silva, A. C. and Azevedo, J. L. (2000). Isolation of endophytic actinomycetes from roots and leaves of maize (*Zea mays* L.). *Brazilian Archives of Biology and Technology*, **43,** 211–221.

Barnett, S., Zhao, S., Ballard, R. and Franco, C. (2017). Selection of microbes for control of *Rhizoctonia* root rot on wheat using a high throughput pathosystem. *Biological Control*, **113**, 45–57.

Beneduzi, A., Peres, D., Vargas, L. K. *et al.* (2008). Evaluation of genetic diversity and plant-growth-promoting activities of nitrogen-fixing bacilli isolated from rice fields in South Brazil. *Applied Soil Ecology*, **39**, 311–320.

Berg, G. (2009). Plant-microbe interactions promoting plant growth and health: perspectives for controlled use of microorganisms in agriculture. *Applied and Environmental Microbiology*, **84**, 11–18.

Boerema, E.B. (1974). *Growth and yield of rice in the Murrumbidgee Valley as influenced by climate, method of sowing, plant density and nitrogen nutrition.* MSc Thesis, Macquarie University, NSW.

Burch, G. and Sarathchandra, U. (2006), Activities and survival of endophytic bacteria in white clover (*Trifolium repens* L.). *Canadian Journal of Microbiology*, **52**, 848–856.

Cao, L., Qiu, Z., You, J. *et al.* (2005). Isolation and characterisation of endophytic streptomycete antagonists of Fusarium wilt pathogen from surface-sterilized banana roots. *FEMS Microbiology Letters*, **247**, 147–152.

Compant, S., Duffy, B., Nowak, J., Clement, C. and Barka, E. A. (2005). Use of plant growth-promoting bacteria for biocontrol of plant diseases: principles, Mechanisms of Action, and Future Prospects. *Applied and Environmental Microbiology*, **71**, 4951–4959.

Conn, V. M. and Franco, C. M. M. (2004a). Analysis of the endophytic actinobacterial population in the roots of wheat (*Triticum aestivum*

L.) by terminal restriction fragment length polymorphism and sequencing of 16S rRNA clones. *Applied and Environmental Microbiology*, **70**, 1787–1794.

Conn, V. M. and Franco, C. M. M. (2004b). Effect of microbial inoculants on the indigenous actinobacterial endophyte population in the roots of wheat as determined by terminal restriction fragment length polymorphism. *Applied and Environmental Microbiology*, **70**, 6407–6413.

Conn, S., Curtin, C., Bézier, A., Franco, C. and Zhang, W. (2008). Purification, molecular cloning, and characterization of glutathione *S*-transferases (GSTs) from pigmented *Vitis vinifera* L. cell suspension cultures as putative anthocyanin transport proteins. *Journal of Experimental Botany*, **59**, 3621–3634.

Coombs, J. T. and Franco, C. M. M. (2003a). Isolation and identification of Actinobacteria from surface-sterilized wheat roots. *Applied and Environmental Microbiology*, **69**, 5603–5608.

Coombs, J. T. and Franco, C. M. M. (2003b). Visualization of an endophytic *Streptomyces* species in wheat seed. *Applied and Environmental Microbiology*, **69**, 4260–4262.

Coombs, J. T., Michelsen, P. P. and Franco, C. M. M. (2003). Evaluation of endophytic Actinobacteria as antagonists of *Gaeumannomyces graminis* var. *tritici* in wheat. *Biological Control*, **29**, 359–366.

Cother, E. J., Reinke, R., McKenzie, C. *et al.* (2004). An unusual stem necrosis of rice caused by *Pantoea ananas* and the first record of this pathogen on rice in Australia. *Australian Plant Pathology*, **33**, 495–503.

Davis, K. E. R., Joseph, S. J. and Janssen, P. H. (2005). Effects of growth medium, inoculum size, and incubation time on culturability and isolation of soil bacteria. *Applied and Environmental Microbiology*, **71**, 826–834.

Dunbar, J., Ticknor, L. O. and Kuske, C. R. (2001). Phylogenetic specificity and reproducibility and new method for analysis of terminal restriction fragment profiles of 16S rRNA genes from bacterial communities. *Applied and Environmental Microbiology*, **67**, 190–197.

Franco, C., Michelsen, P., Percy, N. *et al.* (2007). Actinobacterial endophytes for improved crop performance. *Australasian Plant Pathology*, **36**, 524–531.

Franco, C. M. (2015). Microbispora. In *Bergey's Manual of Systematics of Archaea and Bacteria*, ed. W. B. Whitman, F. Rainey, P. Kämpfer, M. Trujillo, J. Chun, P. DeVos, B. Hedlund and S. Dedysh. doi: 10.1002/9781118960608.gbm00200.

Glickmann, E. and Dessaux, Y. (1995). A critical examination of the specificity of the Salkowski reagent for indolic compound produced by phytopathogenic bacteria. *Applied and Environmental Microbiology*, **61**, 793–796.

Hardoim, P. R., Andreote, F. D., Reinhold-Hurek, B. *et al.* (2011). Rice root-associated bacteria: insight into community structures across 10 cultivars. *FEMS Microbiology Ecology*, **77**, 154–164.

Hartmann, A., Schmid, M., Tuinen, D. and Berg, G. (2009). Plant-driven selection of microbes. *Plant and Soil*, **321**, 235–257.

Hayakawa, M. and Nonomura, H. (1987). Humic acid-vitamin agar, a new medium for the selective isolation of soil actinomycetes. *Journal of Fermentation Technology*, **65,** 501–509.

Hasegawa, S., Meguro, A., Shimizu M. *et al.* (2006). Endophytic actinomycetes and their interactions with host plants. *Actinomycetologica*, **20**, 72–81.

Indananda, C., Matsumoto, A., Inahashi, Y. *et al.* (2010). *Actinophytocola oryzae* gen. nov., sp. nov., isolated from the roots of Thai glutinous rice plants, a new member of the family *Pseudonocardiaceae. International Journal of Systematic and Evolutionary Microbiology*, **60**,1141–1146.

Indananda, C., Thamchaipenet, A., Matsumoto A. *et al.* (2011). *Actinoallomurus oryzae* sp. nov., an endophytic actinomycete isolated from root of Thai jasmine rice plant. *International Journal of Systematic and Evolutionary Microbiology*, **61**, 737–741.

Jenkins, S., Waite, I., Blackburn, A. *et al.* (2009). Actinobacterial community dynamics in long term managed grasslands. *Antonie van Leeuwenhoek*, **95**, 319–334.

Kaewkla, O. and Franco, C. M. M. (2013). *Kribbella endophytica* sp. nov., an endophytic actinobacterium isolated from the surface-sterilized leaf of a native apricot tree. *International Journal of Systematic and Evolutionary Microbiology*, **63**, 1249–1253.

Kawase, T., Saito, A., Sato, T. *et al.* (2004). Distribution and phylogenetic analysis of family 19 chitinases in *Actinobacteria. Applied and Environmental Microbiology*, **70**, 1135–1144.

Kampapongsa, D. and Kaewkla, O. (2016). Biodiversity of endophytic Actinobacteria from jasmine rice (*Oryza sativa* L. KDML 105) grown in Roi-Et Province, Thailand and their antimicrobial activity against rice pathogen. *Annals of Microbiology*, **66**, 587–595.

Khamna, S., Yokota, A. and Lumyong, S. (2009). Actinomycetes isolated from medicinal plant rhizosphere soils: diversity and screening of antifungal compounds, indole-3-acetic acid and siderophore production. *World Journal of Microbiology and Biotechnology*, **25**, 649–655.

Kim, S. T., Cho, K. S., Yu, S. J. and Kang K. Y. (2001). Two-dimensional electrophoretic analysis of rice proteins by polyethylene glycol fractination for protein arrays. *Electrophoresis*, **22**, 2103–2109.

Kizuka, M., Enokita R., Takahashi K. *et al.* (1998). Studies on actnomycetes isolated from plant leaves. *Actinomycetologica*, **12**, 89–91.

Kornerup, A. and Wanscher, J. H. (1978). *Methuen Handbook of Colour*. London: Eyre Methuen Ltd.

Lanoot, B., Vancanneyt, M., Dawyndt, P. *et al.* (2004). BOX-PCR fingerprinting as a powerful tool to reveal synonymous names in the genus *Streptomyces.* Emended descriptions are proposed for the species *Streptomyces cinereorectus, S. fradiae, S. tricolor, S. colombiensis, S. filamentosus, S. vinaceus* and *S. phaeopurpureus. Systematic and Applied Microbiology*, **27**, 84–92.

Lee, S. O., Choi, G. J., Choi, Y. H. *et al.* (2008). Isolation and characterization of endophytic actinomycetes from Chinese cabbage roots as antagonists to *Plasmodiophora brassicae. Journal of Microbiology and Biotechnology*, **18**, 1741–1746.

Loper, J. E. and Buyer, J. S. (1991). Siderophore in microbial interactions on plant surfaces. *Molecular Plant-Microbe Interactions*, **4,** 5–13.

Mano, H. and Morisaki, H. (2008). Endophytic bacteria in rice plant. *Microbes and Environments*, **23**, 109–117.

Marsh, T. L. (2005). Culture-independent microbial community analysis with terminal restriction fragment length polymorphism. In *Methods in Enzymology*, ed. R. L. Jared. San Diego, CA: Academic Press, pp. 308–329.

Nawani, N. N. and Kapadnis, B. P. (2003). Chitin degrading potential of bacteria from extreme and moderate environment. *Indian Journal of Environmental Biology*, **41**, 248–254.

Prakamhang, J., Minamisawa, K., Teamtaisong, K. *et al.* (2009). The communities of endophytic diazotrophic bacteria in cultivated rice (*Oryza sativa* L.). *Applied Soil Ecology*, **42**, 141–149.

Qin, S., Li, J., Chen, H.-H. *et al.* (2009). Isolation, diversity, and antimicrobial activity of rare Actinobacteria from medicinal plants of tropical rain forests in Xishuangbanna, China. *Applied and Environmental Microbiology*, **75**, 6176–6186.

Sarmin, K., Tan, G. Y. A., Franco, C. M. M. *et al.* (2013). *Streptomyces kebangsaanensis* sp. nov. an endophytic actinomycete isolated from a Malaysian ethnomedicinal plant that produces phenazine-1-carboxylic acid. *International Journal of Systematic and Evolutionary Microbiology*, **63**, 3733–3738.

Shimizu, M., Yazawa, S. and Ushijima, Y. (2009). A promising strain of endophytic *Streptomyces* sp. for biological control of cucumber anthracnose. *Journal of General Plant Pathology*, **75**, 27–36.

Shyu, C., Soule, T., Bent, S. *et al.* (2007). MiCA: a web-based tool for the analysis of microbial communities based on terminal-restriction fragment length polymorphisms of 16S and 18S rRNA genes. *Microbial Ecology*, **53**, 562–570.

Stach, J. E. M., Maldonado, L. A., Ward, A. C. *et al.* (2003). New primers for the class Actinobacteria: application to marine and terrestrial environments. *Environmental Microbiology*, **5**, 828–841.

Taechowisan, T., Peberdy, J. F. and Lumyong, S. (2003). Isolation of endophytic actinomycetes from selected plants and their antifungal activity. *World Journal of Microbiology and Biotechnology*, **19**, 381–385.

Tan, H. M., Cao, L. X., He, Z. F. *et al.* (2006). Isolation of endophytic actinomycetes from different cultivars of tomato and their activities against *Ralstonia solanacearum* in vitro. *World Journal of Microbiology and Biotechnology*, **22**, 1275–1280.

Tian, X. L., Cao, L. X., Zeng, Q. G. *et al.* (2004). Study on the communities of endophytic fungi and endophytic actinomycetes from rice and their antipathogenic activities *in vitro*. *World Journal of Microbiology and Biotechnology*, **20**, 303–309.

Tian, X. L., Cao, L. X., Tan, H. *et al.* (2007). Diversity of cultivated and uncultivated actinobacterial endophytes in the stems and roots of rice. *Microbial Ecology*, **53**, 700–707.

Watanabe, T., Kanai, R., Kawase, T. *et al.* (1999). Family 19 chitinases of Streptomyces species: characterization and distribution. *Microbiology*, **145**, 3353–3363.

Watanabe, Y., Nakamura, Y. and Ishii, R. (1997). Relationship between starch accumulation and actvities of the related enzymes in the leaf sheath as a temporary sink organ in rice (*Oryza sativa*). *Australian Journal of Plant Physiology*, **24**, 563–569.

Wensing, A., Braun, S. D., Buttner, P. *et al.* (2010). Impact of siderophore production by *Pseudomonas syringae* pv. *syringae* 22d/93 on epiphytic fitness and biocontrol activity against *Pseudomonas syringae* pv. *glycinea* 1a/96. *Applied and Environmental Microbiology*, **76**, 2704–2711.

9

Isolation, Diversity and Potential Use of Endophytes in the Biomass and Bioenergy Crop *Miscanthus*

JET BEEKWILDER, BRIAN R. MURPHY, EOIN MAC MATHUNA, AARON BARRY AND TREVOR R. HODKINSON

Abstract

Endophytes have the potential to contribute to the sustainable production of bioenergy crops such as the perennial rhizomatous grass *Miscanthus*. They can improve plant growth on marginal land that is otherwise unsuitable for conventional agriculture and can also reduce the need for environmentally damaging chemical inputs including fertilisers and pesticides. This chapter outlines current knowledge of *Miscanthus* endophytes and presents new data on the diversity of root and shoot fungal endophytes isolated from three *Miscanthus* species (*M. sacchariflorus*, *M. sinensis* and *M.* ×*giganteus*). Malt extract, potato dextrose and Czapek Dox media were compared for isolation and growth of the endophytes. The endophytes were then identified using DNA barcoding with three DNA loci (nrITS, nrLSU and TEF). nrITS and nrLSU were found to be the most reliable and consistent barcoding regions. Internal transcribed spacer (ITS) had the highest discriminating potential and is thus recommended for single locus barcoding of endophytes in *Miscanthus*. Most new isolates were Ascomycota belonging to Pezizomycotina with representatives from Dothideomycetes, Eurotiomycetes and Sordariomycetes. One Basidiomycota species was recovered (a known soil yeast *Rhodotorula*). Comparisons between *Miscanthus* endophyte species composition and its better-known sister genus *Saccharum* (including sugarcane) are provided.

9.1 Introduction

9.1.1 *Miscanthus* as a Biomass and Bioenergy Crop

There is a pressing need to develop renewable biomass and bioenergy crops to replace fossil fuels and other fossil-fuel-based products that are contributing to climate change (Hodkinson *et al.*, 2011). Currently, most biomass for producing biofuel (e.g. sugarcane and corn for producing bioethanol, or oil seeds for producing biodiesel) is grown on agricultural land and hence competes with land used for food supply (Perdereau *et al.*, 2017). Therefore, plants that can produce significant amounts of biomass on marginal soils, which have low value for conventional agriculture, have been investigated as alternative biofuel crops (Hodkinson *et al.*, 2015). Since the 1970s, *Miscanthus* species have been under investigation for this purpose (Hodkinson *et al.*, 2001; Jones and Walsh, 2001).

Miscanthus is a perennial panicoid rhizomatous grass that was originally cultivated as an ornamental species for gardens (Teerawatananon *et al.*, 2011; Hodkinson *et al.*, 2015). It is harvested annually by cutting the aboveground plant parts and can produce high yields of up to 37 t ha^{-1} y^{-1} (Lee and Kuan, 2015). The rhizomes, therefore, remain in the soil and support the following season's growth (Hodkinson *et al.*, 2016). *Miscanthus* biomass is rich in carbohydrates and approximately 60% of its dry weight consists of lignocellulose (Lee and Kuan, 2015). After a saccharification pretreatment, the biomass can thus be easily converted into ethanol by a fermentation process (Clifton-Brown *et al.*, 2008).

Miscanthus species have very modest environmental requirements. Climates with a minimum temperature not below -23°C and precipitation higher than 400 mm per year can support continuous biomass production from *Miscanthus* (Xue *et al.*, 2015). Results from a 14-year field study of continuous *Miscanthus* cultivation on a silty clay loam soil, demonstrated that the crop required relatively little nitrogen fertilisation over prolonged periods (Christian *et al.*, 2008). Other studies have indicated that yields on marginal eroded land did not strongly deviate from those on productive soils (Yost *et al.*, 2017). Generally, fertilisation regimes on *Miscanthus* can be minimised to prevent waste of resources and to limit environmental impacts of cultivation. Despite these advantages, *Miscanthus* is still not intensely used as a bioenergy crop. Current limitations mainly relate to the initial establishment of rhizomes in the field and the slow progress of developing new cultivars because of low genetic diversity in breeding material (Clifton-Brown *et al.*, 2008). About 12 different *Miscanthus* species are known, originating from many geographic areas in the world but concentrated in China and South-east Asia (Hodkinson *et al.*, 2015). Three taxa are particularly promising for their high biomass yields, *M. sinensis*, *M. sacchariflorus* and a hybrid of these known as *M.* ×*giganteus* (Hodkinson *et al.*, 2001).

9.1.2 Endophytes in Crop Production

Beneficial microorganisms both in the rhizosphere and stem/leaf are important for stress tolerance and crop performance (Murphy *et al.*, 2015c, 2018). As such they are important for the robustness of a crop like *Miscanthus* and could be partially responsible for its low input requirements with respect to fertilisers and crop protection (Farrar *et al.*, 2014). Many different microorganisms interact with the plant, and a category of these are known as endophytes (Murphy *et al.*, 2015b).

Endophytes are organisms, usually bacteria or fungi, living inside of a plant without causing any signs of disease. They are highly diverse and can be present anywhere in the plant and can in some cases help the plant obtain nutrients or resist pathogens or pests (Wilson, 1995; Murphy *et al.*, 2014). Mycorrhizal fungi are a related category of free-living root symbiotic fungi but are sometimes classified separately from endophytes because endophytes do not form obvious interaction structures with the plant cell such as arbuscules, Hartig net or sheaths. However, they have a role similar to that of endophytic microorganisms in that they are known to recruit nutrients for the plant, especially phosphorus. Mycorrhizal populations in *Miscanthus* monocultures have been studied in a limited number of cases (An *et al.*, 2008; Barnes *et al.*, 2016). Research is required to investigate the role both mycorrhiza and endophytes play in the physiology of *Miscanthus* and how they can be employed to improve robustness and yield. Agronomic traits for *Miscanthus* improvement are discussed in detail by Jones *et al.* (2015).

9.1.3 The Potential for Endophyte-Induced Stress Tolerance and Biomass Improvement in *Miscanthus*

Endophytic bacteria have been isolated from *Miscanthus* (Farrar *et al.*, 2014) and some evidence has been presented indicating their contribution to *Miscanthus* ability to grow on nutrient-poor soils (Straub *et al.*, 2013a). In particular, the endophytic bacterium *Herbaspirillum frisingense* has been studied for its potential nitrogen-fixing capability and its ability to produce effective plant hormones (Straub *et al.*, 2013b). Nitrogen fixation enhances the ability of *Miscanthus* to grow on marginal soils, while plant hormones can promote rapid growth of the plant under certain conditions. *Pseudomonas fluorescens* strains with plant growth-promoting properties have also been isolated from *Miscanthus* and their genomes have been sequenced (Moreira *et al.*, 2016).

In addition to bacteria, fungal endophytes can also contribute to nutrient uptake. Malinowski *et al.* (2000) showed that the leaf endophyte *Neotyphodium coenophialum* modifies nutrient uptake in the forage grass *Festuca arundinacea*. In their study, the growth of *F. arundinacea* shoots increased by 29% in a low phosphorus environment when infected with their naturally occurring endophytes (Malinowski *et al.*, 2000). Besides phosphorus, the same test sample also accumulated more calcium, zinc and copper and the concentrations of iron and boron decreased. Growth-promoting properties of fungal root endophytes have also been demonstrated. For example, a study by Murphy *et al.* (2015b) showed that fungal endophytes recovered from the roots of wall barley, *Hordeum murinum* subsp. *murinum*, could significantly increase the shoot dry weight and grain dry weight of barley cultivars for the plants that received lower nutrient input.

Endophytic fungi from *Miscanthus* have rarely been characterised for abiotic stress resistance or yield improvement. However, a limited number of studies have been published for species within its closely related genus *Saccharum* (including sugarcane). Stuart *et al.* (2010) investigated differences in the population of endophytic fungi in sugarcane due to the presence of transgenes or herbicide treatment, using an rDNA restriction analysis. Furthermore, Leme *et al.* (2013) performed an inventory of endophytic fungi isolated from sugarcane leaves, based on rDNA sequencing and esterase isozyme analysis. The functional role of most of these endophytic fungi for sugarcane cultivation needs to be determined. For example, an endophytic fungus *Neurospora intermedia*, isolated from sugarcane root, was shown to mediate degradation in soil of the herbicide diuron, which is used for pre-emergence weed control (Wang *et al.*, 2017). Thus, this endophytic fungus promotes the establishment of sugarcane in the field after herbicide pretreatment.

9.1.4 The Potential for Endophytic Biocontrol in *Miscanthus*

Miscanthus species are very robust and suffer from few pests or pathogens (Clifton-Brown *et al.*, 2008). In Europe and the USA, hardly any cases of significant harvest losses by biotic agents have been documented. Many microorganisms growing on *Miscanthus* have been isolated from post-harvest storage of *Miscanthus* leaves. However, the main aim of these studies was to investigate their potential as a source for enzymes that can contribute to biomass conversion (Shrestha *et al.*, 2015). Other studies describe isolated cases of pathogenesis. For example, in the 1990s, *Miscanthus* leaf blight caused by the fungus *Stagonospora miscanthi* was found in commercial nurseries and residential ornamental landscapes in Maryland, USA (O'Neill and Farr, 1996). In Europe, on Belgian land used for biomass production, several *Fusarium* species, including *F. avenaceum*, *F. culmorum* and *F. miscanthi* were isolated from fields showing poor emergence of *Miscanthus* (Scauflaire *et al.*, 2013). In pathogenicity tests under greenhouse conditions, isolates displayed the ability to cause rhizome rot disease (Scauflaire *et al.*, 2013). Another pathogen, causing leaf rust, was isolated and characterised in Korea, where the causal agent was identified as the basidiomycete fungus *Puccinia miscanthi* (Kim, 2015; Kim *et al.*, 2012). In general, most of the isolated microbial pathogens are closely related to those found on sugarcane and/or other graminaceous crops, and it has been observed that pathogens isolated from sugarcane are, within *in vitro* tests, pathogens to *Miscanthus*. Also, invertebrate pests have been isolated from *Miscanthus*, including nematodes (Mekete *et al.*, 2011) and aphids (Ameline *et al.*, 2015). However, it is not clear how far these organisms contribute to reductions in crop yield. Therefore, in time, when *Miscanthus* cultivation is scaled up, pathogens and pests from other crops may start causing yield losses and become relevant for this crop. Endophytes can offer biocontrol of such pests and diseases, and so limit the use of pesticides.

Endophytic fungi are known to help plants resist pathogens by several mechanisms (Busby *et al.*, 2016). One way by which fungal endophytes contribute to plant defence is by triggering host resistance, inducing expression of pathogen immunity genes (Mejía *et al.*, 2014). Alternatively, endophytic fungi can directly antagonise pathogens, for example via hyperparasitism, competition or antibiosis (Heather and Sharma, 1987; Pandey *et al.*, 1993; Bailey *et al.*, 2008; Martín *et al.*, 2015). Hence, endophytes can be used as biocontrol agents. In such systems, the endophyte is inoculated/introduced into the plant and once established it either prevents the pathogen from entering the plant or limits its pathogenicity once the plant is infected. Research by Murphy *et al.* (2014) showed that 10 fungal root endophytes isolated from a wild barley species suppressed the development of seed-borne infections in barley. The control group showed reduced growth or died of the seed-borne infections. This indicates that endophytes can be useful as an alternative to pesticides. Another study by Vega *et al.* (2008) showed that endophytes isolated from coffee plants have entomopathogenic properties towards coffee pest insects. For example, the pest insect coffee berry borer is sensitive to entomopathogenic endophytes, and these could be used as biocontrol agents in coffee plants (Vega *et al.*, 2008).

Several endophytic fungi specific for grasses have been described and the mechanisms of their beneficial interactions studied. There are several closely related endophytes known as Balansiaceous endophytes (including family Clavicipitaceae) that specifically inhabit grass. These Balansiaceous endophytes have significant ecological and economic impact, since grasses that have been associated with Balansiaceous endophytes show increased fitness (Bush *et al.*, 1997). See Johnson and Caradus (2019, Chapter 16) for a discussion on the biocontrol applications of the Clavicipitaceae genus *Epichloë*. Balansiaceous endophytes grow systematically and intracellularly and can help the grass survive, by producing secondary metabolites. These secondary metabolites include alkaloids and enzymes (Bush *et al.*, 1997). The enzymes can work as antibiotics and other molecules such as alkaloids can prevent herbivores from eating the host or cause illness once consumed. There are four general groups of alkaloids that have been identified as metabolites in these associations: lolines, peramines, ergot alkaloids and lolitrems. These chemicals show anti-herbivore effects that probably lead to the increased fitness of the plants when associated with endophytes. Peramine and loline mostly act as insect feeding deterrents, while lolitrem B and ergovaline have a pronounced toxicity towards vertebrates (Schardl, 2001).

When exposing herbivores to a purified alkaloid extract, the alkaloids showed clear anti-herbivore results in mammals, like reduced weight gain, restricted blood flow, elevated body temperature, reduced milk production and reduced reproduction (Bush *et al.*, 1997). These alkaloids can also reduce aphid abundance in grasslands in certain environments (Börschig *et al.*, 2014). Kunkel *et al.* (2004) demonstrated that the fungal endophyte *Neotyphodium lolii* produced various ergot alkaloids and caused high mortality in nematodes feeding on the plant. Another example of the use of endophytes as an alternative for pesticides was described by Karpyn Esqueda *et al.* (2017). They showed how endophytes help in resistance to the African black beetle pest, as an alternative to chemical pesticide treatment, because the aboveground symptoms are limited and chemical control in the soil is usually ineffective (Karpyn Esqueda *et al.*, 2017).

A potential role of endophytic fungi for influencing development and enhancing pathogen resistance of *Miscanthus* was suggested by experiments using *Serendipita indica* (= *Piriformospora indica*), a fungus with known growth-promoting properties (Verma *et al.*, 1998; Waller *et al.*, 2005). Inoculation of this endophyte into *Miscanthus* was shown to modulate phytohormone levels (Schmidt *et al.*, 2017). However, no clear effects on plant growth or on pathogens were reported, stressing the need for more comprehensive and extensive studies.

9.1.5 Identification of Endophytic Fungi in *Miscanthus*

Endophytes can be isolated from plants using many different techniques, but usually involves surface sterilisation of plant material and culturing the microbes that subsequently emerge from the plant tissue (Murphy *et al.*, 2015a). DNA barcoding is a technique often used for identifying newly discovered microbial species from host plants. Genetic barcoding uses DNA sequences to identify isolates and can be used to supplement the microscopic morphological examinations of isolates. In this way, it is used in combination with other characters to help identify new species or to assess if existing species names are valid (Moritz and Cicero, 2004). DNA barcoding can be used by non-specialist taxonomists to help identify organisms (Hebert *et al.*, 2003). To barcode a species using conventional Sanger sequencing methods, the DNA has to be extracted, amplified and sequenced. After the sequencing, the sequence of the sample is compared to related accessions in a database such as GenBank to determine the closest matches with any of the sequences of known species (Hodkinson and Parnell, 2007). Phylogenetic analyses might also be necessary to determine the taxonomic identity of a sample, because the most morphologically similar sequences are not always the most closely related (Salamin *et al.*, 2005; Hodkinson *et al.*, 2007b). Identifying the endophyte is the first step in its utilisation in agriculture, forestry or other aspects of biotechnology.

This chapter reports the identification of root and shoot endophytes in three *Miscanthus* species. Fungal endophytes were isolated on potato dextrose, Czapek

Dox and malt extract media to compare the ability of different media to culture the endophytes. DNA barcoding was undertaken using three different genetic loci (TEF, nrLSU, nrITS) that are commonly used for fungal barcoding so that they could be compared for their efficacy in *Miscanthus*. Results are discussed in a taxonomic, phylogenetic and functional context.

9.2 Materials and Methods

Leaves and roots of *Miscanthus sinensis, M. sacchariflorus* and *M.* ×*giganteus* were collected from a private collection in Churchtown, Dublin Ireland (53°17′42.55′N; 6°15′50.71′W). The plants are perennial and had been in the ground for approximately 10 years. They are rhizome propagated duplicates from genetic resource collections at the Trinity College Botanic Garden and Teagasc, Oak Park, Carlow, Ireland.

9.2.1 Isolation of Root Fungal Endophytes

The root samples were washed thoroughly with tap water to remove the majority of microbial surface epiphytes and then rinsed with sterilised distilled water. The samples of each species were separated into tubes containing 70% ethanol, shaken and left for 3 min. Following this, the roots were transferred into 5% sodium hypochlorite for 7 min. The samples were then washed a further time in 70% ethanol, with careful shaking, and finally washed with sterile ultrapure water.

The root segments were cut to 5 mm as evidence suggests that with decreasing size of samples there is a significant increase in the diversity of endophytes recovered (Gamboa *et al.*, 2002). Four root segments were plated out onto one of the following three media: potato dextrose, Czapek Dox and malt extract media, each made up to quarter strength following Murphy *et al.* (2015a). Confirmation of surface sterilisation was carried out by impressing the root tissue onto agar controls. Two control plates per species were plated. The Petri dishes were sealed with Parafilm incubated in the dark for 14 days at 20°C while being monitored daily to examine the growth rates. Seventy-two root segments were plated out in total. The growth rate, quantity and diversity of emergents were compared between the three agar types and species. Czapek Dox was chosen as the most appropriate for subculturing based on growth rate and the most diverse set of emergents.

9.2.2 Isolation of Leaf Fungal Endophytes

Isolation of leaf fungal endophytes followed a similar protocol to the root tissue but samples were cut from separate leaves from each of the species and were sterilised by dipping the leaf into 70% ethanol for 2 s, immersing in 5% sodium hypochlorite for 6 min, dipping in 70% ethanol four times and then rinsing with sterile ultrapure water four times. Samples were then plated out onto the three different media as outlined above for the root isolations.

Table 9.1 Primers used to amplify and DNA barcode TEF, nrLSU and nrITS in *Miscanthus* endophytes.

Barcoding locus	Primer sequence	Reference
TEF 1567R	5' ACHGTRCCRATACCACCRATCTT 3'	Rehner and Buckley (2005)
TEF 983F	5' GCYCCYGGHCAYCGTGAYTTYAT 3'	Rehner and Buckley (2005)
nrLSU LROR	5' ACCCGCTGAACTTAAGC 3'	Vilgalys and Hester (1990)
nrLSU LR5	5' ACCCGCTGAACTTAAGC 3'	Vilgalys and Hester (1990)
nrITS 1F	5' CTTGGTCATTTAGAGGAAGTAA 3'	Gardes and Bruns (1993)
nrITS 4R	5' TCCTCCGCTTATTGATATGC 3'	White *et al*. (1990)
nrITS 5F	5' GGAAGTAAAAGTCGTAACAAGG 3'	White *et al*. (1990)

9.2.3 DNA Extraction, PCR and Sequencing

Approximately 20 mg, fresh weight, of fungal material was scraped from the agar plate and extracted following the 2×CTAB method as outlined in Hodkinson *et al.* (2007b), McGrath *et al.* (2007) and de Cesare *et al.* (2010). PCR used the primers outlined in Table 9.1 for the amplification of three different barcoding loci. The PCR products were cleaned using Exosap and sequenced using Sanger BigDye terminator sequencing using the same primers that were used for the amplification (but individually). Trace files were imported into Geneious 6 (Biomatters) for editing and then analysed using BLAST to find the most similar sequence matches in GenBank (in terms of % similarity).

nrITS is the internal transcribed spacer region of nrDNA (Gardes and Bruns, 1993). It is among the most useful barcoding regions in fungi because it has a clearly defined barcoding gap between the intra- and interspecific variation and can be amplified effectively from small amounts of template DNA (Schoch *et al.*, 2012). The region is usually between 600 and 800 base pairs (bp) in length and is often sufficiently variable, to distinguish closely related fungi. TEF is a translation elongation factor. The TEF sequence is used because it has a conserved region suitable for primers and sequence that can clearly differentiate many taxa (Stielow *et al.*, 2015). nrLSU is the large subunit of nuclear ribosomal DNA and is generally not as variable as the ITS and TEF and may not always differentiate closely related species or genera but it is often useful in combination with the aforementioned loci to align if close matches to other barcoding sequences are not found.

9.3 Results and Discussion

Root and shoot endophytes were retrieved from all three media (malt extract, potato dextrose and Czapek Dox), but with varying success in terms of diversity recovered and growth rate of fungal colonies. The distribution of emergents was even across both the plant species and agar types (an average of six emergents per combination). While virtually all the emergents growing on malt extract and potato dextrose plates reached maximum mycelial growth within a week, Czapek Dox emergents generally grew more slowly. Some other studies support our findings (Sharma and Pandey, 2010). Thus, although potato dextrose and malt extract may be seen as ideal growth media, they can sometimes encourage the faster-growing fungi to dominate. Czapek Dox suppresses the rate of growth in faster-growing species and can also help induce sporulation that is useful for taxon identification (Sharma and Pandey, 2010).

A total of 33 isolates were used for DNA barcoding and 27 of those isolates gave clear and interpretable trace files following Sanger sequencing with three different DNA loci (nrITS, nrLSU and TEF). The nrITS and nrLSU regions were found to be the most reliable and consistent barcoding loci. The universal TEF PCR primers were not always suitable for DNA amplification or produce clear sequencing traces (Table 9.2). The nrITS region had the highest discriminating potential and is thus recommended for single locus barcoding of endophytes in *Miscanthus*. All three of these regions have been generally recommended for DNA barcoding in fungi and their utility is further discussed in Yahr *et al.* (2016).

The sequences were analysed using BLAST and the results shown in Table 9.2, which lists the identified endophyte species extracted from *Miscanthus* roots and shoots. The taxonomic distribution of the endophytes recovered from *Miscanthus* species are shown in Figure 9.1. Figure 9.1a shows the Ascomycota and Figure 9.1b shows the Basidiomycota. Most species belonged to the Ascomycota, and only one belonged to the Basidiomycota (*Rhodotorula mucilaginosa).* Within the Ascomycota, all belonged to the Pezizomycotina with representatives from Dothideomycetes (*Ophiosphaerella,* Phaeosphaeriaceae), Eurotiomycetes (*Penicillium,* Trichocomaceae), and Sordariomycetes (*Roselinia,* Xylariaceae; *Microdochium,* Hyponectriaceae; *Dactylonectria,* Nectriaceae; *Diaporthe,* Diaporthaceae; *Gaeumannomyces* and *Magnaporthe,* Magnaporthaceae).

9.3.1 Comparison of Newly Isolated Endophytes with Other Known Endophytes from *Miscanthus* and *Saccharum*

Chiang *et al.* (2001) isolated and sequenced fungal endophytes of *Miscanthus*; they found that their DNA barcodes from isolates were most similar (percentage sequence similarity in BLAST) to *Cladosporium oxysporum* (86.6%) and *Cladosporium cladosporioides* (98.1%). These species differ from those found in the current study but are in the same class, Dothideomycetes, as two of the species we recovered: *Ophiosphaerella korrae* and *O. hepotricha*. Another endophyte of Chiang *et al.* (2001) showed 79.3% similarity with *Fusarium polyphialidicum*, which is in the same order as the species *Dactylonectria torresensis* isolated here (Hypocreales).

A study by You *et al.* (2011) found that 75% of the endophytes isolated from *Miscanthus sinensis* were *Penicillium* species and the rest were *Aspergillus*. We discovered two taxa of *Penicillium*, namely *P. chrysogenum* and *P. rubens*. The

Table 9.2 Endophyte isolates from root and shoots of *Miscanthus.*

Identified using BLAST analysis and percentage similarity for three barcoding loci, TEF, nrLSU and nrITS. Msa= *M. sacchariflorus*, Msin= *M. sinensis*, Mxg=*M. ×giganteus*

Sample	Tissue source	TEF	Accession (indent. %)	nrLSU	Accession (indent. %)	nrITS	Accession (indent. %)
2EMM	Root Msi	–	–	*Ophiosphaerella korrae*	KP690985 (99%)	*Ophiosphaerella sp.*	KY987527 (99%)
3EMM	Root Msi	–	–	*Ophiosphaerella korrae*	KP690985 (99%)	*Ophiosphaerella sp.*	KY987527 (99%)
4EMM	Root Msi		–	*Ophiosphaerella herpotricha*	KP690992 (99%)	*Ophiosphaerella sp.*	KY987527 (99%)
5EMM	Root Msi	–	–	*Ophiosphaerella korrae*	KP690985 (99%)	*Ophiosphaerella sp.*	KY987527 (99%)
6EMM	Root Msi	–	–	*Diaporthe torresensis*	KP823927 (100%)	*Diaporthe macrodidyma*	KT692589 (100%)
7EMM	Root Msi	–	–	*Microdochium bolleyi*	KP858954 (97%)	*Microdochium bolleyi*	KY365584 (100%)
8EMM	Root Mxg	–	–	*Microdochium bolleyi*	KP858954 (100%)	*Microdochium bolleyi*	KY365584 (100%)
9EMM	Root Mxg	–	–	*Microdochium bolleyi*	KY365584 (100%)	*Microdochium bolleyi*	KP858946 (100%)

(*continued*)

Table 9.2 (*Cont.*)

Sample	Tissue source	TEF	Accession (indent. %)	nrLSU	Accession (indent. %)	nrITS	Accession (indent. %)
10EMM	Root Mxg	*Microdochium paspali*	KJ569504 (96%)	*Microdochium bolleyi*	KP858946 (99%)	*Microdochium bolleyi*	KY365584 (100%)
11EMM	Root Mxg	–	–	*O. herpotricha*	KP690992 (89%)	–	–
12EMM	Root Mxg	*Microdochium paspali*	KJ569504 (96%)	*Microdochium bolleyi*	KY365584 (100%)	*Microdochium bolleyi*	KP858954 (99%)
13EMM	Root Msa	*Gaeumannomyces hyphopodioides*	KX306716 (100%)	*Gaeumannomyces radicicola*	KX306585 (100%)	*Gaeumannomyces hyphopodioides*	KX306515 (99%)
14EMM	Root Msa	*Chaetomium subaffine*	KF001727 (99%)	*Chaetomium globosum*	JN938908 (99%)	*Chaetomium acropullum*	KP994323 (100%)
15EMM	Root Msa	*Alternaria septospora*	JQ672442 (94%)	*Ophiosphaerella korrae*	KP690985 (99%)	*Ophiospaerellasp.*	KY987527 (99%)
16EMM	Root Msa	*Microdochium paspali*	KJ569504 (96%)	*Microdochium bolleyi*	KP858946 (99%)	*Microdochium bolleyi*	KX058046 (100%)
17EMM	Root Msa	*Microdochium paspali*	KJ569504 (96%)	*Microdochium bolleyi*	KP858946 (99%)	*Microdochium bolleyi*	KX058046 (100%)
18EMM	Root Msa	*Microdochium paspali*	KJ569504 (96%)	*Microdochium bolleyi*	KP858946 (99%)	*Microdochium bolleyi*	KY365584 (100%)
1A	Leaf Msi	*Rosellinia sp.*	AB116235 (93%)	*Rosellinia mammiformis*	KF719212 (86%)	*Rosellinia mammiformis*	KT948370 (99%)

2A	Leaf Mxg	*Diaporthe eres*	DQ479931 (97%)	*Phomopsis sp.*	KU973731 (100%)	*Diaporthe phoeniculina*	KP142970 (100%)
3A	Leaf Msa	*Microdochium paspali*	KJ569504 (97%)	*Penicillium chrysogenum*	KX375781 (100%)	*Penicillium rubens*	JX003126 (100%)
1A2	Leaf Msi	*Diaporthe eres*	DQ479931 (97%)	*Diaporthe rhusicola*	JF951166 (99%)	*Podospora theicola*	GQ281798 (100%)
2A2	Leaf Mxg	*Microdochium paspali*	KJ569504 (96%)	*Penicillium chrysogenum*	KY818294 (99%)	*Penicillium chrysogenum*	KY617054 (100%)
4A2	Leaf Msi	*Idriella lunata*	KJ476988 (95%)	*Rhodotorula mucilaginosa*	KY109140 (99%)	*Rhodotorula mucilaginosa*	KX525681 (100%)
5A2	Leaf Msa	*Rosellinia sp.*	AB116235 (98%)	*Rosellinia liminospora*	KF719211 (98%)	*Entoleuca sp.*	KR822145 (97%)
6A2	Leaf Msa	*Microdochium majus*	JX280526 (99%)	*Monographella nivalis* (=*Microdochium*)	AF452030 (99%)	*Microdochium*	KY430576 (100%)
7A2	Leaf Msa	*Podospora anserina*	F0904936 (94%)	*Diaporthe rudis*	KX856061 (99%)	*Diaporthe rudis*	KR909216 (100%)
8A2	Leaf Mxg	*Rosellinia sp.*	AB116235 (96%)	*Rosellinia liminospora*	KF719211 (99%)	*Entoleuca sp.*	KR822145 (97%)

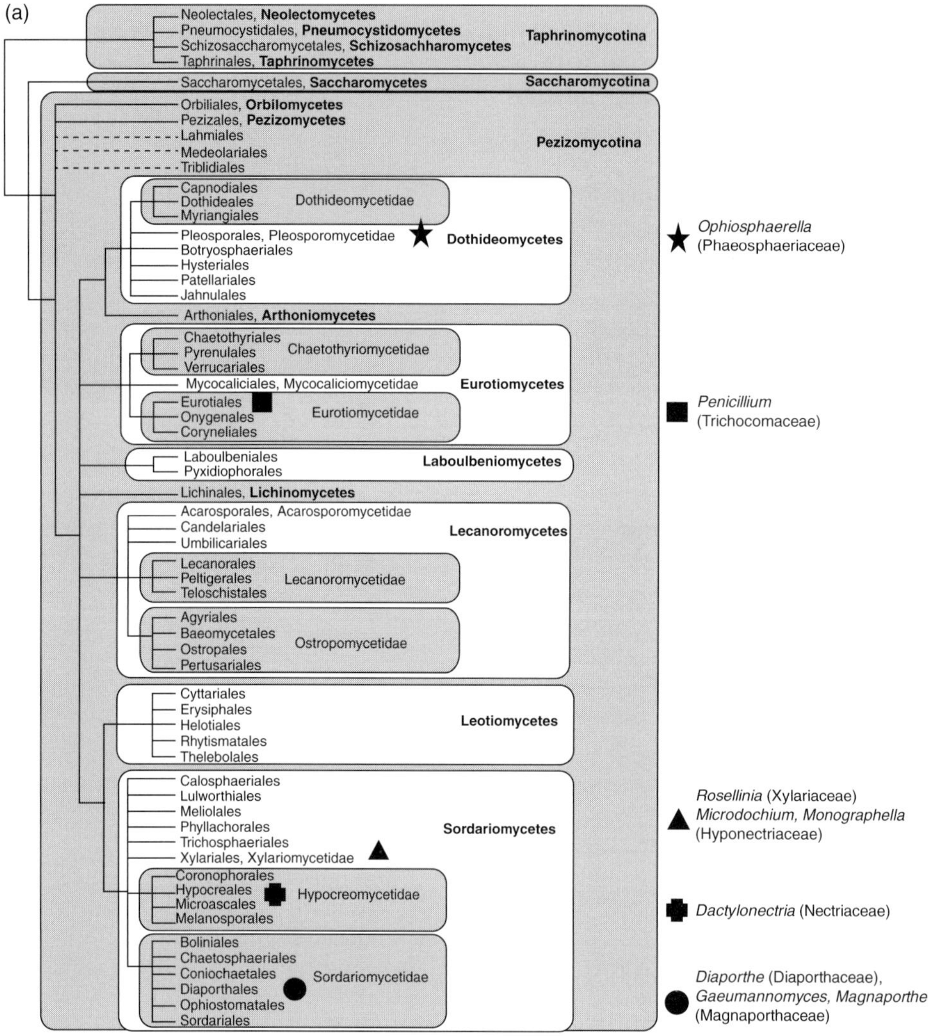

Figure 9.1 Phylogenetic trees of the Dikarya showing position of root and shoot endophytes of *Miscanthus*: (a) Ascomycota; (b) Basidiomycota.
Modified from Hibbett *et al.* (2007).

rest of the endophytes found in our study have never been isolated before as endophytes in *Miscanthus*. It should be noted that the species richness detected here only represents the culturable component of *Miscanthus* endophyte diversity. Next generation sequencing approaches using metagenomic barcoding would be expected to detect far greater species richness than discovered here (Murphy *et al.*, 2015d). Indeed, high diversity of bacterial endophytes has been recorded in mature

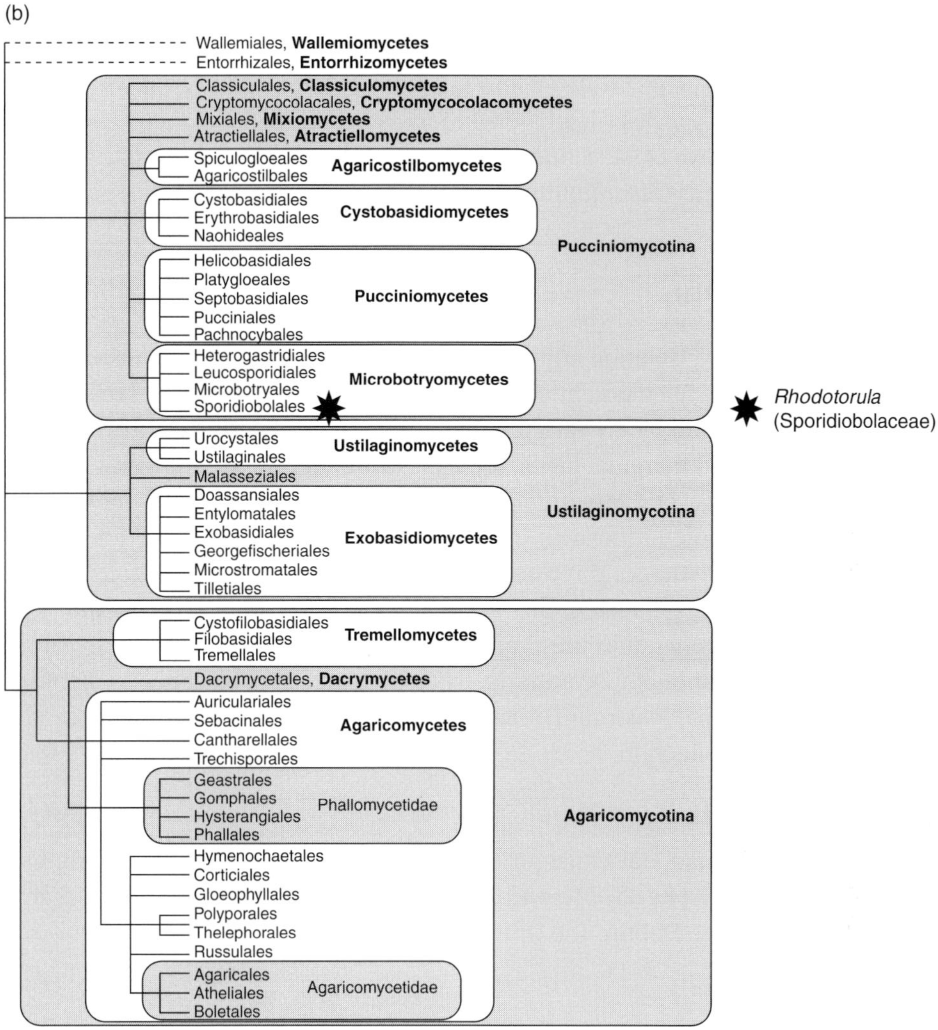

Figure 9.1 (*Cont.*)

Miscanthus plants and also seed using such metagenomic barcoding approaches with the 16S rDNA region (Cope-Selby *et al.*, 2017).

Saccharum is the sister genus to *Miscanthus* (Hodkinson *et al.*, 2002) and is considered an important genetic resource for its improvement. All five species found in *Saccharum* by Leme *et al.* (2013) are from the class Dothideomycetes: *Epicoccum, Phoma, Curvularia, Saccharicola* and *Alternaria.* We also detected Dothideomycetes in *Miscanthus* but none of the genera reported in Leme *et al.* (2013). Nutaratat *et al.* (2014) found nine different endophytes in *Saccharum,* among which were species from *Rhodotorula,* which was also found in our study (4A2). Furthermore, Stuart *et al.* (2010) showed a phylogenetic tree with all their identified samples. They detected members of *Diaporthe* and *Gaeumannomyces*, genera that we also found in *Miscanthus*. Although we detected these genera, species identity within these genera differed and this is likely to be due to co-divergence of host and endophyte species (in a co-evolutionary scenario; Saikkonen *et al.*, 2016; Hodkinson, 2018). Two isolates in this study were identified as *Diaporthe* (2A and 7A2) and one isolate as *Gaeumannomyces* (13EMM).

9.4 Conclusions

Endophyte discovery, isolation and characterisation in *Miscanthus* is very much in its infancy but offers huge potential for sustainable crop production of this highly promising biomass and bioenergy crop by improving yield and stress tolerance on both agricultural and marginal lands. We have outlined how endophytes could be applied to improve several *Miscanthus* traits, have presented methods for optimal fungal endophyte isolation/growth in *Miscanthus* and generated new endophyte species richness statistics for three of its species, *M. sacchariflorus*, *M. sinensis* and *M.* ×*giganteus*. Species detected here are just a subset of total diversity and it is expected that next generation metagenomic barcoding approaches will reveal far greater complexity and allow comparisons to be made between species, habitats and plant tissue so that maximum utilization of the *Miscanthus* microbiome can be achieved in crop production.

Acknowledgements

We thank the International Office of the Hogeschool Utrecht and the European Union for providing Jet Beekwilder with an Erasmus+ Grant to undertake her internship at Trinity College Dublin, The University of Dublin, Ireland.

References

Ameline, A., Kerdellant, E., Rombaut, A. *et al.* (2015). Status of the bioenergy crop *Miscanthus* as a potential reservoir for aphid pests. *Industrial Crops and Products*, **74**, 103–110.

An, G. H., Miyakawa, S., Kawahara, A., Osaki, M. and Ezawa, Z. (2008). Community structure of arbuscular mycorrhizal fungi associated with pioneer grass species *Miscanthus sinensis* in acid sulfate soils: habitat segregation along pH gradients. *Soil Science and Plant Nutrition*, **54**, 517–528.

Bailey, B. A., Bae, H., Strem, M. D. *et al.* (2008). Antibiosis, mycoparasitism, and colonization success for endophytic *Trichoderma* isolates with biological control potential in *Theobroma cacao*. *Biological Control*, **46**, 24–35.

Barnes, C. J., Burns, C. A., Gast, C. J., McNamara, N. P. and Bending, G. D. (2016). Spatio-temporal variation of core and satellite arbuscular mycorrhizal fungus communities in *Miscanthus giganteus*. *Frontiers in Microbiology*, **7**, 1278.

Börschig, C., Klein, A.-M. and Krauss, J. (2014). Effects of grassland management, endophytic fungi and predators on aphid abundance in two distinct regions. *Journal of Plant Ecology*, **7**, 490–498.

Busby, P. E., Ridout, M. and Newcombe, G. (2016). Fungal endophytes: modifiers of plant disease. *Plant Molecular Biology*, **90**, 645–655.

Bush, L. P., Wilkinson, H. H. and Schardl, C. L. (1997). Bioprotective alkaloids of grass-fungal endophyte symbioses. *Plant Physiology*, **114**, 1–7.

Chiang, Y. C., Chou C. H., Lee, P. R. and Chiang, T. Y. (2001). Detection of leaf-associated fungi based on PCR and nucleotide sequence of the ribosomal internal transcribed spacer (ITS) in *Miscanthus*. *Botanical Bulletin of Academia Sinica*, **42**, 39–44.

Christian, D. G., Riche, A. B. and Yates, N. E. (2008). Growth, yield and mineral content of *Miscanthus* × *giganteus* grown as a biofuel for 14 successive harvests. *Industrial Crops and Products*, **28**, 320–327.

Clifton-Brown, J., Chiang, Y.-C. and Hodkinson, T. R. (2008). *Miscanthus*: genetic resources and breeding potential to enhance bioenergy production. In *Genetic Improvement of Bioenergy Crops*, ed. W. Vermerris. New York: Springer Science, pp. 273–290.

Cope-Selby, N., Cookson, A., Squance, M. *et al.* (2017). Endophytic bacteria in *Miscanthus* seed: implications for germination, vertical inheritance of endophytes, plant evolution and breeding. *GCB Bioenergy*, **9**, 57–77.

de Cesare, M., Hodkinson, T. R. and Barth, S. (2010). Chloroplast DNA markers (cpSSRs, SNPs) for *Miscanthus*, *Saccharum* and related grasses (Panicoideae, Poaceae). *Molecular Breeding*, **26**, 539–544.

Farrar, K., Bryant, D. and Cope-Selby, N. (2014). Understanding and engineering beneficial plant–microbe interactions: plant growth promotion in energy crops. *Plant Biotechnology Journal*, **12**, 1193–1206.

Gamboa, M. A., Laureano, S. and Bayman, P. (2002). Measuring diversity of endophytic fungi in leaf fragments: does size matter? *Mycopathologia*, **156**, 41–45.

Gardes, M. and Bruns, T. D. (1993). ITS primers with enhanced specificity for basidiomycetes – application to the identification of mycorrhizae and rusts. *Molecular Ecology*, **2**, 113–118.

Heather, W. A. and Sharma, I. K. (1987). Physiologic specialisation in the hyperparasitism of races of *Melampsora larici-populina* by isolates of *Cladosporium tenuissimum*. *Forest Pathology*, **17**, 185–188.

Hebert, P. D. N., Cywinska, A., Ball, S. L. and deWaard, J. R. (2003). Biological identifications through DNA barcodes. *Proceedings of the Royal Society B: Biological Sciences*, **270**, 313–321.

Hibbett, D. S., Binder, M., Bischoff, J. F. *et al.* (2007). A higher-level phylogenetic classification of the Fungi. *Mycological Research*, **111**, 509–547.

Hodkinson, T. R. and Parnell, J. A. N. (2007). Introduction to the systematics of species rich groups. In *Reconstructing the Tree of Life: Taxonomy and Systematics of Species Rich Taxa*, ed. T. R. Hodkinson and J. A. N. Parnell. Boca Raton, FL: CRC Press, pp. 3–20.

Hodkinson T. R., Chase, M. W. and Renvoize, S. A. (2001). Genetic resources of *Miscanthus*. *Aspects of Applied Biology*, **65,** 239–248.

Hodkinson, T. R., Chase, M. W., Lledó, D. M., Salamin N. and Renvoize, S. A. (2002). Phylogenetics of *Miscanthus*, *Saccharum* and related genera (Saccharinae, Andropogoneae, Poaceae) based on DNA sequences from ITS nuclear ribosomal DNA and plastid *trnL* intron and *trnL-F* intergenic spacers. *Journal of Plant Research*, **115**, 381–392.

Hodkinson, T. R., Salamin, N., Chase, M. W. *et al.* (2007a). Large trees, supertrees, and diversification of the grass family. *Aliso*, **23**, 248–258.

Hodkinson, T. R., Waldren, S., Parnell, J. A. N. *et al.* (2007b). DNA banking for plant breeding, biotechnology and biodiversity evaluation. *Journal of Plant Research*, **120**, 17–29.

Hodkinson, T. R., Jones, M. B., Waldren, S. and Parnell, J. A. N., eds. (2011). *Climate Change Ecology and Systematics*. Cambridge: Cambrige University Press.

Hodkinson, T. R., Klaas, M., Jones, M., Prickett, R. and Barth, S. (2015). *Miscanthus*: a case study for the utilization of natural genetic variation. *Plant Genetic Resources*, **13**, 219–237.

Hodkinson, T. R., Petrunenko, E., Klaas, M. *et al.* (2016). New breeding collections of *Miscanthus sinensis*, *M. sacchariflorus* and hybrids from Primorsky Krai, Far Eastern Russia. In *Perennial Biomass Crops for a Resource-Constrained World*, ed. S. Barth, D. Murphy-Bokern, O. Kalinina, G. Taylor and M. B. Jones. Berlin: Springer, pp. 105–118.

Hodkinson, T. R. (2018). Evolution and taxonomy of the grasses (Poaceae): a model family for the study of species-rich groups. *Annual Plant Reviews Online*, doi: 10.1002/9781119312994.apr0622.

Johnson, L. J. and Caradus, J. R. (2019). The science required to deliver *Epichloë* endophytes to commerce. In *Endophytes for a Growing World*, ed. T. R. Hodkinson, F. M. Doohan, M. J. Saunders and B. R. Murphy. Cambridge: Cambridge University Press, Chapter 16.

Jones, M. B. and Walsh, M., eds. (2001). *Miscanthus for Energy and Fibre*. London: James and James.

Jones, M. B., Finnan, J. and Hodkinson, T. R. (2015). Morphological and physiological traits for higher biomass production in perennial rhizomatous grasses grown on marginal land. *GCB Bioenergy*, **7**, 375–385.

Karpyn Esqueda, K. M., Yen, A. L., Rochfort, S. *et al.* (2017). A review of perennial ryegrass endophytes and their potential use in the management of African black beetle in perennial grazing systems in Australia. *Frontiers in Plant Science*, **8**, 3.

Kim, K. W. (2015). Three-dimensional surface reconstruction and *in situ* site-specific cutting of the teliospores of *Puccinia miscanthi* causing leaf rust of the biomass plant *Miscanthus sinensis*. *Micron*, **73**, 15–20.

Kim, S., Da, K. and Mei, C. (2012). An efficient system for high-quality large-scale micropropagation of *Miscanthus* × *giganteus* plants. *In Vitro Cellular & Developmental Biology-Plant*, **48**, 613–619.

Kunkel, B. A., Grewal, P. S. and Quigley, M. F. (2004). A mechanism of acquired resistance against an entomopathogenic nematode by *Agrotis ipsilon* feeding on perennial ryegrass harboring a fungal endophyte. *Biological Control*, **29**, 100–108.

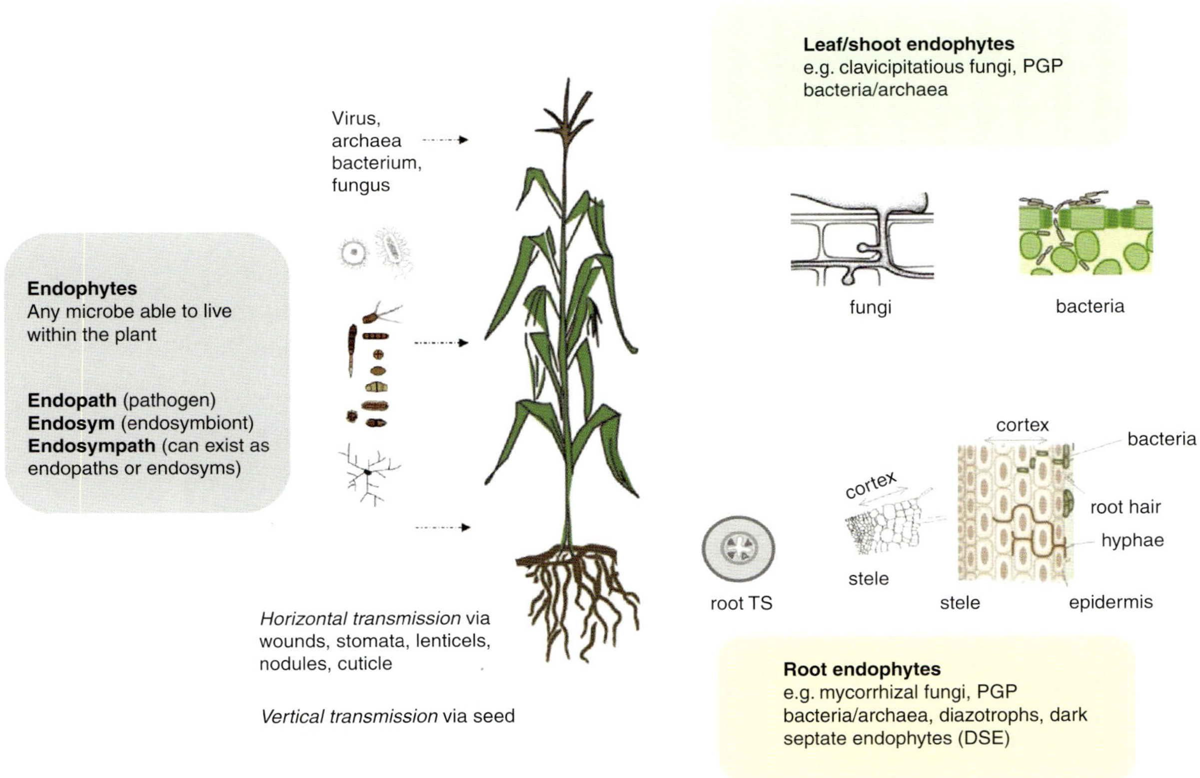

Figure 1.2 Subgroups of endophytes. Endophytes are any microbe that can live within a plant. Subclasses of endophytes can be defined in a hierarchical manner. Endopaths are pathogens, endosymbionts (endosyms) are always mutualistic and endosympaths can exist as both endosymbionts and endopaths. Within endosymbionts, major groups of beneficial microbes can be recognised including plant growth-promoting (PGP) and diazatrophic bacteria/archaea, class 1 (clavicipitatious) fungal endophytes, class 2 fungal endophytes, mycorrhizal fungi and dark septate endophytes (DSEs). They can also be divided by tissue type such as shoot or root. (A black and white version of this figure will appear in some formats.)

Drawn from images in D'Arcy *et al.* (2001), Melotto *et al.* (2008), Hirsch and Mauchline (2012), Gelorini *et al.* (2012), Pickett *et al.* (2014) and Sarah Fogg with permission (www.sarahfogg.co.uk).

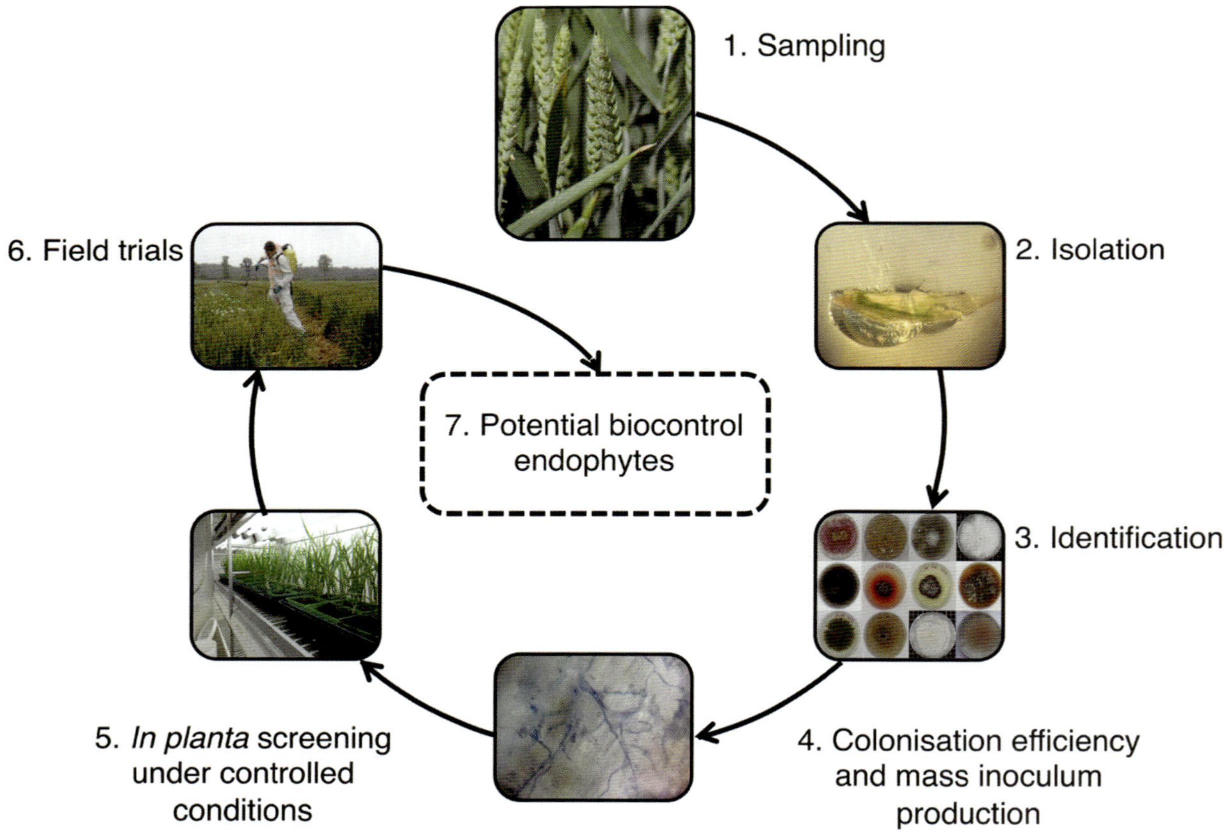

Figure 2.2 The process of discovery of biocontrol endophytes: 1. Sampling healthy plants in areas under disease pressure. 2. Isolation on artificial media from surfaced sterilised tissues (wheat glume). 3. Molecular identification of isolates using ribosomal genes. 4. Colonisation efficiency and mass production tests can reduce the number of potential isolates, as well as detect latent pathogens. 5. Screening of endophytes against diseases in small *in planta* assays provides a better observation of potential candidates. 6. Successful isolates can be tested in field conditions on larger-scale experimental units in order to assess field consistency. 7. Potential biocontrol endophytes can enter an industrial phase or mass production and marketing or continued biological studies about their mode of action and ecological significance. (A black and white version of this figure will appear in some formats.)

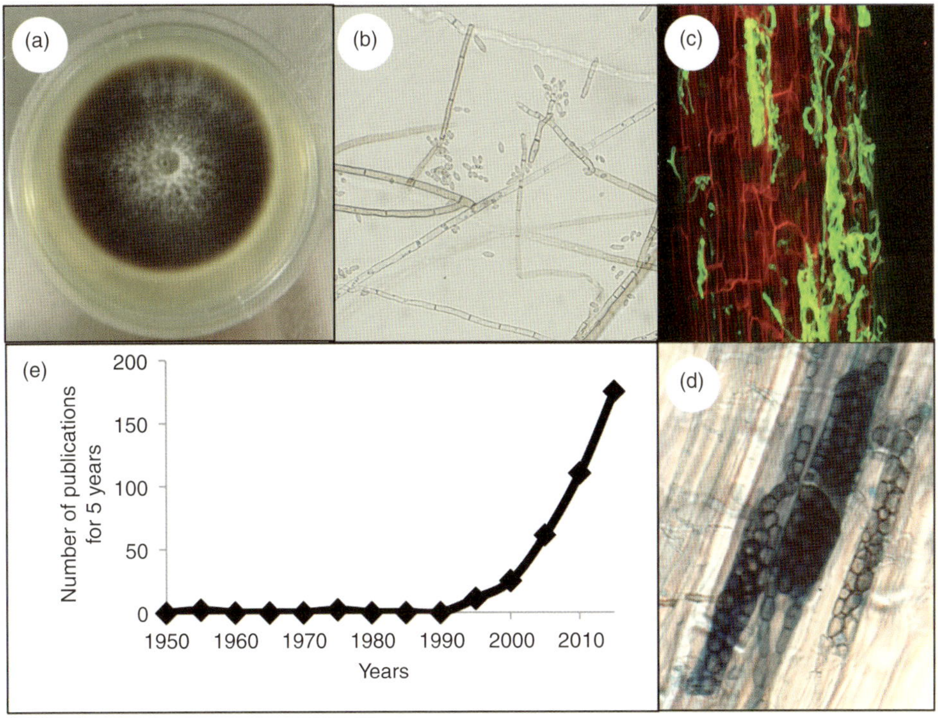

Figure 7.1 Typical characteristics of dark septate endophytes (DSEs) and emergence of DSE-related studies in the literature. (a) Two-week old colony of *Cadophora malorum* CBS 100584 on malt agar medium. (b) Septate hyphae and conidiophores of *Cladosporium cladosporioides* CBS101367. (c) Septate hyphae and microsclerotia of *Phialophora mustea* Pr27 in poplar roots; the fungal structures (green) were labelled with WGA-AF88. (d) Microsclerotia of *Leptodontidium* sp. Me10 in ryegrass (*Lolium*) roots. (e) Increase in the number of DSE-related publications since 1950. The survey was performed using Web of Science tools (http://apps.webofknowledge.com) using the following keywords: 'MRA'; 'DSE'; 'Dark septate endophytes'; and 'Mycelium Radicis Atrovirens.' (A black and white version of this figure will appear in some formats.)

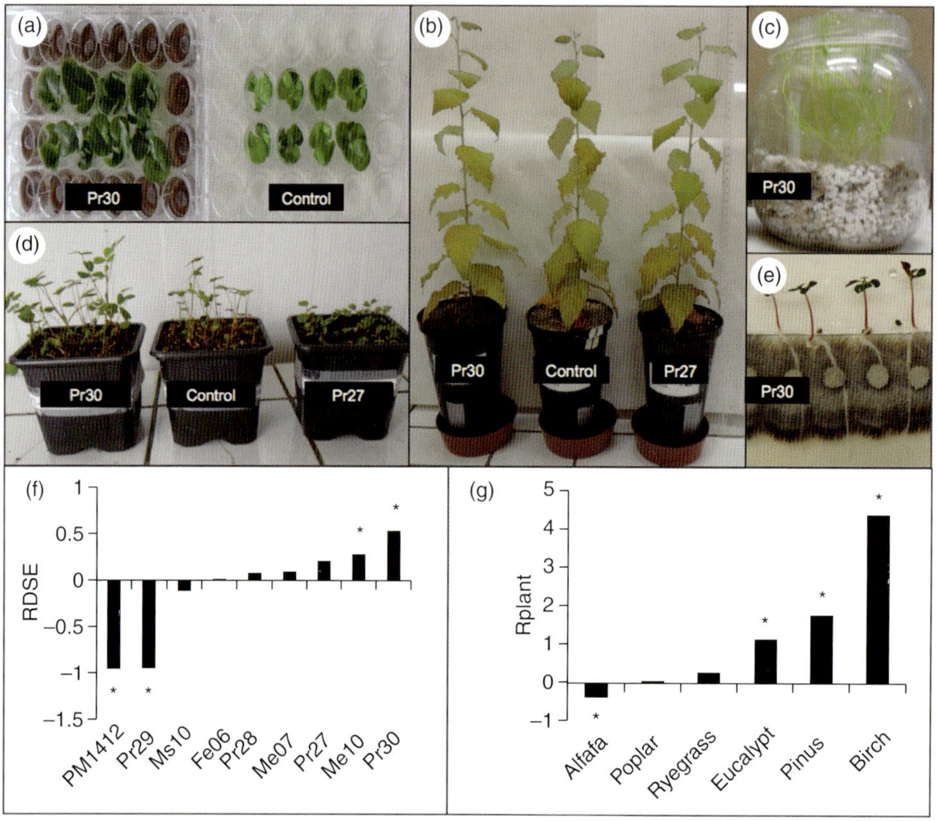

Figure 7.3 Different growth systems used to study DSE–plant interactions and the mutualism– parasitism paradigm. (a) 24-well culture plates used to test the growth-promoting effect of volatile organic compounds from *Leptodontidium* sp. Pr30 on tobacco in the absence of physical contact. (b) Axenic perlite system used to test the effect of *Leptodontidium* sp. Pr30 on ryegrass growth. (c, d) Contrasting effects of *Leptodontidium* sp. Pr30 and *Phialophora mustea* Pr27 on alfalfa (c) and poplar (d) grown in a sterilised trace element-polluted soil. (e) Bi-compartmentalisation plates used to co-culture eucalypt seedlings and *Phialophora mustea* Pr29. (f) Growth responses of ryegrass to colonisation by nine DSE strains. PMI_412, Pr30, Me07, Me10 belong to *Leptodontidium* sp., Pr27 and Pr29 belong to *Phialophora mustea*, Ms10 belongs to *Meliniomyces* sp., Fe06 belongs to *Cadophora* sp., and Pr28 belongs to *Phialocephala fortinii*. (g) Growth responses of six plant species to inoculation with *Phialophora mustea* Pr27. Response (RDSE or Rplant) to fungal inoculation indicates the ratio between the growth value of 'the inoculated minus the control treatments' to that of the control. Values above the *x*-axis indicate a positive response and values below the axis a negative response. Asterisks indicate significant differences between the control and inoculated plants (Kruskal–Wallis, $P < 0.05$). (A black and white version of this figure will appear in some formats.)

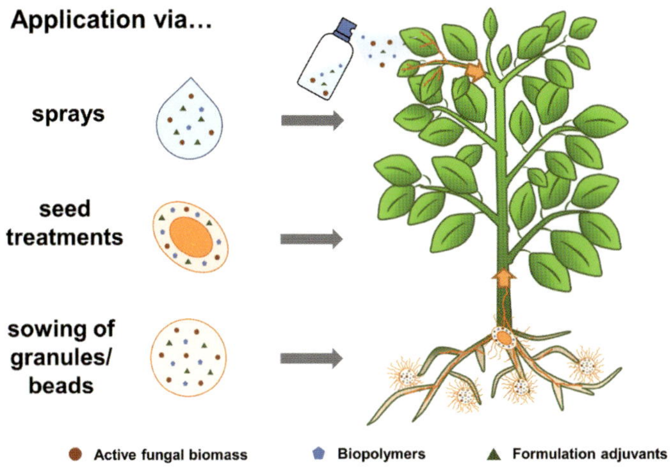

Figure 3.2 Overview of applications of formulated EEPF via sprays, seed treatments or sowing of granules and beads aiming at fungal colonisation of internal plant tissues. (A black and white version of this figure will appear in some formats.)

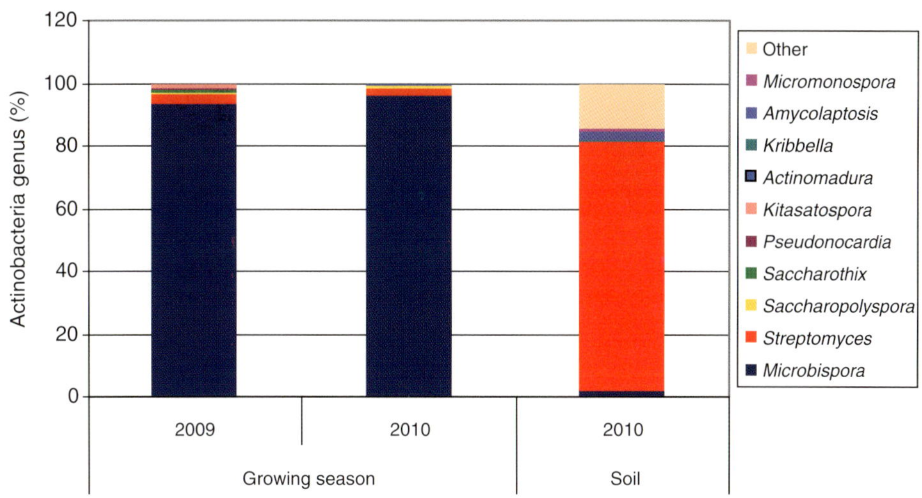

Figure 8.2 Distribution of endophytic Actinobacteria isolated from rice plants in growing season 2009 and 2010 and from soil. (A black and white version of this figure will appear in some formats.)

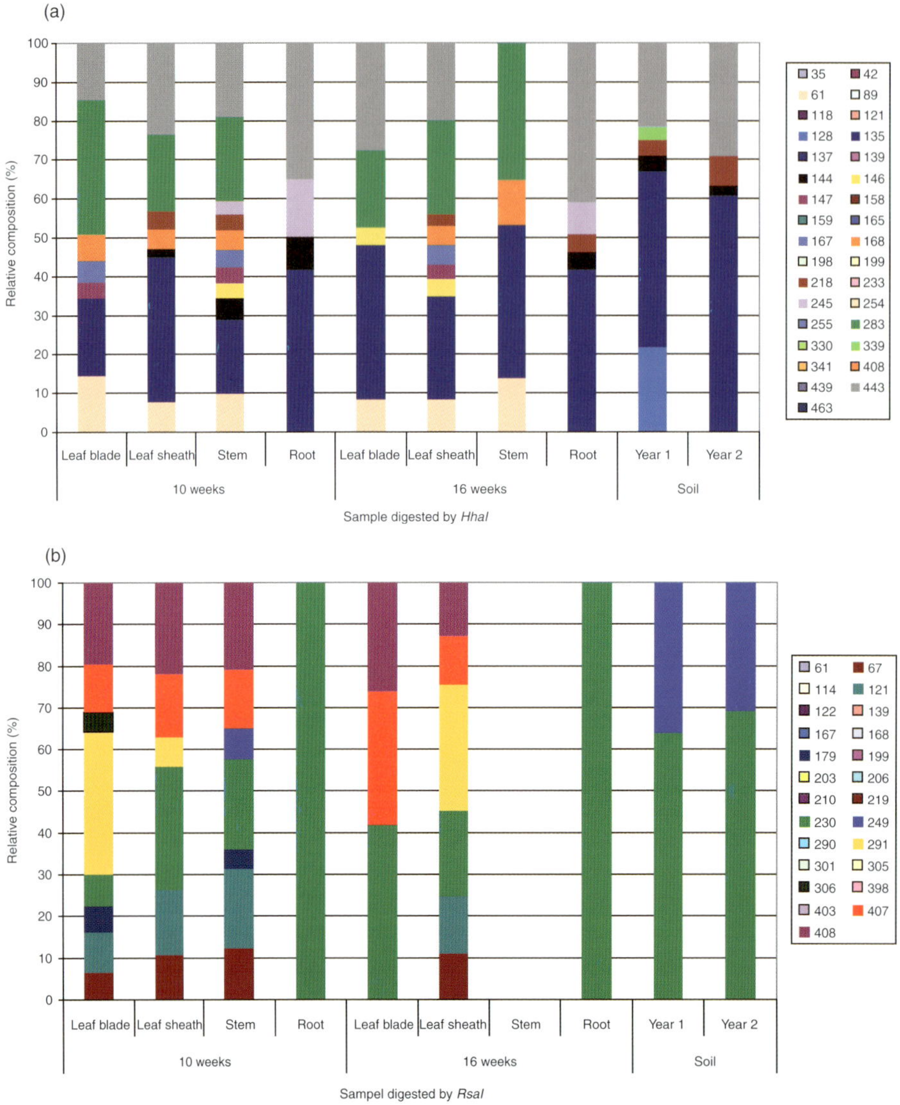

Figure 8.3 Terminal restriction fragments of actinobacterial communities from parts of 10 and 16 week old rice plants and soils after digestion with restriction enzymes: (a) *HhaI* and (b) *RsaI*. (A black and white version of this figure will appear in some formats.)

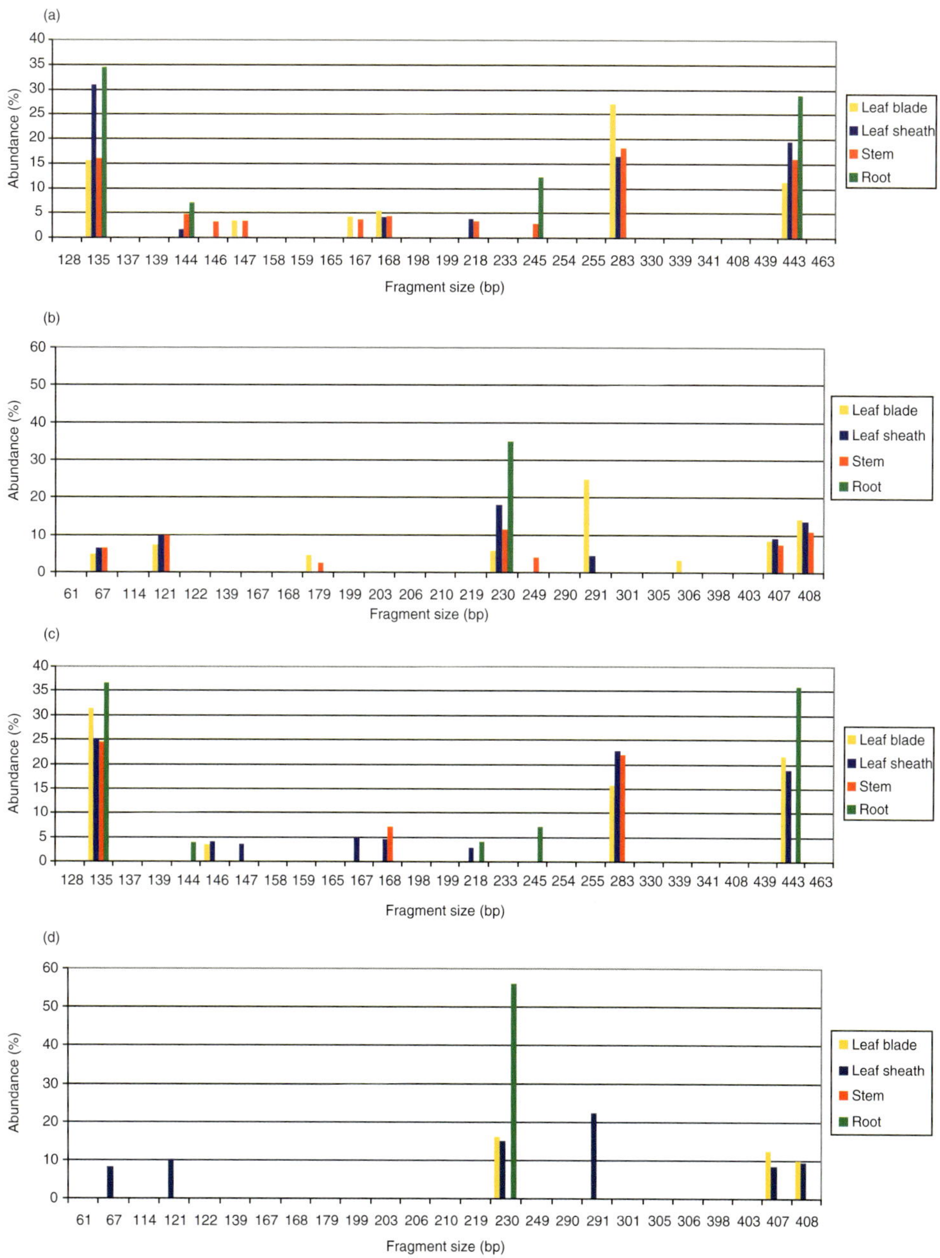

Figure 8.4 Terminal restriction fragments of actinobacterial communities from 10-week-old rice plants after digestion with restriction enzymes (a) *HhaI*, (b) *RsaI* and 16-week-old rice plants after digestion with restriction enzymes (c) *HhaI* and (d) *RsaI*. (A black and white version of this figure will appear in some formats.)

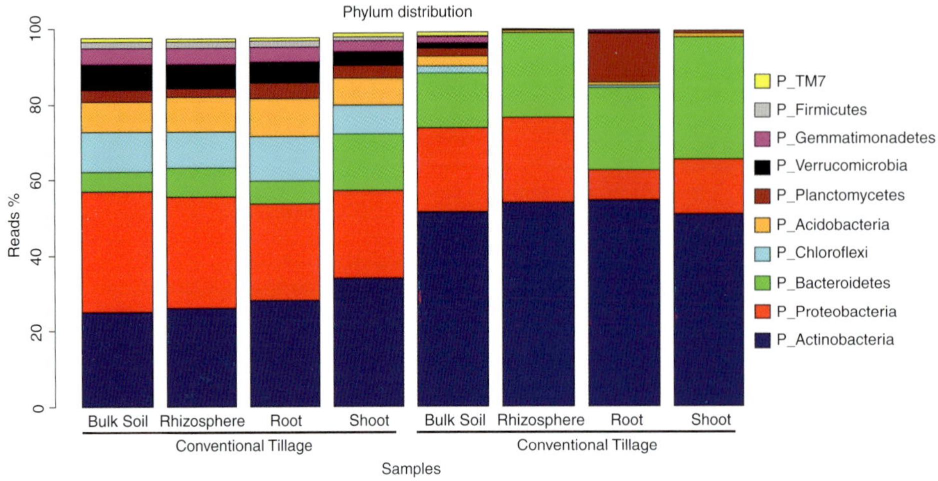

Figure 11.2 Relative abundance of the top 10 phyla in oilseed rape (OSR) in response to tillage system used. Average relative abundance (% of sequencing reads) of prokaryotic phyla associated with bulk soil, rhizosphere, root and shoot microhabitat zones of OSR under conventional tillage and conservation tillage, are displayed in different colours. (A black and white version of this figure will appear in some formats.)
Adapted from Rathore *et al.* (2017).

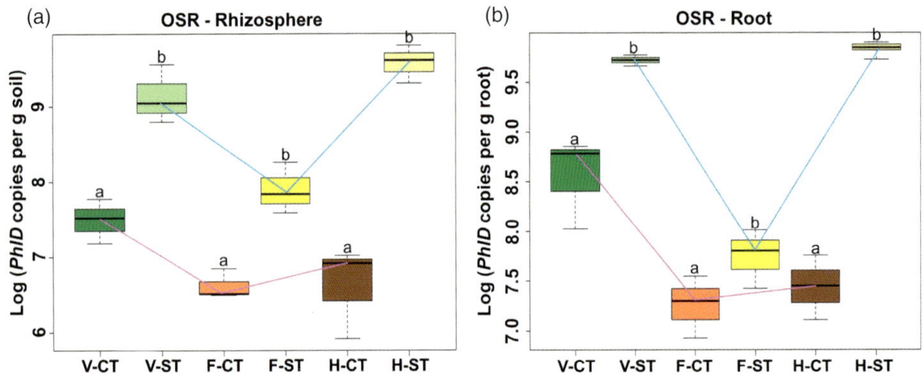

Figure 11.3 Dynamics of *PhlD* + bacterial community in OSR associated to rhizosphere and root microhabitats during the plant growth stages under crop rotation. Quantitative real-time PCR (qPCR) of *PhlD* + population in OSR (a) rhizosphere (b) root, based on the number of *PhlD* gene copies detected per gramme of rhizosphere soil and root sample respectively at three different plant developmental stages; vegetative (V), flowering (F), and harvesting (H) stages under conventional (CT) and conservation (ST) tillage practices. Darker colour defines the samples from CT and lighter colour defines the samples from ST. *Lowercase letters* denote statistically significant differences by Kruskal-Wallis and Dunn's *post hoc* tests, $P < 0.05$, Banjamini-Hochberg (BH) corrected between the tillage systems within the one plant developmental stage. (A black and white version of this figure will appear in some formats.) Rathore *et al.* (unpublished).

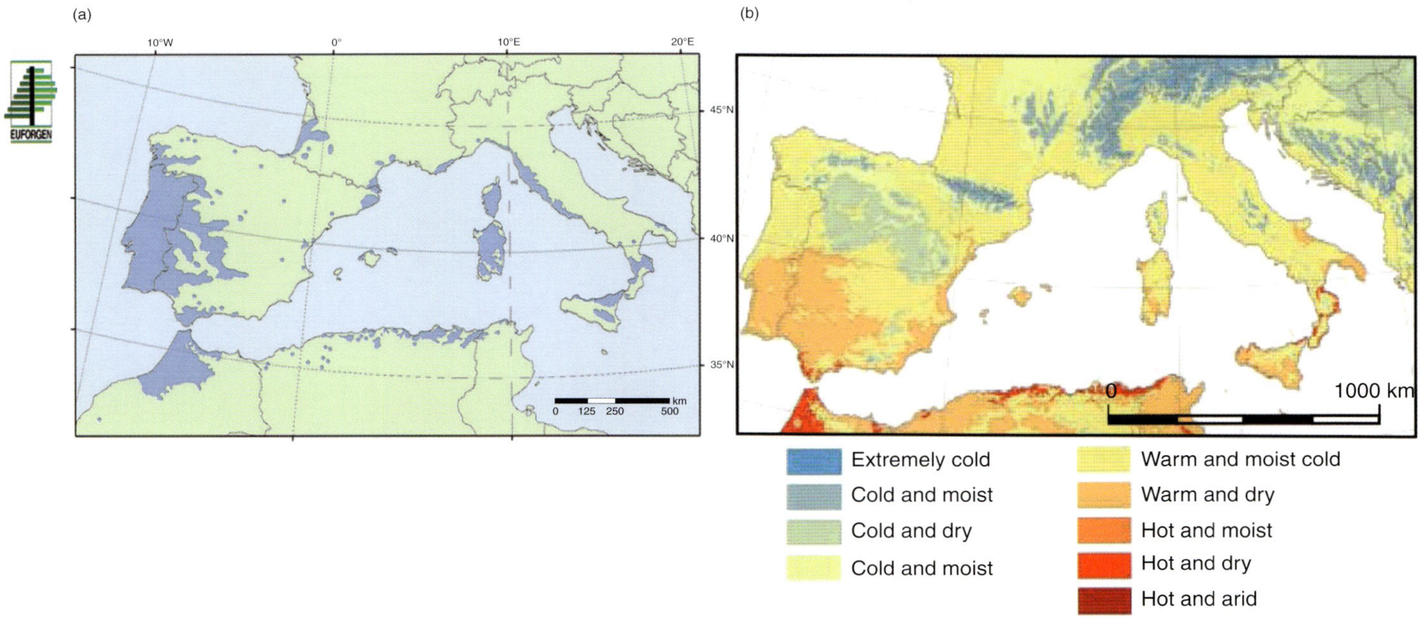

Figure 12.1 Map of Europe with (a) distribution area of *Quercus suber* within the Mediterranean region and (b) environmental zones based on five categories of temperature and two of aridity (adapted from EUFORGEN, 2009). (A black and white version of this figure will appear in some formats.)

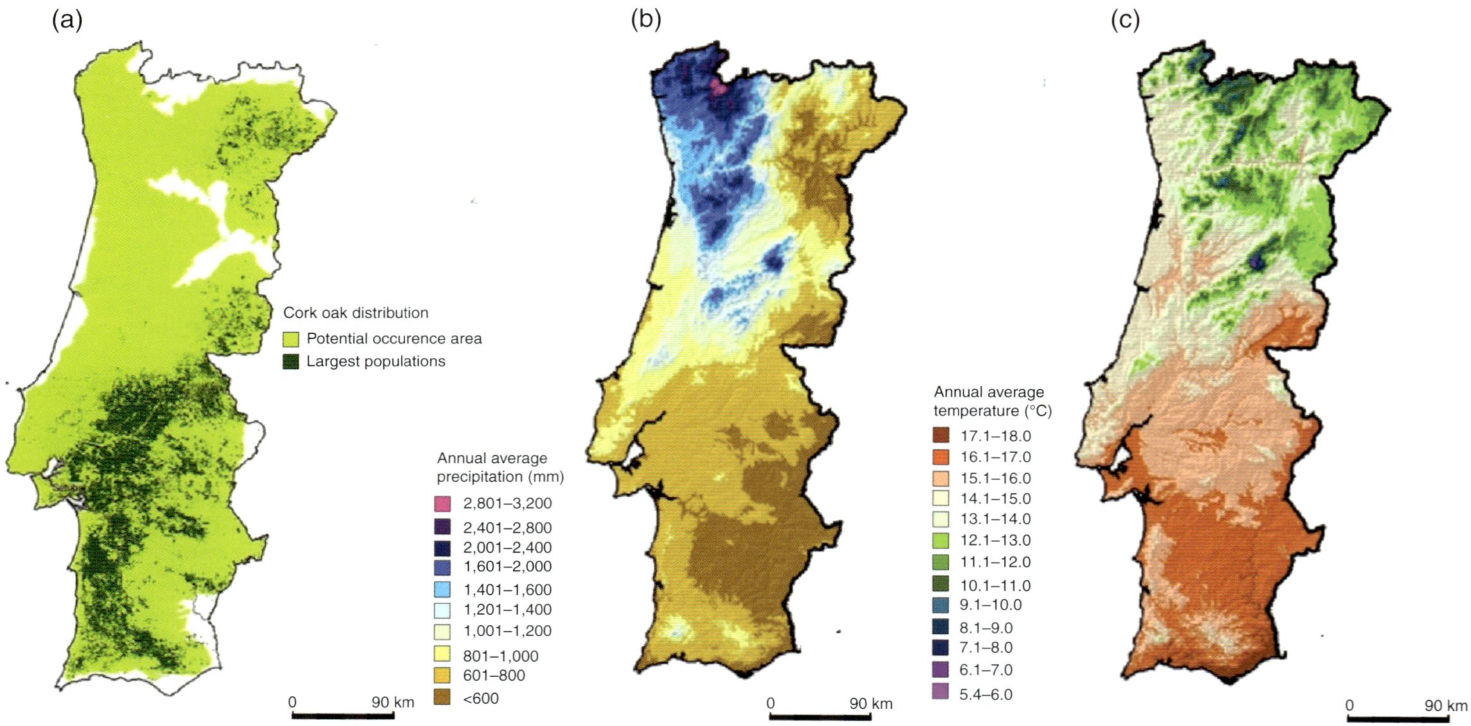

Figure 12.2 (a) Map of cork oak forest distribution in Portugal, (b) total annual precipitation and (c) annual medium temperature. (A black and white version of this figure will appear in some formats.)
(a) was adapted from Capelo and Catry (2007); (b) and (c) were adapted from IPMA (n.d.).

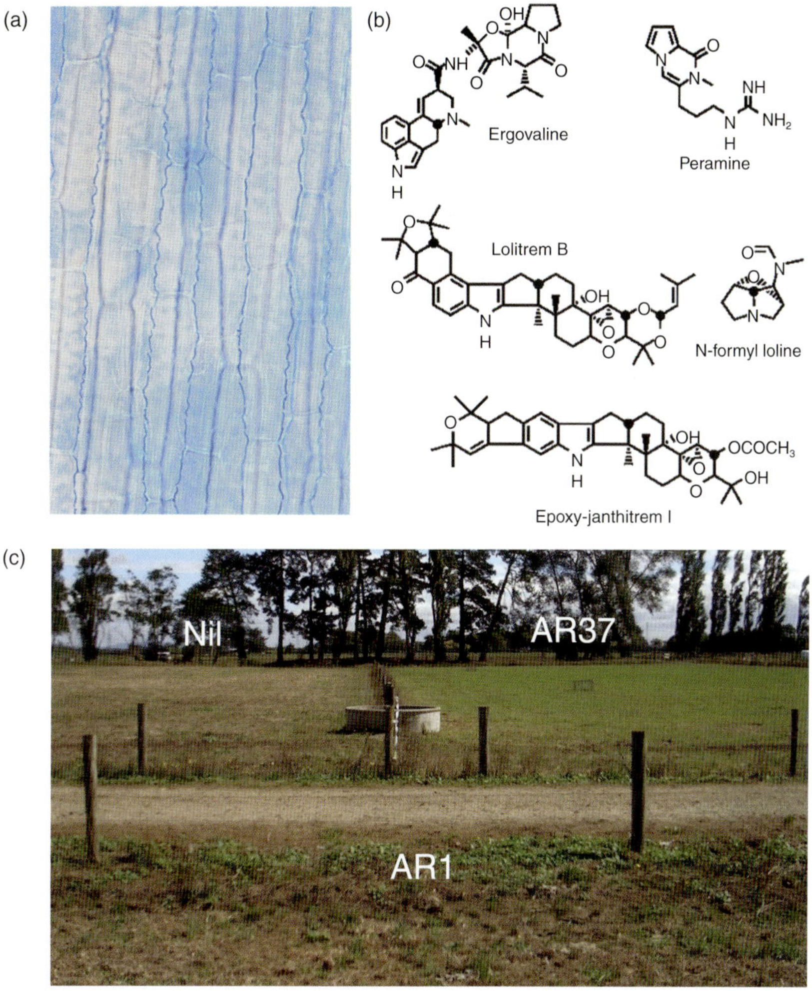

Figure 16.1 (a) An aniline blue-stained microscopy image showing the growth of *Epichloë festucae* in the outer leaf sheaths of perennial ryegrass. (b) Chemical structures of the five alkaloids produced by *Epichloë* endophytes. (c) The difference endophyte makes: pastures sown with perennial ryegrass inoculated without endophyte (Nil), or with AR37 endophyte (AR37) or with AR1 endophyte (AR1). (A black and white version of this figure will appear in some formats.)

Figure 16.2 Major insect pests of pasture grasses in New Zealand. ASW refers to Argentine stem weevil. (A black and white version of this figure will appear in some formats.)
Photos courtesy of AgResearch.

Lee, W. C. and Kuan, W. C. (2015). *Miscanthus* as cellulosic biomass for bioethanol production. *Biotechnology Journal*, **10**, 840–854.

Leme, A. C., Bevilaqua, M. R. R., Rhoden, S. A. *et al.* (2013). Molecular characterization of endophytes isolated from *Saccharum* spp based on esterase and ribosomal DNA (ITS1-5.8S-ITS2) analyses. *Genetics and Molecular Research*, **12**, 4095–4105.

Malinowski, D. P., Alloush, G. A. and Belesky, D. P. (2000). Leaf endophyte *Neotyphodium coenophialum* modifies mineral uptake in tall fescue. *Plant and Soil*, **227**, 115–126.

Martín, J. A., Macaya-Sanz, D. and Witzell, J. (2015). Strong *in vitro* antagonism by elm xylem endophytes is not accompanied by temporally stable in planta protection against a vascular pathogen under field conditions. *European Journal of Plant Pathology*, **142**, 185–196.

McGrath, S., Hodkinson, T. R. and Barth, S. (2007). Extremely high cytoplasmic diversity in natural and breeding populations of *Lolium* (Poaceae). *Heredity*, **99**, 531–544.

Mejía, L. C., Herre, E. A., Sparks, J. P. *et al.* (2014). Pervasive effects of a dominant foliar endophytic fungus on host genetic and phenotypic expression in a tropical tree. *Frontiers in Microbiology*, **5**, 479.

Mekete, T., Sikora, R. A., Kiewnick, S. and Hallmann, S. (2011). Description of plant parasitic nematodes associated with coffee in Ethiopia. *Nematologia Mediterranea*, **36**, 69–77.

Moreira, A. S., Germaine, K. J., Lloyd, A. *et al.* (2016). Draft genome sequence of three endophyte strains of *Pseudomonas fluorescens* isolated from *Miscanthus giganteus*. *Genome Announcements*, **4**(5), e00965-16.

Moritz, C. and Cicero C. (2004). DNA barcoding: promise and pitfalls. *PLoS Biology*, **2**, e354.

Murphy, B. R., Doohan, F. M. and Hodkinson, T. R. (2014). Persistent fungal root endophytes isolated from a wild barley species suppress seed-borne infections in a barley cultivar. *BioControl*, **60**, 281–292.

Murphy, B. R., Batke, S. P., Doohan, F. M. and Hodkinson. T. R. (2015a). Media manipulations and the culture of beneficial fungal root endophytes. *International Journal of Biology*, **7**, 94–102.

Murphy, B. R., Doohan, F. M. and Hodkinson, T. R. (2015b). Fungal root endophytes of a wild barley species increase yield in a nutrient-stressed barley cultivar. *Symbiosis*, **65**, 1–7.

Murphy, B. R., Martin Nieto, L., Doohan, F. M. and Hodkinson, T. R. (2015c). Fungal endophytes enhance agronomically important traits in severely drought-stressed barley. *Journal of Agronomy and Crop Science*, **201**, 419–427.

Murphy, B. R., Martin Nieto, L., Doohan, F. M. and Hodkinson, T. R. (2015d). *Profundae diversitas*: the uncharted genetic diversity in a newly studied group of fungal root endophytes. *Mycology*, **6**, 139–150.

Murphy, B. R., Doohan, F. M. and Hodkinson, T. R. (2018). From concept to commerce: developing a successful fungal endophyte inoculant for agricultural crops. *Journal of Fungi*, **4**, 24, 2–11.

Nutaratat, P., Srisuk, N., Arunrattiyakorn, P. and Limtong, S. (2014). Plant growth-promoting traits of epiphytic and endophytic yeasts isolated from rice and sugar cane leaves in Thailand. *Fungal Biology*, **118**, 683–694.

O'Neil, N. R. and Farr, D. F. (1996). *Miscanthus* blight, a new foliar disease of ornamental grasses and sugarcane incited by *Leptosphaeria* sp. and its anamorphic state *Stagonospora* sp.. *Plant Disease*, **80**, 980–987.

Pandey, R. R., Arora, D. K. and Dubey, R. C. (1993). Antagonistic interactions between fungal pathogens and phylloplane fungi of guava. *Mycopathologia*, **124**, 31–39.

Perdereau, A., Klaas, M., Barth, S. and Hodkinson, T. R. (2017). Plastid genome sequencing reveals biogeographical structure and extensive population genetic variation in wild populations of *Phalaris arundinacea* L. in north-western Europe. *GCB Bioenergy*, **9**, 46–56.

Rehner, S. A. and Buckley, E. (2005). A *Beauveria* phylogeny inferred from nuclear ITS and EF1-α sequences: evidence for cryptic diversification and links to *Cordyceps* teleomorphs. *Mycologia*, **97**, 84–89.

Saikkonen, K., Young, C. A., Helander, M. and Schardl, C. L. (2016). Endophytic Epichloë species and their grass hosts: from evolution to applications. *Plant Molecular Biology*, **90**, 665–675.

Salamin, N., Hodkinson, T. R. and Savolainen, V. (2005). Towards building the tree of life: a simulation study for all angiosperm genera. *Systematic Biology*, **54**, 183–196.

Scauflaire, J., Gourgue, M., Foucart, G. *et al.* (2013). *Fusarium miscanthi* and other *Fusarium* species as causal agents of *Miscanthus* × *giganteus* rhizome rot in Belgium. *European Journal of Plant Pathology*, **137**, 1-3.

Schardl, C. L. (2001). *Epichloë festucae* and related mutualistic symbionts of grasses. *Fungal Genetics and Biology*, **33**, 69–82.

Schmidt, C. S., Mrnka, L., Frantík, T. *et al.* (2017). Combined effects of fungal inoculants and the cytokinin-like growth regulator thidiazuron on growth, phytohormone contents and endophytic root fungi in *Miscanthus* × *giganteus*. *Plant Physiology and Biochemistry*, **120**, 120e131.

Schoch, C. L., Seifert, K. A., Huhndorf, S. *et al.* (2012). Fungal barcoding consortium. *Proceedings of the National Academy of Sciences of the United States of America*, **109**, 6241–6246.

Sharma, G. and Pandey, R. R. (2010). Influence of culture media on growth, colony character and sporulation of fungi isolated from decaying vegetable wastes. *Journal of Yeast and Fungal Research*, **1**, 157–164.

Shrestha, P., Ibáñez, A. B., Bauer, S. *et al.* (2015). Fungi isolated from *Miscanthus* and sugarcane: biomass conversion, fungal enzymes, and hydrolysis of plant cell wall polymers. *Biotechnology for Biofuels*, **8**, 38.

Stielow, J. B., Lévesque, C. A., Seifert, K. A. *et al.* (2015). One fungus, which genes? Development and assessment of universal primers for potential secondary fungal DNA barcodes, *Persoonia*, **22**, 242–263.

Straub, D., Rothballer, M., Hartmann, A. and Ludewig, U. (2013a). The genome of the endophytic bacterium H.frisingense GSF30T identifies diverse strategies in the Herbaspirillum genus to interact with plants. *Frontiers in Microbiology*, **4**,168.

Straub, D., Yang, H., Liu, Y., Tsap, T. and Ludewig, U. (2013b). Root ethylene signalling is involved in *Miscanthus sinensis* growth promotion by the bacterial endophyte *Herbaspirillum frisingense* GSF30T. *Journal of Experimental Botany*, **64**, 4603–4615.

Stuart, R. M., Romao, A. S., Pizzirani-Kleiner, A. A., Azevedo, J. L. and Araujo, W. L. (2010). Culturable endophytic filamentous fungi from leaves of transgenic imidazolinone-tolerant sugarcane and its non-transgenic isolines. *Archives of Microbiology*, **192**, 307–313.

Teerawatananon, A., Jacobs, S. W. L. and Hodkinson, T. R. (2011). Phylogenetics of Panicoideae (Poaceae) based on chloroplast and nuclear DNA sequences. *Telopea*, **13**,115–142.

Vega, F. E., Posada, F., Aime, M. C. *et al.* (2008). Entomopathogenic fungal endophytes. *Biological Control*, **46**, 72–82.

Verma, S., Varma, A., Rexer, K. *et al.* (1998). *Piriformospora indica*, gen. et sp. nov., a new root-colonizing fungus. *Mycologia*, **90**, 896–903.

Vilgalys, R. and Hester, M. (1990). Rapid genetic identification and mapping of enzymatically amplified ribosomal DNA from several *Cryptococcus* species. *Journal of Bacteriology*, **172**, 4238–4246.

Waller, F., Achatz, B., Baltruschat, H. *et al.* (2005). The endophytic fungus *Piriformospora indica* reprograms barley to salt-stress tolerance, disease resistance, and higher yield. *Proceedings of the National Academy of Sciences of the United States of America*, **102**, 13386–13391.

Wang, Y., Li, H., Feng, G., Du, L. and Zeng, D. (2017). Biodegradation of diuron by an endophytic fungus *Neurospora intermedia* DP8-1 isolated from sugarcane and its potential for remediating diuron-contaminated soils. *PLoS One*, **12,** e0182556.

White, T. J., Bruns, T., Lee, S. and Taylor, J. W. (1990). Amplification and direct sequencing of fungal ribosomal RNA genes for phylogenetics. In *PCR Protocols: A Guide to Methods and Applications*, ed. M. A. Innis, D. H. Gelfand, J. J., Sninsky and T. J. White. New York: Academic Press Inc., pp. 315–322.

Wilson, R.T. (1995). *Livestock Production Systems.* Umberleigh, UK: Bartridge Partners.

Xue, S., Kalinina, O. and Lewandowski, I. (2015). Present and future options for the improvement of *Miscanthus* propagation techniques. *Renewable & Sustainable Energy Reviews*, **49**, 1233–1246.

Yahr, R., Schoch, C. L. and Dentinger, B. T. M. (2016). Scaling up discovery of hidden diversity in fungi: impacts of barcoding approaches. *Philosophical Transactions of the Royal Society B*, **371**, 20150336.

Yost, M. A., Randall, B. K., Kitchen, N. R., Heaton, E. A. and Myers, R. L. (2017). Yield potential and nitrogen requirements of *Miscanthus* × *giganteus* on eroded soil. *Agronomy Journal*, **109**, 684–695.

You, Y.-H., Yoon, H.-J., Woo, J.-R. *et al.* (2011). Plant growth-promoting activity of endophytic fungi isolated from the roots of native plants in Dokdo Islands. *Journal of Life Science*, **21**, 1619–1624.

10

Life Within the Leaf: Ecology and Applications of Foliar Bacterial Endophytes

C. Ruth McNees, Isaac V. Greenhut, Audrey D. Law, Muhammad Saleem and Luke A. Moe

Abstract

Endophytic bacteria have evolved to survive within plant leaf tissue while potentially providing benefits to their host. This relationship makes them uniquely applicable as agricultural biocontrol agents, sources of natural chemicals/products, plant growth promoters and mediators of phytoremediation. Foliar bacterial endophytes colonise leaf tissue through vertical transmission (i.e. through seed or pollen) or horizontal transmission (i.e. colonisation through environmental contact of the roots, wounds, stomatal infiltration, insect vectors, and airborne dispersal). The taxonomic diversity of foliar endophytes spans at least seven bacterial classes and they occupy hosts from all taxonomic groups of plants tested. Bacterial leaf endophytes can promote plant health by stimulating and producing plant hormones for growth and preventing pathogenic infection. Plant-pathogenic bacteria can be found residing within leaf tissue asymptomatically raising questions about the relationship between endophyte-host specificity. Similarly, human pathogenic enterobacteria not usually associated with plants have been found to persist as endophytes. Bioactive secondary metabolites produced by these endophytes have been broadly applicable as antifungals, antibiotics, and other compounds used for agricultural and human health. Endophyte research carries unique challenges that require novel and adaptive strategies for separation of plant and bacterial DNA. This chapter will focus on bacteria isolated from within plant leaf tissue with a focus on transmission, diversity, function and challenges associated bacterial leaf endophyte research.

10.1 Introduction

While there have been many definitions of endophytes, translating from the Greek 'endo' meaning 'within' and 'phyte' meaning 'plant' would suggest that any organism living within plant tissues be considered an endophyte (Wilson, 1995). The original use of the term endophyte was applied to fungi found within plant tissues, regardless of whether the microbe was pathogenic (Wennstrom, 1994). More recently, however, the term has been typically applied to non-pathogenic microbes. Identification of a 'true' endophyte usually requires isolation from surface-sterilised plant tissue and visualisation within plant tissues using microscopy, while microorganisms that are not observed by microscopy should be considered 'putative' endophytes (Rosenblueth and Martinez-Romero, 2006). A more modern, general grouping of endophytes would include all microbes (bacterial, archaeal and fungal) capable of colonising plant tissues without causing pathogenic effects and sharing a common dependence on the host plant for some part of their life cycle (Sumner, 2000). It is possible that due to the inadequate methods for identifying endophytes, some of the microbes originally identified as endophytes should not be classified as such (Rosenblueth and Martinez-Romero, 2006).

Analysis of the microbiome associated with the aerial parts of the plant, known as the phyllosphere, has revealed complex biochemical interactions between endophytes and plant hosts (Vorholt, 2012). Plant microbiome analyses support the hypothesis that plants may be considered 'super-organisms' capable of supporting diverse microbial communities, with each type of tissue (e.g. flower, leaf, root, seed, stalk and stem) supporting its own characteristic microbiome (Strobel and Daisy, 2003; Saleem *et al.*, 2016). While most research has been carried out on endophytes with beneficial or deleterious effects on the host, the majority of endophytes have either no known role or elicit no obvious response from the host plant, or can have a range of effects depending on a variety of host factors (Saleem *et al.*, 2017).

10.2 Transmission of Endophytes

Endophytes reside within all plants, though the abundance and diversity of the microbial population varies significantly based on a number of factors, including environmental conditions, plant species and tissue type (Strobel and Daisy, 2003). In a single study by Qin *et al.* (2012) to compare culture-dependent and -independent methods of endophyte identification, it was discovered that DNA sequences obtained by culture-independent means were more abundant from leaf samples, while cultivable isolates were in the greatest number from the roots and stems. The endophyte species composition within each type of tissue sampled (leaf, root, and stem) varied significantly (Qin *et al.*, 2012). Regardless of the plant–bacterial

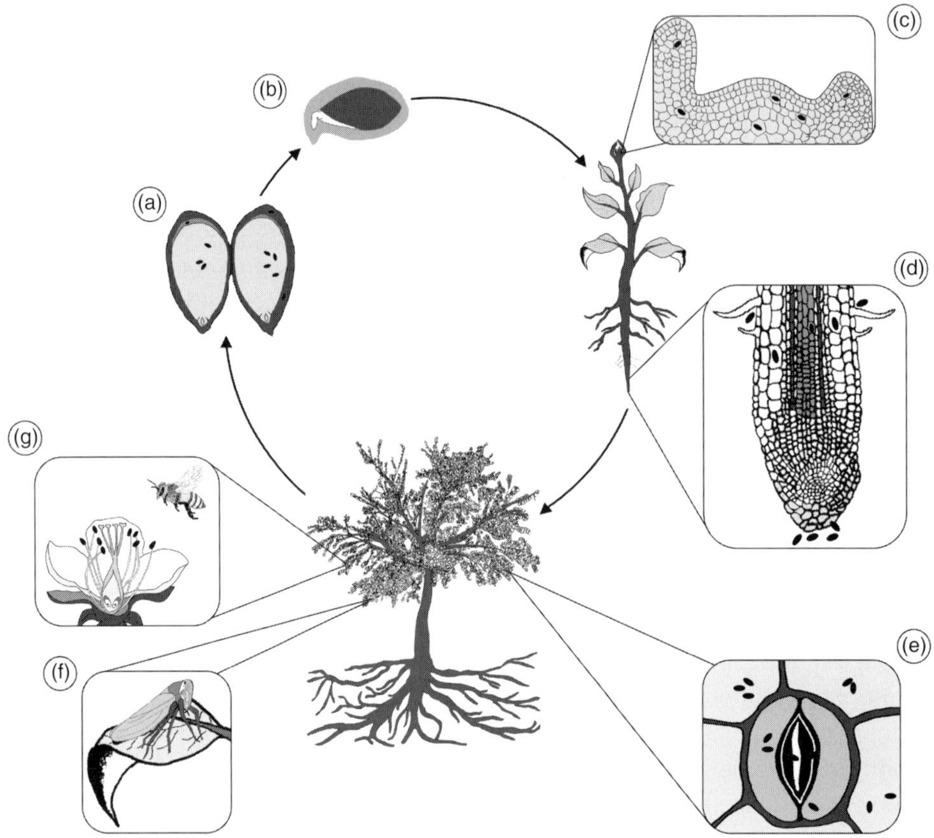

Figure 10.1 Transmission routes for bacterial endophytes: (a) vertical transmission via seed; (b) colonisation of the spermosphere; (c) colonisation of developing reproductive organs via the shoot apical meristem as part of vertical transmission; (d) colonisation of root from soil; (e) colonisation of leaves though stomata; (f) transmission via sap-feeders; (g) transmission to flowers via pollinators. Not drawn to scale.
Figure reprinted with permission from Frank *et al.* (2017).

endophyte relationship, bacterial colonisation of plant tissue occurs in one of two routes: vertical or horizontal transmission (Figure 10.1) (Frank *et al.*, 2017). Vertical transmission relies on inheritance of bacteria from the parent plant, which thus renders a higher probability of selective co-evolution. Alternatively, horizontal transmission relies on bacteria colonising plant tissue after freely living in the environment for some part of their life cycle (Bright and Bulgheresi, 2010).

10.2.1 Vertical Transmission

One early culture-dependent study of the ovules and seeds from 27 species of plants revealed 19 genera of bacteria (Mundt and Hinkle, 1976). Since then, vertical transmission has, for example, been shown in endophytes associated with *Eucalyptus*, where endophytes found in seeds vertically colonise developing seedlings (Ferreira *et al.*, 2008). Analysis of wheatgrass varieties grown in different growth media also confirmed vertical transmission of endophytes, and revealed that the matrix surrounding the roots of the plants had a significant influence on overall endophytic population of the mature plant (Ringelberg *et al.*, 2012). Furthermore, leaf endophytes have been shown to be transmitted to subsequent generations via seed colonisation of switchgrass grown in the field (Gagne-Bourgue *et al.*, 2013). These studies, and others, show that vertical transmission occurs naturally, providing evidence suggestive of bacterial plant host specificity. In seed endophytes, vertical transmission to the pollen, ovule/egg or seed occurs during development (Bright and Bulgheresi, 2010). Using green fluorescent protein (GFP)-labelled strains, it has been demonstrated that endophytes can be transmitted through seeds to roots as key members of the root microbiome (Johnston-Monje and Raizada, 2011; Mitter *et al.*, 2017). It should be noted that not all bacteria associated with seeds are derived from the parent, nor are they guaranteed to colonise the new seedling following germination (Bright and Bulgheresi, 2010). Seed colonisation is influenced by the host plant and soil. Chee-Sanford *et al.* (2006) showed that seeds from different plant species resulted in unique microbial populations following exposure to soil containing a bacterial inoculum.

10.2.2 Horizontal Transmission

Horizontal transmission of endophytes across the leaves of plants can occur via plant exposure to environmental bacteria through airborne transmission or infection through insect or herbivore attack (Bright and Bulgheresi, 2010). The extent of horizontal transmission and the environment in which the plants are grown significantly impacts the endophyte microbial community structure (Bright and Bulgheresi, 2010). Invasion of root tissue by rhizosphere bacteria at the developing lateral root is also an extensively studied route for colonisation (Reinhold-Hurek and Hurek, 2011). Once bacteria colonise a plant, they can penetrate the vascular tissue for movement to other locations using xylem sap, at recorded concentrations of 10^2 to 10^3 colony forming units (CFU)/ml (Bell *et al.*, 1995).

There are several factors that determine the successful establishment of a root endophyte (Ahmad *et al.*, 2011). Motility is required by some endophytes and root derived polysaccharides are considered important chemoattractants which facilitate root colonisation in some species (Bashan and Holguin, 1994; You *et al.*, 1995; Santoyo *et al.*, 2016). Root wounds also release metabolites capable of supporting microbial growth and are attractive sites for microbial entrance (Sprent and Defaria, 1988; Hallmann *et al.*, 1997b; Sørensen, 2007; Hardoim *et al.*, 2015; Santoyo *et al.*, 2016). Some bacteria (e.g. *Enterobacter asburiae* JM22) hydrolyse plant cell wall-bound cellulose, in a process that enables microbial colonisation of the root system (Hallmann *et al.*, 1997b). *Burkholderia* produce cell wall degrading enzymes (endoglucanase and endopolygalacturonase) which can facilitate access to internal root tissue, shown by the accumulation of phenolic compounds at the site of colonisation (Compant *et al.*, 2005).

Inoculation of plants with specific bacteria can lead to internalisation and translocation following colonisation of the root as demonstrated in a study by Ruppel *et al.* (1992) where some, but not all, bacterial species are capable of migrating to leaf tissues. Bacteria can use the transpiration stream within the stem for phyllosphere colonisation (Compant *et al.*, 2005). Analysis of tobacco plants colonised with multiple GFP-tagged rhizobia at the root, revealed migration through the plants and release into the environment through the stomata (Ji *et al.*, 2010). Just as bacteria can be released through stomata, leaf colonisation can also occur through stomatal infiltration. Bacteria can enter the stomata, replicate and emerge later, dispersing in masses (Roos and Hattingh, 1983). Consequently, plants, especially field-grown crops, can be a significant source of airborne bacteria (Lindemann *et al.*, 1982). *Azospirillum*, a plant growth-promoting sporeless rhizosphere bacterium, was shown to be transmitted by wind between infected and non-infected plants in greenhouse and field experiments (Bashan, 1991). As such, weather can have a significant impact on the distance that bacteria can travel in nature (Bovallius *et al.*, 1978).

A number of insects are known to act as transmission vectors through plant herbivory. For example, *Bucephalogonia xanthophis* is capable of transmitting the pathogenic bacterium *Xyllela fastidiosa*, as well as the endophytic *Methylobacterium mesophilicum* in *Catharanthus roseus* (Gai *et al.*, 2009). The American grapevine leafhopper, *Scaphoideus titanus*, is a phloem feeding insect capable of transmitting *Cardinium* sp. to leaf tissues following piercing of the plant tissue (Gonella *et al.*, 2015). *Homalodisca coagulata* has been shown to acquire *Alcaligenes xylosoxidans* subsp. *denitrificans* from plant material and transmit the bacteria to a variety of plant types (Bextine *et al.*, 2005). Furthermore, Lopez-Fernandez *et al.* (2017) showed that microbial communities can be transmitted and altered by insect vectors between host grapevine plants.

10.3 Diversity

There are many factors that determine the structure and diversity of the root endophytic microbiome, such as plant genotype, microbial seed bank, soil physical and chemical properties and climatic changes (Vorholt, 2012; Saleem *et al.*, 2017; Saleem *et al.*, 2018). Monitoring the leaf endophyte population over the course of a growing season has revealed that microbial diversity is most influenced by the rhizosphere microbes at the beginning of the season and shifts to a less diverse, more leaf-specific bacterial population with time (Copeland *et al.*, 2015).

Leaf-specific microbiome analyses have revealed a diverse but limited range of bacteria capable of colonising internal leaf tissue (Ruppel *et al.*, 1992; Kirchhof *et al.*, 2001; Chi *et al.*, 2005; Lin *et al.*, 2005; Reiter and Sessitsch, 2006; Mano *et al.*, 2007; Izumi *et al.*, 2008; Bhore *et al.*, 2010; Lin *et al.*, 2010; Magnani *et al.*, 2010; Costa *et al.*, 2012; Gagne-Bourgue *et al.*, 2013; Ren *et al.*, 2013; Zhang *et al.*, 2017). Culture-dependent and -independent studies have thus far revealed at least 42 bacterial genera within six bacterial classes, including nine genera from *Alphaproteobacteria*, six from *Betaproteobacteria*, and eight from *Gammaproteobacteria*, 12 genera from *Actinobacteria*, and six from *Bacilli* (Figure 10.2) as well as one strain from the *Sphingobacteria* (Costa *et al.*, 2012). Due to the importance of monocots for agricultural production of staple grains, substantial research has been undertaken to determine the microbial community members that may contribute to development, health, and yield of crop species. Bertani *et al.* (2016) found 19 bacterial genera in leaves of rice (*Oryza sativa*), all of which were also found in rhizosphere communities except for *Micrococcus* and *Enterococcus* spp. Comparison of rhizosphere and aerial microbiota showed foliar bacterial loads numbered 10^5–10^6 CFU/g of leaf tissue, with the greatest differences in species richness between leaves and roots caused by *Actinobacteria* genera, which were more diverse in leaves, but in greater abundance in roots (Bertani *et al.*, 2016). A study by Mano *et al.* (2007) examined changes in community dynamics in rice over progressive growth stages, and found that the bacterial load for Gram-negative genera such as *Pantoea* and *Methylobacterium* increased in leaves as the plants matured, but decreased in roots, while conversely Gram-positive genera such as *Curtobacterium* decreased in leaves but increased in roots. Among other examples of foliar endophytes isolated from monocot crops, Magnani *et al.* (2010) isolated 32 strains, including representatives of *Citrobacter, Curtobacterium, Enterobacter, Klebsiella, Pantoea, Pseudomonas,* and *Staphylococcus*, from leaves and stems of sugarcane (*Saccharum*). Four of these were identified as putative, non-nodule forming nitrogen fixers, based on their ability to reduce acetylene to ethylene. The nitrogen-fixing endophyte, *Herbaspirillum frisingense* was isolated from the biomass and bioenergy crop *Miscanthus sacchariflorus* for its diazotrophic capability (Kirchhof *et al.*, 2001), and Gagne-Bourgue *et al.* (2013) found isolates of *Bacillus, Curtobacterium, Microbacterium, Pseudomonas, Sphingomonas* and *Serratia* in leaves of the bioenergy crop switchgrass (*Panicum virgatum*) that produced lipopeptide toxins and phytohormones. Soil inoculation of selected endophytes showed migration into root tissue and subsequently into aerial plant parts (Gagne-Bourgue *et al.*, 2013).

In addition to the monocot hosts described above, several foliar endophytes have been isolated from annual dicots such as *Phaseolus vulgaris* (Figure 10.2), as well as several woody perennial species. Research on bacterial endophytes of poplar

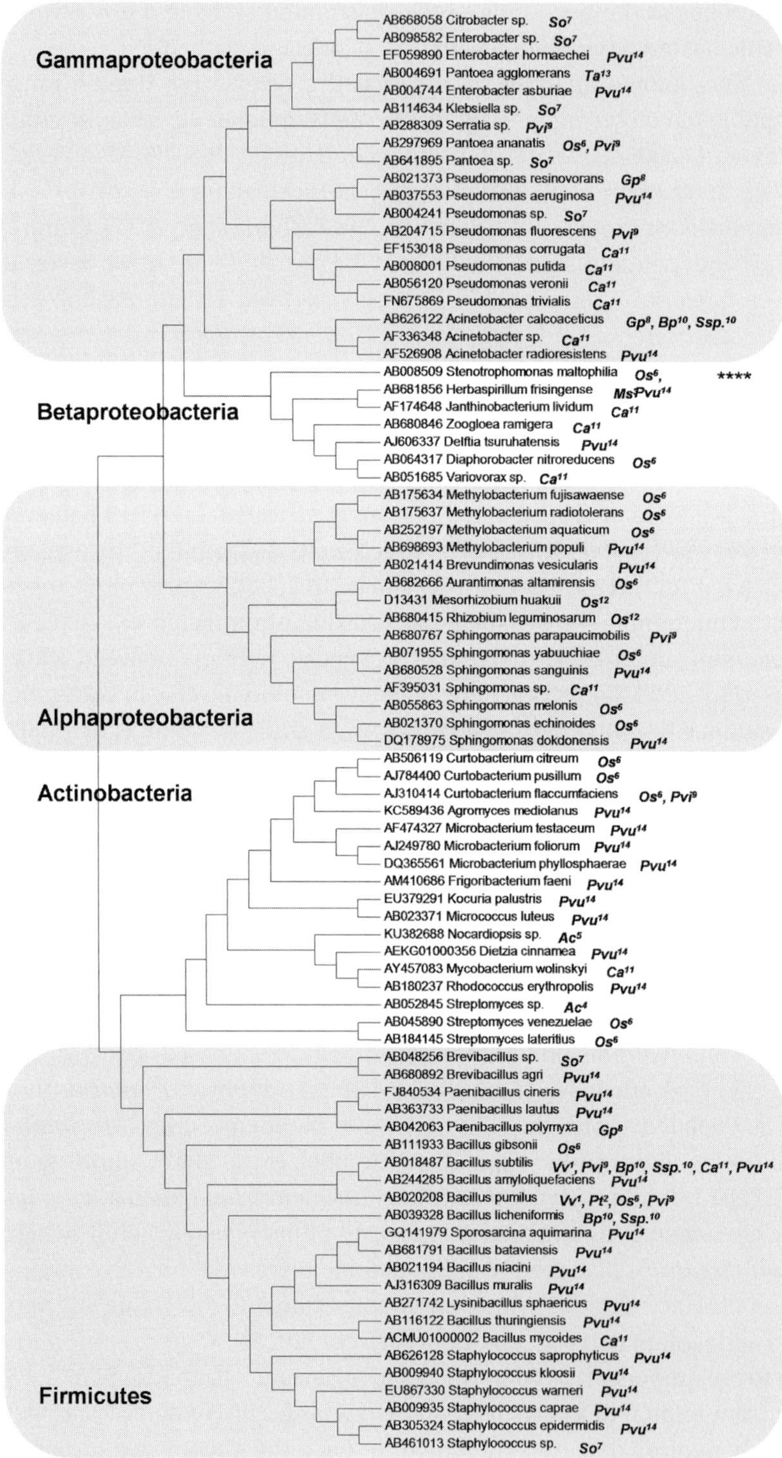
Gammaproteobacteria
AB668058 Citrobacter sp. So[7]
AB098582 Enterobacter sp. So[7]
EF059890 Enterobacter hormaechei Pvu[14]
AB004691 Pantoea agglomerans Ta[13]
AB004744 Enterobacter asburiae Pvu[14]
AB114634 Klebsiella sp. So[7]
AB288309 Serratia sp. Pvi[9]
AB297969 Pantoea ananatis Os[6], Pvi[9]
AB641895 Pantoea sp. So[7]
AB021373 Pseudomonas resinovorans Gp[8]
AB037553 Pseudomonas aeruginosa Pvu[14]
AB004241 Pseudomonas sp. So[7]
AB204715 Pseudomonas fluorescens Pvi[9]
EF153018 Pseudomonas corrugata Ca[11]
AB008001 Pseudomonas putida Ca[11]
AB056120 Pseudomonas veronii Ca[11]
FN675869 Pseudomonas trivialis Ca[11]
AB626122 Acinetobacter calcoaceticus Gp[8], Bp[10], Ssp.[10]
AF336348 Acinetobacter sp. Ca[11]
AF526908 Acinetobacter radioresistens Pvu[14]
AB008509 Stenotrophomonas maltophilia Os[6], ****
AB681856 Herbaspirillum frisingense Ms[3] Pvu[14]
AF174648 Janthinobacterium lividum Ca[11]
Betaproteobacteria
AB680846 Zoogloea ramigera Ca[11]
AJ606337 Delftia tsuruhatensis Pvu[14]
AB064317 Diaphorobacter nitroreducens Os[6]
AB051685 Variovorax sp. Ca[11]
AB175634 Methylobacterium fujisawaense Os[6]
AB175637 Methylobacterium radiotolerans Os[6]
AB252197 Methylobacterium aquaticum Os[6]
AB698693 Methylobacterium populi Pvu[14]
AB021414 Brevundimonas vesicularis Pvu[14]
AB682666 Aurantimonas altamirensis Os[6]
D13431 Mesorhizobium huakuii Os[12]
AB680415 Rhizobium leguminosarum Os[12]
AB680767 Sphingomonas parapaucimobilis Pvi[9]
AB071955 Sphingomonas yabuuchiae Os[6]
AB680528 Sphingomonas sanguinis Pvu[14]
AF395031 Sphingomonas sp. Ca[11]
AB055863 Sphingomonas melonis Os[6]
AB021370 Sphingomonas echinoides Os[6]
DQ178975 Sphingomonas dokdonensis Pvu[14]
Alphaproteobacteria
Actinobacteria
AB506119 Curtobacterium citreum Os[6]
AJ784400 Curtobacterium pusillum Os[6]
AJ310414 Curtobacterium flaccumfaciens Os[6], Pvi[9]
KC589436 Agromyces mediolanus Pvu[14]
AF474327 Microbacterium testaceum Pvu[14]
AJ249780 Microbacterium foliorum Pvu[14]
DQ365561 Microbacterium phyllosphaerae Pvu[14]
AM410686 Frigoribacterium faeni Pvu[14]
EU379291 Kocuria palustris Pvu[14]
AB023371 Micrococcus luteus Pvu[14]
KU382688 Nocardiopsis sp. Ac[5]
AEKG01000356 Dietzia cinnamea Pvu[14]
AY457083 Mycobacterium wolinskyi Ca[11]
AB180237 Rhodococcus erythropolis Pvu[14]
AB052845 Streptomyces sp. Ac[4]
AB045890 Streptomyces venezuelae Os[6]
AB184145 Streptomyces lateritius Os[6]
AB048256 Brevibacillus sp. So[7]
AB680892 Brevibacillus agri Pvu[14]
FJ840534 Paenibacillus cineris Pvu[14]
AB363733 Paenibacillus lautus Pvu[14]
AB042063 Paenibacillus polymyxa Gp[8]
AB111933 Bacillus gibsonii Os[6]
AB018487 Bacillus subtilis Vv[1], Pvi[9], Bp[10], Ssp.[10], Ca[11], Pvu[14]
AB244285 Bacillus amyloliquefaciens Pvu[14]
AB020208 Bacillus pumilus Vv[1], Pt[2], Os[6], Pvi[9]
AB039328 Bacillus licheniformis Bp[10], Ssp.[10]
GQ141979 Sporosarcina aquimarina Pvu[14]
AB681791 Bacillus bataviensis Pvu[14]
AB021194 Bacillus niacini Pvu[14]
AJ316309 Bacillus muralis Pvu[14]
AB271742 Lysinibacillus sphaericus Pvu[14]
AB116122 Bacillus thuringiensis Pvu[14]
ACMU01000002 Bacillus mycoides Ca[11]
AB626128 Staphylococcus saprophyticus Pvu[14]
AB009940 Staphylococcus kloosii Pvu[14]
EU867330 Staphylococcus warneri Pvu[14]
AB009935 Staphylococcus caprae Pvu[14]
AB305324 Staphylococcus epidermidis Pvu[14]
AB461013 Staphylococcus sp. So[7]
Firmicutes

trees (*Populus tremula* × *Populus deltoides*) has yielded interesting findings on the relationship between host genotype and microbiome assembly. Beckers *et al.* (2016) found through comparison of natural and synthetic microbiomes in wild-type and a transgenic poplar line deficient in cinnamoyl-CoA reductase (an enzyme involved in lignin biosynthesis) that the accumulation of metabolites normally directed towards lignin production resulted in selection for bacterial strains capable of utilising these metabolites, especially *Pseudomonas putida*. A more applied study on biocontrol of the fungal pathogens that cause poplar canker reported that *Bacillus pumilus* JK-SX001 colonised poplar through migration from roots to leaves, and once established could promote growth while mitigating disease incidence among seedlings (Ren *et al.*, 2013). Similar research in the more commonly cultivated woody perennial *Vitis vinifera* (grape) also showed that strains of the foliar endophytes *B. pumilus* and *B. subtilis* could exert a biocontrol effect, in this case against the grapevine downy mildew pathogen, *Plasmopara viticola* (Zhang *et al.*, 2017).

In wild specimens of *Arabidopsis thaliana*, a culture-independent orthogonal comparison of epiphytic and endophytic microbial communities in leaves and roots found that species richness, diversity, and evenness were greater for leaf endophytic communities compared to epiphytic communities. This contrasted with root samples which showed higher diversity in epiphytic samples than endophytic samples (Bodenhausen *et al.*, 2013). Culture-dependent studies examining epiphytic and endophytic bacterial communities in *Arabidopsis* have also shown that salicylic acid (SA) and jasmonic acid (JA) defence signalling plays a role in endophyte community structure and abundance, with smaller and less diverse communities found in plants with an elevated defence response (Kniskern *et al.*, 2007). Similarly, a culture-independent study using field experiments of *Arabidopsis*

Figure 10.2 Phylogenetic tree of example species spanning the Firmicutes, Actinobacteria and Proteobacteria phyla demonstrates the taxonomic distribution of foliar endophytes in published culture-dependent and -independent studies. Phylogeny was reconstructed using 16S rRNA gene sequences in MEGA 7.0 using a Maximum Likelihood evolutionary model with 1000 bootstrap replicates. Sequences (see accession numbers) were selected from the Silva database and were not taken directly from cited studies. Abbreviations for host plant: *Vv* = *Vitis vinifera*, *Pt* = *Populus tremula*, *Ms* = *Miscanthus sacchariflorus*, *Ac* = *Aegicerus corniculatum*, *Os* = *Oryza sativa*, *So* = *Saccharum officinarum*, *Gp* = *Gynura procumbens*, *Pvi* = *Panicum virgatum*, *Ca* = *Crocus albiflorus*, *Bp* = *Betula pendula*, *Ssp.* = *Sorbus sp.*, *Ta* = *Triticum aestivum*, *Pvu* = *Phaseolus vulgaris.* Superscripts refer to following references: [1]Zhang *et al.* (2017), [2]Ren *et al.* (2013), [3]Kirchof *et al.* (2001), [4]Lin *et al.* (2005), [5]Lin *et al.* (2010), [6]Mano *et al.* (2007), [7]Magnani *et al.* (2010), [8]Bhore *et al.* (2010), [9]Gagne-Bourgue *et al.* (2013), [10]Izumi *et al.* (2008), [11]Reiter and Sessitch (2006), [12]Chi *et al.* (2005), [13]Ruppel *et al.* (1992), [14]Costa *et al.* (2012).

found correlations between genotype and microbial community structure. Horton *et al.* (2014) used a genome-wide association study to correlate single nucleotide polymorphisms (SNPs) across the genomes of 196 accessions of *Arabidopsis*, with the most prevalent (most sequenced) bacterial and fungal operational taxonomic units (OTUs). The gene ontology terms most associated with the presence or absence of specific taxa within the most abundant 50% of sequenced OTUs include defence response, kinase-related activity and cell wall (including polysaccharide cellulose, callose and pectin biosynthesis) (Horton *et al.*, 2014). This finding agrees with other studies in *Arabidopsis* that use colonisation by wild and synthetic microbial communities to demonstrate the importance of the SA defence signalling pathway in recruitment of specific family-level endophytic root bacterial taxa (Lebeis *et al.*, 2015).

10.4 Function

Beneficial roles of endophytes can include biocontrol of pests and pathogens, phytostimulation, and indirect benefits that occur through competition for the same niche as pathogens. Below we describe examples of aboveground endophytic microbes with known benefits. Endophytes can have a variety of influences on plant health which can be classified as symbiotic, mutualistic and commensal (Ryan *et al.*, 2008). Endophytes have been implicated in improving plant health and crop production through nutrient acquisition, pathogen resistance through antibiosis or competition, removal of contaminants, phytohormone production and inducing plant immune responses (Franco *et al.*, 2007; Murphy *et al.*, 2018). For example, the nitrogen-fixing foliar resident *Methylobacterium* promotes plant growth and seed production in *Jatropha* and other crops (Madhaiyan *et al.*, 2015). Inoculation of tobacco plants with the root endophyte *Paenibacillus polymyxa* C5 reduced infection by the tobacco black shank pathogen through biofilm formation at the root tip (Ren *et al.*, 2012). Additional root endophytes, *Brevibacillus brevis* and *Streptomyces rochei*, have been shown to inhibit the pathogen, *Ralstonia solanacearum*, responsible for tobacco bacterial wilt (Liu *et al.*, 2013). A consortium of endophytes isolated from surface-sterilised *Nicotiana tabacum* seeds from plants cultivated on heavy metal contaminated soil consisted primarily of *Pseudomonas*, but also included *Enterobacter*, *Clostridium* and *Stenotrophomonas*. Furthermore, inoculation with the consortium improved plant biomass when grown in high cadmium soil (Mastretta *et al.*, 2009).

The plant-beneficial trait most commonly ascribed to leaf-dwelling endophytes is phytostimulation, which includes modulation of ethylene levels via 1-aminocyclopropane-1-carboxylate (ACC) deaminase and induced systemic resistance (ISR). A variety of molecules and enzymes have been identified as the result of plant wounding immune responses, altering JA and SA concentrations within the plant tissue, which results in local and global metabolic plant responses. The impact of JA to improve plant resistance was investigated using mutant *Arabidopsis* plants, the results of which indicate that the JA immune response is capable of preventing infection by *Pythium mastophorum*, a common root fungal pathogen present worldwide (Vijayan *et al.*, 1998). Many bacterial endophytes have phytohormone synthetic pathways (Zhang *et al.*, 2006; Gamalero and Glick, 2015). Bacteria capable of producing ACC deaminase can decrease ethylene levels produced by plants during stress, which facilitates promotion of plant health by bacterial indole-3-acetic acid (IAA) under a variety of environmentally stressful situations (Saleem *et al.*, 2007; Glick, 2014).

The defining criterion of bacterial endophytes as non-pathogenic has thus far delineated this group from plant pathogens. Facultative changes in lifestyle, sophisticated metabolic capabilities and factors associated with host genotype (such as primary and secondary metabolite exudates, cell wall composition and defence signalling mechanisms for systemic resistance), all may contribute to the ability of a bacterium to colonise a potential host. Morris *et al.* (2017) discuss the limitations of characterising plant pathogens as such, and their potential to play alternative roles in a host-associated niche. In many cases, pathogenic bacteria may be found in diseased tissue at 5- to 20-fold higher concentrations than in asymptomatic host tissue (Chatterjee *et al.*, 2008; Morris *et al.*, 2017). For example, *Xylella fastidiosa*, the causative agent of Pierce's disease of grapevine, is an obligate pathogen vectored by glassy-winged sharpshooters (*Homalodisca vitripennis*) that occupies and spreads systemically through xylem elements. This pathogen has been well established as a 'generalist', infecting and often persisting for several months asymptomatically in hundreds of alternate host species (Wistrom and Purcell, 2005; Chatterjee *et al.*, 2008). Models by Chatterjee *et al.* (2008) propose that early colonisation utilises hemagglutinin and Type IV pili genes for motility and attachment, but as cell densities increase, diffusible signal factors induce adhesins and gum production, allowing cells to attach to insect vector feeding organs. These sticky metabolites facilitate transport and colonisation to new hosts but result in xylem occlusion and water starvation for the host.

While there is not a wealth of examples of bacterial plant pathogens that also behave as foliar endophytes, any facultative plant pathogen must also have means for survival in the absence of a host, either as a saprophyte in soil or as an epiphyte, until conditions arise for successful infection. In light of the host-specific nature of some bacterial phytopathogens, strong consideration should be given to experimental data showing that formerly pathogenic bacterial populations, which lost the ability to infect their host through mutation to virulence genes or pathogenicity factors, may persist in the host if pathogenic cells of the same species are also present (Kamoun and Kado, 1990; Zhao *et al.*, 2009; Morris *et al.*, 2017). Similarly, non-pathogenic bacteria will utilise the metabolites released during disease by a true pathogen, increasing the severity of damage, and thus earning the designation 'opportunistic pathogen' (Morris *et al.*, 2017). This cooperative or synergistic association blurs the definition of endophyte as a commensal resident that does no harm. Plants are under perpetual colonisation by pathogens and commensals alike, and the inability of a bacterium in the apoplast to cause disease, regardless of taxonomic designation, dictates that by definition these organisms are endophytes.

10.4.1 Potential Post-Harvest Impacts

Bacterial endophytes associated with aerial plant parts intended for raw consumption, such as leafy greens, lettuces and fresh herbs, may have consequences for human health. A body of evidence shows that enteric human pathogens such as *Salmonella* spp. and *E. coli* O157:H7 are capable of living as endophytes in plant tissue, avoiding or suppressing plant defences, in abundance similar to established foliar endophytes (Tyler and Triplett, 2008; Teplitski *et al.*, 2009; Warriner and Namvar, 2010; Kljujev *et al.*, 2018). It is not known whether low levels of endophytic enterobacteria pose a high risk of illness. However, the knowledge that edible plants can serve as an alternate host for human enteropathogens is relatively recent and the mechanisms of plant colonisation and the significance of endophytic enterobacteria to food poisoning outbreaks are areas of ongoing research.

Enterobacteria can persist in the environment, becoming endophytic via foliar entry after leaf contact with soil, contaminated irrigation water, agricultural runoff, damage by phytopathogens, insects and mechanical injury (Brandl, 2006; Simko *et al.*, 2015; Hartmann *et al.*, 2017). After initial contact, the bacteria survive in the phyllosphere by various mechanisms, including biofilm associations and the production of polysaccharides and epitopes that promote adherence to plant surfaces and stomata (Heaton and Jones, 2008; Warriner and Namvar, 2010; Jang and Matthews, 2018). When the bacteria become internalised, a variety of genetic factors often associated with virulence may help them avoid recognition by plant defence mechanisms. However, enterobacteria trigger a defence response in some plants but not others, indicating a host specificity (Warriner *et al.*, 2003; Barak *et al.*, 2005; Brandl, 2006; Teplitski *et al.*, 2009; Berger *et al.*, 2010; Wright *et al.*, 2017). Environmental conditions also affect successful colonisation. For example, *E. coli* O157:H7 cells exposed to soil or manure exhibited greater survival on plants compared to cultures grown solely on LB medium (Seo and Matthews, 2014).

Research on the interaction between enterobacteria and other endophytic microbiota suggest there may be biocontrol potential for microorganisms that decrease the survival and internalisation of enterobacteria (Iniguez *et al.*, 2005; Tyler and Triplett, 2008; Shen *et al.*, 2017). Additionally, a more complete understanding of the genetic mechanisms involved in the specificity between enterobacteria and their plant hosts could lead to the development of cultivars resistant to colonisation by human pathogens (Teplitski *et al.*, 2009). Post-harvest consequences of endophytes for non-food crops are also under investigation. For example, bacteria inhabiting tobacco leaves may be involved in the production of highly carcinogenic tobacco-specific nitrosamines (TSNAs) during the curing process (Law *et al.*, 2016). Many questions remain about the effects of foliar endophytes in pre- and post-harvest crops. Research to advance knowledge in this field could lead to many benefits for human health.

10.4.2 Natural Products

Bacteria are well-known producers of complex chemical entities or secondary metabolites (so-called natural products) that exhibit a variety of plant-beneficial activities. There are a number of examples of bacterial leaf endophytic isolates that produce natural products with bioactivity against plant pathogens *in vitro*. A number of antimycotic compounds are produced by plant-associated bacteria (Raaijmakers and Mazzola, 2012). Ecomycins, lipopeptides produced by the endophytic bacterium *Pseudomonas viridiflava*, show antifungal activity against some of the most destructive crop diseases, including *Fusarium oxysporum*, *Rhizoctonia solani* and *Sclerotinia sclerotiorum*, in addition to a variety of human fungal pathogens (Miller *et al.*, 1998). A *Streptomycete* endophyte isolated from *Kennedia nigriscans*, a medicinal plant used by Aboriginal Australians, was found to produce a number of functionalised peptides with antifungal activity against *Rhizoctonia solani* and *Sclerotinia sclerotiorum*, in addition to other plant pathogens (Castillo *et al.*, 2002). Pseudomycins, another family of peptide compounds with broad-spectrum antifungal activity against a variety of plant pathogens, were isolated from the endophytic bacterium *Pseudomonas syringae*, which also has the ability to act as a pathogen (Harrison *et al.*, 1991; Baltrus *et al.*, 2011; McEvoy *et al.*, 2016). A novel indole derivative produced by *Pantoea ananatis* showed significant activity against *Mycosphaerella musicola*, causative agent of the banana disease Yellow Sigatoka (Aman and Rai, 2016).

Selective isolation of medicinal plant-associated species has revealed a variety of compounds with unique biological activity (Sumner, 2000). In the process of mining for endophytes that synthesise natural products, special attention is given to those plants used for traditional medicinal remedies or that survive in a unique or hostile habitat (Qin *et al.*, 2011). Mangrove endophytes have proven to be a rich source for bacterial secondary metabolites. *Aegicerus corniculatum* was the source for isolation of the Actinobacteria *Streptomyces* sp. GT-20026114 and *Nocardiopsis* sp. A00203, which were found to produce cyclopentonone and 2-pyranone derivatives, respectively (Lin *et al.*, 2005; Lin *et al.*, 2010). An endophytic streptomycete isolated from mangrove produced a number of indolosesquiterpenes with activity against antibiotic-resistant *Staphylococcus aureus* and *Enterococcus faecalis* (Ding *et al.*, 2011). A streptomycete from fern-leaved grevillea produces a variety of kakadumycin compounds with significant activity against multiple bacterial human pathogens and antimalarial activity comparable to chloroquine (Castillo *et al.*, 2003). Coronamycin, a peptide antibiotic isolated from another streptomycete endophyte, also shows activity against the malarial parasite (Ezra *et al.*, 2004). The first indolosesquiterpene isolated from an endophytic bacterium was identified as xiamycin and showed bioactivity against HIV (Ding *et al.*, 2010).

10.5 Challenges in Foliar Endophyte Research

Advances in DNA sequencing technology and data analysis have had an immense impact on the study of environmental microbial populations. However, studying bacterial foliar endophytes poses unique challenges. Bacterial cells living within plant tissue may be present in relatively low abundance; thus, there is no consensus about how much plant tissue is required to adequately capture the diversity of endophytic bacteria, given variations in species, tissue type, and sampling time and conditions (Ding and Melcher, 2016).

In addition to low abundance, there are logistical challenges to investigating the bacterial population within the leaf. Accessing the leaf endophyte population is contingent upon surface sterilisation, usually followed by homogenisation, so that bacteria associated with internal leaf compartments can be cultured or characterised by DNA sequencing. Surface sterilisation is necessary in order to confirm that a cultured or sequenced bacteria was found within the leaf tissue and is not an epiphyte or environmental contaminant (Hallmann, 2001). This process can present a challenge to endophyte research for several reasons. Most protocols involve a series of washes using ethanol (70–95%), dilute hypochlorite solutions and sterile water. Some plant leaves have waxy cuticles that stand up to repeated washes and rinses, but others wilt and tear easily, making it difficult or impossible to have adequate surface sterilisation without affecting internal tissues. The process can be time consuming and difficult to accommodate in high-throughput experiments and large/irregular shaped leaves.

In culture-dependent methods, the endophytic bacteria are grown on general or selective culture media for isolation and further studies. A portion of any microbial population (estimated 50–99%, depending on the environment) will resist cultivation and the remainder will require differing media and culture conditions (Curtis *et al.*, 2006). In addition to homogenisation, specialised methods for accessing endophytes from xylem fluid can be performed, such as vacuum extraction, expulsion using a Scholander pressure bomb or centrifugation to collect intracellular fluid from whole leaves (Bazzi *et al.*, 1987; Dong *et al.*, 1994; Hallmann *et al.*, 1997a). These methods are not appropriate for all leaf tissue types, but different organisms have been recovered when compared with homogenisation methods (Bell *et al.*, 1995; Hallmann *et al.*, 1997a).

Culture-independent methods most often rely on sequencing of PCR-amplified 16S rRNA genes from homogenised leaf tissue metagenomic DNA (Su *et al.*, 2012). This method has the advantage of capturing much of the diversity of the as-yet-uncultured portion of the endophytic community. The use of the 16S rRNA evolutionary marker gene to characterise microbial communities has been developing over decades, with a variety of profiling techniques such as fragment polymorphism (DGGE, T-RFLP, ARISA), as well as cloning and sequencing all or a portion of the 16S rRNA gene by the Sanger method (Nocker *et al.*, 2007; Ding and Melcher, 2016). More recently, next generation sequencing (NGS) technology has made it possible to sequence small fragments (up to 500 base pairs) for several hundred samples at once, using indexed primers that impart a unique nucleotide 'barcode' so that the data can be sorted by sample after sequencing (Caporaso *et al.*, 2011; Kozich *et al.*, 2013). NGS technology used to characterise bacterial communities typically sequences 250 base pairs of the V4 region of the16S rRNA gene, which can be classified up to the genus level of taxonomy (Schloss *et al.*, 2009; Fouhy *et al.*, 2016; Yang *et al.*, 2016). Some environmental/biological samples or tissue types have commercially available kits for DNA extraction which reliably yield DNA in sufficient quantity and purity for PCR amplification. Other environments can be more difficult to extract bacterial DNA from due to either low abundance, the presence of PCR-inhibiting substances, or because the extraction process itself is physically challenging. Extracting endophytic bacterial DNA from leaf tissue can have potentially all of those challenges, as well as one unique to plant environments – plastid DNA of plant cells are invariably extracted along with bacterial DNA at significantly higher relative concentrations. Typical universal primer sets for the bacterial 16S rRNA gene will also amplify from this competing template, in addition to the mitochondrial DNA of the plants, potentially resulting in up to 90% of amplicons derived from the plant rather than bacteria (Wang *et al.*, 2008).

There are several methods for reducing the interference of plant DNA on 16S rRNA gene amplicon sequencing, though none exclude all plastid template, some are incompatible with downstream analyses, and all add time and expense to a project. Physically separating bacterial and plant cells before DNA extraction can successfully reduce plastid/mitochondrial sequences by 80–90%, but is time consuming and requires large amounts of leaf tissue (Jiao *et al.*, 2006; Wang *et al.*, 2008; Ikeda *et al.*, 2009, 2010). Briefly, after mechanical or enzymatic homogenisation of leaf tissue, filtration and differential centrifugations are performed to remove heavier plant particulates, which can be followed by density gradient centrifugation using Nycodenz to further separate bacterial and plant cells (Figure 10.3) (Jiao *et al.*, 2006; Ikeda *et al.*, 2009). Another strategy for reducing the impact of plant DNA in sequencing employs the addition of clamping primers such as locking nucleic acids (LNA) and protein nucleic acids (PNA) to the PCR to exclude amplification of plastid

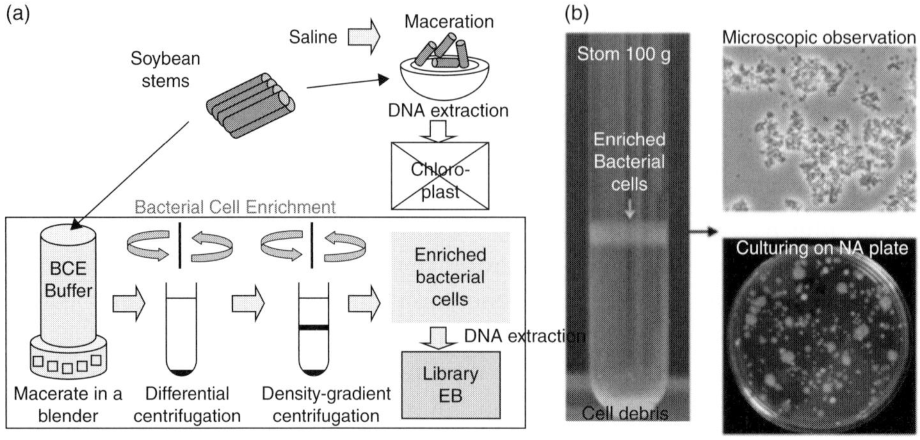

Figure 10.3 Outline of bacterial cell enrichment for community analysis. (a) Comparison of bacterial diversity in bacterial cell enrichment and direct DNA extraction from plant materials. (b) The photographs show the bacterial cells enriched by density gradient centrifugation, their microscopic observation and culture on nutrient agar plates (Ikeda *et al.*, 2009).
Figure reprinted with permission from Ikeda *et al.* (2010).

and mitochondrial DNA (Sakai and Ikenaga, 2013; Ikenaga and Sakai, 2014). These specialised primers are expensive and must be designed for the specific host plant. Alternatively, targeting other regions of the 16S rRNA gene, such as 799F and 1492R primer set, can exclude mitochondrial or plastid DNA. However, variable regions other than V4 can be more difficult to align and classify, and NGS analysis will result in more sequencing errors if the reads do not completely overlap (Sun *et al.*, 2008; Schloss *et al.*, 2009; Kozich *et al.*, 2013; Beckers *et al.*, 2016).

The challenges to studying leaf endophytes are not insurmountable, but are significant enough to impede the pace and scope of research in this area. Method development for improving cultivation of environmental bacteria is an ongoing area of research that is making significant advances (Tyson and Banfield, 2005; Stewart, 2012). The advent of NGS technology has advanced microbiome research in nearly every environment and remains a promising tool for leaf endophyte research. Efforts to develop methods that increase the efficiency and throughput of leaf endophyte extraction will accelerate the study of this unique environment.

10.6 Conclusions

Foliar endophytic bacteria have become a fruitful avenue of research regarding the microbial impact on plants and have led to a variety of pharmaceutical, environmental and agricultural applications. The definition of endophytes will be refined as more information regarding their lifestyle is revealed. Many bacterial endophytes improve the general health of plants and the environment in which they are grown. Improvements in technology to monitor and analyse bacterial populations of endophytes will expand our understanding of plant microbiome diversity and function. Meeting global challenges for food production, such as climate change, increasing population, soil pollution and decreasing resources, will require a holistic paradigm for exploitation of plant–microbe interactions. The unique environmental niche of plant leaves has unlimited potential to aid in addressing these challenges.

References

Ahmad, F., Husain, F. M. and Ahmad, I. (2011). Rhizosphere and root colonization by bacterial inoculants and their monitoring methods: a critical area in PGPR research. In *Microbes and Microbial Technology: Agricultural and Environmental Applications*, ed. I. Ahmad, F. Ahmad and J. Pichtel. New York: Springer, pp. 363–391.

Aman, M. and Rai, V. R. (2016). Antifungal activity of novel indole derivative from endophytic bacteria *Pantoea ananatis* 4G-9 against *Mycosphaerella musicola*. *Biocontrol Science and Technology*, **26,** 476–491.

Baltrus, D. A., Nishimura, M. T., Romanchuk, A. *et al.* (2011). Dynamic evolution of pathogenicity revealed by sequencing and comparative genomics of 19 *Pseudomonas syringae* isolates. *Plos Pathogens*, **7**, e1002132.

Barak, J. D., Gorski, L., Naraghi-Arani, P. *et al.* (2005). *Salmonella enterica* virulence genes are required for bacterial attachment to plant tissue. *Applied and Environmental Microbiology*, **71,** 5685–5691.

Bashan, Y. (1991). Airborne transmission of the rhizosphere bacterium *Azospirillum*. *Microbial Ecology*, **22,** 257–269.

Bashan, Y. and Holguin, G. (1994). Root-to-root travel of the beneficial bacterium *Azospirillum brasilense*. *Applied and Environmental Microbiology*, **60,** 2120–2131.

Bazzi, C., Piazza, C. and Burr, T. J. (1987). Detection of *Agrobacterium tumefaciens* in grapevine cuttings. *EPPO Bulletin*, **17,** 105–112.

Beckers, B., De Beeck, M. O., Thijs, S. *et al.* (2016). Performance of 16s rDNA primer pairs in the study of rhizosphere and endosphere bacterial microbiomes in metabarcoding studies. *Frontiers in Microbiology*, **7**, 650.

Bell, C. R., Dickie, G. A., Harvey, W. L. G. *et al.* (1995). Endophytic bacteria in grapevine. *Canadian Journal of Microbiology*, **41,** 46–53.

Berger, C. N., Sodha, S. V., Shaw, R. K. *et al.* (2010). Fresh fruit and vegetables as vehicles for the transmission of human pathogens. *Environmental Microbiology*, **12,** 2385–2397.

Bertani, I., Abbruscato, P., Piffanelli, P. *et al.* (2016). Rice bacterial endophytes: Isolation of a collection, identification of beneficial strains and microbiome analysis. *Environmental Microbiology Reports*, **8**, 388–398.

Bextine, B., Lampe, D., Lauzon, C. *et al.* (2005). Establishment of a genetically marked insect-derived symbiont in multiple host plants. *Current Microbiology*, **50**, 1–7.

Bhore, S. J., Ravichantar, N. and Loh, C. Y. (2010). Screening of endophytic bacteria isolated from leaves of Sambung Nyawa [*Gynura procumbens* (Lour.) Merr.] for cytokinin-like compounds. *Bioinformation*, **5**, 191–197.

Bodenhausen, N., Horton, M. W. and Bergelson, J. (2013). Bacterial communities associated with the leaves and the roots of *Arabidopsis thaliana*. *PLoS One*, **8**, e56329.

Bovallius, A., Bucht, B., Roffey, R. *et al.* (1978). Long-range air transmission of bacteria. *Applied and Environmental Microbiology*, **35**, 1231–1232.

Brandl, M. T. (2006). Fitness of human enteric pathogens on plants and implications for food safety. *Annual Review of Phytopathology*, **44**, 367–392.

Bright, M. and Bulgheresi, S. (2010). A complex journey: transmission of microbial symbionts. *Nature Reviews Microbiology*, **8**, 218–230.

Caporaso, J. G., Lauber, C. L., Walters, W. A. *et al.* (2011). Global patterns of 16S rRNA diversity at a depth of millions of sequences per sample. *Proceedings of the National Academy of Sciences of the United States of America*, **108**, 4516–4522.

Castillo, U., Harper, J. K., Strobel, G. A. *et al.* (2003). Kakadumycins, novel antibiotics from *Streptomyces* sp NRRL 30566, an endophyte of *Grevillea pteridifolia*. *Fems Microbiology Letters*, **224**, 183–190.

Castillo, U. F., Strobel, G. A., Ford, E. J. *et al.* (2002). Munumbicins, wide-spectrum antibiotics produced by *Streptomyces* NRRL 30562, endophytic on *Kennedia nigriscans*. *Microbiology-Sgm*, **148**, 2675–2685.

Chatterjee, S., Almeida, R. P. P. and Lindow, S. (2008). Living in two worlds: the plant and insect lifestyles of *Xylella fastidiosa*. *Annual Review of Phytopathology*, **46**, 243–271.

Chee-Sanford, J. C., Williams, M. M., Davis, A. S. and Sims, G. K. (2006). Do microorganisms influence seed-bank dynamics? *Weed Science*, **54**, 575–587.

Chi, F., Shen, S.-H., Cheng, H.-P. *et al.* (2005). Ascending migration of endophytic rhizobia, from roots to leaves, inside rice plants and assessment of benefits to rice growth physiology. *Applied and Environmental Microbiology*, **71**, 7271–7278.

Compant, S., Reiter, B., Sessitsch, A. *et al.* (2005). Endophytic colonization of *Vitis vinifera* L. by plant-growth-promoting bacterium *Burkholderia* sp strain PsJN. *Applied and Environmental Microbiology*, **71**, 1685–1693.

Copeland, J. K., Yuan, L. J., Layeghifard, M. *et al.* (2015). Seasonal community succession of the phyllosphere microbiome. *Molecular Plant–Microbe Interactions*, **28**, 274–285.

Costa, L. E. D., De Queiroz, M. V., Borges, A. C. *et al.* (2012). Isolation and characterization of endophytic bacteria isolated from the leaves of the common bean (*Phaseolus vulgaris*). *Brazilian Journal of Microbiology*, **43**, 1562–1575.

Curtis, T. P., Head, I. M., Lunn, M. *et al.* (2006). What is the extent of prokaryotic diversity? *Philosophical Transactions of the Royal Society B: Biological Sciences*, **361**, 2023–2037.

Ding, T. and Melcher, U. (2016). Influences of plant species, season and location on leaf endophytic bacterial communities of non-cultivated plants. *PLoS One*, **11**, e0150895.

Ding, L., Munich, J., Goerls, H. *et al.* (2010). Xiamycin, a pentacyclic indolosesquiterpene with selective anti-HIV activity from a bacterial mangrove endophyte. *Bioorganic & Medicinal Chemistry Letters*, **20,** 6685–6687.

Ding, L., Maier, A., Fiebig, H. H. *et al.* (2011). A family of multicyclic indolosesquiterpenes from a bacterial endophyte. *Organic & Biomolecular Chemistry*, **9,** 4029–4031.

Dong, Z. M., Canny, M. J., McCully, M. E. *et al.* (1994). A nitrogen-fixing endophyte of sugarcane stems: a new role for the apoplast. *Plant Physiology*, **105,** 1139–1147.

Ezra, D., Castillo, U. F., Strobel, G. A. *et al.* (2004). Coronamycins, peptide antibiotics produced by a verticillate *Streptomyces* sp (MSU-2110) endophytic on *Monstera* sp. *Microbiology-Sgm*, **150,** 785–793.

Ferreira, A., Quecine, M. C., Lacava, P. T. *et al.* (2008). Diversity of endophytic bacteria from *Eucalyptus* species seeds and colonization of seedlings by *Pantoea agglomerans*. *Fems Microbiology Letters*, **287,** 8–14.

Fouhy, F., Clooney, A. G., Stanton, C. *et al.* (2016). 16S rRNA gene sequencing of mock microbial populations: impact of DNA extraction method, primer choice and sequencing platform. *BMC Microbiology*, **16**, 123.

Franco, C., Michelsen, P., Percy, N. *et al.* (2007). Actinobacterial endophytes for improved crop performance. *Australasian Plant Pathology*, **36,** 524–531.

Frank, A., Saldierna Guzmán, J. and Shay, J. (2017). Transmission of bacterial endophytes. *Microorganisms*, **5,** E70.

Gagne-Bourgue, F., Aliferis, K. A., Seguin, P. *et al.* (2013). Isolation and characterization of indigenous endophytic bacteria associated with leaves of switchgrass (*Panicum virgatum* L.) cultivars. *Journal of Applied Microbiology*, **114,** 836–853.

Gai, C. S., Lacava, P. T., Quecine, M. C. *et al.* (2009). Transmission of *Methylobacterium mesophilicum* by *Bucephalogonia xanthophis* for paratransgenic control strategy of citrus variegated chlorosis. *Journal of Microbiology*, **47,** 448–454.

Gamalero, E. and Glick, B. R. (2015). Bacterial modulation of plant ethylene levels. *Plant Physiology*, **169,** 13–22.

Glick, B. R. (2014). Bacteria with ACC deaminase can promote plant growth and help to feed the world. *Microbiological Research*, **169,** 30–39.

Gonella, E., Pajoro, M., Marzorati, M. *et al.* (2015). Plant-mediated interspecific horizontal transmission of an intracellular symbiont in insects. *Scientific Reports*, **5**, 15811.

Hallmann, J. (2001). Biotic interactions in plant-pathogen associations. In *Biotic Interactions in Plant-Pathogen Associations*, ed. M. J. Jeger and N. J. Spence. Wallingford, UK: CAB International, pp. 87–119.

Hallmann, J., Kloepper, J. W. and Rodriguez-Kabana, R. (1997a). Application of the Scholander pressure bomb to studies on endophytic bacteria of plants. *Canadian Journal of Microbiology*, **43,** 411–416.

Hallmann, J., Quadthallmann, A., Mahaffee, W. F. *et al.* (1997b). Bacterial endophytes in agricultural crops.

Canadian Journal of Microbiology, **43,** 895–914.

Hardoim, P. R., Van Overbeek, L. S., Berg, G. *et al.* (2015). The hidden world within plants: ecological and evolutionary considerations for defining functioning of microbial endophytes. *Microbiology and Molecular Biology Reviews*, **79,** 293–320.

Harrison, L., Teplow, D. B., Rinaldi, M. *et al.* (1991). Pseudomycins, a family of novel peptides from *Pseudomonas syringae* possessing broad-spectrum antifungal activity. *Journal of General Microbiology*, **137,** 2857–2865.

Hartmann, R., Fricke, A., Stützel, H. *et al.* (2017). Internalization of *Escherichia coli* O157:H7 gfp+ in rocket and Swiss chard baby leaves as affected by abiotic and biotic damage. *Letters in Applied Microbiology*, **65,** 35–41.

Heaton, J. C. and Jones, K. (2008). Microbial contamination of fruit and vegetables and the behaviour of enteropathogens in the phyllosphere: a review. *Journal of Applied Microbiology*, **104,** 613–626.

Horton, M. W., Bodenhausen, N., Beilsmith, K. *et al.* (2014). Genome-wide association study of *Arabidopsis thaliana* leaf microbial community. *Nature Communications*, **5**, 5320.

Ikeda, S., Kaneko, T., Okubo, T. *et al.* (2009). Development of a bacterial cell enrichment method and its application to the community analysis in soybean stems. *Microbial Ecology*, **58,** 703–714.

Ikeda, S., Okubo, T., Anda, M. *et al.* (2010). Community- and genome-based views of plant-associated bacteria: plant-bacterial interactions in soybean and rice. *Plant and Cell Physiology*, **51,** 1398–1410.

Ikenaga, M. and Sakai, M. (2014). Application of locked nucleic acid (LNA) oligonucleotide-PCR clamping technique to selectively PCR amplify the SSU rRNA genes of bacteria in investigating the plant-associated community structures. *Microbes and Environments*, **29,** 286–295.

Iniguez, A. L., Dong, Y. M., Carter, H. D. *et al.* (2005). Regulation of enteric endophytic bacterial colonization by plant defenses. *Molecular Plant-Microbe Interactions*, **18,** 169–178.

Izumi, H., Anderson, I. C., Killham, K. *et al.* (2008). Diversity of predominant endophytic bacteria in European deciduous and coniferous trees. *Canadian Journal of Microbiology*, **54,** 173–179.

Jang, H. and Matthews, K. R. (2018). Influence of surface polysaccharides of *Escherichia coli* O157: H7 on plant defense response and survival of the human enteric pathogen on *Arabidopsis thaliana* and lettuce (*Lactuca sativa*). *Food Microbiology*, **70,** 254–261.

Ji, K. X., Chi, F., Yang, M. F. *et al.* (2010). Movement of rhizobia inside tobacco and lifestyle alternation from endophytes to free-living rhizobia on leaves. *Journal of Microbiology and Biotechnology*, **20,** 238–244.

Jiao, J. Y., Wang, H. X., Zeng, Y. *et al.* (2006). Enrichment for microbes living in association with plant tissues. *Journal of Applied Microbiology*, **100,** 830–837.

Johnston-Monje, D. and Raizada, M. N. (2011). Conservation and diversity of seed associated endophytes in Zea across boundaries of evolution, ethnography and ecology. *PLoS One*, **6,** e20396.

Kamoun, S. and Kado, C. I. (1990). A plant-inducible gene of *Xanthomonas-campestris* pv campestris encodes an exocellular component required for growth in the host and hypersensitivity on nonhosts. *Journal of Bacteriology*, **172,** 5165–5172.

Kirchhof, G., Eckert, B., Stoffels, M. *et al.* (2001). *Herbaspirillum frisingense* sp *nov.*, a new nitrogen-fixing bacterial species that occurs in C4-fibre plants. *International Journal of Systematic and Evolutionary Microbiology*, **51,** 157–168.

Kljujev, I., Raicevic, V., Vujovic, B. *et al.* (2018). *Salmonella* as an endophytic colonizer of plants: a risk for health safety vegetable production. *Microbial Pathogenesis*, **115,** 199–207.

Kniskern, J. M., Traw, M. B. and Bergelson, J. (2007). Salicylic acid and jasmonic acid signaling defense pathways reduce natural bacterial diversity on *Arabidopsis thaliana. Molecular Plant–Microbe Interactions*, **20,** 1512–1522.

Kozich, J. J., Westcott, S. L., Baxter, N. T. *et al.* (2013). Development of a dual-index sequencing strategy and curation pipeline for analyzing amplicon sequence data on the MiSeq Illumina sequencing platform. *Applied and Environmental Microbiology*, **79,** 5112–5120.

Law, A. D., Fisher, C., Jack, A. *et al.* (2016). Tobacco, microbes, and carcinogens: correlation between tobacco cure conditions, tobacco-specific nitrosamine content, and cured leaf microbial community. *Microbial Ecology*, **72,** 120–129.

Lebeis, S. L., Paredes, S. H., Lundberg, D. S. *et al.* (2015). Salicylic acid modulates colonization of the root microbiome by specific bacterial taxa. *Science*, **349,** 860–864.

Lin, C., Lu, C. H. and Shen, Y. M. (2010). Three new 2-pyranone derivatives from mangrove endophytic actinomycete strain *Nocardiopsis* sp A00203. *Records of Natural Products*, **4,** 176–179.

Lin, W. H., Li, L. Y., Fu, H. Z. *et al.* (2005). New cyclopentenone derivatives from an endophytic *Streptomyces* sp isolated from the mangrove plant *Aegiceras comiculatum. Journal of Antibiotics*, **58,** 594–598.

Lindemann, J., Constantinidou, H. A., Barchet, W. R. *et al.* (1982). Plants as sources of airborne bacteria, including ice nucleation-active bacteria. *Applied and Environmental Microbiology*, **44,** 1059–1063.

Liu, Y. X., Shi, J. X., Feng, Y. G. *et al.* (2013). Tobacco bacterial wilt can be biologically controlled by the application of antagonistic strains in combination with organic fertilizer. *Biology and Fertility of Soils*, **49,** 447–464.

Lòpez-Fernàndez, S., Mazzoni, V., Pedrazzoli, F., Pertot, I. and Campisano, A. (2017). A phloem-feeding insect transfers bacterial endophytic communities between grapevine plants. *Frontiers in Microbiology*, **8**, 834.

Madhaiyan, M., Alex, T. H. H., Ngoh, S. T. *et al.* (2015). Leaf-residing *Methylobacterium* species fix nitrogen and promote biomass and seed production in *Jatropha curcas. Biotechnology for Biofuels*, **8**, 222.

Magnani, G. S., Didonet, C. M., Cruz, L. M. *et al.* (2010). Diversity of endophytic bacteria in Brazilian sugarcane. *Genetics and Molecular Research*, **9,** 250–258.

Mano, H., Tanaka, F., Nakamura, C. *et al.* (2007). Culturable endophytic bacterial flora of the maturing leaves and roots of rice plants (*Oryza sativa*) cultivated in a paddy field. *Microbes and Environments*, **22,** 175–185.

Mastretta, C., Taghavi, S., Van Der Lelie, D. *et al.* (2009). Endophytic bacteria from seeds of *Nicotiana tabacum* can reduce cadmium phytotoxicity. *International Journal of Phytoremediation*, **11**, 251–267.

McEvoy, A., O'Regan, F., Fleming, C. C. *et al.* (2016). Bleeding canker of horse chestnut (*Aesculus hippocastanum*) in Ireland: incidence, severity and characterization using DNA sequences and real-time PCR. *Plant Pathology*, **65**, 1419–1429.

Miller, C. M., Miller, R. V., Garton-Kenny, D. *et al.* (1998). Ecomycins, unique antimycotics from *Pseudomonas viridiflava*. *Journal of Applied Microbiology*, **84**, 937–944.

Mitter, B., Pfaffenbichler, N., Flavell, R. *et al.* (2017). A new approach to modify plant microbiomes and traits by introducing beneficial bacteria at flowering into progeny seeds. *Frontiers in Microbiology*, **8**, 11.

Morris, C. E., Barny, M. A., Berge, O. *et al.* (2017). Frontiers for research on the ecology of plant-pathogenic bacteria: Fundamentals for sustainability challenges in bacterial molecular plant pathology. *Molecular Plant Pathology*, **18**, 308–319.

Mundt, J. O. and Hinkle, N. F. (1976). Bacteria within ovules and seeds. *Applied and Environmental Microbiology*, **32**, 694–698.

Murphy, B. R., Doohan, F. M. and Hodkinson, T. R. (2018). From concept to commerce: developing a successful fungal endophyte inoculant for agricultural crops. *Journal of Fungi (Basel)*, **4**, E24.

Nocker, A., Burr, M. and Camper, A. K. (2007). Genotypic microbial community profiling: a critical technical review. *Microbial Ecology*, **54**, 276–289.

Qin, S., Xing, K., Jiang, J.-H. *et al.* (2011). Biodiversity, bioactive natural products and biotechnological potential of plant-associated endophytic Actinobacteria. *Applied Microbiology and Biotechnology*, **89**, 457–473.

Qin, S., Chen, H.-H., Zhao, G.-Z. *et al.* (2012). Abundant and diverse endophytic Actinobacteria associated with medicinal plant *Maytenus austroyunnanensis* in Xishuangbanna tropical rainforest revealed by culture-dependent and culture-independent methods. *Environmental Microbiology Reports*, **4**, 522–531.

Raaijmakers, J. M. and Mazzola, M. (2012). Diversity and natural functions of antibiotics produced by beneficial and plant pathogenic bacteria. *Annual Review of Phytopathology*, **50**, 403–424.

Reinhold-Hurek, B. and Hurek, T. (2011). Living inside plants: bacterial endophytes. *Current Opinion in Plant Biology*, **14**, 435–443.

Reiter, B. and Sessitsch, A. (2006). Bacterial endophytes of the wildflower *Crocus albiflorus* analyzed by characterization of isolates and by a cultivation-independent approach. *Canadian Journal of Microbiology*, **52**, 140–149.

Ren, J. H., Li, H., Wang, Y. F. *et al.* (2013). Biocontrol potential of an endophytic *Bacillus pumilus* JK-SX001 against poplar canker. *Biological Control*, **67**, 421–430.

Ren, X. L., Zhang, N., Cao, M. H. *et al.* (2012). Biological control of tobacco black shank and colonization of tobacco roots by a *Paenibacillus polymyxa* strain C5. *Biology and Fertility of Soils*, **48**, 613–620.

Ringelberg, D., Foley, K. and Reynolds, C. M. (2012). Bacterial endophyte communities of two wheatgrass varieties following propagation in different growing media. *Canadian Journal of Microbiology*, **58**, 67–80.

Roos, I. M. M. and Hattingh, M. J. (1983). Scanning electron-microscopy of *Pseudomonas syringae* pv *morsprunorum* on sweet cherry leaves. *Journal of Phytopathology*, **108,** 18–25.

Rosenblueth, M. and Martinez-Romero, E. (2006). Bacterial endophytes and their interactions with hosts. *Molecular Plant–Microbe Interactions*, **19,** 827–837.

Ruppel, S., Hechtbuchholz, C., Remus, R. *et al.* (1992). Settlement of the diazotrophic, phytoeffective bacterial strain *Pantoea agglomerans* on and within winter-wheat – an investigation using ELISA and transmission electron-microscopy. *Plant and Soil*, **145,** 261–273.

Ryan, R. P., Germaine, K., Franks, A. *et al.* (2008). Bacterial endophytes: recent developments and applications. *Fems Microbiology Letters*, **278,** 1–9.

Sakai, M. and Ikenaga, M. (2013). Application of peptide nucleic acid (PNA)-PCR clamping technique to investigate the community structures of rhizobacteria associated with plant roots. *Journal of Microbiological Methods*, **92,** 281–288.

Saleem, M., Arshad, M., Hussain, S. *et al.* (2007). Perspective of plant-growth-promoting rhizobacteria (PGPR) containing ACC deaminase in stress agriculture. *Journal of Industrial Microbiology & Biotechnology*, **34,** 635–648.

Saleem, M., Law, A. D. and Moe, L. A. (2016). *Nicotiana* roots recruit rare rhizosphere taxa as major root-inhabiting microbes. *Microbial Ecology*, **71,** 469–472.

Saleem, M., Meckes, N., Pervaiz, Z. H. *et al.* (2017). Microbial interactions in the phyllosphere increase plant performance under herbivore biotic stress. *Frontiers in Microbiology*, **8**, 41.

Saleem, M., Law, A. D., Sahib, M. R. *et al.* (2018). Impact of root system architecture on rhizosphere and root microbiome. *Rhizosphere*, **6,** 47–51.

Santoyo, G., Moreno-Hagelsieb, G., Orozco-Mosqueda, M. D. *et al.* (2016). Plant growth-promoting bacterial endophytes. *Microbiological Research*, **183,** 92–99.

Schloss, P. D., Westcott, S. L., Ryabin, T. *et al.* (2009). Introducing MOTHUR: open-source, platform-independent, community-supported software for describing and comparing microbial communities. *Applied and Environmental Microbiology*, **75,** 7537–7541.

Seo, S. and Matthews, K. R. (2014). Exposure of *Escherichia coli* O157:H7 to soil, manure, or water influences its survival on plants and initiation of plant defense response. *Food Microbiology*, **38,** 87–92.

Shen, Z. Y., Mustapha, A., Lin, M. S. *et al.* (2017). Biocontrol of the internalization of *Salmonella enterica* and Enterohaemorrhagic *Escherichia coli* in mung bean sprouts with an endophytic *Bacillus subtilis*. *International Journal of Food Microbiology*, **250,** 37–44.

Simko, I., Zhou, Y. G. and Brandl, M. T. (2015). Downy mildew disease promotes the colonization of romaine lettuce by *Escherichia coli* O157:H7 and *Salmonella enterica*. *BMC Microbiology*, **15**, 19.

Sørensen, J. S. A. (2007). Plant-associated bacteria: lifestyle and molecular interactions. In *Modern Soil Microbiology*, 2 edn, ed. J. D. van Elsas, J. K. Jansson and J. T. Trevors. Boca Raton, FL: CRC Press, pp. 211–236.

Sprent, J. I. and Defaria, S. M. (1988). Mechanisms of infection of plants by nitrogen-fixing organisms. *Plant and Soil*, **110,** 157–165.

Stewart, E. J. (2012). Growing unculturable bacteria. *Journal of Bacteriology*, **194,** 4151–4160.

Strobel, G. and Daisy, B. (2003). Bioprospecting for microbial endophytes and their natural products. *Microbiology and Molecular Biology Reviews*, **67,** 491–502.

Su, C., Lei, L. P., Duan, Y. Q. *et al.* (2012). Culture-independent methods for studying environmental microorganisms: methods, application, and perspective. *Applied Microbiology and Biotechnology*, **93,** 993–1003.

Sumner, J. (2000). *The Natural History of Medicinal Plants*. Portland, OR: Timber Press.

Sun, L., Qiu, F. B., Zhang, X. X. *et al.* (2008). Endophytic bacterial diversity in rice (*Oryza sativa* L.) roots estimated by 16S rDNA sequence analysis. *Microbial Ecology*, **55,** 415–424.

Teplitski, M., Barak, J. D. and Schneider, K. R. (2009). Human enteric pathogens in produce: un-answered ecological questions with direct implications for food safety. *Current Opinion in Biotechnology*, **20,** 166–171.

Tyler, H. L. and Triplett, E. W. (2008). Plants as a habitat for beneficial and/or human pathogenic bacteria. *Annual Review of Phytopathology*, **46**, 53–73.

Tyson, G. W. and Banfield, J. F. (2005). Cultivating the uncultivated: a community genomics perspective. *Trends in Microbiology*, **13,** 411–415.

Vijayan, P., Shockey, J., Levesque, C. A. *et al.* (1998). A role for jasmonate in pathogen defense of *Arabidopsis*. *Proceedings of the National Academy of Sciences of the United States of America*, **95,** 7209–7214.

Vorholt, J. A. (2012). Microbial life in the phyllosphere. *Nature Reviews Microbiology*, **10,** 828–840.

Wang, H. X., Geng, Z. L., Zeng, Y. *et al.* (2008). Enriching plant microbiota for a metagenomic library construction. *Environmental Microbiology*, **10,** 2684–2691.

Warriner, K. and Namvar, A. (2010). The tricks learnt by human enteric pathogens from phytopathogens to persist within the plant environment. *Current Opinion in Biotechnology*, **21,** 131–136.

Warriner, K., Ibrahim, F., Dickinson, M. *et al.* (2003). Internalization of human pathogens within growing salad vegetables. *Biotechnology & Genetic Engineering Reviews*, **20,** 117–134.

Wennstrom, A. (1994). Endophyte: the misuse of an old term. *Oikos*, **71,** 535–536.

Wilson, D. (1995). Endophyte: the evolution of a term, and clarification of its use and definition. *Oikos*, **73,** 274–276.

Wistrom, C. and Purcell, A. H. (2005). The fate of *Xylella fastidiosa* in vineyard weeds and other alternate hosts in California. *Plant Disease*, **89,** 994–999.

Wright, K. M., Crozier, L., Marshall, J. *et al.* (2017). Differences in internalization and growth of *Escherichia coli* O157:H7 within the apoplast of edible plants, spinach and lettuce, compared with the model species *Nicotiana benthamiana*. *Microbial Biotechnology*, **10,** 555–569.

Yang, B., Wang, Y. and Qian, P. Y. (2016). Sensitivity and correlation of hypervariable regions in 16S rRNA genes in phylogenetic analysis. *BMC Bioinformatics*, **17**, 135.

You, C. B., Lin, M., Fang, X. J. *et al.* (1995). Attachment of *Alcaligenes* to rice roots. *Soil Biology & Biochemistry*, **27,** 463–466.

Zhang, H. W., Song, Y. C. and Tan, R. X. (2006). Biology and chemistry of endophytes. *Natural Product Reports,* **23,** 753–771.

Zhang, X., Zhou, Y. Y., Li, Y. *et al.* (2017). Screening and characterization of endophytic *Bacillus* for biocontrol of grapevine downy mildew. *Crop Protection,* **96,** 173–179.

Zhao, Y. F., Sundin, G. W. and Wang, D. P. (2009). Construction and analysis of pathogenicity island deletion mutants of *Erwinia amylovora. Canadian Journal of Microbiology,* **55,** 457–464.

11

Meta-Omics Approach to Unravel the Endophytic Bacterial Communities of *Brassica napus* and Other Agronomically Important Crops in Response to Agricultural Practices

RIDHDHI RATHORE, KIERAN J. GERMAINE, PATRICK D. FORRISTAL, JOHN SPINK AND DAVID N. DOWLING

Abstract

Endophytes are microorganisms that colonise the internal compartments of host plants without harming the host. In many cases, endophytic bacteria have been shown to provide several beneficial effects to their host plant, including growth-promoting activity, modulation of plant metabolism and phytohormone signalling that leads to adaptation to environmental abiotic or biotic stresses. Endophytic bacterial community structure is influenced by plant genotype, soil type, abiotic and biotic factors, such as environmental conditions and microbe–microbe/plant–microbe interactions. In addition, agricultural management practices, such as soil tillage, crop rotation, and fertiliser and pesticide applications have a major effect on the function and structure of the soil, rhizosphere and endophytic bacterial communities. Oilseed rape (*Brassica napus* L.) is an important break crop in cereal crop rotation and can significantly reduce the rate of 'take-all' fungal disease (*Gaeumannomyces graminis* var. *tritici*) and, as a result, improves the yield of subsequent cereal crops. Additionally, oilseed rape is the world's third largest source of vegetable oil used for human nutrition and as a source of oil for biodiesel production. Therefore, the promotion of agricultural practices that maintain the natural diversity of *B. napus* endophytic bacteria is receiving attention as an important element for a sustainable agricultural system that ensures crop productivity and quality while reducing inputs. This chapter reviews the existing literature on the role of endophytic bacteria in oilseed rape crop production, agricultural factors influencing oilseed rape bacterial diversity and discusses how meta-omics is enhancing our understanding of the endophytic bacteria and their function.

11.1 Introduction

In the twenty-first century, modern agriculture is facing the significant challenge of producing high-quality nutritious food for a continuously increasing global population, while also conserving the natural resources that are so critical to its ability to maintain productivity (Godfray *et al.*, 2010; Tilman *et al.*, 2011). In general, food crops are the most important aspect of the food system. They directly provide edible produce for human consumption, for livestock (ultimately supporting the meat and dairy industries), the drinks industry and even biofuel production (Steinfeld *et al.*, 1998; Trostle, 2008). To accomplish these increasing food demands, intensive agricultural practices, and excessive use of agrochemicals and heavy machinery, are often employed. These can, in turn, have negative impacts, both on soil quality and the wider environment. For example, poor agronomic practices can lead to soil degradation/erosion and reduce soil fertility, further inducing a loss of biodiversity and increasing greenhouse gas emissions linked to climate change (Tilman *et al.*, 2002). Moreover, major international research initiatives such as the European Union's Horizon 2020 programme, aims to improve both the productivity and sustainability of agriculture. Advancements in modern agriculture are multifaceted, and include the use of plant growth-promoting bacteria to alleviate the dependency on chemical inputs and have also been shown to result in significant yield increases (Bottini *et al.*, 2004). Beneficial plant–microbe interactions are thus hot topics in agricultural research (Farrar *et al.*, 2014; George *et al.*, 2016).

11.2 Importance of Oilseed Rape and its Endophytic Bacteria in Agriculture

Brassica napus L. (oilseed rape, OSR) is the most economically important species of Brassicaceae worldwide (Etesami and Alikhani, 2016). OSR, also known as rapeseed, swede rape, Argentine rape, oil rape and canola, is in high demand due to its uses in a variety of sectors including: for human nutrition as the oil has a high proportion of unsaturated fatty acids (linoleic acid 15–20% and linolenic acid 8–12%); as a valuable animal feed with high energy and protein content; the production of renewable biodiesel fuel and environmentally friendly lubricating oil (Cardone *et al.*, 2003; Delourme *et al.*, 2006). OSR is also an important break crop in cereal crop rotation and can significantly reduce the rate of 'take-all' fungal disease (*Gaeumannomyces graminis* var. *tritici*). As a result, it can increase the yield of subsequent cereal crops by 20–30% (Angus *et al.*, 1991; Kirkegaard *et al.*, 2008). Long-term studies have reported that OSR improves the yield of wheat grown in rotation by 10–26% compared to wheat in monoculture or wheat in rotation with other cereal crops, such as barley and oats (Schonhammer and Fischbeck, 1987; Christen *et al.*, 1992). Moreover, the deep tap root system of OSR can improve soil structure and tilth by improving soil aggregates, porosity and increasing water and nitrogen uptake, which is important for the healthy growth of a subsequent cereal crop (Kirkegaard *et al.*, 2008). OSR is grown as winter or spring varieties. The winter varieties are established in early autumn and can take up considerable quantities of nitrogen and reduce the nitrate leaching from the soil. In addition, it covers the ground due to its large biomass and reduces the risk of surface run-off, soil erosion and soil acidification (Teagasc, 2009). The total area of OSR cultivation is increasing rapidly and in 2016 the FAO reported that OSR cultivation has increased by 79% from 1989 with over 71 million tons of rapeseed being produced worldwide.

A number of recent microbiome studies have significantly advanced our understanding of the microbial assemblages of OSR. Figure 11.1 shows the most abundant bacterial families found in OSR calculated from recent studies (Gkarmiri *et al.*, 2017; Rathore *et al.*, 2017; Rybakova *et al.*, 2017; Zhao *et al.*, 2017). Some of these studies have reported the beneficial attributes of endophytic bacteria associated with OSR from bacterial phyla such as Proteobacteria, Actinobacteria and Firmicutes. Many members of these phyla have been reported to be involved in plant growth promotion (PGP), heavy metal and antibiotic resistance and suppression of various fungal pathogens (Table 11.1). For instance, Gkarmiri *et al.* (2017) used RNA stable isotopes ($^{12}CO_2$, $^{13}CO_2$) and high-throughput sequencing to characterise the active bacterial microbiome that colonise the rhizosphere and roots of OSR. The authors reported that the bacterial phyla Acidobacteria, Actinobacteria, Chloroflexi, Gemmatimonadetes, Planctomycetes, Proteobacteria and Verrucomicrobia were the most active in the rhizosphere, whereas Proteobacteria, Actinobacteria and Bacteroidetes were active in roots of OSR. The root genera with the highest relative abundance and metabolic activity were *Actinoplanes*, *Agrobacterium*, *Devosia*, *Flavobacterium*, *Rhizobium* and *Streptomyces*. *Kaistobacter* and *Rhodoplanes* were highly active in rhizosphere fraction. Members of the mentioned genera are well known for their PGP and biocontrol activity. Also, *Rhizobium* and *Rhodoplanes* are potential nitrogen fixers (Gkarmiri *et al.*, 2017). Rybakova *et al.* (2017) studied the *Brassica napus* seed microbiome through amplicon 16S metagenomic analysis and reported the most abundant taxa found within the OSR seed were *Acetobacteraceae*, Acidobacteria, *Bacillus*, *Bradyrhizobiaceae*, *Burkholderiaceae*, *Burkholderia*,

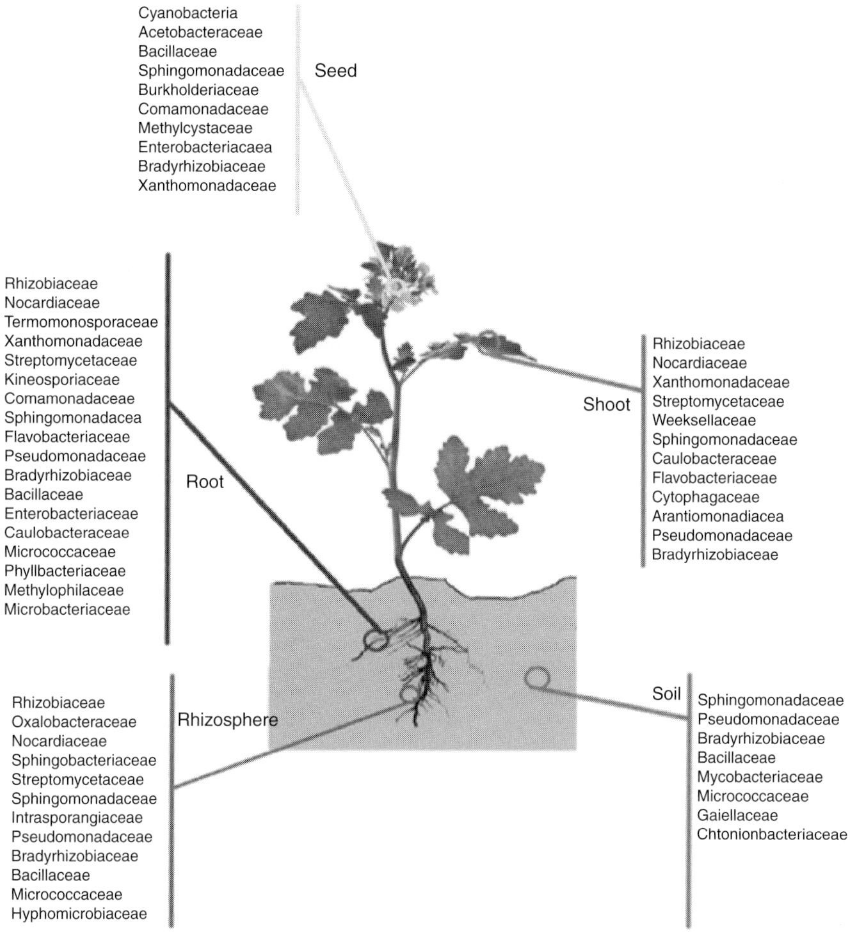

Figure 11.1 The most abundant bacterial families associated with *Brassica napus* in relation to shoot, root, rhizosphere and soil.
Compiled from data in Gkarmiri *et al.* (2017), Rathore *et al.* (2017), Rybakova *et al.* (2017) and Zhao *et al.* (2017).

Comamonadaceae, Cyanobacteria, *Methylobacterium, Mesorhizobium, Pseudomonas, Rhizobiaceae, Sphingomonas* and *Xanthomonadaceae.* These bacterial taxa were already known for their beneficial activity in plants.

Granér *et al.* (2003) selected four cultivars of *Brassica napus* that were genetically distinct from each other and known to possess different sensitivities towards the plant pathogen *Verticillium longisporum.* They studied the endophytic bacterial populations of OSR through agar plating assays, fatty acid methyl ester (FAME) profiling and functional characterisation. They isolated bacterial strains of *Bacillus* and *Pseudomonas* which were shown to express PGP properties and significant

Table 11.1 Endophytic bacteria associated with *Brassica napus* and their activity.

Microbial species	Beneficial activity	References
Bacillus spp.	Heavy metal resistance, antagonism towards *Leptosphaeria maculans* and *Sclerotinia sclerotiorum*	Xing *et al.*, 2005; Rajesh and Fern, 2006
Bacillus subtilis	Growth promotion, antagonism towards *Xanthomonas campestris*, *S. sclerotiorum* and *Colletotrichum higginsianum*	He *et al.*, 2013; Jiang *et al.*, 2007; Zhang *et al.*, 2009
Bacillus subtilis CEN3 *Bacillus cereus CEN5*	Antifungal activity towards fungal rice pathogens and plant growth promotion traits	Etesami and Alikhani, 2016
Bacillus subtilis 205	Biocontrol of *Rhizoctonia solani*	Fiddaman and Rossall, 1995
Bacillus amyloliquefaciens *Bacillus endophyticus*	Antagonism towards fungal pathogens	Danielsson *et al.*, 2007
Microbacterium sp. G16 *Pseudomonas fluorescens G10*	Heavy metal and antibiotic resistance, plant growth promotion under heavy metal contaminated soils	Sheng *et al.*, 2008
Paenibacillus polymyxa	Antagonism towards *Verticillium longisporum*	Granér, 2002; Granér *et al.*, 2003
Alcaligenes spp., *Pseudomonas chlororaphis*, *Serratia proteamaculans*, *Stenotrophomonas spp.*	Inhibition of *Verticillium dahliae*	Alström, 2001
Pseudomonas brassicacearum	ACC deaminase activity, promoted root elongation, increased root and shoot biomass	Achouak *et al.*, 2000
Azorhizobium caulinodans	Potential for plant growth promotion	O'Callaghan *et al.*, 2000
Enterobacter sp.	Inhibition of *Verticillium dahliae* and plant growth promotion	Alström, 2001; Wang *et al.*, 2013
Pseudomonas fluorescens	Heavy metal and antibiotic resistance, plant growth promotion, and antagonism towards *Xanthomonas campestris*	Sheng *et al.*, 2008
Pseudomonas fluorescens L321. L111, L228 and F113	ACC deaminase activity, phosphate solubilisation, biocontrol activity. Enhanced OSR growth in field trials	Lally *et al.*, 2017
Rhanella sp.	Plant growth promotion and heavy metal uptake	He *et al.*, 2013
Serratia plymuthica	Improved seed germination	Abuamsha *et al.*, 2011

inhibition of *V. longisporum*. Germida *et al.* (1998) also studied the bacterial diversity and composition associated with OSR roots through FAME profiling. They observed a higher proportion of *Bacillus*, *Flavobacterium*, *Micrococcus* and *Rathayibacter* genera living inside the root compared to those in the rhizoplane populations. Montalban *et al.* (2016) found that in zinc-contaminated soil, endophytic bacterial strains *Pseudomonas* sp. 228 & 256 and *Serratia* sp. strain 246 facilitated the root development of *B. napus*. While in a similar study, endophytic bacterial strains from *B. napus* cultivated in a cadmium (Cd), lead (Pb) and zinc (Zn)-contaminated field were capable of phosphorus solubilisation, nitrogen fixation, indole acetic acid (IAA) production and demonstrated ACC deaminase activity (Croes *et al.*, 2013). Lally *et al.* (2017) performed field trials that involved the inoculation of OSR with bacterial endophytes expressing plant growth promotion traits such as phosphate solubilisation, phytohormone production and ACC deaminase activity. They found significant enhancement of plant growth parameters such as plant height and biomass in endophyte-inoculated plots compared to uninoculated plots. Clearly, these microbiomes have a significant role to play in the health and development of OSR. By gaining a deeper understanding of the factors that modulate the microbial community structure and population sizes of the OSR microbiome, we can devise strategies and practices that can positively select for these beneficial microbes in the OSR microbiome and so reduce our reliance on chemical fertiliser and fungicide inputs in OSR production.

11.3 Agricultural Management Practices as Drivers of the OSR Microbiome

Agricultural land management greatly alters the physical, chemical and biological properties of soils (García-Orenes *et al.*, 2013). Soil tillage may lead to reductions in soil organism diversity due to mechanical destruction, soil compaction, changes in pore volume, desiccation and access to food resources (Giller, 1996). The excessive use of pesticides and fertilisers can drastically modify the function and structure of microbial communities, thereby altering the normal functioning of terrestrial ecosystems (Pampulha and Oliveira, 2006). Agricultural management alters the quantity and quality of plant residues entering the soil and their spatial distribution, through changes in nutrients and inputs. A number of sustainable agricultural approaches can be used to improve or maintain the soil quality, encourage biodiversity and reduce soil erosion such as conservation tillage (no tillage or reduced tillage), crop rotation, use of plant growth-promoting bacteria (PGPB) inoculants, mineral/organic fertilisers, and use of drought and disease resistant crop varieties (García-Orenes *et al.*, 2010). It is very important to understand the impact of these management practices on the microbial diversity structure and composition to develop a sustainable agricultural model, based not only on crop productivity but also on ecological principles.

Beneficial microbes often originate from soil, initially infecting roots of the host plant and colonising the plant apoplast. Therefore it could be presumed that the endophytic community represents a subset of the wider rhizospheric microbial population and it would be influenced by agronomic practices. However, research on the effect of agricultural practices on endophyte population dynamics is limited.

11.3.1 Effect of Soil Types and Plant Genotypes on OSR Microbiota

Many research studies have suggested that soil is the major starting inoculum of microbiota from which plants select their microbiome. Therefore, soil types play a vital role in microbiota selection as distinct soil types can have very different microbial community structures (Bulgarelli *et al.*, 2012; Lundberg *et al.*, 2012; Bulgarelli *et al.*, 2013). However, the assessment of the extent to which soil type influences the plant microbial community under field conditions is difficult to measure, as other factors such as the climate, cropping history and other agricultural management practices can also influence the soil microbiome (Costa *et al.*, 2006; Costa *et al.*, 2007). Schreiter *et al.* (2014) studied the rhizosphere microbiome in three distinct soil types and observed that it differed depending on the soil type and ultimately influenced the formation of distinct bacterial endophytic compositions in lettuce. Bulgarelli *et al.* (2012) reported, through pyrosequencing of 16S rRNA gene fragments amplified from the total community DNA of *Arabidopsis thaliana*, that soil type explains the composition of root-inhabiting bacterial communities and that the plant genotype is also important.

Other studies have also found that plant community composition is a major driving factor for microbiota selection. Different plant species, and even different genotypes of one species, can possess different root architecture and root exudation patterns that strongly influence plant-associated bacterial community structure and composition (Bais *et al.*, 2006, Bulgarelli *et al.*, 2013). Rybakova *et al.* (2017) studied the endophytic *Brassica napus* seed microbiome and reported that three different cultivars of OSR shared only one-third of the total 8362 operational taxonomic units (OTUs) which shows that the cultivar dependency and specificity on OSR endophytic microbiome. Moreover, Lundberg *et al.* (2012) studied the microbiome of more than 600 *Arabidopsis thaliana* plants and reported that rhizosphere and root endophytic microbiota are dependent on host genotypes. They also observed different rhizosphere and endophytic bacterial communities in plants grown in two geochemically distinct bulk soils.

11.3.2 Soil Tillage

Soil tillage is a major agricultural practices influencing the soil. Soil tillage modifies aeration, temperature and permeability, which have significant effects on soil

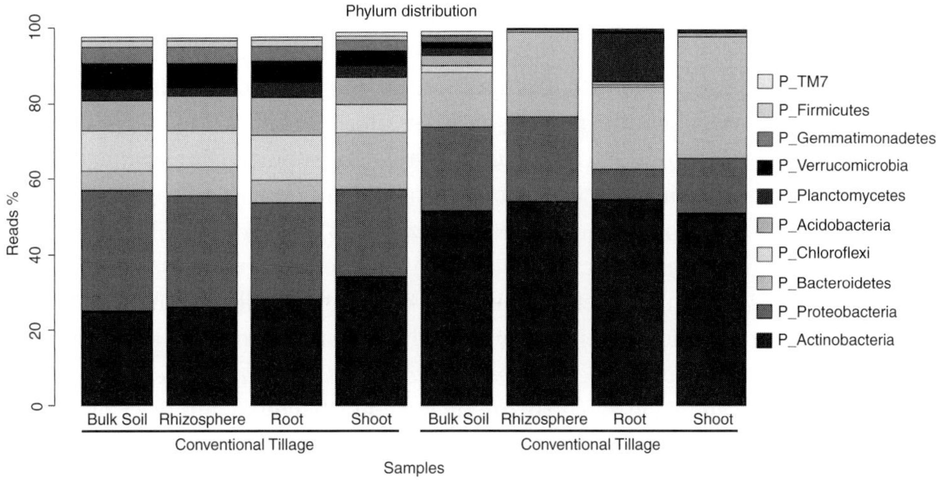

Figure 11.2 Relative abundance of the top 10 phyla in oilseed rape (OSR) in response to tillage system used. Average relative abundance (% of sequencing reads) of prokaryotic phyla associated with bulk soil, rhizosphere, root and shoot microhabitat zones of OSR under conventional tillage and conservation tillage, are displayed in different colours. (A black and white version of this figure will appear in some formats. For the colour version, please refer to the plate section.)
Adapted from Rathore *et al.* (2017).

microorganisms and their functions in the worked layer. In general, two types of tillage are used for farming: (1) mouldboard plough-based conventional tillage, and (2) conservation reduced to no tillage. Conventional tillage refers to soil inversion down to approximately 25–30 cm using a mouldboard plough, whereas conservation reduced tillage disturbs only the top 10 cm of soil in order to improve the conditions for seed germination.

Rathore *et al.* (2017) studied the impact of conventional versus conservation tillage on bulk soil, rhizosphere and endophytic (root and shoot) microbiota of OSR by using Illumina 16S amplicon sequencing (Figure 11.2). They reported that the root and shoot associated bacterial communities displayed markedly distinct profiles as a result of tillage practices. The root microbiota was highly sensitive to soil tillage. The phyla Proteobacteria, Bacteroidetes, Actinobacteria, Acidobacteria, Verrucomicrobia, Firmicutes and Chloroflexi largely dominated the OSR microbiota. In general, under conservation tillage, the enrichment of phylum Bacteroidetes and depletion of phylum Actinobacteria significantly differentiated the conservational tillage from the conventional tillage regime. Moreover, the root microbiota showed greater OTU diversity under conventional tillage. Conservation tillage displayed a remarkable abundance of bacterial members from the phyla Proteobacteria, Bacteroidetes and Firmicutes.

Dorr de Quadros *et al.* (2012) demonstrated through 16S amplicon metagenomic sequencing that soil microbial diversity, including the anaerobic community such as Chlamydiae, Chlorobi, Crenarchaeota, Euryarchaeota, Firmicutes and Verrucomicrobia were significantly higher in no tillage compared to conventionally tillage plots of a subtropical Acrisol. Exchangeable phosphorus (P) and magnesium (Mg) concentration, total organic carbon, nitrogen, and mineral nitrogen were also higher in the no-tillage plots. Navarro-Noya *et al.* (2013) observed similar results in a semiarid agroecosystem. Under reduced tillage the microbial biomass, plant organic matter and nutrient content were higher than conventional tillage. Yin *et al.* (2017) reported through amplicon metagenomic analysis that copiotrophic bacterial families such as *Cytophagaceae*, *Pseudomonadaceae* and *Oxalobacteriaceae* were more abundant in reduced tillage of wheat and oligotrophic bacterial families were abundant in conventional tillage. However, Degrune *et al.* (2017) observed higher abundance of copiotrophic bacteria under conventional tillage and oligotrophic bacteria under reduced tillage. Therefore, tillage alters soil microbial community structure but the response of individual groups of bacteria appears to be very context-specific and cannot be generalised across various agroecosystems. The response is largely dependent on the soil physical and chemical conditions, soil types and climatic conditions. Sampling depth can lead to different results and conclusions.

11.3.3 Effects of Crop Rotation on OSR Microbiota

Another agriculture management practice that impacts on microbial diversity is crop rotation, which is the practice of growing different crops in succession on the same land. Benefits from diversified crop rotations include increased soil nutrients and organic matter content (McDaniel *et al.*, 2014; Lehman *et al.*, 2017), enhanced soil water use efficiency (Larkin *et al.*, 2010), stabilised soil temperature, improved soil structure and tilth (Ball *et al.*, 2005), enhanced soil microbial biomass, diversity and activity (McDaniel *et al.*, 2014) and disruption of pests, weeds and disease cycles (Peters *et al.*, 2003; Hopkins *et al.*, 2004). Some of these rotational effects result from variation in root architecture and rhizodeposition, biogeochemical cycles, host plant identity and genetics, soil characteristics and climatic conditions. These collectively influence the size, structure and dynamics of soil and plant-associated microbial communities (Garbeva *et al.*, 2004; White *et al.*, 2013; Lehman *et al.*, 2015). Hooper *et al.* (2000) stated that increasing aboveground crop biodiversity can result in corresponding increases in diversity belowground. Additionally, crop rotation may be the best, most widely practised and most cost-effective method for reducing soil-borne fungal pathogen populations.

For instance, *Brassica napus* rotated with tobacco has been shown to supress the tobacco black shank (*Phytophthora parasitica* var. *nicotianae*) fungal disease by secreting antimicrobial compounds such as 2-butenoic acid, benzothiazole, 2-(methylthio) benzothiazole, 1-(4-ethylphenyl)-ethanone, and 4-methoxyindole from the roots (Fang *et al.*, 2016). Plant roots can continuously secrete many compounds into the rhizosphere to mediate the interactions between the roots and pathogens (Bais *et al.*, 2004). Soil microbes can recognise the signals in the root exudates and subsequently colonise the host plant (Bais *et al.*, 2006). Oilseed rape is an effective break crop for cereals to inhibit take-all fungus infection in wheat (Angus *et al.*, 1991). Rathore *et al.* (unpublished) studied the 2,4-diacetylphloroglucinol (2,4-DAPG) antibiotic producing bacterial populations in the rhizosphere and root of OSR at different plant growth stages under crop rotation. They quantified the copy number of the *PhlD* gene in the soil, rhizosphere and roots at the vegetative, flowering and harvesting stage of the OSR growth cycle. They found that the root microbiota had the highest abundance of the *PhlD* gene at harvesting stage followed by vegetative stage and then flowering stage. Moreover, overall 2,4-DAPG producing bacterial abundance was higher under conservation tillage compared to conventional tillage practices at all plant growth stages (Figure 11.3). 2,4-DAPG is key component of the natural biological control against the take-all fungus. Therefore, these results suggest that under crop rotation with conservation tillage, OSR may reduce the rate of fungal infection by favouring the beneficial bacterial population in its rhizosphere and root at all plant growth stages.

Zhao *et al.* (2017) compared the endophytic microbiome of symptomatic and asymptomatic *Brassica napus* roots infected with *Plasmodiophora brassicae* (a causal agent of clubroot disease) through 16S amplicon sequencing. They observed that the bacterial population and its relative abundance were far higher in the asymptomatic roots than symptomatic OSR root samples. For instance, the bacterial families *Comamonadaceae, Flavobacteriaceae, Methylophilaceae, Oxalobacteraceae, Pseudomonadaceae, Rhizobiaceae, Sphingobacteriaceae* and *Xanthomonadaceae* were significantly higher in asymptomatic roots under *P. brassicae* infection. Moreover, many of these microorganisms have biocontrol and PGP functions. It was shown several decades ago that disease-suppressive properties of soil were largely induced by long-term cultivation of monoculture leading to build up of a host-specific microbial community (Whipps, 1997; Kwak and Weller, 2013; Hegewald *et al.*, 2017). To date, the specific involvement of endophytic bacteria in development of disease-suppressive soils has not been adequately studied. However, *Actinomyces, Bacillus, Pseudomonas* and *Streptomyces* that lead endophytic lifestyles were shown to contribute to the disease-suppressive traits of soils (Raaijmakers and Weller, 1998; Weller *et al.*, 2002; Mendes *et al.*, 2011).

11.3.4 Effects of Fertiliser and Pesticide Application on OSR Microbiota

Modern agriculture largely depends on the widespread application of agrochemicals, including inorganic fertilisers and pesticides. Indiscriminate, long-term and over-application of these chemicals have severe effects on soil ecology that may lead to alterations in, or the erosion of, beneficial or plant and soil probiotic microflora

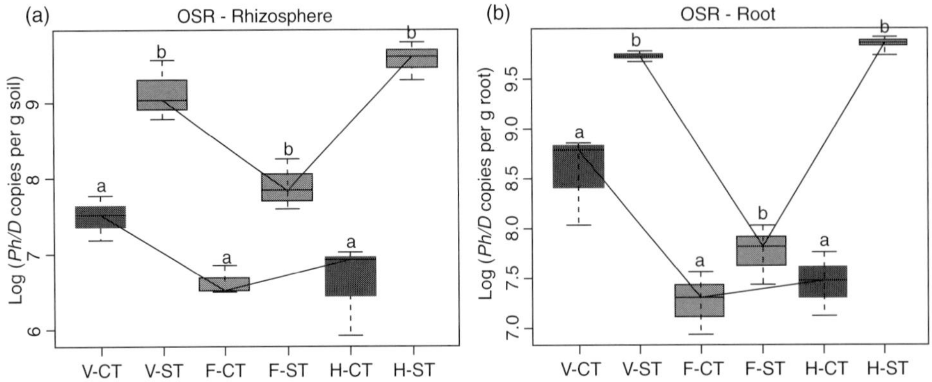

Figure 11.3 Dynamics of *PhlD*+ bacterial community in OSR associated to rhizosphere and root microhabitats during the plant growth stages under crop rotation. Quantitative real-time PCR (qPCR) of *PhlD*+ population in OSR (a) rhizosphere (b) root, based on the number of *PhlD* gene copies detected per gramme of rhizosphere soil and root sample respectively at three different plant developmental stages; vegetative (V), flowering (F), and harvesting (H) stages under conventional (CT) and conservation (ST) tillage practices. Darker colour defines the samples from CT and lighter colour defines the samples from ST. *Lowercase letters* denote statistically significant differences by Kruskal–Wallis and Dunn's *post hoc* tests, $P < 0.05$, Banjamini-Hochberg (BH) corrected between the tillage systems within the one plant developmental stage. (A black and white version of this figure will appear in some formats. For the colour version, please refer to the plate section.) Rathore *et al.* (unpublished).

(Kalia and Gosal, 2011). Field-grown plants could be intentionally exposed to these detrimental chemicals (e.g. by spraying) or indirectly from residues remaining from previous applications (Khan *et al.*, 2004). These chemicals or their degraded products interact with soils and plant-associated microbes, and cause DNA, protein, oxidative or membrane damage (Pham *et al.*, 2004), they can adversely affect N_2 fixation (Anderson *et al.*, 2004), protein synthesis and metabolic enzyme activity (Boldt and Jacobsen, 1998). They can disrupt the signalling between the microbes and plant phytochemicals (Fox *et al.*, 2007), reduce the photosynthate allocation (Datta *et al.*, 2009), or can restrict root growth (Eberbach and Douglas, 1989). Additionally, pesticides that persist in soils may have a long-lasting impact on rhizobia survival and function (Eliason *et al.*, 2004, Fox *et al.*, 2007).

Fox *et al.* (2007) reported that the pesticides disrupted the signalling between legume plant-derived phytochemicals (luteolin, apigenin) and *Rhizobium* Nod D receptors which are essential for nodulation and symbiotic nitrogen fixation. However, PGPB, including symbiotic nitrogen fixers, can affect plant development either indirectly by circumventing the toxic effects of pesticides or directly by synthesising plant growth-regulating substances (Jeon *et al.*, 2003). The fipronil and pyriproxyfen insecticide-tolerant and phytohormone-producing *R. leguminosarum* strain MRP1 promoted the growth of pea when it was applied as seed inoculant and plants were cultivated in fipronil and pyriproxyfen treated soil (Ahemad and Khan, 2010). Germaine *et al.* (2006) inoculated pea with a *Pseudomonas* endophyte capable of degrading the organo-chlorine herbicide, 2,4-dichlorophenoxyacetic acid (2,4-D). When inoculated plants were exposed to 2,4-D, they showed no accumulation of the herbicide into their tissues and experienced little or no signs of phytotoxicity. In contrast, uninoculated plants showed significant accumulation of 2,4-D and displayed signs of toxicity including a reduction in biomass, leaf abscission and callus development on their roots. Diazotrophic microorganisms perform biological nitrogen fixation using the genes involved in the synthesis and functioning of nitrogenase (*nif*) (Souza *et al.*, 2015). It was demonstrated that the colonisation ability of the nitrogen-fixing endophytic bacterium *Acetobacter diazotrophicus* was significantly decreased in sugarcane plants fertilised with high levels of nitrogen (Fernandez *et al.*, 2012). Analysis of the endophytic population of maize roots under treatment with herbicides and different fertiliser types revealed that differentiated microbial communities developed in the maize plants cultivated using mineral fertiliser, compared to plants cultivated using organic fertiliser (Seghers *et al.*, 2004). These studies did not, however, reveal if the changes in endophyte population were a consequence of changes in overall soil microbial population upon the fertiliser and pesticide treatment.

11.4 Conclusions: Current Challenges and Future Prospects

Endophytes have become a research focus in agriculture due to their potential application to plant production and stress resistance (Murphy *et al.*, 2018). However, the interaction between different endophytes and their hosts is still poorly understood. For instance, several plant growth-promoting microbes are commercialised for their application as microbial inoculants such as biofertiliser, plant straighteners, biocontrol agents, biopesticides and phytostimulants (Berg, 2009). However, the efficacy of many microbial inoculants is often not observed at field scale due to issues with the establishment of the biological agent, as they have to face competition with the native soil microbiota (Le Cocq *et al.*, 2017). Moreover, several other environmental factors shape the endophytic microbiome. Therefore, further research is necessary into agricultural management practices which stimulate the indigenous bacterial community to respond and support host plant growth and defence. The advancement of meta-omic approaches has opened the doors for detailed analysis of the plant microbiome, and will allow researchers to investigate practices to manipulate the plant microbiota and their intrinsic traits in order to test interactions.

References

Abuamsha, R., Salman, M. and Ehlers, R.-U. (2011). Improvement of seed bio-priming of oilseed rape (*Brassica napus* ssp. oleifera) with *Serratia plymuthica* and *Pseudomonas chlororaphis*. *Biocontrol Science and Technology*, **21**, 199–213.

Achouak, W., Sutra, L., Heulin, T. *et al.* (2000). *Pseudomonas brassicacearum* sp. nov. and *Pseudomonas thivervalensis* sp. nov., two root-associated bacteria isolated from *Brassica napus* and *Arabidopsis thaliana*. *International Journal of Systematic and Evolutionary Microbiology*, **50**, 1, 9–18.

Ahemad, M. and Khan, M. S. (2010). Comparative toxicity of selected insecticides to pea plants and growth promotion in response to insecticide-tolerant and plant-growth-promoting *Rhizobium leguminosarum*. *Crop Protection*, **29**, 325–329.

Alström, S. (2001). Characteristics of bacteria from oilseed rape in relation to their biocontrol activity against *Verticillium dahliae*. *Journal of Phytopathology*, **149**, 57–64.

Anderson, A., Baldock, J. A., Rogers, S. L., Bellotti, W. and Gill, G. (2004). Influence of chlorsulfuron on rhizobial growth, nodule formation, and nitrogen fixation with chickpea. *Australian Journal of Agricultural Research*, **55**, 1059–1070.

Angus, J., Herwaarden, A., Howe, G. and Van, H. A. (1991). Productivity and break crop effects of winter-growing oilseeds. *Australian Journal of Experimental Agriculture*, **31**, 669–677.

Bais, H. P., Park, S. W., Weir, T. L., Callaway, R. M. and Vivanco, J. M. (2004). How plants communicate using the underground information superhighway. *Trends in Plant Science*, **9**, 26–32.

Bais, H. P., Weir, T. L., Perry, L. G., Gilroy, S. and Vivanco, J. M. (2006). The role of root exudates in rhizosphere interations with plants and other organisms. *Annual Review of Plant Biology*, **57**, 233–266.

Ball, B. C., Bingham, I., Rees, R. M., Watson, C. A. and Litterick, A. (2005). The role of crop rotations in determining soil structure and crop growth conditions. *Canadian Journal of Soil Science*, **85**, 557–577.

Berg, G. (2009). Plant-microbe interactions promoting plant growth and health, perspectives for controlled use of microorganisms in agriculture. *Applied Microbiology and Biotechnology*, **84**, 11–18.

Boldt, T. S. and Jacobsen, C. S. (1998). Different toxic effects of the sulfonylurea herbicides metsulfuron methyl, chlorsulfuron and thifensulfuron methyl on fluorescent pseudomonads isolated from an agricultural soil. *FEMS Microbiology Letters*, **161**, 29–35.

Bottini, R., Cassan, F. and Piccoli, P. (2004). Gibberellin production by bacteria and its involvement in plant growth promotion and yield increase. *Applied Microbiology and Biotechnology*, **65**, 497–503.

Bulgarelli, D., Schlaeppi, K., Spaepen, S., Ver Loren van Themaat, E. and Schulze-Lefert, P. (2013). Structure and functions of the bacterial microbiota of plants. *Annual Review of Plant Biology*, **64**, 807–838.

Bulgarelli, D., Rott, M., Schlaeppi, K. *et al.* (2012). Revealing structure and assembly cues for *Arabidopsis* root-inhabiting bacterial microbiota. *Nature*, **488**, 91–95.

Cardone, M., Mazzoncini, M., Menini, S. *et al.* (2003). *Brassica carinata* as an alternative oil crop for the production of biodiesel in Italy: agronomic evaluation, fuel production by transesterification and characterization. *Biomass and Bioenergy*, **25**, 623–636.

Christen, O., Sieling, K. and Hanus, H. (1992). The effect of different preceding crops on the development, growth and yield of winter wheat. *European Journal of Agronomy*, **1**, 21–28.

Costa, R., Gotz, M., Mrotzek, N. *et al.* (2006). Effects of site and plant species on rhizosphere community structure as revealed by molecular analysis of microbial guilds. *FEMS Microbiology Ecology*, **56**, 236–249.

Costa, R., Gomes, N. C., Krogerrecklenfort, E. *et al.* (2007). *Pseudomonas* community structure and antagonistic potential in the rhizosphere: insights gained by combining phylogenetic and functional gene-based analyses. *Environmental Microbiology*, **9**, 2260–2273.

Croes, S., Weyens, N., Janssen, J. *et al.* (2013). Bacterial communities associated with *Brassica napus* L. grown on trace element-contaminated and non-contaminated fields: a genotypic and phenotypic comparison. *Microbial Biotechnology*, **6**, 371–384.

Danielsson, J., Reva, O. and Meijer, J. (2007). Protection of oilseed rape (*Brassica napus*) toward fungal pathogens by strains of plant-associated *Bacillus amyloliquefaciens*. *Microbial Ecology*, **54,** 134–140.

Datta, A., Sindel, B. M., Kristiansen, P., Jessop, R. S. and Felton, W. L. (2009). Effect of isoxaflutole on the growth, nodulation and nitrogen fixation of chickpea (*Cicer arietinum* L.). *Crop Protection*, **28**, 923–927.

Degrune, F., Theodorakopoulos, N., Colinet, G. *et al.* (2017). Temporal dynamics of soil microbial communities below the seedbed under two contrasting tillage regimes. *Frontiers in Microbiology*, **8**, 1127.

Delourme, R., Falentin, C., Huteau, V. *et al.* (2006). Genetic control of oil content in oilseed rape (*Brassica napus* L.). *Theoretical and Applied Genetics*, **113**, 1331–1345.

Dorr de Quadros, P., Zhalnina, K., Davis-Richardson, A. *et al.* (2012). The effect of tillage system and crop rotation on soil microbial diversity and composition in a subtropical acrisol. *Diversity*, **4**, 375.

Eberbach, P. L. and Douglas, L. A. (1989). Herbicide effects on the growth and nodulation potential of *Rhizobium trifolii* with *Trifolium subterraneum* L. *Plant and Soil*, **119**, 15–23.

Eliason, R., Schoenau, J. J., Szmigielski, A. M. and Laverty, W. M. (2004). Phytotoxicity and persistence of flucarbazone-sodium in soil. *Weed Science*, **52**, 857–862.

Etesami, H. and Alikhani, H. A. (2016). Rhizosphere and endorhiza of oilseed rape (*Brassica napus* L.) plant harbor bacteria with multifaceted beneficial effects. *Biological Control*, **94**, 11–24.

Fang, Y., Zhang, L., Jiao, Y. *et al.* (2016). Tobacco rotated with rapeseed for soil-borne *Phytophthora* pathogen biocontrol: mediated by rapeseed root exudates. *Frontiers in Microbiology*, **7**, 894.

Farrar, K., Bryant, D. and Cope-Selby, N. (2014). Understanding and engineering beneficial plant–microbe interactions: plant growth promotion in energy crops. *Plant Biotechnology Journal*, **12**, 1193–1206.

Fernandez, O., Theocharis, A., Bordiec, S. *et al.* (2012). *Burkholderia phytofirmans* PsJN acclimates grapevine to cold by modulating carbohydrate metabolism. *Molecular Plant–Microbe Interactions*, **25**, 496–504.

Fiddaman, P. J. and Rossall, S. (1995). Selection of bacterial antagonists for the biological control of *Rhizoctonia solani* in oilseed rape (*Brassica napus*). *Plant Pathology*, **44**, 695–703.

Fox, J. E., Gulledge, J., Engelhaupt, E., Burow, M. E. and McLachlan, J. A. (2007). Pesticides reduce symbiotic efficiency of nitrogen-fixing rhizobia and host plants. *Proceedings of the National Academy of Sciences of the United States of America*, **104**, 10282–10287.

Garbeva, P., van Veen, J. A. and van Elsas, J. D. (2004). Microbial diversity in soil: selection microbial populations by plant and soil type and implications for disease suppressiveness. *Annual Review of Phytopathology*, **42**, 243–270.

García-Orenes, F., Guerrero, C., Roldán, A. *et al.* (2010). Soil microbial biomass and activity under different agricultural management systems in a semiarid Mediterranean agroecosystem. *Soil and Tillage Research*, **109**, 110–115.

García-Orenes, F., Morugán-Coronado, A., Zornoza, R. and Scow, K. (2013). Changes in soil microbial community structure influenced by agricultural management practices in a Mediterranean agro-ecosystem. *PLoS One*, **8**, e80522.

George, T. S., Dou, D. and Wang, X. (2016). Plant-microbe interactions: manipulating signals to enhance agricultural sustainability and environmental security. *Plant Growth Regulation*, **80**, 1–3.

Germaine, K. J., Liu, X., Cabellos, G. G., Hogan, J. P., Ryan, D. and Dowling, D. N. (2006). Bacterial endophyte-enhanced phytoremediation of the organochlorine herbicide 2,4-dichlorophenoxyacetic acid. *FEMS Microbiology Ecology*, **57**, 302–310.

Germida, J. J., Siciliano, S. D., Renato de Freitas, J. and Seib, A. M. (1998). Diversity of root-associated bacteria associated with field-grown canola (*Brassica napus* L.) and wheat (*Triticum aestivum* L.). *FEMS Microbiology Ecology*, **26**, 43–50.

Giller, P. S. (1996). The diversity of soil communities, the 'poor man's tropical rainforest'. *Biodiversity and Conservation*, **5**, 135–168.

Gkarmiri, K., Mahmood, S., Ekblad, A., Alstrom, S., Hogberg, N. and Finlay, R. (2017). Identifying the active microbiome associated with roots and rhizosphere soil of oilseed rape. *Applied and Environmental Microbiology*, **83**, e01938-17.

Godfray, H. C., Beddington, J. R., Crute, I. R. *et al.* (2010). Food security: the challenge of feeding 9 billion people. *Science*, **327**, 812–818.

Granér, G. (2002). *Studies on Fungal Pathogens on Oilseed Rape* (Brassica napus). Uppsala, Sweden: Sveriges Lantbruksuniversitet (Swedish University of Agricultural Sciences).

Granér, G., Persson, P., Meijer, J. and Alstrom, S. (2003). A study on microbial diversity in different cultivars of *Brassica napus* in relation to its wilt pathogen, *Verticillium longisporum*. *FEMS Microbiology Letters*, **224**, 269–276.

He, H., Ye, Z., Yang, D. *et al.* (2013). Characterization of endophytic *Rahnella* sp. JN6 from *Polygonum pubescens* and its potential in promoting growth and Cd, Pb, Zn uptake by *Brassica napus*. *Chemosphere*, **90**, 1960–1965.

Hegewald, H., Koblenz, B., Wensch-Dorendorf, M. and Christen, O. (2017). Yield, yield formation, and blackleg disease of oilseed rape cultivated in high-intensity crop rotations. *Archives of Agronomy and Soil Science*, **63**, 1785–1799.

Hooper, D. U., Bignell, D. E., Brown, V. K. *et al.* (2000). Interactions between aboveground and belowground biodiversity in terrestrial ecosystems: patterns, mechanisms, and feedbacks. *BioScience*, **50**, 1049–1061.

Hopkins, B. G., Hutchinson, P. J., Patterson, P. *et al.* (2004). Cropping sequence and rotation: impact on potato production and soil condition. *Proceedings of the University of Idaho Winter Commodity Schools*, **35**, 97-109.

Jeon, J. S., Lee, S. S., Kim, H. Y. *et al.* (2003). Plant growth promotion in soil by some inoculated microorganisms. *The Journal of Microbiology*, **41**, 271–276.

Jiang, M. L., Zhao, R., Hu, X. J., Zhang, Y. B. and Wang, G. P. (2007). Colonization of antifungal endobacterium BY-2 in oilcrop rape and its control effect on disease caused by *Sclerotinia sclerotiorum*. *Acta Phytopathologica Sinica*, **37**, 192–196.

Kalia, A. and Gosal, S. K. (2011). Effect of pesticide application on soil microorganisms. *Archives of Agronomy and Soil Science*, **57**, 569–596.

Khan, M. S., Zaidi, A. and Aamil, M. (2004). Influence of herbicides on chickpea-*Mesorhizobium* symbiosis. *Agronomie* **24**, 123–127.

Kirkegaard, J., Christen, O., Krupinsky, J. and Layzell, D. (2008). Break crop benefits in temperate wheat production. *Field Crops Research*, **107**, 185–195.

Kwak, Y.-S. and Weller, D. M. (2013). Take-all of wheat and natural disease suppression: a review. *The Plant Pathology Journal*, **29**, 125–135.

Lally, R. D., Galbally, P., Moreira, A. S. *et al.* (2017). Application of endophytic *Pseudomonas fluorescens* and a bacterial consortium to *Brassica napus* can increase plant height and biomass under greenhouse and field conditions. *Frontiers in Plant Science*, **8**, 2193.

Larkin, R. P., Griffin, T. S. and Honeycutt, C. W. (2010). Rotation and cover crop effects on soilborne potato diseases, tuber yield, and soil microbial communities. *Plant Disease*, **94**, 1491–1502.

Le Cocq, K., Gurr, S. J., Hirsch, P. R. and Mauchline, T. H. (2017). Exploitation of endophytes for sustainable agricultural intensification. *Molecular Plant Pathology*, **18**, 469–473.

Lehman, M. R., Cambardella, A. C., Stott, E. D. *et al.* (2015). Understanding and enhancing soil biological health: the solution for reversing soil degradation. *Sustainability*, **7**, 988–1027.

Lehman, R. M., Osborne, S. L. and Duke, S. E. (2017). Diversified no-till crop rotation reduces nitrous oxide emissions, increases soybean yields, and promotes soil carbon accrual. *Soil Science Society of America Journal*, **81**, 76–83.

Lundberg, D. S., Lebeis, S. L., Paredes, S. H. *et al.* (2012). Defining the core *Arabidopsis thaliana* root microbiome. *Nature*, **488**, 86–90.

McDaniel, M. D., Tiemann, L. K. and Grandy A. S. (2014). Does agricultural crop diversity enhance soil microbial biomass and organic matter dynamics? A meta-analysis. *Ecological Applications*, **24**, 560–570.

Mendes, R., Kruijt, M., de Bruijn, I. *et al.* (2011). Deciphering the rhizosphere microbiome for disease-suppressive bacteria. *Science*, **332**, 1097–1100.

Montalban, B., Croes, S., Weyens, N. *et al.* (2016). Characterization of bacterial communities associated with *Brassica napus* L. growing on a Zn-contaminated soil and their effects on root growth. *Internation Journal of Phytoremediation*, **18**, 985–993.

Murphy, B. R., Doohan, F. M. and Hodkinson, T. R. (2018). From concept to commerce: developing a successful fungal endophyte inoculant for agricultural crops. *Journal of Fungi*, **4**, 2–11.

Navarro-Noya, Y. E., Gómez-Acata, S., Montoya-Ciriaco, N. *et al.* (2013). Relative impacts of tillage, residue management and crop-rotation on soil bacterial communities in a semi-arid agroecosystem. *Soil Biology and Biochemistry*, **65**, 86–95.

O'Callaghan, K. J., Stone, P. J., Hu, X. *et al.* (2000). Effects of glucosinolates and flavonoids on colonization of the roots of *Brassica napus* by *Azorhizobium caulinodans* ORS571. *Applied Environmental Microbiology*, **66**, 2185–2191.

Pampulha, M. E. and Oliveira, A. (2006). Impact of an herbicide combination of bromoxynil and prosulfuron on soil microorganisms. *Current Microbiology*, **53**, 238–243.

Peters, R. D., Sturz, A. V., Carter, M. R. and Sanderson, J. B. (2003). Developing disease-suppressive soils through crop rotation and tillage management practices. *Soil and Tillage Research*, **72**, 181–192.

Pham, C., Min, J. and Gu, M. (2004). Pesticide induced toxicity and stress response in bacterial cells. *Bulletin of Environmental Contamination and Toxicology*, **72**, 380–386.

Raaijmakers, J. M. and Weller, D. M. (1998). Natural plant protection by 2,4-diacetylphloroglucinol-producing *Pseudomonas* spp. in take-all decline soils. *Molecular Plant–Microbe Interactioins*, **11**, 144–152.

Rajesh, R. and Fern, W. G. D. (2006). Preliminary phenotypic and molecular screening for potential bacterial biocontrol agents of *Leptosphaeria maculans*, the blackleg pathogen of canola. *Biocontrol Science and Technology*, **16**, 567–582.

Rathore, R., Dowling, D. N., Forristal, P. D. *et al.* (2017). Crop establishment practices are a driver of the plant microbiota in winter oilseed rape (*Brassica napus* L.). *Frontiers in Microbiology*, **8**, 1489.

Rybakova, D., Mancinelli, R., Wikstrom, M. *et al.* (2017). The structure of the *Brassica napus* seed microbiome is cultivar-dependent and affects the interactions of symbionts and pathogens. *Microbiome*, **5**, 104.

Schonhammer, A. and Fischbeck, G. (1987). Investigations on cereal crop rotations and monocultures. III. Changes in soil properties. *Bayerisches Landwirt Jahrb*, **64**, 681–694.

Schreiter, S., Ding, G.-C., Heuer, H. *et al.* (2014). Effect of the soil type on the microbiome in the rhizosphere of field-grown lettuce. *Frontiers in Microbiology* **5**, 144.

Seghers, D., Wittebolle, L., Top, E. M., Verstraete, W. and Siciliano, S. D. (2004). Impact of agricultural practices on the *Zea mays* L. endophytic community. *Applied Environmental Microbiology*, **70**, 1475–1482.

Sheng, X.-F., Xia, J.-J., Jiang, C.-Y., He, L.-Y. and Qian, M. (2008). Characterization of heavy metal-resistant endophytic bacteria from rape (*Brassica napus*) roots and their potential in promoting the growth and lead accumulation of rape. *Environmental Pollution*, **156**, 1164–1170.

Souza, R., Ambrosini, A. and Passaglia, L. M. P. (2015). Plant growth-promoting bacteria as inoculants in agricultural soils. *Genetics and Molecular Biology*, **38**, 401–419.

Steinfeld, H., de Haan, C. and Blackburn, H. (1998). Livestock and the environment, issues and options. In *Agriculture and the Environment Perspectives on Sustainable Development,* ed. E. Lutz. New York: World Bank, pp. 283–301.

Teagasc (2009). Winter oilseed rape. Fact Sheet. Carlow, Ireland: Teagasc Crop Research Centre.

Tilman, D., Cassman, K. G., Matson, P. A., Naylor, R. and Polasky, S. (2002). Agricultural sustainability and intensive production practices. *Nature*, **418**, 671–677.

Tilman, D., Balzer, C., Hill, J. and Befort, B. L. (2011). Global food demand and the sustainable intensification of agriculture. *Proceedings of the National Academy of Sciences of the United States of America*, **108**, 20260–20264.

Trostle, R. (2008). *Global Agricultural Supply and Demand: Factors Contributing to the Recent Increase in Food Commodity Prices*. Washington, DC: US Department of Agriculture, Economic Research Service.

Wang, W., Deng, Z., Tan, H. and Cao, L. (2013). Effects of Cd, Pb, Zn, Cu-resistant endophytic *Enterobacter* sr CBSB1 and R*hodotorula* sp. CBSB79 on the growth and phytoextraction of Brassica plants in multimetal contaminated soils. *International Journal of Phytoremediation*, **15**, 488–497.

Weller, D. M., Raaijmakers, J. M., Gardener, B. B. and Thomashow, L. S. (2002). Microbial populations responsible for specific soil suppressiveness to plant pathogens. *Annual Review of Phytopathology*, **40**, 309–348.

Whipps, J. M. (1997). Developments in the biological control of soil-borne plant pathogens. *Advances in Botanical Research*, **26**, 1–134.

White, P. J., George, T. S., Gregory, P. J. *et al.* (2013). Matching roots to their environment. *Annals of Botany*, **112**, 207–222.

Xing, K., Han, J. and Liu, H (2005). A preliminary study on the inhibitory effect of an endophytic bacterium yc8 isolated from rape. *Acta Agriculturae Universitatis Jiangxiensis*, **27**, 852–856.

Yin, C., Mueth, N., Hulbert, S. *et al.* (2017). Bacterial communities on wheat grown under long-term conventional tillage and no-till in the Pacific northwest of the United States. *Phytobiomes*, **1**, 83–90.

Zhang, H., Sun, Y., Xie, X. *et al.* (2009). A soil bacterium regulates plant acquisition of iron via deficiency-inducible mechanisms. *The Plant Journal*, **58**, 568–577.

Zhao, Y., Gao, Z., Tian, B. *et al.* (2017). Endosphere microbiome comparison between symptomatic and asymptomatic roots of *Brassica napus* infected with *Plasmodiophora brassicae*. *PLoS One*, **12**, e0185907.

12

The Influence of Endophytes on Cork Oak Forests Under a Changing Climate

Daniela Costa, Rui M. Tavares, Paula Baptista and Teresa Lino-Neto

Abstract

Quercus suber L. is an evergreen tree species with high economic, ecological and social importance within the Mediterranean Basin. Cork oak forests occupy more than 2 million hectares worldwide, being mainly located in Algeria, Morocco, Portugal and Spain. As in other Mediterranean ecosystems, cork oak forests have been reported as important reservoirs of biological diversity, including endemic species that are currently under threat due to abiotic and biotic stress. Despite the adaptation of cork oak to the Mediterranean climate, which is characterised by warm, dry summers and wet winters, the growth and productivity of this species is sensitive to climatic change and variability. Extended periods of high temperature and/or low precipitation leading to low level of available water in the soil, can trigger the decline of cork oak and increase vulnerability to pathogen attack. Plant microbiomes are major factors for preserving plant health and productivity under challenging climates and their endophytic components can have dual ecological function, both as detrimental microbes or as beneficial symbionts. Endophytes can play beneficial roles for plant health and productivity but some can become opportunistic pathogens that take advantage of weakened plants that are stressed by environmental conditions. This review discusses endophytes in the context of Mediterranean bioclimates, the geographic distribution of cork oaks and the spread of opportunistic disease-causing agents. Some studies have begun to characterise and isolate endophytes from cork oak, which represents the first steps towards understanding how cork oak endophytes might help ameliorate the negative impacts of climate change for this tree species.

12.1 Introduction

Cork oak (*Quercus suber* L.) is an evergreen broad-leaved tree species with slow growth rates (reaching 15–20 m in height at maturity) and long-lived exemplars (200 years or more) (Gil and Varela, 2008). The most economically important feature of the species is cork, a thick bark that covers the trunk and branches, which is a natural sealing and a thermally insulating tissue (Pereira, 2007). For this reason, *Q. suber* are mainly exploited for cork production that is mostly used for bottle stoppers, but also other products such as indoor pavements, insulation materials and clothes (Allard *et al.*, 2013). A new cork growth ring is produced each year, which is believed to be an evolutionary adaptation to cope with fire, common in the Mediterranean region (Pausas, 1997). Cork can be harvested for the first time (virgin cork) when the tree is approximately 25 years old and subsequently at intervals of 9 to 15 years. Cork harvesting represents a stressful practice for the tree, as an immediate water loss occurs (through evaporation in debarked trunk and branches) and the plant can become more vulnerable to pathogen attacks (e.g. *Diplodia corticola*) through wounds caused during cork harvest (Oliveira and Costa, 2012; Costa *et al.*, 2015).

Quercus suber is well adapted to the Mediterranean-type climate characterised by warm, dry summers and wet winters. Cork oak thrives within a temperature range of 13 to 20°C but is sensitive to temperatures below 0°C (Aranda *et al.*, 2005a). Some abiotic factors affecting this tree are related with local conditions, such as extreme drought conditions (Costa *et al.*, 2016), soil properties or slope (Costa *et al.*, 2008). Cork oak prefers soils with coarse and medium textures that display good drainage conditions (Paulo *et al.*, 2015). Although other regions of the world have a similar climate to the Mediterranean, such as California, central Chile, southern Africa and southwestern and southern Australia (Di Castri, 1991), this review focuses on the Mediterranean Basin ecosystem from which the cork oak is endemic and has the largest populations of the species.

Future climate models predict drier conditions for the Mediterranean region, manifested in a decrease in the rainy season period coupled with higher intensity precipitation events, and a longer dry summer season (Solomon *et al.*, 2007; Giorgi and Lionello, 2008; Hodkinson *et al.*, 2011). Although cork oak is well adapted to Mediterranean climates, the predicted changes in climate could affect its growth as well as the cork yield and quality (Costa *et al.*, 2001; Mendes *et al.*, 2016; Acácio *et al.*, 2017). For example, cork-ring width is higher in years with increased precipitation (>500 mm) and moderate low temperatures during spring (5–6°C) (Caritat *et al.*, 1996, 2000). Also, a decrease in cork growth rates (25% to 50%) was reported under prolonged drought conditions (more than 2 months) (Costa *et al.*, 2016; Oliveira *et al.*, 2016). Taking advantage of plant-beneficial microorganisms, such as ectomycorrhizal fungi or other endophytes, could be an effective way to alleviate the consequences of a drought environment on cork oak forests. This review addresses the relationship between Mediterranean bioclimates and cork oak distribution as well as the role of endophytes for cork oak sustainability under a changing climate. The potential use of cork oak endophytes as a protection/mitigation measure against biotic and abiotic stress will be explored.

12.1.1 Cork Oak Forests Distribution and Their Economic and Ecological Importance

Cork oak forests cover more than 2 million hectares worldwide, with the largest areas located in the Mediterranean region, where cork oaks play an important socio-economic and ecological role (Allard *et al.*, 2013). The most extensive worldwide forest area is situated in the Iberian Peninsula, with the greatest total land cover in Portugal and Spain (715 922 and 574 248 ha respectively), followed by Morocco (383 120 ha), Algeria (230 000 ha), Tunisia (85 771 ha), France (65 228 ha) and Italy (64 800 ha) (Figure 12.1a; Allard *et al.*, 2013). Approximately 200 thousand tonnes of raw cork are harvested worldwide each year (APCOR, 2016). The Iberian Peninsula is responsible for 80% of the annual global cork production, with ~50% of this produced in Portugal. Besides being the primary producer of raw cork, Portugal is also one of the world's greatest importers of raw cork that will be processed and later exported. The export of cork generates an income of over 1400 million euros worldwide, 62.7% of which is allocated to Portugal, highlighting the economic importance of the crop to this region (APCOR, 2016). In the Iberian Peninsula, cork oak forests can be managed using two different silvicultural systems (González-García *et al.*, 2013). The most common is the savannah-type cork oak forest, known as *montado* in Portugal and as *dehesa* in Spain. Such cork oak stands are characterised by an agro-silvo-pastoral system, where cork is produced using a low planting density (about 60–100 trees ha^{-1}) and is associated with crop production (cereals such as wheat, barley or oats and legumes such as *Lupinus luteus*) and pasture. Higher density stands (about 400 trees ha^{-1}) are less common and are characterised by mixed forests where cork oaks usually grows with other oaks or pines. Silvicultural or agricultural practices can significantly effect biodiversity, including endophytic fungal diversity as described for *Coffea arabica* (Saucedo-García *et al.*, 2014) and *Zea mays* (Seghers *et al.*, 2004).

Despite their economic importance, cork oak forests are also important ecosystems rich in biodiversity with flora and fauna typical of Mediterranean ecosystems (Pereira *et al.*, 2008). The Mediterranean Basin provides a habitat to many endemic plant and animal species and has been classified as a global biodiversity hotspot with conservation priority, due to its vulnerability under predicted climatic changes (Myers *et al.*, 2000; Malcolm *et al.*, 2006). Cork oak forests have been considered important for biodiversity conservation and are protected by the 92/43/EEC Habitat Directive (Council of European Union, 1992). They include the habitat of species with high protection status such as the Iberian lynx (*Lynx pardina*),

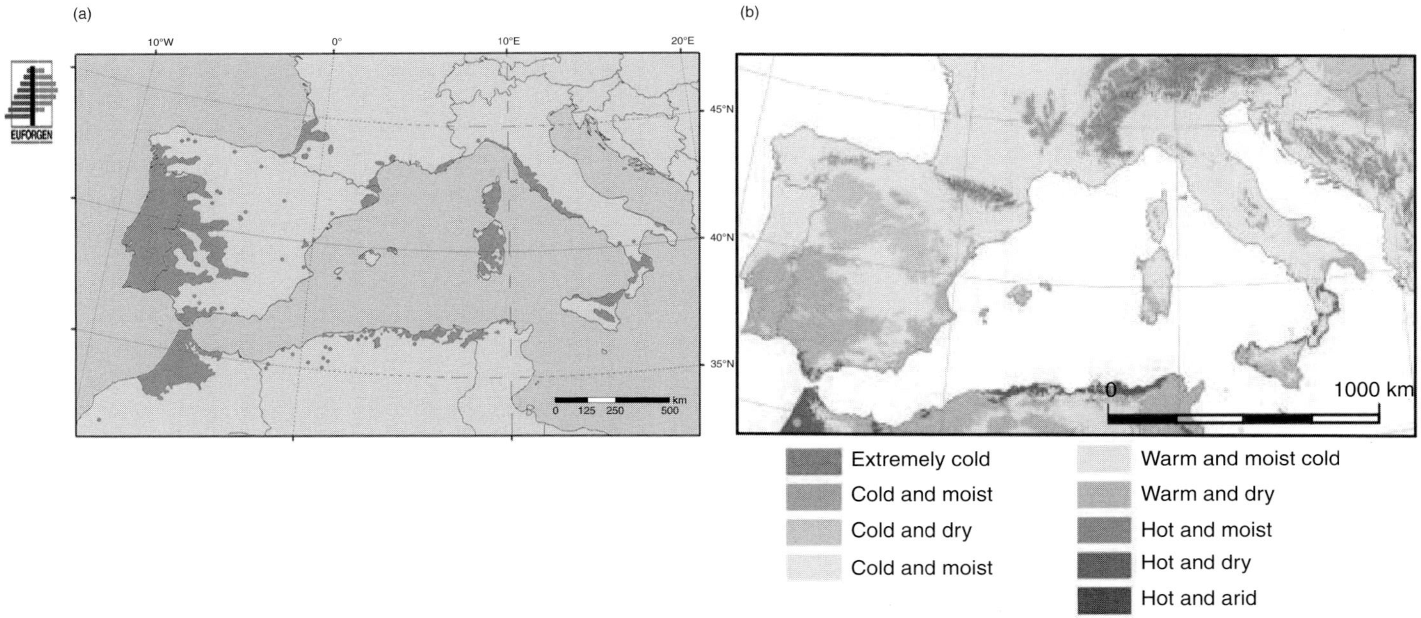

Figure 12.1 Map of Europe with (a) distribution area of *Quercus suber* within the Mediterranean region and (b) environmental zones based on five categories of temperature and two of aridity (adapted from EUFORGEN, 2009). (A black and white version of this figure will appear in some formats. For the colour version, please refer to the plate section.)

the imperial eagle (*Aquila adalberti*) and the black vulture (*Aegypius monachus*) (Carrete and Donázar, 2005; Rigueiro-Rodríguez *et al.*, 2009). In addition, cork oak forests play an important role for the regulation of the local water balance, due to greater soil capacity to hold water when under canopy (Joffre and Rambal, 1993), while deep roots allow groundwater redistribution to shallower soil layers (Besson *et al.*, 2006). Also, carbon stocks in soil are almost two times higher (50.2 Mg ha^{-1}) under cork oak tree canopy, compared to a 15 m distance from the tree (Howlett *et al.*, 2011). Cork harvesting practices by manual extraction has a small environmental impact and the anthropogenic management of these ecosystems is essential for the conservation of biodiversity and their ecosystem services (Pereira and Fonseca, 2003; Bugalho *et al.*, 2011).

12.1.2 *Quercus suber* Distribution According to Bioclimate

Cork oak is native to the Mediterranean region and is well adapted to climatic conditions in the region. The Mediterranean climate generally exhibits pronounced seasonality in air temperature and precipitation, characterised by a hot dry period during summer (average temperature of the hottest month is ~30°C and precipitation is below 30 mm) and a cool wet period during winter (average temperature of 0°C and precipitation between 100 mm and 500 mm, depending on region) (Lionello *et al.*, 2006). The local climate is also affected by regional factors, such as the proximity of Mediterranean Sea and the complex physiography of the region, leading to a variety of sub-climates and significant spatial variability (Lionello *et al.*, 2006). The combination of climatic indicators associated with vegetation maps (species or communities) can be used to define bioclimatic classes (Ozenda and Borel, 2000). A few decades ago, Emberger (1930) created a pluviothermic indicator (Emberger index) that allowed the classification of Mediterranean climates in the context of aridity and temperature. This classification was then used to establish and identify different bioclimatic zones (e.g. Holdridge, 1947; Bagnouls and Gaussen, 1957). Taking into consideration the broader-scale adaptation of forest trees, a map of environmental zones in Europe was recently created by de Vries *et al.* (2015), resulting in eight environmental zones (from extremely cold, cold, cool, warm, hot and extremely hot regions, either moist, dry or arid; Figure 12.1b). Comparing these environmental zones with the distribution map of cork oak within the Mediterranean region (Figure 12.1a), the preference of *Q. suber* for bioclimates displaying warm temperatures (that can be either moist or dry) is illustrated. Cork oak distribution in regions with warmer temperatures is expected as this tree species exhibits a high sensitivity to low temperatures (below 0°C) (Aranda *et al.*, 2005a). When considering the distribution of cork oak in Portugal (Figure 12.2a), a relationship between cork oak distribution and total annual precipitation was observed (Figure 12.2b). Regions with limited rainfall (600–800 mm) have a higher density of cork oak populations than the most

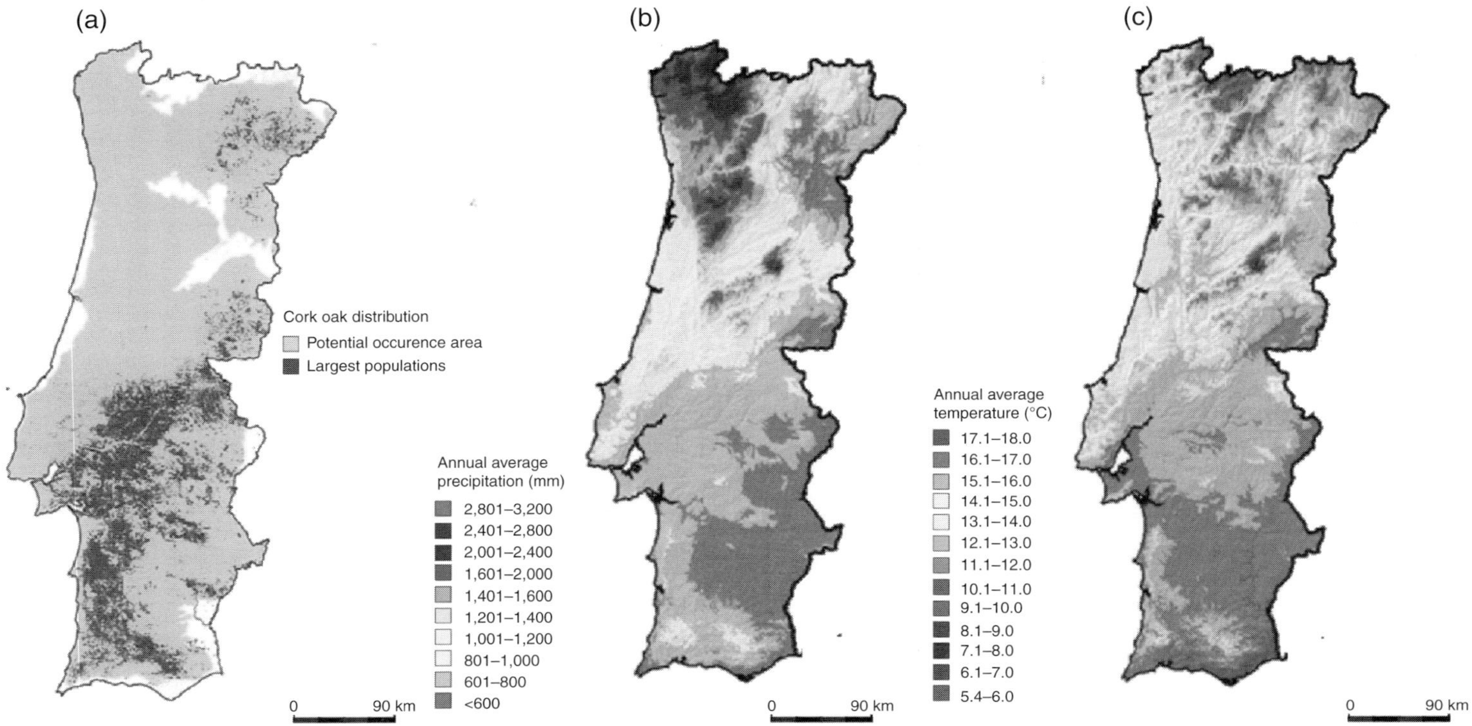

Figure 12.2 (a) Map of cork oak forest distribution in Portugal, (b) total annual precipitation and (c) annual medium temperature. (A black and white version of this figure will appear in some formats. For the colour version, please refer to the plate section.)
(a) was adapted from Capelo and Catry (2007); (b) and (c) were adapted from IPMA (n.d.).

humid or arid regions in the country (Paulo *et al.*, 2015). This relationship was not so obvious when considering the average annual temperature (Figure 12.2c), as the temperature range in Portugal is among the optimal temperatures for cork oak (13°C to 20°C; Aranda *et al.*, 2005a). Forest distribution confirms that cork oak represents a good adaptation to an environment with reduced water availability conditions. Besides a deep root system, cork oak leaf physiology and plasticity allow the more efficient use of water (e.g. leaf area decreases, and thickness increases under arid conditions) (Gouveia and Freitas, 2009; Ramírez-Valiente *et al.*, 2011; Besson *et al.*, 2014). Furthermore, cork oak phenology provides good adaptation to environmental conditions (e.g. earlier bud bursts as a consequent of warmer springs) (Pinto *et al.*, 2011). Despite these adaptations, water availability is a major limiting factor for cork oak growth, as revealed by cork-ring chronologies of mature cork oaks in response to climate, dating back 9–10 years (Costa *et al.*, 2016). Water limitation tends to be higher during the early stages of development, as cork oak seedlings were found to be highly sensitive to drought (Aranda *et al.*, 2005b).

The climate change predictions for the Mediterranean region project an increase in annual temperature (up to +4°C) and a decrease in annual precipitation (up to -25%) (Giorgi and Lionello, 2008), associated with a higher risk of the occurrence of extreme events, such as flooding and heat waves (Solomon *et al.*, 2007). A regional climate change index for identifying the most afflicted worldwide regions to climate change revealed that the Mediterranean region will be one of the most affected (Giorgi, 2006). Models that take into consideration environmental and geographic indicators to predict climatic patterns within the Mediterranean region suggest a clear climatic pattern shift with temperature increases and precipitation decreases (Rego and Rocha, 2014). The prediction of a drier and warmer climate suggests significant vulnerability of the Iberian Peninsula to vegetation shifts (Gonzalez *et al.*, 2010). For example, the evergreen oak forests are expected to transition into shrublands, due to the combination of drought and more frequent wildfires (Acácio *et al.*, 2009, 2017). The mortality of cork oak trees has been also associated with water-limited sites (Costa *et al.*, 2010). The extensive exposure of individuals to drought environments is expected to increase the frequency of cork oak decline, which is one of the tree death syndromes characterised by gradual loss of foliage, die back from the top of the tree, detachment of bark and sprouting of epicormic shoots (Camilo-Alves *et al.*, 2017), and facilitate further pathogen attacks (reviewed by La Porta *et al.*, 2008; Gentilesca *et al.*, 2017). In addition, the increased mortality rate of dominant *Quercus suber* trees, present on mixed Mediterranean oak forests, was recently described to affect the recovery and self-replacement of cork oaks, reducing the dominance of *Q. suber* in favour of other coexisting woody species including shrubs (e.g. *Pistacia lentiscus* and *Arbutus unedo*) (Ibáñez *et al.*, 2017). All these results predict a decrease in cork oak growth and distribution under the expected climatic change in this region and further suggest the shift of Mediterranean forests towards more open systems with lower tree cover and higher shrub dominance.

12.2 Endophytes and *Quercus suber* Sustainability

Besides the increasing adverse environmental conditions that cork oak forests are facing (water limitation and high temperatures), cork oaks are also increasingly threatened by attacks of pathogens and pests (Moricca *et al.*, 2016). A growing number of studies have highlighted the importance of fungal pathogens on cork oak decline and mortality. *Phytophthora cinnamomi* has been reported to have had a major impact in *Q. suber* decline within the Mediterranean region (Brasier *et al.*, 1993; Braisier, 1996). This oomycete is a root pathogen that causes sudden death to the trees due to root system loss (Camilo-Alves *et al.*, 2013). Leaf pathogens, such as *Discula quercina*, cause leaf necrosis and can reduce cork oak productivity through a decrease in photosynthesis and modification of stomatal conductance (Linaldeddu *et al.*, 2009). More recently, *Diplodia corticola* and *Biscogniauxia mediterranea* have received considerable attention because they are endophytes widely found on cork oak trees that can become opportunistic pathogens (Franceschini *et al.*, 2005). *Diplodia corticola* is described as a more virulent pathogen than *B. mediterranea* and causes disease symptoms including dieback, branch and trunk cankers (sunken necrotic lesions) and cork bark cracks (Linaldeddu *et al.*, 2009). *Biscogniauxia mediterranea* causes extensive inner bark and xylem necrosis, associated with blackish exudation from the outer bark, typical of charcoal disease (Linaldeddu *et al.*, 2005; Linaldeddu *et al.*, 2009). A common feature of all these cork oak pathogens is their opportunistic behaviour. They are endophytes that live within plant tissues but have a negative impact on the sustainability of cork oak forests, especially when cork oaks are weakened by stress factors.

Several microorganisms displaying an endophytic lifestyle (endophytes) can live within any plant tissue (roots, stems and leaves) and are usually considered non-pathogenic, as they do not cause visible disease symptoms in the plant host (Schulz and Boyle, 2005). Several fungal endophytes have been described as being beneficial to the plant host and their ecological functions have been widely described (White *et al.*, 2014; Hardoim *et al.*, 2015). According to their colonisation characteristics, fungal endophytes can be divided into classes and grouped as clavicipitaceous and non-clavicipitaceous (Rodriguez *et al.*, 2009). Clavicipitaceous, class 1 endophytes live in grass hosts, whereas non-clavicipitaceous, classes 2, 3 and 4 have a broad range of hosts (Rodriguez *et al.*, 2009). Endophytes from all classes can provide benefits for drought tolerance and growth improvement (Murphy *et al.*, 2015) but class 2 endophytes can also benefit the host by enhancing their tolerance to pH or salinity extremes (Rodriguez *et al.*, 2009). Despite the beneficial effects to plant hosts, the endophytic community can also include latent pathogens, which can cause disease symptoms when the corresponding plant host is facing an environmental stress (Desprez-Loustau *et al.*, 2006; Saikkonen, 2007; Sieber, 2007). For example, endophytic species belonging to the Botryosphaeriaceae family are frequently found in woody plants and are able to cause disease when plant hosts are weakened by environmental stress (Slippers and Wingfield, 2007). The same opportunistic behaviour is described for many cork oak endophytes, like *Discula quercina* found in oak forests (Moricca and Ragazzi, 2008), *D. corticola* (Botryosphaeriaceae) and *B. mediterranea* (Franceschini *et al.*, 2005). Henriques *et al.* (2012) found high rates of *B. mediterranea* in diseased young cork oak forest; it is also has a high colonisation frequency in declined cork oak trees (Franceschini *et al.*, 2005; Martín *et al.*, 2005). The colonisation frequency of this endophyte also increased when trees were facing environmental stresses such as drought (Desprez-Loustau *et al.*, 2006; Linaldeddu *et al.*, 2011).

Due to the opportunistic behaviour of cork oak endophytes, the predicted changes in climate could represent a significant threat to the sustainability of cork oak trees in the future (Moricca *et al.*, 2016). The combination of increased abiotic stresses that weakens the host plant and the existence of fungal species that can rapidly adapt to a changing environment, as is the case of *B. mediterranea* (Henriques *et al.*, 2016), suggests an intensification of disease occurrence in the future. Climate change can also affect the ecological role of the microbial community by contributing to more aggressive pathogens (La Porta *et al.*, 2008) and also by disturbing the beneficial interactions occurring between plant and microorganisms (Compant *et al.*, 2010). An increase in temperature can enhance the colonisation of existing fungal genotypes (Fujimura *et al.*, 2008). However, shifts in the endophytic community structure and composition can occur due to extreme event occurrence such as fires (Huang *et al.*, 2016). Such community shifts could have a negative impact on tree health, especially if the beneficial symbiosis cannot be re-established or maintained.

To our knowledge, very few studies describing cork oak endophytic communities have been reported and most are from Italy (Table 12.1). Using culture-based methods for studying the fungal endophytic community present on twigs of healthy and declining cork oak stands in Sardinia, Franceschini *et al.* (2005) identified 32 fungal taxa based on their morphological features. The common species isolated from all stands were *Aureobasidium pullulans*, *B. mediterranea*, *D. quercina* and *Penicillium purpurogenum*, revealing once more the opportunistic behaviour of *B. mediterranea* and *D. quercina*. A similar approach was used by Linaldeddu *et al.* (2011) to describe the fungal endophytic community in healthy and water-stressed cork oak trees. The authors assessed cork oak twigs, branches and wood trunk communities, resulting in the identification of 18 species with higher colonisation rates in asymptomatic and non-water-stressed trees than water-stressed trees. They also found *B. mediterranea* in all trees but higher isolation frequency was found in trees under water stress. Also, a study of the endophytic community in branches and leaves of healthy and declining cork oak trees resulted in the identification of 21 fungal species, in which *B. mediterranea* was more frequently isolated in declining trees (Pulina *et al.*, 2006). Among others, this pathogen was also found in declined cork oak trees in Tunisia (Hasnaoui *et al.*, 2017). Recently, we studied the fungal endophytic community in roots, twigs and leaves of four Portuguese cork oak stands with differences in water availability, using culture-based methods and morphotype grouping (unpublished results). This work resulted in the identification of 23 different morphotypes. The driest cork oak stand exhibited the most rich and diverse endophyte community of all forests. The use of DNA-based technologies for the identification of microbial diversity, such as the most recent high-throughput sequencing approaches, could provide further insights on endophyte communities.

The fungal community associated with Moroccan cork oak roots (including the endophytic community) was also recently described, but the endophytic community was not discriminated (Maghnia *et al.*, 2017). Overall, 32 fungal species were identified as endophytes of cork oak trees (Table 12.1). Some species, such as *Alternaria alternata* and *A. pullulans*, were also described as endophytes of others *Quercus* spp. while species like *Botryosphaeria corticola*, *Coryneum depressum* and *Stereum gausapatum* have been only found in *Q. suber*. In addition, some of these species were described as having potential to control diseases (e.g. *Epicoccum nigrum* (Madrigal *et al.*, 1994; Larena *et al.*, 2005) and *Aereobasidium pullulans* (Lima *et al.*, 1997)) and to promote plant growth (e.g. *Trichoderma* spp. (Naseby *et al.*, 2000) and *Sordaria fimicola* (Dewan *et al.*, 1994)).

12.3 Endophytes as Contributors to Climate Change Adaptation in Plants

The use of beneficial endophytic microorganisms to cope with abiotic and biotic stress could be a suitable strategy for plants to help mitigate the effects of climate change. Several studies describe abiotic stress tolerance and plant growth promotion by using endophytes (Murphy *et al.*, 2015; Dastogeer and Wylie, 2017; Rho *et al.*, 2017). For example, a *Trichoderma hamatum* isolate was described to improve *Theobroma cacao* root growth under drought stress (Bae *et al.*, 2009) and several fungal endophytic species were reported to promote tomato growth under drought stress (Azad and Kaminskyj, 2016). Furthermore, Redman *et al.* (2011) detected an enhanced tolerance to drought stress of rice plants through symbiosis with class 2 fungal endophytes, suggesting that these endophytes are able to increase rice drought and cold tolerance. The production of phytohormones by the fungal endophyte *Paecilomyces formosus* also protected rice plants from the negative effects

Table 12.1 Endophytic fungal species found in *Quercus suber* tissues.

Endophytic fungal species	Tree status	Organ	Reference	Quercus spp.	Reference
Alternaria alternata	Ht/ Dt	Bd/Tw	Franceschini *et al.*, 2005	*Q. acuta*	Hashizume *et al.*, 2008
				Q. cerris	Ragazzi *et al.*, 2001; Ragazzi *et al.*, 2003; Boncaldo *et al.*, 2008; Moricca *et al.*, 2012
				Q. ilex	Fisher *et al.*, 1994; Collado *et al.*, 1999
				Q. liaotungensis	Sun *et al.*, 2012
				Q. pubescens	Ragazzi *et al.*, 2003; Boncaldo *et al.*, 2008; Moricca *et al.*, 2012
				Q. pyrenaica	Martínez-Álvarez *et al.*, 2012
				Q. robur	Gennaro *et al.*, 2003; Ragazzi *et al.*, 2003; Kwaśna *et al.*, 2016
Alternaria spp.	Ht/ Dt	Bd/Tw	Franceschini *et al.*, 2005	*Q. ilex*	Fisher *et al.*, 1994; Collado *et al.*, 1999; Pulina *et al.*, 2006
		Tw/Br/ Wd	Linaldeddu *et al.*, 2011	*Q. variabilis*	Wang *et al.*, 2007
		Lv/Br	Pulina *et al.*, 2006		
Aureobasidium pullulans	Ht/ Dt	Bd/Tw	Franceschini *et al.*, 2005	*Q. cerris*	Boncaldo *et al.*, 2008; Moricca *et al.*, 2012
		Tw/Br/ Wd	Linaldeddu *et al.*, 2011	*Q. ilex*	Fisher *et al.*, 1994; Pulina *et al.*, 2006
		Lv/Br	Pulina *et al.*, 2006	*Q. pubescens*	Boncaldo *et al.*, 2008; Moricca *et al.*, 2012
				Q. pyrenaica	Martínez-Álvarez *et al.*, 2012
				Q. robur	Kwaśna *et al.*, 2016; Hasnaoui *et al.*, 2017

Table 12.1 (*Cont.*)

Endophytic fungal species	Tree status	Organ	Reference	Quercus spp.	Reference
Bionectria solani	Ht/ Dt	Bd/Tw	Franceschini *et al.*, 2005	*Q. ilex*	Pulina *et al.*, 2006
		Lv/Br	Pulina *et al.*, 2006		
Biscogniauxia mediterranea	Ht/ Dt	Bd/Tw	Franceschini *et al.*, 2005	*Q. cerris*	Boncaldo *et al.*, 2008; Moricca *et al.*, 2012
		Tw/Br/Wd	Linaldeddu *et al.*, 2011	*Q. ilex*	Pulina *et al.*, 2006
		Lv/ Br	Pulina *et al.*, 2006	*Q. pubescens*	Boncaldo *et al.*, 2008; Moricca *et al.*, 2012
	Dct	Br	Hasnaoui *et al.*, 2017		
Botryosphaeria corticola	Dct	Tw/Br/Wd	Linaldeddu *et al.*, 2011		
		Br	Hasnaoui *et al.*, 2017		
Botrytis cinerea	Ht/ Dt	Bd/Tw	Franceschini *et al.*, 2005	*Q. ilex*	Pulina *et al.*, 2006
		Tw/Br/Wd	Linaldeddu *et al.*, 2011		
		Lv/Br	Pulina *et al.*, 2006		
Cladosporium spp.	Ht/ Dt	Bd/Tw	Franceschini *et al.*, 2005	*Q. cerris*	Ragazzi *et al.*, 2001; Ragazzi *et al.*, 2003; Kwaśna *et al.*, 2016
		Tw/Br/Wd	Linaldeddu *et al.*, 2011	*Q. ilex*	Fisher *et al.*, 1994; Pulina *et al.*, 2006
		Lv/Br	Pulina *et al.*, 2006	*Q. pubescens*	Ragazzi *et al.*, 2003
				Q. robur	Ragazzi *et al.*, 2003 Kwaśna, H. *et al.*, 2016; Kwaśna, H. *et al.*, 2016; Gonthier *et al.*, 2006
Coryneum depressum	Ht/ Dt	Bd/Tw	Franceschini *et al.*, 2005		
		Tw/Br/Wd	Linaldeddu *et al.*, 2011		

(*continued*)

Table 12.1 (*Cont.*)

Endophytic fungal species	Tree status	Organ	Reference	Quercus spp.	Reference
Cryphonectria sp.	Ht/ Dt	Lv/Br	Pulina *et al.*, 2006		
Cytospora spp.	Ht/ Dt	Bd/Tw	Franceschini *et al.*, 2005	*Q. cerris*	Boncaldo *et al.*, 2008
				Q. ilex	Fisher *et al.*, 1994; Collado *et al.*, 1999; Pulina *et al.*, 2006
		Tw/Br/ Wd	Linaldeddu *et al.*, 2011		
		Br	Pulina *et al.*, 2006	*Q. pubescens*	Ragazzi *et al.*, 2003; Boncaldo *et al.*, 2008
	Dct	Br	Hasnaoui *et al.*, 2017		
Dictyochaeta parva	Ht/ Dt	Bd/Tw	Franceschini *et al.*, 2005		
Diplodia corticola	Ht/ Dt	Bd/Tw	Franceschini *et al.*, 2005	*Q. cerris*	Boncaldo *et al.*, 2008
				Q. ilex	Pulina *et al.*, 2006
		Br	Pulina *et al.*, 2006	*Q. pubescens*	Boncaldo *et al.*, 2008
Diplodia spp.	Ht/ Dt	Bd/Tw	Franceschini *et al.*, 2005	*Q. cerris*	Ragazzi *et al.*, 2001, 2003
				Q. ilex	Collado *et al.*, 1999
				Q. pubescens	Ragazzi *et al.*, 2003
				Q. robur	Ragazzi *et al.*, 2003
Discula quercina	Ht/ Dt	Bd/Tw	Franceschini *et al.*, 2005	*Q. cerris*	Ragazzi *et al.*, 2001; Boncaldo *et al.*, 2008; Moricca *et al.*, 2012
		Tw/Br/ Wd	Linaldeddu *et al.*, 2011		
		Lv/Br	Pulina *et al.*, 2006	*Q. garryana*	Wilson and Carroll, 1994
	Dct	Br	Hasnaoui *et al.*, 2017	*Q. ilex*	Pulina *et al.*, 2006
				Q. pubescens	Boncaldo *et al.*, 2008; Moricca *et al.*, 2012
Dothiorella iberica	Ht/ Dt	Tw/Br/ Wd	Linaldeddu *et al.*, 2011	*Q. ilex*	Pulina *et al.*, 2006
		Br	Pulina *et al.*, 2006		

Table 12.1 (*Cont.*)

Endophytic fungal species	Tree status	Organ	Reference	Quercus spp.	Reference
Epicoccum nigrum	Ht/ Dt	Bd/Tw	Franceschini *et al.*, 2005	*Q. cerris*	Ragazzi *et al.*, 2001, 2003; Gennaro *et al.*, 2003; Boncaldo *et al.*, 2008; Moricca *et al.*, 2012
		Tw/Br/ Wd	Linaldeddu *et al.*, 2011		
		Lv/Br	Pulina *et al.*, 2006	*Q. ilex*	Pulina *et al.*, 2006
				Q. pubescens	Ragazzi *et al.*, 2003; Boncaldo *et al.*, 2008; Moricca *et al.*, 2012
				Q. robur	Gonthier *et al.*, 2006
Myrioconium spp.	Ht/ Dt	Bd/Tw	Franceschini *et al.*, 2005	*Q. ilex*	Pulina *et al.*, 2006
		Tw/Br/ Wd	Linaldeddu *et al.*, 2011		
		Lv/Br	Pulina *et al.*, 2006		
Penicillium sp.	Ht	Lv/Br	Pulina *et al.*, 2006	*Q. robur*	Kwaśna *et al.*, 2016
				Q. ilex	Collado *et al.*, 1999
				Q. pyrenaica	Martínez-Álvarez *et al.*, 2012
Penicillium purpurogenum	Ht/ Dt	Bd/Tw	Franceschini *et al.*, 2005		
		Tw/Br/ Wd	Linaldeddu *et al.*, 2011		
Pestalotia spp.	Ht/ Dt	Bd/Tw	Franceschini *et al.*, 2005	*Q. ilex*	Pulina *et al.*, 2006
Phoma spp.	Ht/ Dt	Bd/Tw	Franceschini *et al.*, 2005	*Q. acuta*	Hashizume *et al.*, 2008
				Q. cerris	Ragazzi *et al.*, 2001, 2003; Gennaro *et al.*, 2003; Boncaldo *et al.*, 2008
				Q. ilex	Fisher *et al.*, 1994
				Q. liaotungensis	Sun *et al.*, 2012

(*continued*)

Table 12.1 (*Cont.*)

Endophytic fungal species	Tree status	Organ	Reference	Quercus spp.	Reference
				Q. pubescens	Ragazzi *et al.*, 2003; Boncaldo *et al.*, 2008
				Q. robur	Gennaro *et al.*, 2003; Ragazzi *et al.*, 2003; Petrini and Fisher, 1990; Gonthier *et al.*, 2006
				Q. variabilis	Wang *et al.*, 2007
Pleurophoma cava	Ht/ Dt	Bd/Tw	Franceschini *et al.*, 2005	*Q. ilex*	Pulina *et al.*, 2006
				Q. liaotungensis	Sun *et al.*, 2012
		Tw/Br/Wd	Linaldeddu *et al.*, 2011		
	Dct	Br	Pulina *et al.*, 2006; Hasnaoui *et al.*, 2017		
Preussia aemulans	Ht/ Dt	Bd/Tw	Franceschini *et al.*, 2005	*Q. ilex*	Pulina *et al.*, 2006
		Tw/Br/Wd	Linaldeddu *et al.*, 2011		
	Ht	Br	Pulina *et al.*, 2006		
Seimatosporium glandigenum	Ht/ Dt	Bd/Tw	Franceschini *et al.*, 2005	*Q. ilex*	Pulina *et al.*, 2006
	Ht/ Dct	Lv	Pulina *et al.*, 2006		
S. kriegerianum	Ht/ Dt	Bd/Tw	Franceschini *et al.*, 2005		
Sordaria fimicola	Ht/ Dt	Bd/Tw	Franceschini *et al.*, 2005	*Q. ilex*	Fisher *et al.*, 1994; Pulina *et al.*, 2006
		Tw/Br/Wd	Linaldeddu *et al.*, 2011	*Q. pyrenaica*	Martínez-Álvarez *et al.*, 2012
	Ht	Lv/Br	Pulina *et al.*, 2006		

Table 12.1 (*Cont.*)

Endophytic fungal species	Tree status	Organ	Reference	Quercus spp.	Reference
Sporormiella minima	Ht/ Dt	Bd/Tw	Franceschini *et al.*, 2005		
Stereum gausapatum	Ht/ Dt	Bd/Tw	Franceschini *et al.*, 2005		
		Tw/Br/ Wd	Linaldeddu *et al.*, 2011		
	Ht	Br	Pulina *et al.*, 2006		
Trichoderma citrinoviride	Ht/ Dt	Tw/Br/ Wd	Linaldeddu *et al.*, 2011		
T. fertile	Ht/ Dt	Bd/Tw	Franceschini *et al.*, 2005	*Q. ilex*	Pulina *et al.*, 2006
	Ht	Lv/Br	Pulina *et al.*, 2006	*Q. robur*	Kwaśna, H. *et al.*, 2016
T. harzianum	Ht/ Dt	Bd/Tw	Franceschini *et al.*, 2005	*Q. cerris*	Moricca *et al.*, 2012
	Ht	Br	Pulina *et al.*, 2006	*Q. ilex*	Pulina *et al.*, 2006
				Q. pubescens	Ragazzi *et al.*, 2003
				Q. robur	Ragazzi *et al.*, 2003; Kwaśna *et al.*, 2016

Notes: From healthy (Ht), declined (Dt) or declining (Dct) trees. Indication is given whenever the same endophyte was found in other *Quercus* species. Organs: Bd: Buds; Tw: Twigs; Br: Branches; Wd: wood cores; Lv: Leaves.

of heat stress (Waqas *et al.*, 2015a). The rapid adaptation ability of endophytes to changing environments, as described by Henriques *et al.* (2016) for the opportunistic endophyte *B. mediterranea*, could be a useful trait to manipulate for plant protection in climate changing scenarios. Fungal endophytes adapted to habitats with high temperatures can thus promote heat tolerance (Rodriguez *et al.*, 2008).

Despite conferring tolerance to abiotic stress, endophytes could also improve plant responses towards biotic stress. Fungal endophytes have the potential to promote disease suppression in cereal species (O'Hanlon *et al.*, 2012), in addition to other grass species. For example, when *Lolium perenne* is infected with *Neotyphodium* spp. enhanced production of seeds and reproductive tillers can occur (Hesse *et al.*, 2003). The beneficial effect of endophytes has also been studied on other plant hosts. For example, in sunflower, the negative effects of a stem rot pathogen were attenuated by the endophytic fungi *Penicillium citrinum* and *Aspergillus terreus* (Waqas *et al.*, 2015b). Studies on the positive role of endophytes on plant tree species are scarcer. Fungal endophytes were described to play an important role for *Theobroma cacao* defence against *Phytophthora* spp., reducing leaf damage and loss (Arnold *et al.*, 2003; Mejía *et al.*, 2008). Furthermore, Aleppo pine endophytic fungi were recently reported to promote a reduction of necrosis damage caused by the pathogen *Gremmeniella abietina* (Romeralo *et al.*, 2015).

The use of endophytic species for the mitigation of abiotic stresses or control of diseases in oaks, including cork oak, has not yet been adequately assessed or reported. Only the beneficial effects of mycorrhizal associations to enhance water uptake, nutrient acquisition and plant growth have been highlighted (Reis *et al.*, 2017). Although more studies are needed to fully understand the cork oak endophytic community and possible endophyte community shifts in response to environmental changes, the exploitation of the potential endophyte protection role seems a sustainable way to mitigate the consequences of expected climatic changes.

12.4 Conclusions

Quercus suber is an important forest species in the Mediterranean Basin that is vulnerable to the predicted changes in climate. The increasing occurrence and severity of drought stress, makes this plant species more susceptible to diseases, including those caused by opportunistic endophytic pathogens. While vegetation shifts due to a changing climate could result in the loss of cork oak dominance in mixed forests, the endophytic communities can adapt to the new environmental conditions and develop more aggressive pathogens. On the other hand, endophytes could comprise beneficial microorganisms that could help cork oak to deal with the changing environment. Our knowledge on the endophytic community of cork oak forests is limited in comparison to other *Quercus* species. A better understanding of these communities in their natural environment and their behaviour under different climatic scenarios would be the first step in promoting any beneficial interactions between the plant and endophyte for cork oak survival and production. These studies would create the opportunity to identify appropriate strategies to cope with increasing abiotic and biotic stress, including the use of endophytic species able to improve plant growth and tolerance against pathogens.

Acknowledgements

This work was funded by FEDER through the Operational Competitiveness Program (COMPETE) and by Portuguese national funds through the Foundation for Science and Technology (FCT) within the scope of the project POCI-01-0145-FEDER-028635 – 'SuberControl – Influence of endosphere microbiome to control diseases in cork oak (*Quercus suber* L.)'. The work was also supported by Fundação Ciência e Tecnologia (FCT/MCTES/PIDDAC, Portugal), under the project (PEst- OE/BIA/UI4046/2014; UID/MULTI/04046/2013) and PhD grant to D.C. (SFRH/BD/120516/2016).

References

Acácio, V., Holmgren, M., Rego, F., Moreira, F. and Mohren, G. M. (2009). Are drought and wildfires turning Mediterranean cork oak forests into persistent shrublands? *Agroforestry Systems*, **76**, 389–400.

Acácio, V., Dias, F. S., Catry, F. X., Rocha, M. and Moreira, F. (2017). Landscape dynamics in Mediterranean oak forests under global change: understanding the role of anthropogenic and environmental drivers across forest types. *Global Change Biology*, **23**, 1199–1217.

Allard, G., Berrahmouni, N., Besacier, C. *et al.* (2013). State of forest resources in the Mediterranean Region. In *State of Mediterranean Forests 2013*. Rome: FAO, pp. 27–114.

Alves, A., Correia, A., Luque, J. and Phillips, A. (2004). *Botryosphaeria corticola*, sp. nov. on *Quercus species*, with notes and description of *Botryosphaeria stevensii* and its anamorph, *Diplodia mutila*. *Mycologia*, **96**, 598–613.

APCOR (2016). *Cork Yearbook 2016*. Santa Maria de Lamas, Portugal: Portuguese Cork Association.

Aranda, I., Castro, L., Alía, R., Pardos, J. A. and Gil, L. (2005a). Low temperature during winter elicits differential responses among populations of the Mediterranean evergreen cork oak (*Quercus suber*). *Tree Physiology*, **25**, 1085–1090.

Aranda, I., Castro, L., Pardos, M., Gil, L. and Pardos, J. A. (2005b). Effects of the interaction between drought and shade on water relations, gas exchange and morphological traits in cork oak (*Quercus suber* L.) seedlings. *Forest Ecology and Management*, **210**, 117–129.

Arnold, A. E., Mejía, L. C., Kyllo, D. *et al.* (2003). Fungal endophytes limit pathogen damage in a tropical tree. *Proceedings of the National Academy of Sciences of the United States of America*, **100**, 15649–54.

Aronson, J., Pereira, J. S. and Pausas, J. G., eds. (2012). *Cork Oak Woodlands On The Edge: Ecology, Adaptive Management, and Restoration*. Washington, DC: Island Press.

Aschmann, H. (1973). Distribution and peculiarity of Mediterranean ecosystems. In *Mediterranean Type Ecosystems. Ecological Studies (Analysis and Synthesis)*, ed. F. di Castri and H. A. Mooney. Berlin: Springer, pp. 11–19.

Azad, K. and Kaminskyj, S. (2016). A fungal endophyte strategy for mitigating the effect of salt and drought stress on plant growth. *Symbiosis*, **68**, 73–78.

Bae, H., Sicher, R. C., Kim, M. S. *et al.* (2009). The beneficial endophyte *Trichoderma hamatum* isolate DIS 219b promotes growth and delays the onset of the drought response in *Theobroma cacao*. *Journal of Experimental Botany*, **60**, 3279–3295.

Bagnouls, F. and Gaussen, H. (1957). Les climats biologiques et leur classification. *Annales de Géographie*, **66**, 193–220.

Besson, C. K., Otieno, D., Do Vale, R. L. *et al.* (2006). Hydraulic lift in cork oak trees in a savannah-type Mediterranean ecosystem and its contribution to the local water balance. *Plant and Soil*, **282**, 361–378.

Besson, C. K., Do Vale R. L., Rodrigues, M. L. *et al.* (2014). Cork oak physiological responses to manipulated water availability in a Mediterranean woodland. *Agricultural and Forest Meteorology*, **184**, 230–242.

Boncaldo, E., Sicoli, G., Mannerucci, F. and Luisi, N. (2008). Characterisation of fungal endophytic communities of deciduous oak species in Southern Italy. *Italian Journal of Forest and Mountain Environments*, **63**, 321–332.

Braisier, C. M. (1996). *Phytophthora cinnamomi* and oak decline in southern Europe. Environmental constraints including climate change. *Annales des Sciences Forestieres*, **53**, 347–358.

Brasier, C. M., Robredo, F. and Ferraz, J. F. P. (1993). Evidence for *Phytophthora cinnamomi* involvement in Iberian oak decline. *Plant Pathology*, **42**, 140–145.

Bugalho, M. N., Caldeira, M. C., Pereira, J. S., Aronson, J. and Pausas, J. G. (2011). Mediterranean cork oak savannas require human use to sustain biodiversity and ecosystem services. *Frontiers in Ecology and the Environment*, **9**, 278–286.

Camilo-Alves, C. S. P., Clara, M. I. E. and Ribeiro, N. M. C. A. (2013). Decline of Mediterranean oak trees and its association with *Phytophthora cinnamomi*: a review. *European Journal of Forest Research*, **132**, 411–432.

Camilo-Alves, C. S. P., Vaz, M., Da Clara, M. I. E. and Ribeiro, N. M. A. (2017). Chronic cork oak decline and water status: new insights. *New Forests*, **48**, 753–772.

Capelo, J. and Catry, F. (2007). A distribuição do sobreiro em Portugal. In *Os Montados – Muito para além das árvores*. Vol. 3, ed. J. S. Silva. Colecção Árvores e Florestas de Portugal. *Jornal Público/Fundação Luso-Americana para o Desenvolvimento/Liga para a Protecção da Natureza*. Lisboa. 9 vols, pp. 107–113.

Caritat, A., Molinas, M. and Gutiérrez, E. (1996). Annual cork-ring width variability of *Quercus suber* L. in relation to temperature and precipitation (Extremadura, southwestern Spain). *Forest Ecology and Management*, **86**, 113–120.

Caritat, A., Gutierrez, E. and Molinas, M. (2000). Influence of weather on cork-ring width. *Tree Physiology*, **20**, 893–900.

Carrete, M. and Donázar, J. A. (2005). Application of central-place foraging theory shows the importance of Mediterranean dehesas for the conservation of the cinereous vulture, *Aegypius monachus*. *Biological Conservation*, **126**, 582–590.

Collado, J., Platas, G., González, I. and Peláez, F. (1999). Geographical and seasonal influences on the distribution of fungal endophytes in *Quercus ilex*. *New Phytologist*, **144**, 525–532.

Compant, S., Heijden, M. G. A. and Sessitsch, A. (2010). Climate change effects on beneficial plant–microorganism interactions. *FEMS Microbiology Ecology*, **73**, 197–214.

Costa, A., Pereira, H. and Oliveira, A. (2001). A dendroclimatological approach to diameter growth in adult cork-oak trees under production.

Trees - Structure and Function, **15**, 438–443.

Costa, A., Madeira, M. and Oliveira, Â. C. (2008). The relationship between cork oak growth patterns and soil, slope and drainage in a cork oak woodland in Southern Portugal. *Forest Ecology and Management*, **255**, 1525–1535.

Costa, A., Pereira, H. and Madeira, M. (2010). Analysis of spatial patterns of oak decline in cork oak woodlands in Mediterranean conditions. *Annals of Forest Science*, **67**, 204–204.

Costa, A., Nunes, L. C., Spiecker, H. and Graça, J. (2015). Insights into the responsiveness of cork oak (*Quercus suber* L.) to bark harvesting. *Economic Botany*, **69**, 171–184.

Costa, A., Barbosa, I., Roussado, C., Graça, J. and Spiecker, H. (2016). Climate response of cork growth in the Mediterranean oak (*Quercus suber* L.) woodlands of southwestern Portugal. *Dendrochronologia*, **38**, 72–81.

Council of European Union (1992). Council directive 92/43/EEC. *Official Journal of the European Communities*, **L 269**, 1–15.

Dastogeer, K. M. G. and Wylie, S. J. (2017). Plant-fungi association: Role of fungal endophytes in improving plant tolerance to water stress. In *Plant-Microbe Interactions in Agro-Ecological Perspectives*, ed. D. P. Singh, H. B. Singh and R. Prabha. Singapore: Springer Singapore, pp. 161–176.

Desprez-Loustau, M. L., Marçais, B., Nageleisen, L. M., Piou, D. and Vannini, A. (2006). Interactive effects of drought and pathogens in forest trees. *Annals of Forest Science*, **63**, 597–612.

de Vries, S. M. G., Alan, M., Bozzano, M. *et al.* (2015). Pan-European strategy for genetic conservation of forest trees and establishment of a core network of dynamic conservation units. *European Forest Genetic Resources Programme (EUFORGEN), Biodiversity International, Rome, Italy.* **xii**, 3.

Dewan, M. M., Ghisalbertib, E. L., Rowland, C. and Sivasithamparam, K. (1994). Reduction of symptoms of take-all of wheat and rye-grass seedlings by the soil-borne fungus *Sordaria fimicola*. *Applied Soil Ecology*, **1**, 45–51.

Di Castri, F. (1991). An ecological overview of the five regions of the world with Mediterranean climate. In *Biogeography of Mediterranean Invasions*, ed. R. H. Groves and F. Di Castri. Cambridge: Cambridge University Press, pp. 3–15.

Emberger, L. (1930). Sur une formule climatique applicable en géographie botanique. *Comptes Rendus de l'Académie des Sciences*, **191**, 389–390.

EUFORGEN (2009). Distribution map of cork oak (*Quercus suber*). www.euforgen.org.

Fisher, P. J., Petrini, O., Petrini, L. E. and Sutton, B. C. (1994). Fungal endophytes from the leaves and twigs of *Quercus ilex* L. from England, Majorca and Switzerland. *New Phytologist*, **127**, 133–137.

Franceschini, A., Linaldeddu, B. T. and Marras, F. (2005). Occurrence and distribution of fungal endophytes in declining cork oak forests in Sardinia (Italy). *IOBC-WPRS Bulletin*, **28**, 67–74.

Fujimura, K. E., Egger, K. N. and Henry, G. H. R. (2008). The effect of experimental warming on the root-associated fungal community of *Salix arctica*. *ISME Journal*, **2**, 105–114.

Gennaro, M., Gonthier, P. and Nicolotti, G. (2003). Fungal endophytic communities in healthy and declining *Quercus robur* L. and *Q. cerris* L. Trees in Northern Italy. *Journal of Phytopathology*, **151**, 529–534.

Gentilesca, T., Camarero, J. J., Colangelo, M., Nolè, A. and Ripullone, F. (2017). Drought-induced oak decline in the western Mediterranean region: an overview on current evidences, mechanisms and management options to improve forest resilience. *iForest – Biogeosciences and Forestry*, **10**, 796–806.

Gil, L. and Varela, M. C. (2008). EUFORGEN Technical guidelines for genetic conservation and use for cork oak (*Quercus suber*). *Biodiversity International*, 6.

Giorgi, F. (2006). Climate change hot-spots. *Geophysical Research Letters*, **33**, L08707.

Giorgi, F. and Lionello, P. (2008). Climate change projections for the Mediterranean region. *Global and Planetary Change*, **63**, 90–104.

Gonthier, P., Gennaro, M. and Nicolotti, G. (2006). Effects of water stress on the endophytic mycota of *Quercus robur*. *Fungal Diversity*, **21**, 69–80.

Gonzalez, P., Neilson, R. P., Lenihan, J. M. and Drapek, R. J. (2010). Global patterns in the vulnerability of ecosystems to vegetation shifts due to climate change. *Global Ecology and Biogeography*, **19**, 755–768.

González-García, S., Dias, A. C. and Arroja, L. (2013). Life-cycle assessment of typical Portuguese cork oak woodlands. *Science of the Total Environment*, **452**, 355–364.

Gouveia, A. C. and Freitas, H. (2009). Modulation of leaf attributes and water use efficiency in *Quercus suber* along a rainfall gradient. *Trees – Structure and Function*, **23**, 267–275.

Hardoim, P. R., van Overbeek, L. S., Berg, G. *et al.* (2015). The hidden world within plants: ecological and evolutionary considerations for defining functioning of microbial endophytes. *Microbiology and Molecular Biology Reviews*, **79**, 293–320.

Hashizume, Y., Sahashi, N. and Fukuda, K. (2008). The influence of altitude on endophytic mycobiota in *Quercus acuta* leaves collected in two areas 1000km apart. *Forest Pathology*, **38**, 218–226.

Hasnaoui, F., Zouaoui, I., Seghaeir, W. O. and Abbes, C. (2017). Identification and pathogenicity of Fungi associated with decline of cork oak in the north west of Tunisia. *Journal of New Sciences, 40*, 2164–2168.

Henriques, J., Inácio, M. L., Lima, A. and Sousa, E. (2012). New outbreaks of charcoal canker on young cork oak trees in Portugal. *IOBC/WPRS Bulletin*, **76**, 85–88.

Henriques, J., Nóbrega, F., Sousa, E. and Lima, A. (2016). Analysis of the genetic diversity and phylogenetic relationships of *Biscogniauxia mediterranea* isolates associated with cork oak. *Phytoparasitica*, **44**, 19–34.

Hesse, U., Schöberlein, W., Wittenmayer, L. *et al.* (2003). Effects of *Neotyphodium* endophytes on growth, reproduction and drought-stress tolerance of three *Lolium perenne* L. genotypes. *Grass and Forage Science*, **58**, 407–415.

Hodkinson, T. R., Jones, M. B., Waldren, S. and Parnell, J. A. N., eds. (2011). *Climate Change Ecology and Systematics*. Cambridge: Cambrige University Press.

Holdridge, L. R. (1947). Determination of world plant formations from simple climatic data. *Science*, **105**, 367–368.

Howlett, D. S., Moreno, G., Losada, M. R. M., Nair, P. R. and Nair, V. D. (2011). Soil carbon storage as influenced by tree cover in the dehesa cork oak silvopasture of central-western Spain. *Journal of Environmental Monitoring*, **13**, 1897.

Huang, Y. L., Devan, M. N., U'Ren, J. M., Furr, S. H. and Arnold, A. E. (2016). Pervasive effects of wildfire on foliar endophyte communities in montane forest trees. *Microbial Ecology*, **71**, 452–468.

Ibáñez, B., Gómez-Aparicio, L., Stoll, P. *et al.* (2015). A neighborhood analysis of the consequences of *Quercus suber* decline for regeneration dynamics in Mediterranean forests. *PLoS One*, **10**, e0117827.

Ibáñez, B., Gómez-Aparicio, L., Ávila, J. M., Pérez-Ramos, I. M. and Marañón, T. (2017). Effects of *Quercus suber* decline on woody plant regeneration: potential implications for successional dynamics in Mediterranean forests. *Ecosystems*, **20**, 630–644.

IPMA (n.d.) Clima de Portugal Continental. www.ipma.pt/pt/educativa/tempo.clima/

Joffre, R. and Rambal, S. (1993). How tree cover influences the water balance of Mediterranean rangelands. *Ecology*, **74**, 570–582.

Kwaśna, H., Szewczyk, W. and Behnke-Borowczyk, J. (2016). Fungal root endophytes of *Quercus robur* subjected to flooding. *Forest Pathology*, **46**, 35–46.

Larena, I., Torres, R., De Cal, A. *et al.* (2005). Biological control of postharvest brown rot (*Monilinia* spp.) of peaches by field applications of *Epicoccum nigrum*. *Biological Control*, **32**, 305–310.

La Porta, N., Capretti, P., Thomsen, I. M. *et al.* (2008). Forest pathogens with higher damage potential due to climate change in Europe. *Canadian Journal of Plant Pathology*, **30**, 177–195.

Lima, G., Ippolito, A., Nigro, F. and Salerno, M. (1997). Effectiveness of *Aureobasidium pullulans* and *Candida oleophila* against postharvest strawberry rots. *Postharvest Biology and Technology*, **10**, 169–178.

Linaldeddu, B. T., Franceschini, A. and Pulina, M. A. (2005). Epidemiological aspects of *Biscogniauxia mediterranea* in declining cork oak forest in Sardinia (Italy). *IOBC/WPRS Bulletin*, **28**, 308.

Linaldeddu, B. T., Sirca, C., Spano, D. and Franceschini, A. (2009). Physiological responses of cork oak and holm oak to infection by fungal pathogens involved in oak decline. *Forest Pathology*, **39**, 232–238.

Linaldeddu, B. T., Sirca, C., Spano, D. and Franceschini, A. (2011). Variation of endophytic cork oak-associated fungal communities in relation to plant health and water stress. *Forest Pathology*, **41**, 193–201.

Lionello, P., Malanotte-Rizzoli, P. and Boscolo, R. (2006). The Mediterranean climate: an overview of the main characteristics and issues. In *Mediterranean Climate Variability*, ed. P. Lionello, P. Malanotte-Rizzoli, Paola and R. Boscolo. Amsterdam: Elsevier, pp. 1–26.

Madrigal, C., Pascual, S. and Melgarejo, P. (1994). Biological control of peach twig blight (*Monilinia laxa*) with *Epicoccum nigrum*. *Plant Pathology*, **43**, 554–561.

Maghnia, F. Z., Abbas, Y., Mahé, F. *et al.* (2017). Habitat- and soil-related drivers of the root-associated fungal community of *Quercus suber* in the Northern Moroccan forest. *PLoS One*, **12**, e0187758.

Malcolm, J. R., Liu, C., Neilson, R. P., Hansen, L. and Hannah, L. E. E. (2006). Global warming and extinctions of endemic species from biodiversity hotspots. *Conservation Biology*, **20**, 538–548.

Martín, J., Cabezas, J., Buyolo, T. and Patón, D. (2005). The relationship between

Cerambyx spp. damage and subsequent *Biscogniauxia mediterranum* infection on *Quercus suber* forests. *Forest Ecology and Management*, **216**, 166–174.

Martínez-Álvarez, P., Rodríguez-Ceinós, S., Martín-García, J. and Diez, J. J. (2012). Monitoring endophyte populations in pine plantations and native oak forests in Northern Spain. *Forest Systems*, **21**, 373–382.

Mejía, L. C., Rojas, E. I., Maynard, Z. *et al.* (2008). Endophytic fungi as biocontrol agents of *Theobroma cacao* pathogens. *Biological Control*, **46**, 4–14.

Mendes, M. P., Ribeiro, L., David, T. S. and Costa, A. (2016). How dependent are cork oak (*Quercus suber* L.) woodlands on groundwater? A case study in southwestern Portugal. *Forest Ecology and Management*, **378**, 122–130.

Moricca, S. and Ragazzi, A. (2008). Fungal endophytes in Mediterranean oak forests: a lesson from *Discula quercina*. *Phytopathology*, **98**, 380–386.

Moricca, S., Ginetti, B. and Ragazzi, A. (2012). Species- and organ-specificity in endophytes colonizing healthy and declining Mediterranean oaks. *Phytopathologia Mediterranea*, **51**, 587–598.

Moricca, S., Linaldeddu, B. T., Ginetti, B. *et al.* (2016). Endemic and emerging pathogens threatening cork oak trees: Management options for conserving a unique forest ecosystem. *Plant Disease*, **100**, 2184–2193.

Murphy, B. R., Martin Nieto, L., Doohan, F. M. and Hodkinson, T. R. (2015). Fungal endophytes enhance agronomically important traits in severely drought-stressed barley. *Journal of Agronomy and Crop Science*, **201**, 419–427.

Myers, N., Mittermeier, R. A., Mittermeier, C. G., Da Fonseca, G. A. and Kent, J. (2000). Biodiversity hotspots for conservation priorities. *Nature*, **403**, 853–858.

Naseby, D. C., Pascual, J. A. and Lynch, J. M. (2000). Effect of biocontrol strains of *Trichoderma* on plant growth, *Pythium ultimum* populations, soil microbial communities and soil enzyme activities. *Journal of Applied Microbiology*, **88**, 161–169.

O'Hanlon, K. A., Knorr, K., Jørgensen, L. N., Nicolaisen, M. and Boelt, B. (2012). Exploring the potential of symbiotic fungal endophytes in cereal disease suppression. *Biological Control*, **63**, 69–78.

Oliveira, G. and Costa, A. (2012). How resilient is *Quercus suber* L. to cork harvesting? A review and identification of knowledge gaps. *Forest Ecology and Management*, **270**, 257–272.

Oliveira, V., Lauw, A. and Pereira, H. (2016). Sensitivity of cork growth to drought events: insights from a 24-year chronology. *Climatic Change*, **137**, 261–274.

Oses, R., Valenzuela, S., Freer, J., Sanfuentes, E. and Rodriguez, J. (2008). Fungal endophytes in xylem of healthy Chilean trees and their possible role in early wood decay. *Fungal Diversity*, **33**, 77–86.

Ozenda, P. and Borel, J. L. (2000). An ecological map of Europe: why and how? *Comptes Rendus de l'Académie des Sciences-Series III-Sciences de la Vie*, **323**, 983–994.

Paulo, J. A., Palma, J. H. N., Gomes, A. A. *et al.* (2015). Predicting site index from climate and soil variables for cork oak (*Quercus suber* L.) stands in Portugal. *New Forests*, **46**, 293–307.

Pausas, J. G. (1997). Resprouting of *Quercus suber* in NE Spain after fire. *Journal of Vegetation Science*, **8**, 703–706.

Pereira, H. (2007). *Cork: Biology, Production and Uses*. Amsterdam: Elsevier.

Pereira, J. S., Bugalho, M. N. and Caldeira, M. D. C. (2008). *From the Cork Oak to Cork*. Portugal: APCOR - Portuguese Cork Association.

Pereira, P. M. and Fonseca, M. P. (2003). Nature vs. nurture: The making of the montado ecosystem. *Ecology and Society*, **7**, 7.

Petrini, O. and Fisher, P. J. (1990). Occurrence of fungal endophytes in twigs of *Salix fragilis* and *Quercus robur*. *Mycological Research*, **94**, 1077–1080.

Pinto, C. A., Henriques, M. O., Figueiredo, J. P. *et al.* (2011). Phenology and growth dynamics in Mediterranean evergreen oaks: effects of environmental conditions and water relations. *Forest Ecology and Management*, **262**, 500–508.

Pulina, M. A., Linaldeddu, B. T. and Franceschini, A. (2006). Topoclimats et communautés des champignons endophytiques dans des bois de chênes-lièges dépéris et non dépéris en Sardaigne (Italie). *Les risques liés au temps et au climat*, 474.

Ragazzi, A., Mancini, F., Dellavalle, I., Capretti, P. and Moricca, S. (2001). Endophytic fungi in Quercus cerris: isolation frequency in relation to phenological phase, tree health and the organ affected. *Phytopathologia Mediterranea*, **40**, 165–171.

Ragazzi, A., Moricca, S., Capretti, P., Dellavalle, I. and Turco, E. (2003). Differences in composition of endophytic mycobiota in twigs and leaves of healthy and declining *Quercus* species in Italy. *Forest Pathology*, **33**, 31–38.

Ramírez-Valiente, J. A., Valladares, F., Huertas, A. D., Granados, S. and Aranda, I. (2011). Factors affecting cork oak growth under dry conditions: Local adaptation and contrasting additive genetic variance within populations. *Tree Genetics and Genomes*, **7**, 285–295.

Redman, R. S., Kim, Y. O., Woodward, C. J. *et al.* (2011). Increased fitness of rice plants to abiotic stress via habitat adapted symbiosis: a strategy for mitigating impacts of climate change. *PLoS One*, **6**, e14823.

Rego, F. C. and Rocha, M. S. (2014). Climatic patterns in the Mediterranean region. *Ecologia Mediterranea*, **40**, 49–59.

Reis, F., Tavares, R. M., Baptista, P. and Lino-Neto, T. (2017). Mycorrhization of Fagaceae forests within Mediterranean ecosystems. In *Mycorrhiza: Function, Diversity, State of the Art*, 4th edn, ed. A. Varma. Berlin: Springer International Publishing, pp. 75–97.

Rho, H., Hsieh, M., Kandel, S. L. *et al.* (2017). Do endophytes promote growth of host plants under stress? A meta-analysis on plant stress mitigation by endophytes. *Microbial Ecology*, **75**, 407–418.

Rigueiro-Rodríguez, A., McAdam, J. and Mosquera-Losada, M. R., eds. (2009). *Agroforestry in Europe. Current Status and Future Prospects*. Berlin: Springer Science & Business Media B. V.

Rodriguez, R. J., Henson, J., Van Volkenburgh, E. *et al.* (2008). Stress tolerance in plants via habitat-adapted symbiosis. *The ISME Journal*, **2**, 404–416.

Rodriguez, R. J., White Jr., J. F., Arnold, A. E. and Redman, A. R. A. (2009). Fungal endophytes: diversity and functional roles. *New Phytologist*, **182**, 314–330.

Romeralo, C., Santamaría, O., Pando, V. and Diez, J. J. (2015). Fungal endophytes reduce necrosis length produced by *Gremmeniella abietina* in *Pinus halepensis* seedlings. *Biological Control*, **80**, 30–39.

Saikkonen, K. (2007). Forest structure and fungal endophytes. *Fungal Biology Reviews*, **21**, 67–74.

Saucedo-García, A., Anaya, A. L., Espinosa-García, F. J. and González, M. C. (2014). Diversity and communities of foliar endophytic fungi from different agroecosystems of *Coffea arabica* L. in two regions of Veracruz, Mexico. *PloS One*, **9**, e98454.

Schulz, B. and Boyle, C. (2005). The endophytic continuum. *Mycological Research*, **109**, 661–686.

Seghers, D., Wittebolle, L., Top, E. M., Verstraete, W. and Siciliano, S. D. (2004). Impact of agricultural practices on the *Zea mays* L. endophytic community. *Applied and Environmental Microbiology*, **70**, 1475–1482.

Sieber, T. N. (2007). Endophytic fungi in forest trees: are they mutualists? *Fungal Biology Reviews*, **21**, 75–89.

Slippers, B. and Wingfield, M. J. (2007). Botryosphaeriaceae as endophytes and latent pathogens of woody plants: diversity, ecology and impact. *Fungal Biology Reviews*, **21**, 90–106.

Solomon, S., Qin, D., Manning, M. *et al.* (2007). Climate change 2007: The physical science basis. In *Contribution of Working Group I to the Fourth Assessment Report of the Intergovernmental Panel on Climate Change*. Geneva: IPCC, pp. 235–337.

Sun, X., Ding, Q., Hyde, K. D. and Guo, L. D. (2012). Community structure and preference of endophytic fungi of three woody plants in a mixed forest. *Fungal Ecology*, **5**, 624–632.

Wang, F. W., Jiao, R. H., Cheng, A. B., Tan, S. H. and Song, Y. C. (2007). Antimicrobial potentials of endophytic fungi residing in *Quercus variabilis* and brefeldin A obtained from *Cladosporium* sp. *World Journal of Microbiology and Biotechnology*, **23**, 79–83.

Waqas, M., Khan, A. L., Shahzad, R. *et al.* (2015a). Mutualistic fungal endophytes produce phytohormones and organic acids that promote japonica rice plant growth under prolonged heat stress. *Journal of Zhejiang University. Science. B*, **16**, 1011.

Waqas, M., Khan, A. L., Hamayun, M. *et al.* (2015b). Endophytic fungi promote plant growth and mitigate the adverse effects of stem rot: an example of *Penicillium citrinum* and *Aspergillus terreus*. *Journal of Plant Interactions*, **10**, 280–287.

White, J. F., Torres M. S., Johnson H., Irizarry I. and Tadych M. (2014). A functional view of plant microbiomes: endosymbiotic systems that enhance plant growth and survival. In *Advances in Endophytic Research*, ed. V. Verma and A. Gange. New Delhi: Springer, pp. 425–439.

Wilson, D. and Carroll, G. C. (1994). Infection studies of *Discula quercina*, an endophyte of *Quercus garryana*. *Mycologia*, **86,** 635–647.

Part IV

Endophytes for Novel Biomolecules and In Vitro Methods

13

Endophytic Fungi: A Quintessential Source of Potential Bioactive Compounds

Vineet Meshram and Mahiti Gupta

Abstract

The fortuitous discovery of penicillin from *Penicillium chrysogenum* heralded the golden era of antibiotics. Since then, fungi have significantly contributed to the welfare of humans by producing bioactive compounds which have been used as antibacterial, anticancer, antioxidant and immunomodulatory agents. However, in recent years, microorganisms associated with plants have emerged as fountainheads of bioactive molecules with high therapeutic potential. In general terms, endophytes are an extremely diverse and ubiquitous group of microorganisms that resides within the living internal tissues of a host plant in a non-invasive manner. Endophytes communicate with their host plant through metabolic interactions which enables them to produce signal molecules with interesting biological activities. Further, the genetic recombination of endophytes with the host plant enables them to mimic the biological properties of the host and produce analogous bioactive compounds. Thus, they start producing the host plant phytochemicals when cultured independently. The endless need for potent drugs has prompted researchers to explore alternative avenues for finding novel bioactive molecules, and endophytes appear to be a plausible target for drug discovery. This chapter reviews the current research trends with these promising organisms.

13.1 Introduction

Since time immemorial, natural products have been used for the treatment of various kinds of human ailments. Plants have served as sources of phytochemicals for drug discovery and development (Kaul *et al.*, 2012). However, in the recent times, there has been a paradigm shift and the microorganisms associated with the plants rather than plants themselves have emerged as a lucrative source of bioactive compounds with high therapeutic potential (Gouda *et al.*, 2016; Omeje *et al.*, 2017). Endophytic microorganisms asymptomatically live together with plants in mutualistic alliance (Bacon and White, 2000; Strobel and Daisy, 2003; Kusari and Spiteller, 2016). They produce a plethora of structurally diverse bioactive compounds, such as enzymes, enzyme inhibitors, and volatile organic compounds, with significant antibacterial, antifungal, antioxidant, anticancer, antidiabetic, anti-inflammatory, anti-obesity, cardioprotective and neuroprotective properties – many of which are undergoing clinical trials (Aly *et al.*, 2011; Kaul *et al.*, 2012; Gupta *et al.*, 2015; Saxena *et al.*, 2015; Meshram *et al.*, 2016). Furthermore, recent discoveries have revealed that endophytic microorganisms could synthesise plant analogous compounds which reflect an adaptive functional role of endophytes in biocenosis (Zhao *et al.*, 2010; Aly *et al.*, 2013; Ludwig-Muller, 2015; Nicoletti and Fiorentino, 2015). This intrinsic nature of endophytes to produce analogous and non-analogous bioactive metabolites in most of the plants of therapeutic or commercial importance has given an impetus to the possibilities of exploiting these microorganisms for production of several life-saving drugs (Kusari *et al.*, 2012, 2013; Venugopalan and Srivastava, 2015). Even, if the metabolites produced by endophytes do not directly exhibit desired biological functions, they might still provide lead structures which could act as parent molecules for deriving several drugs via combinatorial synthesis (Kusari *et al.*, 2009a; Kusari and Spiteller, 2016). Several reports have suggested that the metabolites produced by endophytes have promising applications in the food, agrochemical and pharmaceutical industries (Strobel and Daisy, 2003; Aly *et al.*, 2011; Kharwar *et al.*, 2011; Kaul *et al.*, 2012; Nicoletti and Fiorentino, 2015; Gouda *et al.*, 2016; Vasundhara *et al.*, 2016). The current chapter highlights the potential use of medicinally important plant secondary metabolites produced by fungal endophytes and outlines their possible applications in pharmaceutical industry.

13.1.1 What Is an Endophyte?

Endophytes are a ubiquitous, polyphyletic group of microorganisms, that spend their entire or a significant portion of their life cycle within the living tissues of their host plant without causing negative or overt symptoms (Petrini, 1991; Bacon and White, 2000; Kusari *et al.*, 2012; Meshram *et al.*, 2016). During the alliance, neither of the interacting partners is harmed and the benefits obtained are dependent on the interacting partners. Thus endophytism is a novel, evolutionarily cost-effective plant–microbe association driven by location and not defined by function (Kusari *et al.*, 2012, 2013; Saxena *et al.*, 2015). Endophytes exhibit a variety of relationships with their host plant including symbiotic partners, benign commensals, decomposers and latent pathogens (Promputha *et al.*, 2007). Endophytes can provide fitness benefits to their host plants by mediating abiotic and biotic stress tolerance, decreasing water consumption and defending from pests and microbial infections. Thus, endophytes play a significant role in plant symbiosis, protecting their host from pathogens, pests and abiotic stresses (Kusari *et al.*, 2014; Meshram *et al.*, 2016; Hodkinson and Murphy, 2019, Chapter 1).

Endophytes are hyperdiverse and are widely distributed in various ecological niches including the Arctic and Antarctic, deserts, mangroves, rainforests, marine and coastal environments (Strobel and Daisy, 2003; Chowdhary *et al.*, 2012, Stierle and Stierle, 2015). Each plant species is inhabited by a large community of endophytes. For example, 25–33 culturable endophytes are expected to colonise a single leaf of a tropical tree *Manilkara bidentata* or *Guarea Guidonia*, and more than 50 different genera in the roots of an arid grassland species *Bouteloua gracilis* (Zhao *et al.*, 2010; Porras-Alfararo and Bayman, 2011). The fossil records of plant-associated microorganisms date back to more than 400 million years ago during the Devonian period, suggesting that plant–endophyte alliance may have evolved from at least the time when land plants first appeared on the earth (Kusari and Spiteller, 2012; Venugopalan and Srivastava, 2015; Hodkinson, 2018).The term 'endophyte' was coined by a German botanist Anton de Bary in 1866 to describe any organisms asymptomatically living inside the plants, whereas endophytes were first discovered from a Eurasian ryegrass Darnel (*Lolium temulentum*) in 1904 (De Bary, 1866; Freeman, 1904; Porras-Alfaro and Bayman, 2011; Chowdhary *et al.*, 2012). However, except for some infrequent studies, endophytes did not receive much of attention from the bioactive chemical research community until the discovery of a taxol-producing endophytic fungus *Taxomyces andreanea* from the Pacific yew, *Taxus brevifolia*, in 1993. This prompted researchers to further explore endophytic fungi for potential bioactive compounds (Stierle *et al.*, 1993; Strobel and Daisy, 2003; Kusari and Spiteller, 2012, Stierle and Stierle, 2015) and spurred a worldwide search for novel endophytes with unique biological properties (Kharwar *et al.*, 2011; Aly *et al.*, 2013; Kusari *et al.*, 2014; Nicoletti and Fiorentino, 2015). On one hand, endophytes were investigated from an ecological prospective to understand their colonisation, transmission, tissue specificity, responses to biotic and abiotic stresses and evolutionary relatedness (Schulz *et al.*, 1999; Arnold *et al.*, 2000; Schulz and Boyle, 2005; Arnold, 2007, Rodriguez *et al.*, 2009). On the other hand, a significant number of discoveries were made that highlighted the potential of endophytes for *de novo* synthesis of bioactive compounds which may directly or indirectly be used as therapeutic agents against a plethora of maladies (Strobel and Daisy, 2003; Aly *et al.*, 2013; Kusari *et al.*, 2014; Venugopal and Srivastava, 2015; Nicoletti and Fiorentino, 2015; Gupta *et al.*, 2015; Meshram *et al.*, 2016).

13.1.2 Rationale for Plant Selection

There are about 300 000 higher plant species existing on earth (Aly *et al.*, 2011). Each plant harbours a suite of endophytic microorganisms (Meshram *et al.*, 2016) and fungal endophytes are more frequently encountered than their bacterial counterparts (Saxena *et al.*, 2015). Given the high plant species richness, it is crucial to use innovative and imaginative selection strategies to narrow down the target species and search for novel endophytic fungi displaying bioactivities (Strobel and Daisy, 2003; Kusari and Spiteller, 2012). Several plant selection strategies can be adopted for isolation of fungal endophytes equipped to produce host related compounds (Kusari and Spiteller, 2016). The classes of plants selected include, but are not limited to, the following:

Traditional Medicinal Plants: Plants used by native people in traditional medicinal preparations, or plants with ethnobotanical history, are bio-prospected for endophytes, especially for the ones that are proficient in producing one or more plant related bioactive compounds (Strobel and Daisy, 2003; Kusari and Spiteller, 2012, 2016). This is one of the most commonly employed selection strategies.

Plants from Biodiversity Hotspots: Plants dwelling in areas of abundant biodiversity where a profusion of abiotic and biotic factors play a vital role in generation of bioactive compounds and a rich diversity of endophytes (Strobel and Daisy, 2003; Kusari and Spiteller, 2012).

Plants from Unique Ecological Niches: Plants from inimitable ecological niches, especially those with an atypical biology, and possessing unusual strategies for existence are an exclusive source of endophytes with useful biological activities (Strobel and Daisy, 2003; Kusari and Spiteller, 2012, 2016).

Endemic and Endangered Plants: Plants that are endemic, that have occupied a particular geographical niche for long periods of time, are likely to harbour unique endophytes with profound biological activities (Strobel and Daisy, 2003; Kusari and Spiteller, 2012).

Phylogenetically Related Plants: Plants are selected on the basis of their phylogenetic relatedness to a previously recognised or traditional medicinal plant of high importance. The endophytes inhabited in these plants are more likely to produce analogous compounds similar to their host (Kusari and Spiteller, 2016).

Plants Selected by Data Mining: This approach can be used in combination and complementary to the abovementioned strategies. The information collected using various bioprospecting strategies can be assembled together, systematically investigated and stored for further use by applying appropriate data mining approaches (Kusari and Spiteller, 2016).

13.2 Endophytic Fungi as Hidden Producers of Bioactive Compounds to Their Host Plant

Endophytic fungi are chemical synthesisers inside the plants with par excellence (Kaul *et al.*, 2012). The unobtrusive union of endophytic fungi with plant enables them to co-evolve (Kusari *et al.*, 2012, 2013). This co-evolution between endophytic fungi and its host plant determines the production of bioactive compounds which often contributes in a variety of ways in the plant–microbe interaction and can endow a variety of fitness benefits to the host plant (Saxena *et al.*, 2015; Venugopalan and Srivastava, 2015; Kusari and Spiteller, 2016). Endophytic fungi communicate with their host plant through metabolic interactions (Krohn *et al.*, 2007; Kusari *et al.*, 2012). According to the xenohormesis hypothesis (Howitz and Sinclair, 2008), heterotrophic organisms like fungi, under selective evolutionary pressure, develop the ability to sense the stress induced chemical cues from host plant and start producing analogous chemicals themselves. In essence they mimic the biological properties of the host plant (Kusari *et al.*, 2014; Venugopalan and Srivastava, 2015).

A milestone in the area of bioprospecting for endophytic fungi was undoubtedly the discovery of *Taxomyces andreanea* from the Pacific yew, *Taxus brevifolia*, which exhibited the *de novo* synthesis of the first billion dollar anticancer drug, taxol, in its culture broth similar to its host plant (Stierle *et al.*, 1993; Stierle and Strobel, 1995). Since then, endophytic fungi have been harnessed for production of an array of plant related compounds such as camptothecin, podophyllotoxin, hypericin, huperzine, vinblastin and vincristine (Puri *et al.*, 2005, 2006; Kusari *et al.*, 2008; Zhu *et al.*, 2010; Kumar *et al.*, 2013).This section discusses the production of medicinally important plant secondary metabolites by their fungal endophytes, with special emphasis on the source microorganisms and its biological activities.

13.2.1 Taxol

We are still a long way from achieving bona fide cures of several major cancers (Omeje *et al.*, 2017) which are responsible for over 8.8 million deaths annually (Stierle and Stierle, 2015; WHO, 2017). Therefore, the development of new and effective anticancer agents is urgently required. Among various anticancer drugs, taxol is a highly reactive diterpenoid, originally isolated from the inner bark of the Pacific yew which was later found in all yew species (Wani *et al.*, 1971; Stierle *et al.*, 1993, Vasundhara *et al.*, 2016). Taxol is an exceptionally potent anticancer drug that renders its anticancerous property by preventing tubulin molecules from depolymerisation during the process of cell division (Stierle and Strobel, 1995; Stierle and Stierle, 2015). Today, taxol is in clinical use for treatment of range of cancers such as ovarian and breast cancers (Vasundhara *et al.*, 2016; Omeje *et al.*, 2017). However, within a few years of its application to cancer treatment, issues of limited supply of plant tissue became evident. There was thus a pressing need to find alternative sources which led to the isolation of an endophytic fungus *T. andreanae*, from the inner bark of *T. brevifolia*, that is able to produce taxol in under axenic monoculture conditions (Stierle *et al.*, 1993; Kharwar *et al.*, 2011; Stierle and Stierle, 2015). Ever since this landmark discovery, several other endophytic fungi showing taxol production from yew and other plant species have been discovered, including *Alternaria, Bartalinia, Fusarium, Lasiodiplodia, Metarhizium, Monochaetia, Pestalotiopsis, Penicillium, Phoma, Pithomyces, Seimatoantlerium, Sporormia, Trichothecium, Tubercularia* and *Truncatella* spp. (Table 13.1, Figure 13.1) (Zhao *et al.*, 2010; Ludwig-Muller, 2015; Vasundhara *et al.*, 2016; Omeje *et al.*, 2017). The taxol yield varies (from nanograms to milligrams per litre media) among these fungal species and their productivity is often lost within a few generations of *in vitro* culture (Gangadevi and Muthumary, 2008; Gond *et al.*, 2014). This has triggered researchers to question whether the source of the fungal taxol is in fact the host plant pool (Nicoletti and Fiorentino, 2015). However, isolation of taxol from non-taxol-producing tress such as *Aegle marmelos, Morinda citrifolia* and *Terminalia arjuna* has been also reported which contradicts the idea of fungal taxol being a host plant adduct (Gangadevi and Muthumary, 2008, 2009a, b). Whether such strains can be efficiently used by the pharmaceutical industry is still under debate. Nevertheless, there is no doubt that the issue of taxol production by endophytic fungi has prompted a wide ranging surge in research activity concerning pharmacological drugs and other bioactive compounds obtained from plants and their endophytes (Nicoletti and Fiorentino, 2015; Venugopalan and Srivastava, 2015).

13.2.2 Vinblastine and Vincristine

Vinblastine and vincristine are vinca alkaloids that were serendipitously discovered during the development of an oral hypoglycaemic agent from the extract of the plant *Catharanthus roseus* (Cragg and Pezzuto, 2016). Vinblastine and vincristine were the first plant-based chemotherapeutic agents that acquired a market place for clinical use. Vinblastine was introduced in 1960s for the treatment of breast and testicular cancers, and Hodgkin's disease, whereas its oxidised form vincristine was introduced in 1963 and for the treatment of various kinds of leukaemia (Ayob *et al.*, 2017). These compounds bind irreversibly to microtubule and spindle proteins causing arrest of the tumour cells in metaphase, ultimately leading to apoptotic cell death (Zhao *et al.*, 2010; Kharwar *et al.*, 2011; Cragg and Pezzuto, 2016). This sparked a worldwide hunt for novel alternative sources of vinblastine and vincristine. Guo *et al.* (1998) first described the production of vinblastine from endophytic *Alternaria* sp., whereas Lingqi *et al.* (2000) discovered an endophytic *Fusarium oxysporum* from *C. roseus* that successfully produced vincristine. Four years later, Yang *et al.* (2004) also reported a vincristine-producing endophytic mycelia sterilia from the Yunnan province of China. Vinblastine is more abundant in nature

Table 13.1 Endophytic fungi producing bioactive compounds (and their derivatives) and their host plants.

Compound	Source endophyte	Host Plant	Bioactivity	Yield µg/l unless stated otherwise	Reference(s)
Taxol	*Taxomyces andreanae*	*Taxus brevifolia*	Cytotoxic	0.02–0.05	Stierle *et al.*, 1993
	Pestalotiopsis microspora	*T. wallichiana*	Cytotoxic	60–70	Strobel *et al.*, 1996
	Tubercularia sp TF5	*T. mairei*	Cytotoxic	185.4	Wang *et al.*, 2001
	Botrytis sp. HD181-23	*T. cuspidata*	Cytotoxic	206.34	Hu *et al.*, 2006
	Alternaria alternata TPF6	*T. chinensis*	–	84.5	Tian *et al.*, 2006
	Bartalinia robillardoides AMB9	*Aegle marmelos*	Cytotoxic	187.6	Gangadevi and Muthumary, 2008
	Fusarium mairei UH23	*T. chinensis*	–	286.4	Dai *et al.*, 2008
	F. solanil Tax-3	*T. chinensis*	–	163.35	Deng *et al.*, 2009
	Metarhizium anisopliae H-27	*T. chinensis*	–	846.1	Liu *et al.*, 2009
	Cladosporium cladosporioides MD2	*T. media*	–	800	Zhang *et al.*, 2009
	Chaetomella raphigera TAC15	*Terminalia arjuna*	–	79.6	Gangadevi and Muthumary, 2009b
	Lasiodiplodia theobromae	*Morinda citrifolia*	Cytotoxic	245	Pandi *et al.*, 2011
	Phoma betae	*Ginkgo biloba*	Cytotoxic	795	Kumaran *et al.*, 2012
	F. redolens	*Taxus baccata*	Antimitotic	66	Garyali *et al.*, 2013
	P. medicaginis	*T. wallichiana*	–	1125	Zaiyou *et al.*, 2017

(continued)

Table 13.1 (*Cont.*)

Compound	Source endophyte	Host Plant	Bioactivity	Yield µg/l unless stated otherwise	Reference(s)
Baccatin III	*Didmyostilbe* sp. DF110	*T. chinensis*	Cytotoxic	–	Wang andTang, 2011
	Diaporthe phaseolorum	*T. wallichiana*	–	219	Zaiyou *et al.*, 2013
	Trichoderma sp. IRB54[a]	*T. wallichiana*	–	187.56	Li *et al.*, 2015
Vinblastine	*Alternaria* sp.	*Catheranthus roseus*	–	–	Guo *et al.*, 1998
	F. oxysporum	*C. roseus*	–	76	Kumar *et al.*, 2013
	Talaromyces radicus	*C. roseus*	–	70	Palem *et al.*, 2015
	N. sphaerica	*C. roseus*	Cytotoxic	860	Ayob *et al.*, 2017
Vincristine	*F. oxysporum*	*C. roseus*	–	–	Lingqi *et al.*, 2000
	Mycelia sterilia 97CY_(3)	*C. roseus*	–	0.21	Yang *et al.*, 2004
	F. oxysporum AA-CRL-6	*C. roseus*	–	67	Kumar *et al.*, 2013
	T. radicus CrP20	*C. roseus*	Cytotoxic	670	Palem *et al.*, 2015
	Eutypella sp. CrP14	*C. roseus*	Cytotoxic	53	Kuriakose *et al.*, 2016
Camptothecin	*E. infrequens*	*Nothapodytes foetida*	Cytotoxic	–	Puri *et al.*, 2005
	Neurospora crassa	*N. foetida*	Cytotoxic	–	Rehman *et al.*, 2008
	Nodulisporium sp.	*N. foetida*	–	45	Rehman *et al.*, 2009
	Mycelia sterilia_XK001[a]	*Camptotheca acuminata*	–	677	Duan *et al.*, 2009
	F. solani[a]	*C. acuminata*	–	ND	Kusari *et al.*, 2009c

	F. solani (MTCC 9667, 9668)[a]	*Apodytes dimidiata*	–	0.37–0.53	Shweta *et al.*, 2010
	Trichoderma atroviride LY357	*C. acuminata*	–	142.15	Pu *et al.*, 2013
	F. solani S-019	*C. acuminata*	Cytotoxic	150	Ran *et al.*, 2017
Podophyllotoxin	*Phialocephala fortinii* (PPE5, PPE7)	*Podophyllum peltatum*	Cytotoxicity	0.5–189	Eyberger *et al.*, 2006
	Trametes hirsute	*Sinopodophyllum hexandrum*	Cytotoxic	30[b]	Puri *et al.*, 2006
	F. oxysporum JRE1	*Juniperus recurva*	–	28[b]	Kour *et al.*, 2008
	Aspergillus fumigates[a]	*J. communis*	Antibacterial	0.04	Kusari *et al.*, 2009b
	Mucor fragilis TW5	*S. hexandrum*	–	49.3[b]	Huang *et al.*, 2014
	F. solani P1	*Podophyllum hexandrum*	–	29 μg/g	Mohammad *et al.*, 2012
	Alternaria tenuissima	*Sinopodophyllum emodi*	–	–	Liang *et al.*, 2016
Huperzine A	*Acremonium* sp.2F09P03B	*Huperzia serrata*	–	8.32	Li *et al.*, 2007
	Blastomyces sp.	*Phlegmariurus cryptomerianus*	–	20–30 μg/g[b]	Zan *et al.*, 2009
	Botrytis sp.	*P. cryptomerianus*	–	20–30 μg/g[b]	Zan *et al.*, 2009
	Shiraia sp. Slf14	*H. serrata*	AChEi	327.8	Zhu *et al.*, 2010
	Cladosporium cladosporioides LF70	*H. serrata*	AChEi	56.84	Zhang *et al.*, 2011
	Colletotrichum gloeosporioides ES026	*H. serrata*	–	1 μg/g[b]	Shu *et al.*, 2014

(continued)

Table 13.1 (*Cont.*)

Compound	Source endophyte	Host Plant	Bioactivity	Yield µg/l unless stated otherwise	Reference(s)
	Paecilomyces tenuis YS-13	*H. serrata*	AChEi	21	Su and Yang, 2014
	Trichoderma sp. L44	*H. serrata*	AChEi	37.63 µg/g[b]	Dong *et al.*, 2014
	Ceriporia lacerate MY311	*Phlegmariurus phlegmaria*	AChEi	40.53	Zhang *et al.*, 2015
Resveratrol	*Alternaria* sp. MG1	*Vitis vinifera*	–	224.4	Shi *et al.*, 2012; Che *et al.*, 2016a
	Alternaria sp. R6[a]	*Myoporum bontioides*	Antioxidant, Cytotoxic	–	Wang *et al.*, 2014
	Aspergillus niger	*V. vinifera 'C. sauvignon'*	–	1480	Liu *et al.*, 2016
Hypericin	*T. subthermophila*	*H. perforatum*	Antimicrobial, Cytotoxic	0.35 µg/g[b]	Kusari *et al.*, 2008
Piperine	*Periconia* sp.	*Piper longum*	Antimycobacterial	~750,000 µg	Verma *et al.*, 2011
	C. gloeosporioides PF12	*P. nigrum*	–	–	Chithra *et al.*, 2014a
	Mycosphaerella sp. PF13	*P. nigrum*	–	–	Chithra *et al.*, 2014b
	Phomopsis sp. PF5	*P. nigrum*	Growth promotion	880	Chithra *et al.*, 2017
Berberine	*Alternaria* sp. BBH6	*Phellodendron amurense*	–	–	Duan *et al.*, 2009
	F. solani SDV01	*Coscinium fenestratum*	Cytotoxic	196	Vinodhini and Agastian, 2013
	Alternaria sp. HL-Y-3	*Coptis chinensis*	–	9.31 µg/g	Zhang *et al.*, 2016

Emodin	*Penicillium janthinellum*	*Melia azedarach*	Antibacterial	–	Marinho *et al.*, 2005
	Thielavia subthermophila	*Hypericum perforatum*	Antimicrobial	1.10 μg/g[b]	Kusari *et al.*, 2008
	Talaromyces sp. YE3016	*Aconitum carmichaeli*	–	8000 μg	Xie *et al.*, 2016
Rohitukine	*F. proliferatum*	*Dysoxylum binectariferum*	Cytotoxic	1.86 μg/g[b]	Kumara *et al.*, 2012
	F. oxysporum (MTCC-11383), *F. oxysporum* (MTCC-11384), *F. solani* (MTCC-11385),	*D. binectariferum*	–	2.96–3.59 μg/g[b]	Kumara *et al.*, 2014
	Gibberella fujikuroi MTCC11382	*Amoora rohituka*	–	1.92 μg/g[b]	Kumara *et al.*, 2014
Vincamine	Unidentified strain Vm-J2	*Vinca minor*	–	10000	Yin and Sun, 2011
	Geomyces sp. CH1	*Nerium indicum*	AChEi	1279	Na *et al.*, 2016
Aconitine	*Cladosporium cladosporioides*XJ-AC03	*Aconitum leucostomum*	–	236.4 μg/g	Yang *et al.*, 2013
Gymnemagenin	*Penicillium oxalicum*	*Gymnema sylvestre*	–	–	Parthasarathy and Sathiyabama, 2014
Curcumin	*Gibberella fujikuroi* M7226	*Curcuma wenyujin*	–	–	Yan *et al.*, 2014
Bilobalide	*Pestalotiopsis uvicola*GZUYX13	*Ginkgo biloba*	–	106	Qian *et al.*, 2016
Peimisine	*Fusarium* sp. WBS007	*Fritillaria unibracteata*	–	90.9	Pan *et al.*, 2014
	F. redolens 6WBY3	*F. unibracteata*	–	16.02	Pan *et al.*, 2015

[a] refers to derivatives of the bioactive compounds, [b] refers to dry weight of mycelium

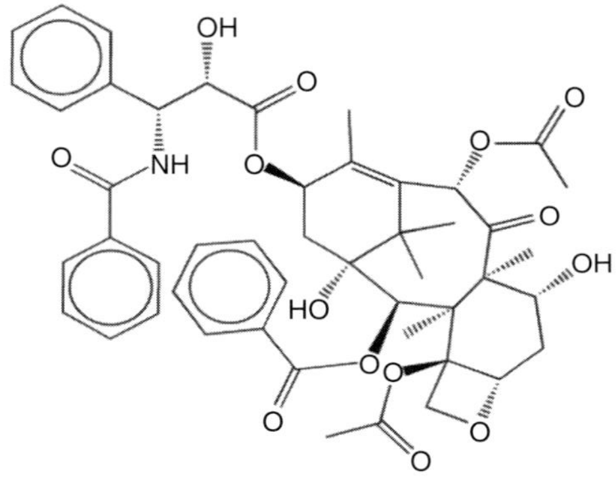

Taxol

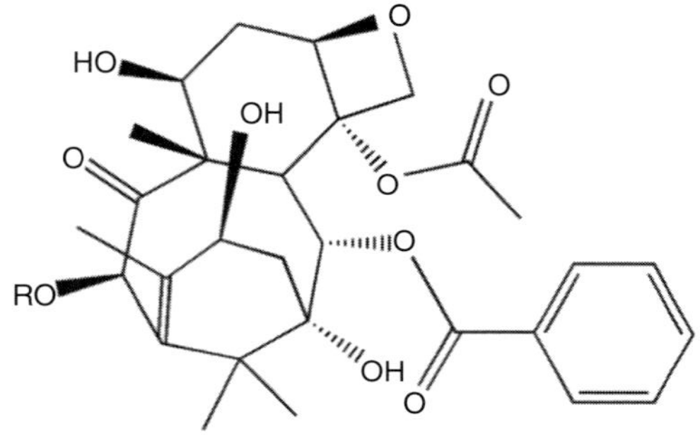

Baccatin

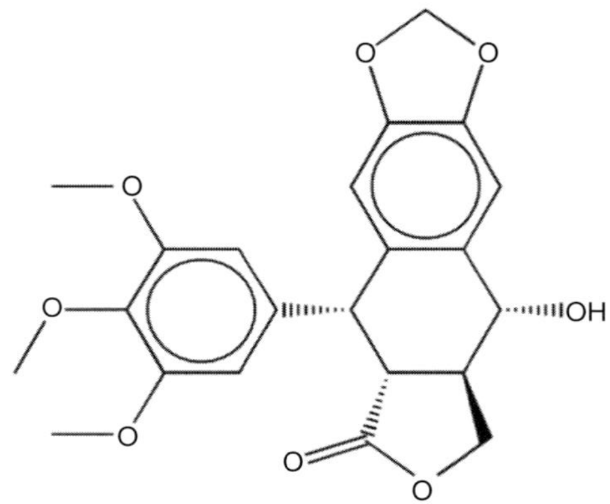

Podophyllotoxin

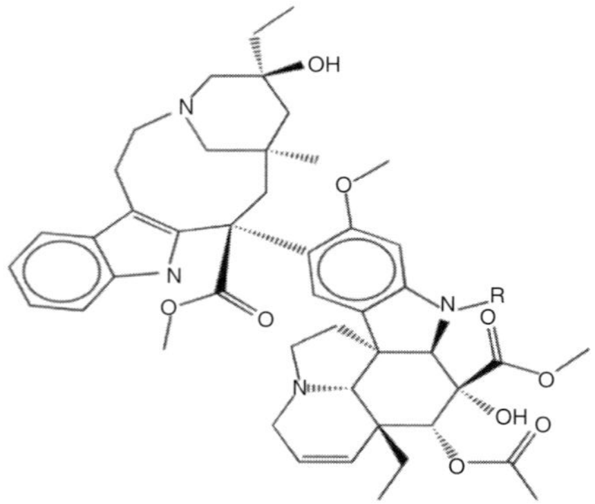

Vinblastine $R=CH_3$
Vincristine $R=CHO$

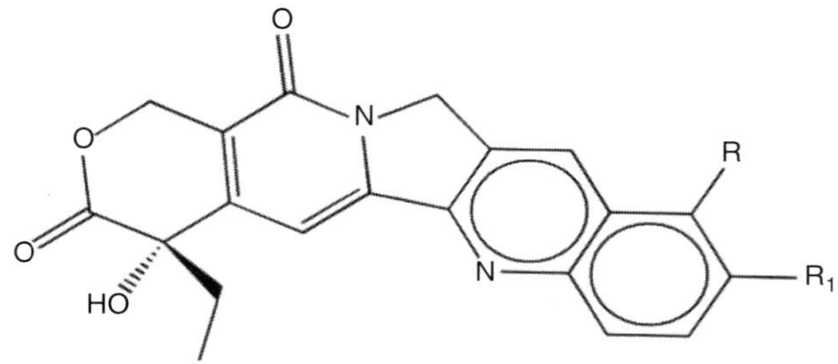

Campthothecin	**$R=R_1=H$**
9 methoxycamptothecin	**$R=OCH_3$, $R_1=H$**
10 methoxycamptothecin	**$R=H, R_1=OH$**

Figure 13.1 Structures of taxol, baccatin, vinblastine and vincristine, camptothecin and podophyllotoxin.

compared to vincristine but vincristine is more valued as an anticancerous agent. Endophytic *F. oxysporum* was found to successfully carry out the biotransformation of vinblastine into vincristine (Kumar and Ahmed, 2013). Later in the same year, Kumar *et al.* (2013) also reported vinblastine and vincristine production by endophytic *F. oxysporum* isolated from *C. roseus*. Similarly, Palem *et al.* (2015) isolated an endophytic *Talaromyces radicus* from *C. roseus* that could produce vinblastine and vincristine. The partially purified vincristine from the fungus exhibited potential dose dependent cytotoxic activity towards HeLa, MCF7, A549, U251, A431 cancer cell lines. Furthermore, Kuriakose *et al.* (2016) also reported a vincristine-producing endophytic fungus, *Eutypella* sp. from stems of *C. roseus*. Fungal vincristine from *Eutypella* sp. is highly active and induces apoptosis against A431 cells by generating a reactive oxygen species and activating the intrinsic pathway leading to loss of mitochondrial membrane potential. Recently, Ayob *et al.* (2017) isolated an endophytic *Nigrospora sphaerica* from *C. roseus* that could produce vinblastine. The fungus produced vinblastine with 10-fold better cytotoxicity towards breast cancer cell line MDA-MB 231compared to vinblastine extracted from *C. roseus* (Table 13.1, Figure 13.1).

13.2.3 Camptothecin

Camptothecin, a pentacyclic quinoline alkaloid, is predominantly isolated from the wood of the Chinese ornamental plant *Camptotheca acuminata* and roots of *Nothapodytes foetida* (Puri *et al.*, 2005; Kusari *et al.*, 2009a; Ran *et al.*, 2017). Camptothecin is the third largest plant-based antineoplastic agent that executes its cytotoxic property by selectively inhibiting topoisomerase I, an enzyme which plays a vital role in DNA replication (Kharwar *et al.*, 2011; Cragg and Pezzuto, 2016). Several reports have described the therapeutic potential of camptothecin and its derivatives against colon, cervical, lung, ovarian and uterine cancer (Pu *et al.*, 2013; Zhao *et al.*, 2010). The anticancerous efficacy of camptothecin is prominently hindered by its low solubility in aqueous medium, toxicity and short half-life period (Rehman *et al.*, 2008; Shweta *et al.*, 2010). Furthermore, there is inconsistent and inadequate amounts of camptothecin present in plant species which makes it non-viable for satisfying global market demand (Puri *et al.*, 2005; Amna *et al.*, 2006). Although, camptothecin is not directly used as an anticancer drug, two of its major derivatives 9-methoxycamptothecin and 10-hydroxycamptothecin have shown promising non-toxic anticancerous activity (Kusari *et al.*, 2009a; Shao *et al.*, 2017). Further, extensive studies on stucture-activity relationship of camptothecin has led to the development of several semi-synthetic derivatives such as Topotecan and Irinotecan which are now being prescribed for treatment of various kinds of cancers (Shweta *et al.*, 2010). However, production of semi-synthetic derivatives is a complex process and suffers from yield issues which makes the production cost high (Cragg and Pezzuto, 2016). This has prompted researchers to look for sustainable alternative sources of camptothecin and its derivatives including endophytes (Rehman *et al.*, 2009; Zhao *et al.*, 2010). Puri *et al.* (2005) first reported production of camptothecin from endophytic *Entrophosphora infrequens* isolated from the inner bark of *Nothapodytes foetida* and showed that it exhibited potential cytotoxic activity against A-549, HEP-2 and OVCAR-5 cell lines. Furthermore, comparative studies on production of camptothecin under shake flask and bioreactor revealed that camptothecin production was enhanced 8.7 fold under bioreactor conditions as compared to shake flask fermentation methodology. A decrease in production time from 216 h to 120 h was also noted during the studies (Amna *et al.*, 2006). Endophytic *Neurospora crassa* and *Nodulisporium* sp. isolated from *N. foetida* were also found to produce camptothecin in liquid culture medium. Camptothecin obtained from *N. crassa* showed cytotoxic activity against lung and ovarian cancer cell lines (Rehman *et al.*, 2008, 2009). Furthermore, endophytic fungal isolate XK001 isolated from *C. acuminata* produced 10-hydroxycamptothecin in its liquid culture (Duan *et al.*, 2009). Similarly, Kusari *et al.* (2009a) isolated camptothecin and its two derivatives 9-methoxycamptothecin and 10-hydroxycamptothecin from endophytic *Fusarium solani* isolated from the bark of *C. acuminata*. Later, two more strains of endophytic *F. solani* (MTCC 9667 and MTCC 9668) recovered from *Apodytes dimidiata* also produced camptothecin derivatives 9-methoxycamptothecin and 10-hydroxycamptothecin (Shweta *et al.*, 2010). Furthermore, endophytic *Aspergillus* sp. LY341, LY355, and *Trichoderma atroviride* LY357 were also found to produce camptothecin (Pu *et al.*, 2013). More recently, camptothecin produced by endophytic *F. solani* S-019 was shown to exhibit strong cytotoxic activity against Vero cells (Table 13.1, Figure 13.1) (Ran *et al.*, 2017). The above reports clearly suggest that endophytic fungi could produce ample amounts of camptothecin. Furthermore, endophyte production will also eliminate the need to harvest rare medicinal plants from their natural environment and help with their conservation.

13.2.4 Podophyllotoxin

Podophyllotoxin is a highly valued aryltetralin lignin that acts as a precursor for synthesis of anticancer drugs like etoposide, teniposide and etopophos phosphate (Eyberger *et al.*, 2006; Puri *et al.*, 2006; Karwar *et al.*, 2011). Podophyllotoxin was originally isolated from resins of the plant *Podophyllum emodi* for the treatment of genital warts (Leiter *et al.*, 1950). Podophyllotoxin acts as a potent inhibitor of microtubule assembly, whereas its derivatives etoposide and teniposide inhibit topoisomerase enzyme II causing cell arrest in the S phase (Huang *et al.*, 2014). Etoposide, etopophos and teniposide are in clinical use for the treatment of bronchial and testicular cancers (Stierle and Stierle, 2015; Cragg and Pezzuto, 2016). The phenomenal success of these drugs has increased the global demand of podophyllotoxin. However, there are several limitations in meeting the worldwide demand of these compounds. The podophyllotoxin content in the host plant is quite low and their over harvesting from natural habitat has contributed to the endangered status of the species (Liang *et al.*, 2016). Agricultural production has been unsuccessful due to limiting environmental conditions (Puri *et al.*, 2006). Further, metabolic engineering and cell tissue culture techniques have not yielded desirable results (Kour *et al.*, 2008). Therefore, a viable, cost-effective, inexhaustible alternative avenue needs to be explored for reproducible and dependable productivity (Kusari *et al.*, 2009b, 2013). Recent biotechnological advancements, including genetic manipulations have made microbial fermentation a feasible technology for commercial production of bioactive compounds (Kusari *et al.*, 2014; Venugopalan and Srivastava, 2015). Yang *et al.* (2003) first reported the production of podophyllotoxin from six endophytic fungi isolated from *Sinopodophyllum hexandrum*, *Diphylleia sinensis* and *Dysosma veitchii*. Three years later, Eyberger *et al.* (2006) isolated two strains of endophytic *Phialocephala fortinii* PPE5 and *P. fortinii* PPE7 from the rhizome of *P. peltatum* that possessed the ability to produce podophyllotoxin under axenic culture conditions. Similarly, podophyllotoxin and podophyllotoxin glycoside production was also detected in the Sabouraud broth culture of endophytic *Trametes hirsute* isolated from *S. hexandrum* (Puri *et al.*, 2006). Furthermore, endophytic *F. oxysporum* and *Aspergillus fumigates* isolated from *Juniperus recurva* and *J. communis*, respectively, produced podophyllotoxin and deoxypodophyllotoxin (Kour *et al.*, 2008; Kusari *et al.*, 2009b). Huang *et al.* (2014) also reported the production of podophyllotoxin from *Mucor fragilis* TW5. Additionally, it also produced kaempferol, a medicinally important compound with potential anticancer, antioxidant and anti-inflammatory properties. Recently, *Alternaria tenuissima* isolated from *Sinopodophyllum emodi* was found to produce podophyllotoxin under *in vitro* conditions (Table 13.1, Figure 13.1) (Liang *et al.*, 2016). The studies suggest that endophytic fungi could be promising starting candidates for the production of podophyllotoxin, which could be further genetically modified to achieve enhanced production. This will also reduce the mass destruction of endogenous plants and aid with their conservation.

13.2.4 Huperzine A

Huperzine A (HupA) is a naturally occurring lycopodium alkaloid originally isolated from the Chinese club moss *Huperzia serrata* (Zhu *et al.*, 2010; Dong *et al.*, 2014; Su and Yang, 2014). HupA has drawn worldwide attention because of its potential in treatment of Alzheimer's disease (Zhang *et al.*, 2011; Ying *et al.*, 2014). Currently, there is no known cure available for Alzheimer's disease; however, inhibition of acetylcholinesterase appears to be among the most accepted strategies for its treatment (Zhu *et al.*, 2010; Wang *et al.*, 2016). HupA is a second-generation, highly selective, reversible inhibitor of acetylcholinesterase which blocks the activity of the enzyme thereby increasing the availability of acetylcholine in the central cholinergic synapses (Zhang *et al.*, 2011; Dong *et al.*, 2014; Su and Yang, 2014). HupA was discovered by Chinese researchers in 1986 and was later approved for treatment of Alzheimer's disease in the 1990s, whereas in the USA, it is used as a dietary supplement for preventing further memory deprivation. Furthermore, ZT-1, a semi-synthetic analogue of HupA has successfully completed phase II clinical trials in China and Europe (Wang *et al.*, 2011; Dong *et al.*, 2014). HupA possess better acetylcholinesterase inhibitory activity, longer duration of action, higher oral bioavailability, superior therapeutic index and minimal side effects than its counterparts tacrine, donepezil, rivastigmine and galanthamine (Zhu *et al.*, 2010; Shu *et al.*, 2014; Zhang *et al.*, 2015). With the demographic trend of ageing societies, there has been an increased demand of HupA. The majority of HupA is sourced from the lycopod Huperziaceae family. However, *H. serrata* has a narrow geographic distribution, slow growth rate and an extremely low content of HupA (*c.* 0.007%). Moreover, the germinated spores of *H. serrata* require approximately 15 years to develop into mature sporophytes which can then be harvested for HupA extraction (Zhang *et al.*, 2011; Su and Yang, 2014). The other species of Huperziaceae are even more difficult to obtain and are much rarer in nature than *H. serrata* which makes them less desirable candidates for commercial extraction (Zhu *et al.*, 2010). Increasing efforts have been made to develop an alternative means of HupA production. HupA can also be produced by *in vitro* tissue culture methods or by chemical synthesis. However, chemical synthesis of HupA is a cumbersome and expensive process because of the cost of catalysts thiophenol and diphenylphosphorylazide (Su and Yang, 2014). Furthermore, *in vitro* propagated tissues could produce higher amount of HupA but is not sufficient to meet the global demand. Thus, bioprospecting novel natural resources including endophytes for HupA production has become a research hotspot (Dong *et al.*, 2014; Su and Yang, 2014; Wang *et al.*, 2016).

Endophytic fungi colonising Huperziaceae have been extensively studied. Li *et al.* (2007) first reported production of HupA from endophytic *Acremonium* sp. 2F09P03B isolated from *H. serrata*. Similarly, endophytic *Shiraia* sp. Slf14 and *Cladosporium cladosporioides* LF70 isolated from leaves of *H. serrata* produced HupA which exhibited complete inhibition of acetylcholinesterase activity under *in vitro* conditions (Zhu *et al.*, 2010; Wang *et al.*, 2011; Zhang *et al.*, 2011). In addition, endophytic fungi such as *Colletotrichum gloeosporioides* ES026, *Paecilomyces tenuis* YS-13 and *Trichoderma* sp. L44 isolated from *H. serrata* were also found to produce HupA under *in vitro* conditions (Dong *et al.*, 2014; Shu *et al.*, 2014; Su and Yang, 2014). Besides this, endophytic fungal strains such as *Blastomyces* sp. HA15, *Botrytis* sp. HA23, *Ceriporia* lacerate MY311 and *Hypoxylon investiens* MY183, isolated from the lycopods *Phlegmariurus cryptomerianus* and *P. phlegmaria*, respectively, also produced Hup A (Table 13.1, Figure 13.2) (Zan *et al.*, 2009; Zhang *et al.*, 2015). HupA has a rigid molecular configuration and its structural modification via chemical methods to obtain derivatives is a challenging task. However, endophytic *Ceriporia lacerate* HS-ZJUT-C13A successfully biotransformed HupA into five different

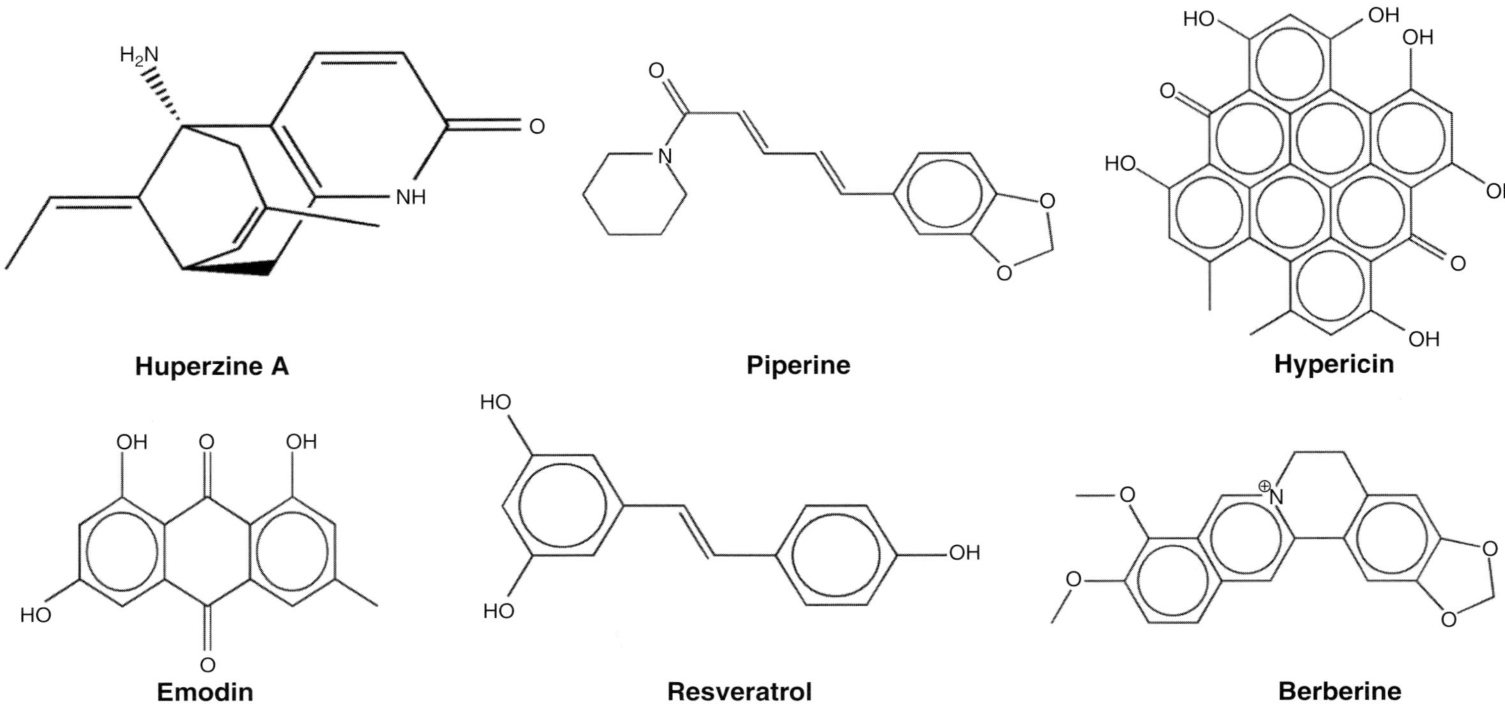

Figure 13.2 Structures of huperzine A, resveratrol, hypericin, emodin, piperine and berberine.

compounds namely huptremules A–D, 8α,15α- epoxyhuperzine A, tremulenediol A, conocenol B and 11,12-dihydroxy-1-tremulen-5-one. All the biotransformed compounds exhibited potential acetylcholinesterase inhibitory activity. In addition, huptremules share similar structural features with ZT-1. Thus, biotransformation using fungal endophytes can also be a valuable approach for producing derivatives of HupA (Ying *et al.*,2014). Microbial HupA production will have positive economic and environmental impacts. It seems to be a feasible strategy to meet global market demand via microbial fermentation and genetic manipulation of the source fungi. In addition, it will also safeguard the depletion of rare source plants from their natural habitats and will thus preserve biodiversity. However, before taking endophytic fungi into consideration for industrial-scale production of HupA, the molecular basis of HupA synthesis in fungi and the genes involved in biosynthetic pathways needs to be identified.

13.2.5 Resveratrol

Resveratrol (3,5,4'-trihydoxy-trans-stilbene) is an extraordinary natural product known to provide multifarious therapeutic effects such as cardiovascular protection, prevention against cancer, treatment of ischaemic injuries, enhancement of stress resistance and extension of lifespan from yeasts to vertebrates (Shi *et al.*, 2012; Che *et al.*, 2016a). The therapeutic effect of wine has been highly attributed to the transfer of resveratrol from grapes to wine during winemaking. Today, resveratrol is a constituent of health supplements and wellness products apart from being studied as a pharmacophore for drug development (Saxena and Srivastava, 2014). Currently, resveratrol is produced from plant cell suspension culture of *Polygonum* sp., *Vitis vinifera* and *Arachis hypogeal* which have shortcomings such as low yield, instability and cost effectiveness (Zhang *et al.*, 2013; Wang *et al.*, 2014). The growing market demand has instigated researchers to look for plausible alternative sources that could efficiently produce resveratrol (Saxena and Srivastava, 2014; Che *et al.*, 2016b).

Shi *et al.* (2012) screened endophytic fungi isolated from *V. vinifera*, *V. quinquangularis*, and *P. cuspidatum* for production of resveratrol. Among 65 screened isolates, 21 isolates produced resveratrol (6–123 μg/l) under axenic culture medium conditions. Endophytic *Alternaria* sp. MG1 isolated from *V. vinifera* exhibited highest and stable resveratrol production. Furthermore, a reaction system was proposed for efficient production of resveratrol from resting cells of a non-genetically modified endophytic *Alternaria* sp. MG1 (Zhang *et al.*, 2013). In addition, a qRT-PCR approach was used to identify key genes involved in production of resveratrol from the endophytic *Alternaria* sp. MG1. Among the evaluated nine TUBA, EF1, EF2, UBC, UFD, RPS5, RPS24, ACTB and 18S reference genes, a combination of TUBA and EF1 was found to be most suitable for normalisation of *Alternaria* sp. MG1, whereas 18S appeared as least stable (Che *et al.*, 2016c).

Moreover, *de novo* transcriptome sequencing was also performed to reveal the genetic basis of resveratrol biosynthesis pathway in endophytic *Alternaria* sp. MG1. It was found that 84 genes were involved in pathways for conversion of glucose to resveratrol which codes for several enzymes involved in glycolysis (20), phenylalanine biosynthesis (10), phenylpropanoid biosynthesis (4) and stilbenoid biosynthesis (4). It was also observed that chalcone synthase is responsible for resveratrol biosynthesis in endophytic *Alternaria* sp. MG1 rather than stilbene synthase (Che *et al.*, 2016c). Apart from this, a novel and competent enzymatic reaction system was developed (either by using free enzyme solution or immobilised enzymes for successful bioconversion of resveratrol). The system developed provided a way to produce resveratrol within 2 hours (Che *et al.*, 2016a). Furthermore, a mangrove endophytic *Alternaria* sp. R6 isolated from *Myoporum bontioides* was found to produce three resveratrol derivatives resveratrodehydes A–C which exhibited a broad spectrum of activities against human breast MDA-MB-435, liver HepG2, and colon HCT-116 cancer cell lines. Additionally, compounds A and C showed moderate antioxidant activity (Wang *et al.*, 2014). Recently, endophytic *Aspergillus niger* C2J6 isolated from the grape variety cabernet sauvignon also demonstrated stable resveratrol production upon subsequent subculturing (Table 13.1, Figure 13.2) (Liu *et al.*, 2016). Resveratrol production is a classic example of xenohormesis, where on sensing the stress cues from the host, the symbiont organisms produce bioactive compounds like resveratrol to counteract biotic or abiotic stresses.

13.2.6 Hypericin

Hypericin is a naphthodianthrone derivative and characteristic constituent of *Hypericum* plant species, first isolated from the medicinal herb *H. perforatum*, commonly known as St. John's wort (Karioti and Bilia, 2010). Hypericin has been long used as an antidepressant owing to its monoamine oxidase inhibiting property. Furthermore, it also exhibited wound healing, anti-inflamatory, antimicrobial, antiviral and antitumour activities (Kusari *et al.*, 2008; Kusari and Spiteller, 2012). Hypericin significantly inhibited a plethora of enveloped viruses including HIV-1, HSV-1, HSV-2, BVDV, BIV and influenza A virus (Kusari *et al.*, 2009c). It is one of the most powerful naturally occurring photodynamic agents, which in combination with light significantly induces apoptosis/necrosis of cancer cells (Nicoletti and Fiorentino, 2015). Hypericin has been found as a metabolic product of endophytic *Thielavia subthermophila* INFU/Hp/KF/34B isolated from stems of *H. perforatum* in India. In addition, the endophyte also produced emodin, which is considered to be the chief precursor in production of fungal hypericin. Both hypericin and emodin showed antimicrobial activity against a spectrum of pathogenic fungi and bacteria under *in vitro* conditions (Table 13.1, Figure 13.2) (Kusari *et al.*, 2008). Furthermore, the metabolic profiling of endophytic *T. subthermophila* under light and dark conditions revealed that hypericin production is not a light driven phenomenon in case of endophytic *T. subthermophila*. The fungus also lacks the *hyp-1* gene which is responsible for production of hypericin in the host plant. Thereafter, they also proposed a new biosynthetic pathway for production of hypericin by endophytic *T. subthermophila*. In addition, fungal hypericin and emodin also exhibited photodynamic cytotoxicity against human acute monocytic leukaemia cell lines (Kusari *et al.*, 2009c). Hypericin production from endophytic fungi is thus a nascent area of study with very scant and preliminary data. However, the available reports suggest that endophytes are prospective candidates for hypericin production owing to their novel mechanism for biosynthesis.

13.2.7 Piperine

Piperine is a naturally occurring pungent alkaloid from *Piper longum* and *P. nigrum* (pepper). The medicinal properties of these plants are largely attributed to the presence of piperine (Gorgani *et al.*, 2016), which in the dried fruits of *P. nigrum* is 3–9% (Chithra *et al.*, 2014a,b). Piperine exhibits an array of pharmacological properties such as analgesic, hepatoprotective, neuroprotective, antibacterial, anticonvulsant, antidepressant, antihyperlipidemic, anti-inflammatory, antipyretic, antioxidant and antitumour activities (Verma *et al.*, 2011). It stimulates the production of digestive enzymes of the pancreas and intestine and also stimulates biliary secretions (Chithra *et al.*, 2014a). Furthermore, piperine acts as a bioenhancer, thereby increasing the bioavailability of the important substances such as curcumin, β-carotene, amino acids and several pharmacological drugs like sulfadiazine, tetracycline, streptomycin, rifampicin, pyrazinamide, isoniazid, ethambutol and phenytoin (Verma *et al.*, 2011; Chithra *et al.*, 2014b). Several endophytic fungi have been reported to produce piperine in their culture broth. Verma *et al.* (2011) first reported the production of piperine under liquid culture conditions by an endophytic *Periconia* sp. isolated from *Piper longum*. Fungal piperine from endophytic *Periconia* sp. also exhibited promising anti-mycobacterial activity against *Mycobacterium tuberculosis* and *M. smegmetis*. Furthermore, endophytic *Colletotrichum gloeosporioides* PF12 and *Mycosphaerella* sp. PF13 isolated from stems of *P. nigrum* were also found to produce piperine in their liquid cultures (Chithra *et al.*, 2014a,b). Recently, endophytic *Phomopsis* sp. PF 5 isolated from *P. nigrum* also showed piperine production. Furthermore, when the fungus colonised *Oryza sativa*, a significant enhancement in the growth of the crop was observed (Table 13.1, Figure 13.2) (Chithra *et al.*, 2017).

13.2.8 Berberine

Berberine is an isoquinoline alkaloid isolated from several unrelated medicinal plants such as *Berberis aquifolium*, *B. vulgaris*, *Coptis chinensis*, *Coscinium fenestratum*, *Hydrastis canadensis*, *Phellodendron amurense* and *P. chinense* (Vinodhini and Agastian, 2013; Wojtyczka *et al.*, 2014; Nicoletti and Fiorentino, 2015). In China, it has been traditionally used for treatment of diabetes. Furthermore, it is also an important over the counter drug for curing gastrointestinal infections (Yin *et al.*, 2008). In addition, berberine also possesses a wide range of pharmacological properties including cardioprotective, anticancer, anti-inflammatory and antimicrobial activities (Vinodhini and Agastian, 2013). Berberine has been isolated as a secondary metabolite of several endophytic fungal species. Endophytic *Alternaria* sp. BBH6 isolated from *P. amurense* showed berberine production (Duan *et al.*, 2009). Similarly, endophytic *Fusarium solani* SDV01 isolated from the roots of *C. fenestratum* also demonstrated berberine production in its culture broth. The fungal extract containing berberine also exhibited cytotoxicity against HT-29 cell lines (Vinodhini and Agastian, 2013). Recently, endophytic *Alternaria* sp. HL-Y-3 isolated from *Coptis chinensis* also showed berberine production (Table 13.1, Figure 13.2) (Zhang *et al.*, 2016). Furthermore, Agusta *et al.* (2014) carried out the biotransformation of protoberberine into berberine 7-*N*-oxide via endophytic *Coelomycetes* AFKR-3. The biotransformed berberine demonstrated similar antimicrobial activity to standard berberine. The above reports suggest that endophytic fungi are putative source of berberine, however more detailed, rationalised and target based studies are required before moving to preclinical trials.

13.2.9 Miscellaneous Compounds

Apart from the above-described compounds, there are some other important host related bioactive compounds produced by different endophytic fungi. Rohitukine is a chromane alkaloid possessing anticancer, anti-inflammatory and immunomodulatory properties. Flavopiridol and P-276-00, the semi-synthetic derivatives of rohitukine are in advanced stages of clinical trials for cancer treatment. Endophytic *Fusarium proliferatum*, *F. oxysporum*, *F. solani* and *Gibberella fujikuroi* isolated from *Dysoxylum binectariferum* and *Amoora rohituka*, respectively, produced rohitukine (Table 13.1, Figure 13.3) (Kumara *et al.*, 2012, 2014). Similarly, vincamine, a monoterpenoid indole alkaloid used for treatment of cerebral insufficiencies was found to be produced by an unidentified endophytic fungal strain Vm-J2 from *Vinca minor* and a *Geomyces* sp. CH1 isolated from *Nerium indicum* under *in vitro* conditions (Yin and Sun, 2011; Na *et al.*, 2016) (Table 13.1, Figure 13.3). Furthermore, endophytic *Cladosporium cladosporioides* isolated from *Aconitum leucostomum* produced 'aconitine', a C19 diterpene alkaloid that is known to have strong analgesic, anti-inflammatory and anticancer properties (Table 13.1, Figure 13.3) (Yang *et al.*, 2013). Gymnemagenin is considered a wonder drug for diabetes and cancer, isolated from leaves and roots of *Gymnema sylvestre*. Endophytic *Penicillium oxalicum* isolated from *G. sylvestre* have been reported to produce gymnemagenin similar to its host plant (Table 13.1, Figure 13.3) (Parthasarathy and Sathiyabama, 2014).

Curcumin is a food additive mainly isolated from *Curcuma longa* (turmeric) with several medicinal properties and has been found as a fermentation product of endophytic *Gibberella fujikuroi* isolated from *Curcuma wenyujin* (Table 13.1, Figure 13.3)

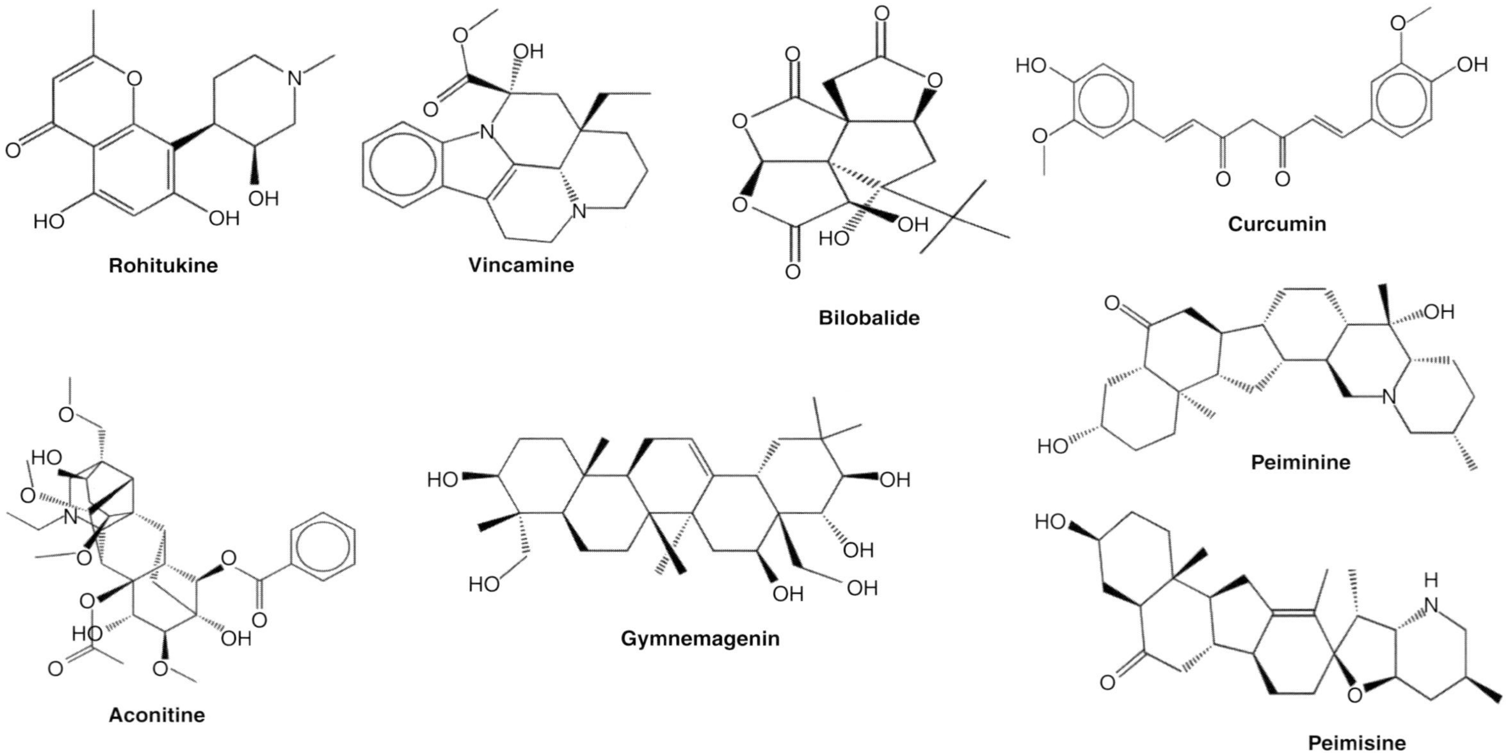

Figure 13.3 Structures of rohitukine, vincamine, aconitine, gymnemagenin, curcumin, bilobalide, peimisine and peiminine.

(Yan *et al.*, 2014). Similarly, endophytic *Fusarium* sp. WBS007 and *F. redolens* isolated from *Fritillaria unibracteata* have been found to produce peimisine and peiminine, steroidal alkaloids originally characterised from the host plant (Table 13.1, Figure 13.3) (Pan *et al.*, 2014, 2015). Recently, endophytic *Pestalotiopsis uvicola* isolated from *Ginkgo biloba* also produced bilobalide similar to its host plant (Table 13.1, Figure 13.3) (Qian *et al.*, 2016).

Apart from being a putative source of plant related phytochemicals, endophytic fungi also produce several medicinally important precursor molecules like emodin and baccatin. Emodin serves as a precursor molecule for synthesis of hypericin, whereas baccatin is involved in commercial production of anticancer drug taxol. Endophytic *Penicillium janthinellum*, *T. subthermophilia* and *Talaromyces* sp. isolated from *Melia azedarach*, *H. perforatum* and *Aconitum carmichaeli*, respectively, have been reported to produce emodin and its derivatives under *in vitro* conditions (Table 13.1, Figure 13.2) (Marinho *et al.*, 2005; Kusari *et al.*, 2008; Xie *et al.*, 2016). Similarly, endophytic *Didmyostilbe* sp. DF110, *Diaporthe phaseolorum* and *Trichoderma* sp. IRB54 isolated from *T. chinensis* and *T. wallichiana* produced baccatin and its derivatives as their fermentation product (Table 13.1, Figure 13.1) (Wang and Tang, 2011; Zaiyou *et al.*, 2013; Li *et al.*, 2015).

13.3 Conclusions

Endophytic fungi are a rich source of structurally diverse compounds with promising therapeutic applications. Structural diversity is not the only reason endophytic fungi are of great interest for drug discovery and development: they often also provide highly selective and specific biological activities based on their unique mode of action. Taxol, camptothecin and hypericin are some very good examples of this. The production of bioactive compounds exclusive to their host by endophytic fungi has nurtured expectations of exploiting them as alternative and plausible sources of these compounds. However, the commercial production of these compounds by endophytic fungi still remains a future goal. A major obstacle preventing the commercial production of these compounds using endophytic fungi is the diminution in ability to produce secondary metabolites on repeated subculturing under axenic culture conditions. However, this problem can be overcome by adopting strategies such as strain improvement (via mutagenesis or genetic transformations), optimisation of culture medium and supplementation of precursors, elicitors and/or inhibitors. Considering the fact, that endophytic fungi reside inside the host plant along with other endophytes, it is compelling that they continuously interact with each other and have profound impact on the production of each other's secondary metabolites. Thus, a co-cultivation method can also be adopted for enhanced production of desired bioactive compounds. Furthermore, owing to the fact that endophytic fungi are involved in multifarious interactions with their host plant and other endophytes, even the slightest of variation in the *in vitro* culture conditions can greatly impact the kind and range of secondary metabolite they produce. Moreover, a systematic study on the plant–endophyte and endophyte–endophyte interspecies crosstalk is desirable for better understanding and sustainable production of bioactive compounds using endophytic fungi. Apart from this, there is need for microbiologists, chemists, toxicologists and pharmacologists to work in a coordinated manner for the discovery of the target lead molecules. If we will be able to achieve the abovementioned targets, the bioactive compounds from endophytic fungi will emerge as future medicines for the treatment of major health problems.

Acknowledgements

Authors appreciate the kind help of Dr Nadeem Akhtar, Post Doctoral Fellow, Department of Animal Biosciences University of Guelph, Guelph, Ontario, Canada, Dr Sumit Jaiswal, Research Associate, School of Energy and Environment, Thapar University, Patiala, Punjab, India and Dr Surbhi Goel, Research Scholar, Enzyme Engineering Lab, Department of Biochemical Engineering and Biotechnology, Indian Institute of Delhi, New Delhi, India for their unconditional support and cooperation.

References

Agusta, A., Wulansari, D., Nurkanto, A. and Fathoni, A. (2014). Biotransformation of protoberberine alkaloids by the endophytic fungus *Coelomycetes* AFKR-3 isolated from yellow moonsheed plant (*Archangelisia flava* (L.) Merr.). *Procedia Chemistry*, **13**, 38–43.

Aly, A. H., Debbab, A. and Proksch, P. (2011). Fungal endophytes: unique plant inhabitants with great promises. *Applied Microbiology and Biotechnology*, **90**, 1829–1845.

Aly, A. H., Debbab, A. and Proksch, P. (2013). Fungal endophytes-secret producers of bioactive plant metabolites. *Pharmazie*, **68**, 499–505.

Amna, T., Puri, S. C., Verma, V. *et al.* (2006). Bioreactor studies on the endophytic fungus *Entrophospora infrequens* for the production of an anticancer alkaloid camptothecin. *Canadian Journal of Microbiology*, **52**, 189–196.

Arnold, A. E. (2007). Understanding the diversity of foliar endophytic fungi: progress, challenges, and frontiers. *Fungal Biology Reviews*, **21**, 51–66.

Arnold, A. E., Maynard, Z., Gilbert, G. S., Coley, P. D. and Kursar, T. A. (2000). Are tropical fungal endophytes hyperdiverse? *Ecology Letters*, **3**, 267–274.

Ayob, F. W., Simarani, K., Abidin, N. Z. and Mohamad, J. (2017). First report on a novel *Nigrospora sphaerica* isolated from *Catharanthus roseus* plant with anticarcinogenic properties. *Microbial Biotechnology*, **10**, 926–932.

Bacon, C. W. and White, J. F. (2000). Microbial endophytes. New York, NY: Marcel Dekker.

Che, J. X., Shi, J. L., Gao, Z. H. and Zhang, Y. (2016a). A new approach to produce Resveratrol by enzymatic bioconversion. *Journal of Microbiology and Biotechnology*, **26**, 1348–1357.

Che, J., Shi, J., Gao, Z. and Zhang, Y. (2016b). Transcriptome analysis reveals the genetic basis of the resveratrol biosynthesis pathway in an endophytic fungus (*Alternaria* sp. MG1) isolated from *Vitis vinifera*. *Frontiers in Microbiology*, **7**, 1257.

Che, J. X., Shi, J. L., Lu, Y. and Liu, Y. L. (2016c). Validation of reference genes for normalization of gene expression by qRT-PCR in a resveratrol-producing endophytic fungus (*Alternaria* sp. MG1). *AMB Express*, **6**, 1–10.

Chithra, S., Jasim, B., Sachidanandan, P., Jyothis, M. and Radhakrishnan, E. K. (2014a). Piperine production by endophytic fungus *Colletotrichum gloeosporioides* isolated from *Piper nigrum*. *Phytomedicine*, **21**, 534–540.

Chithra, S., Jasim, B., Anisha, C., Mathew, J. and Radhakrishnan, E. K. (2014b). LC-MS/MS based identification of piperine production by endophytic *Mycosphaerella*sp. PF13 from *Piper nigrum*. *Applied Biochemistry and Biotechnology*, **173**, 30–35.

Chithra, S., Jasim, B., Mathew, J. and Radhakrishnan, E. K. (2017). Endophytic *Phomopsis* sp. colonization in *Oryza sativa* was found to result in plant growth promotion and piperine production. *Physiologia Plantarum*, **160**, 437–446.

Chowdhary, K., Kaushik, N., Coloma, A. G. and Raimundo, C. M. (2012). Endophytic fungi and their metabolites isolated from Indian medicinal plant. *Phytochemistry Reviews*, **11**, 467–485.

Cragg, G. M. and Pezzuto, J. M. (2016). Natural products as a vital source for the discovery of cancer chemotherapeutic and chemopreventive agents. *Medical Principles and Practice*, **25**, 41–59.

Dai, W. and Tao, W. (2008). Preliminarily study on fermentation conditions of taxol-producing endophytic fungus. *Chemical Industry and Engineering Progress*, **27**, 883–886.

De Bary, A. (1866). Morphologie und physiologie der Pilze, Flechten and myxomyceten. In *Hofmeister's Handbook of Physiological Botany*, ed. W. Hofmeister, Vol. II. Leipzig, Germany: Engelmann, pp. 1–335.

Deng, B. W., Liu, K. H., Chen, W. Q., Ding, X. W. and Xie, X. C. (2009). *Fusarium solani*, Tax-3, a new endophytic taxol-producing fungus from *Taxus chinensis*. *World Journal of Microbiology and Biotechnology*, **25**, 139–143.

Dong, L. H., Fan, S. W., Ling, Q. Z., Huang, B. B. and Wei, Z. J. (2014). Identification of huperzine A-producing endophytic fungi isolated from *Huperzia serrata*. *World Journal of Microbiology and Biotechnology*, **30**, 1011–1017.

Duan, L. I., Liwei, G. and Hong, Y. (2009). Isolation and identification of producing endophytic fungi of berberine from the plant *Phellodendron amurense*. *Journal of Anhui Agricultural Sciences*, 2009, **22**.

Eyberger, A. L., Dondapati, R. and Porter, J. R. (2006). Endophyte fungal isolates from *Podophyllum peltatum* produce podophyllotoxin. *Journal of Natural Products*, **69**, 1121–1124.

Freeman, E. M. (1904). The seed fungus of *Lolium temulentum* L. *Philosophical*

Transactions of the Royal Society of London B, **196**, 1–27.

Gangadevi, V. and Muthumary, J. (2008). Taxol, an anticancer drug produced by an endophytic fungus *Bartalinia robillardoides* Tassi, isolated from a medicinal plant, *Aegle marmelos* Correa ex Roxb. *World Journal of Microbiology Biotechnology*, **24**, 717–724.

Gangadevi, V. and Muthumary, J. (2009a). A novel endophytic taxol-producing fungus *Chaetomella raphigera* isolated from a medicinal plant, *Terminalia arjuna*. *Applied Biochemistry and Biotechnology*, **158**, 675–684.

Gangadevi, V. and Muthumary, J. (2009b). Taxol production by *Pestalotiopsis terminaliae*, an endophytic fungus of *Terminalia arjuna* (arjun tree). *Biotechnology and Applied Biochemistry*, **52**, 9–15.

Garyali, S. and Reddy, M. S. (2013). Taxol production by an endophytic fungus, *Fusarium redolens*, isolated from Himalayan yew. *Journal of Microbiology and Biotechnology*, **23**, 1372–1380.

Gond, S. K., Kharwar, R. N. and White, J. F. (2014). Will fungi be the new source of the blockbuster drug taxol? *Fungal Biology Reviews*, **28**, 77–84.

Gorgani, L., Mohammadi, M., Najafpour, G. D. and Nikzad, M. (2016). Piperine: the bioactive compound of black pepper: from isolation to medicinal formulations. *Comprehensive Reviews in Food Science and Food Safety*, **16**, 124–140.

Gouda, S., Das, G., Sen, S. K., Shin, H. S. and Patra, J. K. (2016). Endophytes: a treasure house of bioactive compounds of medicinal importance. *Frontiers in Microbiology*, **7**, 1538.

Guo, B., Li, H. and Zhang, L. (1998). Isolation of a fungus producing vinblastine. *Journal of Yunnan University (Natural Science)*, **20**, 214–215.

Gupta, M., Saxena, S. and Goyal, D. (2015). Potential pancreatic lipase inhibitory activity of an endophytic *Penicillium* species. *Journal of Enzyme Inhibition Medicinal Chemistry*, **30**, 15–21.

Hodkinson, T. R. (2018). Evolution and taxonomy of the grasses (Poaceae): a model family for the study of species-rich groups. *Annual Plant Reviews Online*, doi: 10.1002/9781119312994.apr0622.

Hodkinson, T. R. and Murphy, B. R. (2019). Endophytes for a growing world. In *Endophytes for a Growing World*, ed. T R. Hodkinson, F. M. Doohan, M. J. Saunders and B. R. Murphy. Cambridge: Cambridge University Press, Chapter 1.

Howitz, K. T. and Sinclair, D. A. (2008). Xenohormesis: sensing the chemical cues of other species. *Cell*, **133**, 387–391.

Hu, K., Tan, F., Tang, K., Zhu, S. and Wang, W. (2006). Isolation and screening of endophytic fungi synthesizing taxol from *Taxus chinensis* var. *mairei*. *Journal of Southwest China Normal University (Natural Science Edition)*, **31**, 134–137.

Huang, J. X., Zhang, J., Zhang, X. R. *et al.* (2014). *Mucor fragilis* as a novel source of the key pharmaceutical agents podophyllotoxin and kaempferol, *Pharmaceutical Biology*, **52**, 1237–1243.

Karioti, A. and Bilia, A. R. (2010). Hypericins as potential leads for new therapeutics. *International Journal of Molecular Sciences*, **11**, 562–594.

Kaul, S., Gupta, S., Ahmed, M. and Dhar, M. K. (2012). Endophytic fungi from medicinal plants: a treasure hunt for bioactive metabolites. *Phytochemistry Reviews*, **11**, 487–505.

Kharwar, R. N., Mishra, A., Gond, S. K., Stierle, A. and Stierle, D. (2011). Anticancer compounds derived from fungal endophytes: their importance and future challenges. *Natural Product Reports*, **28**, 1208–1228.

Kour, A., Shawl, A. S., Rehman, S. *et al.* (2008). Isolation and identification of an endophytic strain of *Fusarium oxysporum* producing podophyllotoxin from *Juniperus recurva*. *World Journal of Microbiology and Biotechnology*, **24**, 1115–1121.

Krohn, K., Farooq, U., Florke, U. *et al.* (2007). Secondary metabolites isolated from an endophytic *Phoma* sp.: absolute configuration of tetrahydropyre-nophorol using the solid-state TDDFT CD methodology. *European Journal of Organic Chemistry*, **19**, 3206–3211.

Kumar, A. and Ahmad, A. (2013). Biotransformation of vinblastine to vincristine by the endophytic fungus *Fusarium oxysporum* isolated from *Catharanthus roseus*. *Biocatalysis and Biotransformation*, **31**, 89–93.

Kumar, A., Patil, D., Rajamohanan, P. R. and Ahmad, A. (2013). Isolation, purification and characterization of vinblastine and vincristine from endophytic fungus *Fusarium oxysporum* isolated from *Catharanthus roseus*. *PLoS One*, **8**, e71805.

Kumara, P. M., Zuehlke, S., Priti, V. *et al.* (2012). *Fusarium proliferatum*, an endophytic fungus from *Dysoxylum binectariferum* Hook. f, produces rohitukine, a chromane alkaloid possessing anti-cancer activity. *Antonie Van Leeuwenhoek*, **101**, 323–329.

Kumara, P. M., Soujanya, K. N., Ravikanth, G. *et al.* (2014). Rohitukine, a chromone alkaloid and a precursor of flavopiridol, is produced by endophytic fungi isolated from *Dysoxylum binectariferum* Hook. f and *Amoora rohituka* (Roxb). Wight & Arn. *Phytomedicine*, **21**, 541–546.

Kumaran, R. S., Choi, Y. K., Lee, S. *et al.* (2014). Isolation of taxol, an anticancer drug produced by the endophytic fungus, *Phoma betae*. *African Journal of Biotechnology*, **11**, 950–960.

Kuriakose, G. C., Palem, P. P. C. and Jayabaskaran, C. (2016). Fungal vincristine from *Eutypella*spp – CrP14 isolated from *Catharanthus roseus* induces apoptosis in human squamous carcinoma cell line -A431. *BMC Complementary and Alternative Medicine*, **16**, 302.

Kusari, S. and Spiteller, M. (2012). Metabolomics of endophytic fungi producing associated plant secondary metabolites: progress, challenges and opportunities. In *Metabolomics*, ed. U. Roessner. London: InTech, pp. 241–266.

Kusari, S. and Spiteller, M. (2016). The promise of endophytic fungi as sustainable resource of biologically relevant pro-drugs: a focus on Cameroon. In *Fungi Applications and Management Strategies*, ed. S. K. Deshmuk, J. K. Misra, J. P. Tewari and T. Papp, Boca Raton, FL: CRC Press, pp. 1–13.

Kusari, S., Lamshoft, M., Zuhlke, S. and Spiteller, M. (2008). An endophytic fungus from *Hypericum perforatum* that produces hypericin. *Journal of Natural Products*, **71**, 159–162.

Kusari, S., Zuhlke, S. and Spiteller, M. (2009a). An endophytic fungus from *Camptotheca acuminata* that produces camptothecin and analogues. *Journal of Natural Products*, **72**, 2–7.

Kusari, S., Lamsho, M. and Spiteller, M. (2009b). *Aspergillus fumigatus* Fresenius, an endophytic fungus from

Juniperus communis L. Horstmann as a novel source of the anticancer pro-drug deoxypodophyllotoxin. *Journal of Applied Microbiology*, **107**, 1019–1030.

Kusari, S., Zuhlke, S., Kosuth, J., Cellarova, E. and Spiteller, M. (2009c). Light-independent metabolomics of endophytic *Thielavia subthermophila* provides insight into microbial hypericin biosynthesis. *Journal of Natural Products*, **72**, 1825–1835.

Kusari, S., Hertweck, C. and Spiteller, M. (2012). Chemical ecology of endophytic fungi: origins of secondary metabolites. *Chemistry and Biology*, **19**, 792–798.

Kusari, S., Pandey, S. P. and Spiteller, M. (2013). Untapped mutualistic paradigms linking host plant and endophytic fungal production of similar bioactive secondary metabolites. *Phytochemistry*, **91**, 81–87.

Kusari, S., Singh, S. and Jayabaskaran, C. (2014). Biotechnological potential of plant-associated endophytic fungi: hope versus hype. *Trends in Biotechnology*, **32**, 297–303.

Leiter, J., Downing, V., Hartwell, J. L. and Shear, J. J. (1950). Damage induced in sarcoma 37 with podophyllin, podophyllotoxin alpha-peltatin, beta-peltatin, and quercetin. *Journal of the National Cancer Institute*, **10**, 1273–1293.

Li, M. C. and Wang, X. (2009). Isolation and identification of the 10-hydroxycamptothecin-producing endophytic fungi from *Camptotheca acuminata* Decne. *Acta Botanica Boreali- Occidentalia Sinica*, **29**, 0614–0617.

Li, W., Zhou, J, Lin Z. and Hu, Z. (2007). Study on fermentation condition for production of huperzine A from endophytic fungus 2F09P03B of *Huperzia serrata*. *Chinese Medicinal Biotechnology*, **2**, 254–259.

Li, Y., Yang, J., Zhou, X., Zhao, W. and Jian, Z. (2015). Isolation and identification of a 10-deacetyl baccatin-III-producing endophyte from *Taxus wallichiana*. *Applied Biochemistry and Biotechnology*, **175**, 2224–2231.

Liang, Z., Zhang, J., Zhang, X., Li, J., Zhang, X. and Zhao, C. (2016). Endophytic fungus from *Sinopodophyllum emodi* (Wall.) Ying that produces podophyllotoxin. *Journal of Chromatographic Science*, **54**, 175–178.

Lingqi, Z., Bo, G., Haiyan, L. *et al.* (2000). Preliminary study on the isolation of endophytic fungus of *Catharanthus roseus* and its fermentation to produce products of therapeutic value. *Chinese Traditional Herbal Drugs*, **31**, 805–807.

Liu, K., Ding, X., Deng, B. and Chen, W. (2009). Isolation and characterization of endophytic taxol-producing fungi from *Taxus chinensis*. *Journal of Industry Microbiology and Biotechnology*, **36**, 1171–1177.

Liu, Y., Nan, L., Liu, J. *et al.* (2016). Isolation and identification of resveratrol-producing endophytes from wine grape *Cabernet Sauvignon*. *Springer Plus*, **5**, 1029.

Ludwig-Muller, J. (2015). Plants and endophytes: equal partners in secondary metabolite production? *Biotechnology Letters*, **37**, 1325–1334.

Marinho, A. M. S., Rodrigues-Filho, E., Moitinh, M. L. R. and Santos, L. S. (2005). Biologically active polyketides produced by *Penicillium janthinellum* isolated as an endophytic fungus from fruits of *Melia azedarach*. *Journal of the Brazilian Chemical Society*, **16**, 280–283.

Meshram, V., Saxena, S. and Paul, K. (2016). Xylarinase: a novel clot busting enzyme from an endophytic fungus *Xylaria curta*. *Journal of Enzyme Inhibition Medicinal Chemistry*, **31**, 1–10.

Mohammad, N., Mauji, R., Pravej, A. and Grothaus, P. (2012). *Fusarium solani*, P1, a new endophytic podophyllotoxin-producing fungus from roots of *Podophyllum hexandrum*. *African Journal of Microbiology Research*, **6**, 2493–2499.

Na, R., Jiajia, L., Dongliang, Y. *et al.* (2016). Indentification of vincamine indole alkaloids producing endophytic fungi isolated from *Nerium indicum*, Apocynaceae. *Microbiological Research*, **192**, 114–121.

Nicoletti, R. and Fiorentino, A. (2015). Plant bioactive metabolites and drugs produced by endophytic fungi of spermatophyta. *Agriculture*, **5**, 918–970.

Omeje, E. O., Ahomafor, J. E., Onyekaba, T. U. *et al.* (2017). Endophytic fungi as alternative and reliable sources for potent anticancer agents. In *Natural Products and Cancer Drug Discovery*, ed. F. A. Badria. London: IntechOpen, pp. 142–157.

Palem, P. P. C., Kuriakose, G. C. and Jayabaskaran, C. (2015). An endophytic fungus, *Talaromyces radicus*, isolated from *Catharanthus roseus*, produces vincristine and vinblastine, which induce apoptotic cell death. *PLoS One*, **10**, e0144476.

Pan, B. F., Su, X., Hu, B. *et al.* (2015). *Fusarium redolens* 6WBY3, an endophytic fungus isolated from *Fritillaria unibracteata* var. wabuensis, produces peimisine and imperialine-3β-D-glucoside. *Fitoterapia*, **103**, 213–221.

Pan, F., Hou, K., Gao, F. *et al.* (2014). Peimisine and peimininen production by endophytic fungus *Fusarium* sp. isolated from *Fritillaria unibracteata* var. wabuensis. *Phytomedicine*, **21**, 1104–1109.

Pandi, M., Kumaran, R. S., Choi, Y.-K., Kim, H. J. and Muthumary, J. (2011). Isolation and detection of taxol, an anticancer drug produced from *Lasiodiplodia theobromae*, an endophytic fungus of the medicinal plant *Morinda citrifolia*. *African Journal of Biotechnology*, **10**, 1428–1435.

Parthasarathy, R. and Sathiyabama, M. (2014). Gymnemagenin-producing endophytic fungus isolated from a medicinal plant *Gymnema sylvestre* R.Br. *Applied Biochemistry and Biotechnology*, **172**, 3141–3152.

Petrini, O. (1991). Fungal endophytes of tree leaves. In *Microbial Ecology of Leaves*, ed. N. J. Fokkema and V. D. Heuvel. Cambridge: Cambridge University Press, pp. 185–187.

Porras-Alfaro, A. and Bayman, P. (2011). Hidden fungi, emergent properties: endophytes and microbiomes. *Annual Review of Phytopathology*, **49**, 291–315.

Promputtha, I., Lumyong, S. and Dhanasekaran, V. (2007). Phylogenetic evaluation of whether endophytes become saprotrophs at host senescence. *Microbial Ecology*, **53**, 579–590.

Pu, X., Qu, X., Chen, F. *et al.* (2013). Camptothecin-producing endophytic fungus *Trichoderma atroviride* LY357: Isolation, identification, and fermentation conditions optimization for camptothecin production. *Applied Microbiology and Biotechnology*, **97**, 9365–9375.

Puri, S. C., Verma, V., Amna, T., Qazi, G. N. and Spiteller, M. (2005). An endophytic fungus from *Nothapodytes foetida* that produces camptothecin. *Journal of Natural Products*, **68**, 1717–1719.

Puri, S. C., Nazir, A., Chawla, R. *et al.* (2006). The endophytic fungus *Trametes hirsuta* as a novel alternative source of podophyllotoxin and related aryl tetralin ligans. *Journal of Biotechnology*, **122**, 494–510.

Qian, Y. X., Kang, J. C., Luo, Y. K. *et al.* (2016). A bilobalide-producing endophytic fungus, *Pestalotiopsis uvicola* from medicinal plant *Ginkgo biloba*. *Current Microbiology*, **73**, 280–286.

Ran, X., Zhang, G., Li, S. and Wang, J. (2017). Characterization and antitumor activity of camptothecin from endophytic fungus *Fusarium solani* isolated from *Camptotheca acuminate*. *African Health Sciences*, **17**, 566–574.

Rehman, S., Shawl, A. S., Kour, A. *et al.* (2008). An endophytic *Neurospora* sp. from *Nothapodytes foetida* producing camptothecin. *Applied Biochemistry and Microbiology*, **44**, 203–209.

Rehman, S., Shawl, A. S., Kour, A. *et al.* (2009). Comparative studies and identification of camptothecin produced by an endophyte at shake flask and bioreactor. *Natural Product Research*, **23**, 1050–1057.

Rodriguez, R. J., White, J. F. J., Arnold, A. E. and Redman, R. S. (2009). Fungal endophytes: diversity and functional roles. *New Phytologist*, **182**, 314–330.

Saxena, S. and Srivastava, A. (2014). Resveratrol: biological activities and therapeutic potential. *Journal of Pharmaceutical Technology, Research and Management*, **2**, 145–157.

Saxena, S., Meshram, V. and Kapoor, N. (2015). *Muscodor tigerii* sp. nov. -volatile antibiotic producing endophytic fungus from north eastern Himalayas. *Annals of Microbiology*, **65**, 47–57.

Schulz, B. and Boyle, C. (2005). The endophytic continuum. *Mycological Research*, **109**, 661–686.

Schulz, B., Rommert, A. K., Dammann, U., Aust, H. J. and Strack, D. (1999). The endophyte–host interaction: a balanced antagonism? *Mycological Research*, **103**, 1275–1283.

Shao, C., Zheng, J., Chi, J. *et al.* (2017). Excretion Study of 10-methoxycamptothecin and its metabolite 10-hydroxycamptothecin in rats by RP-HPLC method with fluorescence detection. *Acta Chromatographica*, **29**, 453–458.

Shi, J., Zeng, Q., Liu, Y. and Pan, Z. (2012). *Alternaria* sp. MG1, a resveratrol-producing fungus: isolation, identification, and optimal cultivation conditions for resveratrol production. *Applied Microbiology and Biotechnology*, **95**, 369–379.

Shu, S., Zhao, X., Wang, W. *et al.* (2014). Identification of a novel endophytic fungus from *Huperzia serrata* which produces huperzine A. *World Journal of Microbiology and Biotechnology*, **30**, 3101–3109.

Shweta, S., Zuehlke, S., Ramesha, B. T. *et al.* (2010). Endophytic fungal strains of *Fusarium solani*, from *Apodytes dimidiata* E. Mey. ex Arn (Icacinaceae) produce camptothecin, 10-hydroxycamptothecin and 9-methoxycamptothecin. *Phytochemistry*, **71**, 117–122.

Stierle, A. and Strobel, G. A. (1995). The search for a taxol-producing microorganism among the endophytic fungi of the Pacific yew, *Taxus brevifolia*. *Journal of Natural Products*, **58**, 1315–1324.

Stierle, A., Strobel, G. A. and Stierle, D. (1993). Taxol and taxane production by *Taxomyces andreanae*. *Science*, **260**, 214–216.

Stierle, A. A. and Stierle, D. B. (2015). Bioactive secondary metabolites produced by the fungal endophytes of conifers. *Natural Product Communication*, **10**, 1671–1682.

Strobel, G. and Daisy, B. (2003). Bioprospecting for microbial endophytes and their natural products. *Microbiology and Molecular Biology Reviews*, **67**, 491–502.

Strobel, G. A., Hess, W. M., Ford, E., Sidhu, R. S. and Yang, X. (1996). Taxol from fungal endophytes and issue of biodiversity. *Journal of Industrial Microbiology*, **17**, 417–423.

Su, J. and Yang, M. (2014). Huperzine A production by *Paecilomyces tenuis* YS-13, an endophytic fungus isolated from *Huperzia serrata*. *Natural Product Research*, **29**, 1035–1041.

Tian, R., Yang, Q., Zhou, G. *et al.* (2006). Taxonomic study on a taxol producing fungus isolated from bark of *Taxus chinensis* var. *mairei*. *Journal of Wuhan Botanical Research*, **24**, 541–545.

Vasundhara, M., Kumar, A. and Reddy, M. S. (2016). Molecular approaches to screen bioactive compounds from endophytic fungi. *Frontiers in Microbiology*, **7**, 1774.

Venugopalan, A. and Srivastava, S. (2015). Endophytes as in vitro production platforms of high value plant secondary metabolites. *Biotechnology Advances*, **1**, 873–887.

Verma, V. C., Lobkovsky, E., Gange, A. C., Singh, S. K. and Prakash, S. (2011). Piperine production by endophytic fungus *Periconia* sp. isolated from *Piper longum* L. *The Journal of Antibiotics*, **64**, 427–431.

Vinodhini, D. and Agastian, P. (2013). Berberine production by endophytic fungus *Fusarium solani* from *Coscinium fenestratum*. *International Journal of Biological and Pharmaceutical Research*, **4**, 1239–1245.

Wang, B., Li, A. and Wang, X. (2001). An endophytic fungus for producing taxol. *Science in China (Series C)*, **31**, 271–274.

Wang, J., Cox, D. G., Ding, W., Huang, G., Lin, Y. and Li, C. (2014). Three new resveratrol derivatives from the mangrove endophytic fungus *Alternaria* sp. *Marine Drugs*, **12**, 2840–2850.

Wang, Y. and Tang, K. (2011). A new endophytic taxol- and baccatin III-producing fungus isolated from *Taxus chinensis* var. *mairei*. *African Journal of Biotechnology*, **10**, 16379–16386.

Wang, Y., Zeng Q. G., Zhang, Z. B. *et al.* (2011). Isolation and characterization of endophytic huperzine A-producing fungi from *Huperzia serrata*. *Journal of Industrial Microbiology and Biotechnology*, **38**, 1267–1278.

Wang, Y. A., Lai, Z., Li, X. X. *et al.* (2016). Isolation, diversity and acetylcholinesterase inhibitory activity of the culturable endophytic fungi harboured in *Huperzia serrata* from Jinggang Mountain, China. *World Journal of Microbiology and Biotechnology*, **32**, 20.

Wani, M. C., Talor, H. L., Wall, M. E., Coggon, P. and McPhail, A. T. (1971). Plant antitumor agents. VI. Isolation and structure of taxol, a novel antileukemic and antitumor agent from *Taxus brevifolia*. *Journal of the American Chemical Society*, **93**, 2325–2327.

Wojtyczka, R. D., Dziedzic, A., Kępa, M. *et al.* (2014). Berberine enhances the antibacterial activity of selected antibiotics against coagulase-negative *Staphylococcus* strains *in vitro*. *Molecules*, **19**, 6583–6596.

World Health Organization (WHO). (2017). Cancer. WHO fact sheet. Geneva: World Health Organization.

Xie, X. S., Fang, X. W., Huang, R. *et al.* (2016). A new dimeric anthraquinone from endophytic *Talaromyces* sp. YE3016. *Natural Product Research*, **30**, 1706–1711.

Yan, J., Qi, N., Wang, S., Gadhave, K. and Yang, S. (2014). Characterization of secondary metabolites of an endophytic fungus from *Curcuma wenyujin*. *Current Microbiology*, **69**, 740–744.

Yang, X., Guo, S., Zhang, L. and Shao, H. (2003). Selection of producing podophyllotoxin endophytic fungi from podophyllin plant. *Natural Product Research and Development*, **15**, 419–422.

Yang, X., Zhang, L., Guo, B. and Guo, S. (2004). Preliminary study of a vincristine-producing endophytic fungus isolated from leaves of *Catharanthus roseus*. *Chinese Traditional and Herbal Drugs*, **35**, 79–81.

Yang, K., Liang, J., Li, Q. *et al.* (2013). *Cladosporium cladosporioides*XJ-AC03, an aconitine-producing endophytic fungus isolated from *Aconitum leucostomum*. *World Journal of Microbiology and Biotechnology*, **29**, 933–938.

Yin, H. and Sun, Y. H. (2011). Vincamine-producing endophytic fungus isolated from *Vinca minor*. *Phytomedicine*, **18**, 802–805.

Yin, J., Xing, H. and Ye, J. (2008). Efficacy of berberine in patients with type 2 diabetes mellitus. *Metabolism*, **57**, 712–717.

Ying, Y. M., Shan, W. G. and Zhan, Z. J. (2014). Biotransformation of Huperzine a by a fungal endophyte of *Huperzia serrata* furnished sesquiterpenoid–alkaloid hybrids. *Journal of Natural Products*, **77**, 2054–2059.

Zaiyou, J., Li, M., Guifang, X. and Xiuren, Z. (2013). Isolation of an endophytic fungus producing baccatin III from *Taxus wallichiana* var. *mairei*. *Journal of Industrial Microbiology and Biotechnology*, **40**, 1297–1302.

Zaiyou, J., Li, M. and Xiqiao, H. (2017). An endophytic fungus efficiently producing Paclitaxel isolated from *Taxus wallichiana* var. mairei. *Medicine*, **96**, e7406.

Zan, J., Jun, W. and Sheng-li, P. (2009). Isolation and preliminary identification of the endophytic fungi which produce Hupzine A from four species in Hupziaceae and determination of huperzine A by HPLC. *Fudan University Journal of Medical Sciences*, **36**, 0–449.

Zhang, F. F., Wang, M. Z., Zheng, Y. X. *et al.* (2015). Isolation and characterzation of endophytic HuperzineA producing fungi from *Phlegmariurus phlegmaria*. *Microbiology*, **84**, 701–709.

Zhang, F. H., Xiang, J. H., Cui, W. X. *et al.* (2016). Isolation and identification of berberine from endophytic fungi HL-Y-3]. *Zhongguo Zhong Yao Za Zhi*, **41**, 2998–3001.

Zhang, J., Shi, J. and Liu, Y. (2013). Bioconversion of resveratrol using resting cells of non-genetically modified *Alternaria* sp. *Biotechnology and Applied Biochemistry*, **60**, 236–243.

Zhang, L., Guo, B., Li, H. *et al.* (1999). Preliminary study on the isolation of endophytic fungus of *Catharanthus roseus* and its fermentation to produce products of therapeutic value. *Chinese Traditional and Herbal Drugs*, **31**, 805–807.

Zhang, P., Zhou, P. and Yu, L. (2009). An endophytic taxol-producing fungus from *Taxus media*, *Cladosporium cladosporioides* MD2. *Current Microbiology*, **59**, 227–232.

Zhang, Z. B., Zeng, Q. G., Yan, R. M., Wang, Y., Zou, Z. R. and Zhu, D. (2011). Endophytic fungus *Cladosporium cladosporioides* LF70 from *Huperzia serrata* produces huperzine A. *World Journal of Microbiology and Biotechnology*, **27**, 479–486.

Zhao, J., Zhou, L., Wang, J. *et al.* (2010). Endophytic fungi for producing bioactive compounds originally from their host plants. *Current Research, Technology and Education Tropics in Applied Microbiology and Microbial Biotechnology*, **1**, 567–576.

Zhu, D., Wang, J. X., Zeng, Q. G., Zhang, Z. B. and Yan, R. M. (2010). A novel endophytic huperzine A-producing fungus, *Shiraia* sp. Slf14, isolated from *Huperzia serrata*. *Journal of Applied Microbiology*, **109**, 1469–1478.

14

Enhancing Secondary Metabolite Production in Medicinal Plants Using Endophytic Elicitors: A Case Study of *Centella asiatica* (Apiaceae) and Asiaticoside

SHUBHPRIYA GUPTA AND PREETI CHATURVEDI

Abstract

Fungal endophytes are a vital component of the plant microbiome. The symbiotic relationship between endophytic fungi and medicinal plants can considerably influence plant secondary metabolism pathways, thus affecting their metabolite production and the quality and quantity of crude drugs produced. This chapter focuses on how fungal endophytic symbiosis can increase production of secondary metabolites during *in vitro* culture. Other than promoting secondary metabolite accumulation, endophytic fungi can also promote the growth of host plants and improve their resistance to abiotic and biotic stresses. Therefore, an understanding of the relationship between endophytic fungi and their host medicinal plants is of utmost importance. This knowledge can be applied in the production of novel and improved drugs from medicinally valued plants. *In vitro* elicitation of secondary metabolites by endophytic fungi has been reported in several medicinal plants. A case study is documented in this chapter which shows enhancement of asiaticoside by an endophytic fungal (*Colletotrichum gloeosporioides*) elicitor isolated from *in vivo* grown plants of *Centella asiatica*. These findings may motivate further exploitation of fungal endophytes as an effective and beneficial way to enhance the production of pharmacologically important compounds from medicinal plants.

14.1 Introduction

The term 'endophyte' (Greek *endon*, within; *phyton*, plant) was first coined by de Bary in 1866. Endophytes can be described as microorganisms that live within plants for at least part of their life cycle without causing any visible appearance of disease (Bacon and White, 2000). The existence of endophytic fungi has been known since the end of the nineteenth century (Guerin, 1898). Freeman (1904) was the first to describe a fungal endophyte and its mode of infection in the grass Darnel, *Lolium temulentum*. Since then, several investigators have isolated endophytes from many different plant species. Endophytes are rich and consistent components of fungal diversity and many new discoveries may represent previously undescribed species (Murphy *et al.*, 2015). They can have profound effects on plant ecology, fitness, evolution and the composition of plant communities (Clay and Holah, 1999). Endophytic fungi can form mutualistic relationships with plants to get nutrition and protection. In return, endophytes can enhance the fitness of the host plant, for example, improving nutrient uptake, producing a variety of bioactive metabolites, antibiotics, plant growth promoters, plant growth inhibitors and enzymes (Furumai *et al.*, 2003; Igarashi *et al.*, 2003). The quality and quantity of crude drugs extracted from medicinal plants is largely affected by genetic, biotic and abiotic factors (Sherameti *et al.*, 2005). Endophytic fungi can play an important role in influencing and improving the production of crude drugs through a particular fungus–host interaction (Strobel *et al.*, 1996; Keith, 1998). Endophyte inoculation can also positively influence *in vitro* propagation and large-scale production of medicinal plants. Besides enhancing plant growth, fungal endophytes may also aid in increasing their secondary metabolite production. *In vitro* experiments can help to establish the biochemical and physiological relationship of the host plant and their respective endophytic fungi, facilitating the understanding of mutualistic associations of plant and endophyte. In this chapter, we describe plant–endophyte interactions and the factors influencing the production of secondary metabolites in the co-cultivated medicinal plants under *in vitro* conditions. A case study presented here shows that a fungal endophyte *Colletotrichum gloeosporioides* improves the production of a host secondary metabolite, viz asiaticoside, by co-cultivation with *in vitro* cultures of *C. asiatica*, a medicinally valuable plant species.

14.2 Host and Endophyte Relationship

Fungi are frequently encountered endophytes (Staniek *et al.*, 2008) and the fossil record indicates that plants have been associated with endophytic fungi for more than 400 million years (Krings *et al.*, 2007). Endophytes are likely to have associated with plants when they first colonised land, thus playing a long and important role in driving the evolution of life on land (Hodkinson, 2018). It is likely that each of the approximate 300 000 species of higher plants that exist on earth contains endophytes (Strobel and Daisy, 2003) and endophytes are also present in lower plants such as mosses, liverworts and ferns (Boullard, 1957). It is estimated that there may be as many as 1 million different endophyte species (Dreyfuss and Chapela, 1994).

An increasing demand of plant active metabolites and their paucity in natural sources, has led to diverse strategies of bioprospecting endophytic fungi as an alternative source of bioactive compounds. Some of the popular examples include paclitaxel, also known as taxol (Stierle *et al.*, 1993), vinblastine (Guo *et al.*, 1998), cryptocandin (Strobel *et al.*, 1999), vincristine (Zhang *et al.*, 2000), pestacin and isopestacin (Strobel *et al.*, 2002; Harper *et al.*, 2003), camptothecin (Puri *et al.*, 2005), rohitukine (Kumara *et al.*, 2012) and azadirachtin (Kusari *et al.*, 2012). Interestingly, a single endophyte may be able to produce or elicit several bioactive metabolites. As a result, the role of endophytes in the production of novel metabolites for exploitation in modern medicine, agriculture and industry is receiving high attention (Wang *et al.*, 2000; Gunatilaka, 2006). To date, only a small number of plants have been investigated for the endophyte diversity and their bioactive potential (Andrews and Hirano, 1991).

The initial step, in dealing with endophytic organisms, is the selection of a suitable and promising plant for the study, followed by isolation of the endophytes. Selection often focuses on (1) plants with an unusual biology; (2) plants that have medicinal and ethnobotanical history; (3) plants that are endemic or endangered; or (4) plants growing in areas of enormous biodiversity. Different parts of the plant are then cultured and assessed for endophytic diversity. After isolation, the endophytes are encouraged to sporulate on specific plant materials and are eventually identified via standard morphological and molecular methods such as DNA barcoding (Murphy *et al.*, 2015). Eventually, when an endophyte is acquired in pure culture, it is then tested for its ability to grow in a shake or still culture using various media and growth conditions. It can also be immediately placed for storage in media, such as 15% glycerol at -70°C. Once appropriate growth conditions are achieved, the microbe is fermented, extracted and bioactive compound(s) are isolated and characterised.

14.3 Co-Culturing Experiments for Metabolite Production

There is a growing demand to optimise the quality and yield of medicinal plants. This has led to the advancement of biotechnological techniques such as plant cell and tissue culture, use of biotic and abiotic elicitors, genetic manipulation and metabolic engineering (Hussain *et al.*, 2012). Plant cell/tissue/organ culture technology has resulted in the selection of high-yielding cell lines, ultimately leading to higher production and accumulation of valuable secondary metabolites. The elicitation of plant cells is an approach which involves the use of substances to induce or enhance biosynthesis of a specific compound. Biotic and abiotic elicitors, such as fungal carbohydrates, yeast extract, methyl jasmonate and chitosan, have been used to enhance the production of secondary metabolites in cell cultures (Karuppusamy *et al.*, 2009). Genetic manipulation (e.g. *Agrobacterium*-mediated transformation), alteration of the metabolic pathway and manipulation of cellular activities have also been used to increase the yield of target compounds. However, these methods are expensive and, therefore, new and cost-effective methods have been investigated.

The symbiotic relationship between the host plant and microbial endophytes can be explored to investigate the accumulation of secondary metabolites under *in vitro* conditions (Yuan *et al.*, 2007, 2010). To date, most of the known fungal secondary metabolites have been isolated from endophytic fungi (Ghisalberti, 2002). These fungi have been extensively studied over the last several years as prolific sources of bioactive metabolites including alkaloids, terpenoids, flavonoids, steroids, phenylpropanoids, phenolics and volatile organic compounds. Some of these compounds have anticancer, antioxidant, antibiotic and other biological activities. Interactions between endophytic fungi and their host plant involve a balanced antagonism when compared to the interactions of a plant pathogen. Researchers have tried to elucidate the molecular mechanisms involved in the establishment of endophyte colonisation (Sherameti *et al.*, 2005). Nevertheless, there is little literature about the secondary metabolism of plants induced by their endophytes (Mucciarelli *et al.*, 2003). However, reduction of secondary metabolite production on repeated subculturing of endophytic fungi under axenic monoculture conditions is a key challenge that needs to be addressed in order to establish, restore and sustain *in vitro* biosynthetic potential of endophytes. The loss of production could be due to the lack of host stimulus in the culture media and/or silencing of genes in axenic cultures (Sachin *et al.*, 2013).

Co-cultivation experiments could be an answer to overcome such limitations, where it is possible to cultivate endophytic fungi by reproducing naturally occurring conditions. Another advantage of co-cultivation is that interference from external microbes is removed or minimised when plants are cultivated aseptically. Co-culturing of endophytic fungi may enhance the accumulation of constitutively present natural products up to many fold or may trigger the expression of silent biosynthetic pathways yielding new compounds. Some findings have shown that intimate physical interaction of the endophytic fungi during co-culture triggers the expression of some gene clusters in *in vitro* plants that are kept silent under standard laboratory conditions (Ola *et al.*, 2013). Naoumkina *et al.*, (2008) have reported that the use of yeast extract in cell cultures of *Medicago truncatula* induced the accumulation of flavonoids, triterpenes, and other defence-related molecules. In another study, the use of yeast extract, jasmonic acid, *Aspergillus niger* and *A. flavus* fungal mycelium homogenates, chitosan and glucan in cell suspensions of *Morinda elliptica* had different effects on anthraquinone production (Chong *et al.*, 2005). It was also found that the addition of an endophytic fungal elicitor into plant cell suspension culture medium can improve the ability of plant cells to produce secondary metabolites. The oligosaccharides from the endophytic fungus *Fusarium oxysporum* when added to 26-day-old *Dioscorea zingiberensis* cell cultures, significantly enhanced diosgenin production (Li *et al.*, 2012). Wang *et al.* (2012) observed that the effect of the fungus in promoting the growth of the host plant was much higher than that of its elicitor and hence proposed that the fungus could perhaps be more effective in improving the quality of herbal medicines. Co-culture of *Rumex gmelini* seedlings and its endophytic fungus *Aspergillus* sp. resulted in significant improvement in production of bioactive secondary metabolites (Ding *et al.*, 2017). Thus, co-cultivation system of plant and fungus under *in vitro* conditions helps to elucidate the relationships between both partners and improve the production of the desired compounds as well.

14.4 Endophytic Fungi and Host Plant Metabolism

The endophytic fungi isolated from a healthy host plant when reintroduced into tissue-cultured plants can affect the growth of the inoculated plants in different ways depending on the fungal strain. Both the plant and the endophyte influence each other's physiology and alter their secondary metabolism. This can disturb the equilibrium between microorganism virulence and plant defence that is important for asymptomatic colonisation (Sherameti *et al.*, 2005). Sometimes, the fungal strain can become pathogenic, leading to wilting, leaf-drop and darkness in the roots, in spite of being an endophyte. Wilting is often caused by the colonisation of xylem vessels by pathogenic microorganisms and can lead to leaf fall. There are reports where, out of many endophytic fungal strains isolated from the host plant, only a few strains colonised the root system asymptomatically in *in vitro* plants (Junker *et al.*, 2012). The successful symbiosis between the endophytic fungus and *in vitro* plant is directly related to environmental conditions and the physiological state of the host.

Under adverse conditions, fungal endophtyes may manifest themselves as pathogens. It might be possible that the experimental conditions used for co-cultivation act as a stress factor to the plants. In this case, the fungal endophyte can be seen as a pathogen under *in vitro* conditions (Nery-Silva *et al.*, 2007). Therefore, strain selection is a crucial step during *in vitro* colonisation of endophytic fungi in tissue-cultured plants. During the pre-symbiotic phase, *in vitro* plant and fungal endophyte secrete several signals, mostly secondary metabolites, subsequently perceived by roots and mycelium inducing morphological and physiological changes. These signal molecules mediate and enhance the symbiotic interaction between both the partners (Akiyama and Hayashi, 2006). Once a fungal endophyte has established successfully with *in vitro* plants, it triggers a metabolic response of the plant for the production of secondary metabolites. The fungal endophytes need to maintain a balanced antagonism to grow asymptomatically within the *in vitro* plant. The major concern after the asymptomatic establishment of endophytic fungus is the overgrowth of fungus that can cause an inhibition in growth of the plant due to exhaustion of nutrients from the medium or it can lead to the death of the plant cell/plant (Li and Tao, 2009). This can be regulated by taking relatively small inoculums of endophytic fungi and/or by replacing the medium frequently to ensure sufficient nutrients for seedlings.

14.5 Secondary Metabolite Accumulation in Co-Culture Systems

The fungal endophytes can be isolated from host medicinal plant and can be reintroduced into *in vitro* medicinal plants, which can provide an effective strategy for increasing biomass and secondary metabolite content under co-culture systems. There are several reports of endophyte-induced biosynthesis of secondary metabolites in host plants. The case study presented in this chapter serves as an example of *in vitro* induction of secondary metabolites by plant endofungal co-culture system.

14.5.1 Endophyte-Enhanced Terpenoid Production in *Centella*: A Case Study

Centella asiatica (L.) Urban is an important medicinal plant and has drawn scientific attention due to its remarkable concentration of terpenoids viz. asiaticoside, asiatic acid, madecassoside and madecassic acid. The major active terpenoid component of the plant, asiaticoside, demonstrates a number of biological activities such as wound healing (Shukla *et al.*, 1999), memory enhancement (Nalini *et al.*, 1992), immunomodulatory (Plohman *et al.*, 1994) and antidepressant activity (Chen *et al.*, 2005). Asiaticoside also plays an important role in stabilising connective tissue, and imparts a balancing effect on the connective and cardiovascular tissue (Darnis

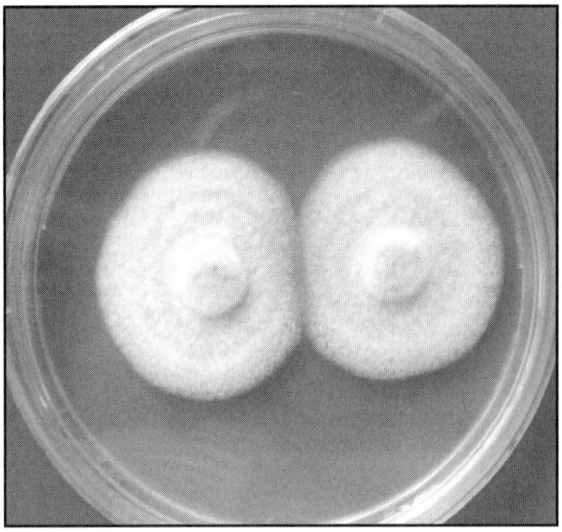

Figure 14.1 Morphology of *Colletotrichum gloeosporioides*.

et al., 1979; Montecchio *et al.*, 1991). Asiaticoside possesses anticancerous properties (Babykutty *et al.*, 2009) and is, therefore, an important constituent of many medicinal formulations. Asiaticoside is obtained from the leaves of *C. asiatica*, which are also colonised by endophytic fungi. Different functional traits that are important to promote plant growth can be associated with these endophytes. In the present study, the endophytic fungus, *Colletotrichum gloeosporioides*, isolated from *in vivo* plants of *C. asiatica* was able to produce asiaticoside under axenic conditions which was confirmed by LC-MS analysis (Gupta, 2015; Gupta *et al.*, 2018). The endophytic fungal isolate was subjected to genomic DNA isolation and sequencing, targeting the internal transcribed spacer (ITS) region (partial 18S ribosomal RNA gene, internal transcribed spacer 1, 5.8S ribosomal RNA gene, internal transcribed spacer 4, partial 28S ribosomal RNA gene) using the standard ITS 1/4 primers. A BLAST search showed that the isolated fungal strain had a similarity of 99% and 99% query coverage with *C. gloeosporioides* in the National Center for Biotechnology Information (NCBI) GenBank database. A phylogenetic analysis of the isolated strain of *C. gloeosporioides* was also carried out by the neighbour joining (NJ) method and a consensus tree was constructed from 74 aligned ITS sequences with 1000 bootstrap replications. It also confirmed the strain was *C. gloeosporioides* (Gupta *et al.*, 2018). The amplified ITS sequence of *C. gloeosporioides* was submitted to NCBI GenBank under accession number KP682510. The morphology of the cultured *C. gloeosporioides* strain is presented in Figure 14.1. In order to optimise the yield of asiaticoside, *C. gloeosporioides* was reintroduced into the *in vitro* plants of *C. asiatica* through co-cultivation (Figure 14.2a).

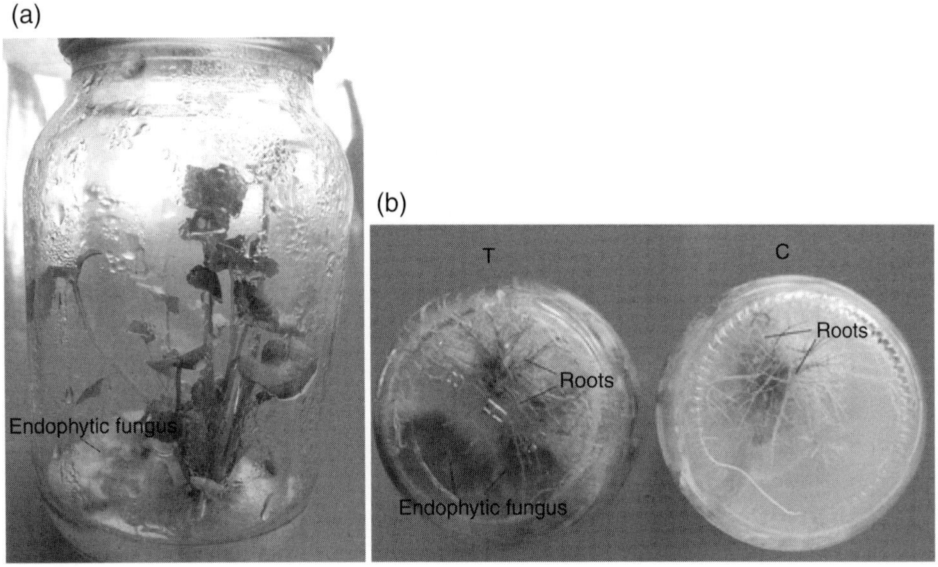

Figure 14.2 (a) Co-cultivation of *Colletotrichum gloeosporioides* with *in vitro Centella asiatica* plants. (b) Increased number of roots in co-cultivated (T) plants and control (C) plants.

14.5.2 Inoculation of *In Vitro* Grown Plants with *Colletotrichum gloeosporioides*

In the present study, plants of *Centella asiatica* were raised *in vitro* for co-cultivation with the endophytic fungus *Colletotrichum gloeosporioides.* To establish *in vitro* plants of *C. asiatica,* leaf explants were cultured aseptically in Murashige and Skoog (1962) medium (MS). MS + 0.75 mg L^{-1} 6-benzylaminopurine (BA) was found to be most proficient for shoot induction and proliferation, with a shoot induction factor of 86.66%. Profuse rooting was observed on half-strength MS medium supplemented with 0.5 mg L^{-1} naphthalene acetic acid (NAA) with a root induction factor of 100%. Plants of *C. asiatica* were inoculated with *C. gloeosporioides* after 45 days of establishment in MS medium. 5 mm diameter disc of the fungal isolate was inoculated near the base of *C. asiatica* plants grown in 50% MS basal medium. Plants without the fungal inoculant served as controls. After inoculation, culture bottles containing the plants and fungal isolate were kept in a culture room with 15 μmol m^{-2} s^{-1} photosynthetically active radiation (PAR), at an average temperature of 28°C, with a photoperiod of 16 hours. Plant health was evaluated on a daily basis. The plants were collected on the 20th day after inoculation for assessment of changes in morphological parameters (Figure 14.3). The absence of disease symptoms in co-cultivated micropropagated plants indicate the presence of *C. gloeosporioides* in non-antagonistic form.

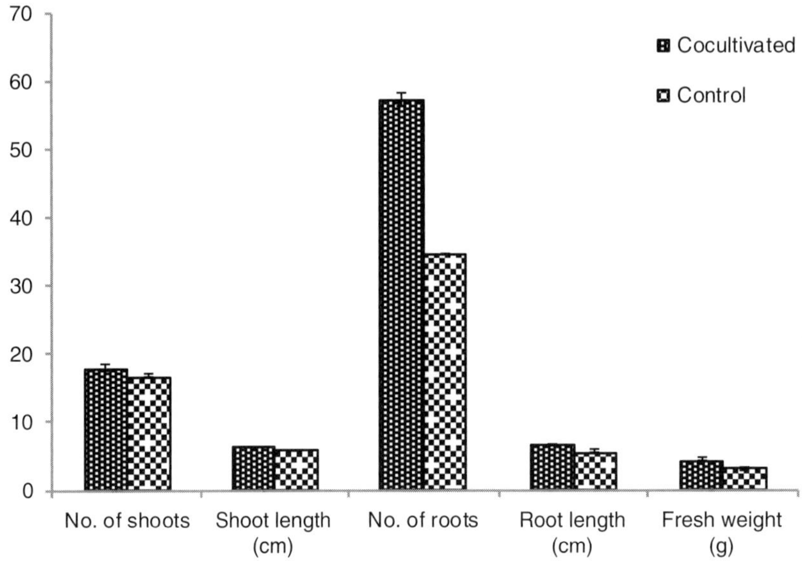

Figure 14.3 Growth responses in co-cultivated plants.

14.5.3 Morphological Parameters of *Centella asiatica*

Growth parameters of *Centella asiatica* were measured to assess the growth promotion capability of endophytic *Colletotrichum gloeosporioides*. The number of roots were counted, shoot and root lengths were measured in centimetres, and fresh weight was measured in grams. There was no significant difference in the number of shoots of co-cultivated (18.0 ± 0.57) and control (16.66 ± 0.33) plants wherein the shoot length was 5.8 cm and 6.23 cm, respectively. In contrast with the control plants, *C. gloeosporioides* inoculation led to a sustained extension of the root system of *in vitro* grown plants. The root length of the co-cultivated plants (6.56 ± 0.17 cm) was significantly higher ($P < 0.05$) than control plants (5.4 ± 0.03 cm). The number of roots in the co-cultivated plants (57.33 ± 0.88) was also significantly higher than the control plants (34.66 ± 0.33) (Figure 14.2 b). There was a 1.65-fold increase in the number of roots in the co-cultivated plants. Similarly, fresh weight (4.26 ± 0.17 g) of the co-cultivated plants was also significantly higher ($P < 0.05$) in comparison to the control plants (3.0 ± 0.13 g). The significant increase in number of roots and fresh weight clearly reflected stimulation by endophytic fungus.

Plant host stimulation by foliar endophytic fungi has been variously documented consisting of larger and more numerous tillers (Hill *et al.*, 1991, 1996), greater leaf elongation (Belesky and Fedders, 1996), increased leaf area, leaf thickness, stem length (Malinowski and Belesky, 2000) and occasionally altered root architecture (Malinowsky *et al.*, 1999). Pocasangre (2000) and Niere (2001) reported higher root and shoot weights of tissue-cultured banana plants co-cultivated with *Fusarium oxysporum* endophytes. Prasad *et al.* (2004) co-cultivated *Bacopa monnieri* and *Piriformospora indica* and found that plants co-cultivated with *P. indica* showed significant increases in the number of leaves, number of branches, and root and shoot length. Likewise, *in vitro Euphorbia pekinensis* plants co-cultivated with two endophytic fungal strains of *Fusarium* spp. exhibited significant increase in growth compared to the control (Dai *et al.*, 2008).

In the present study, gas chromatography–mass spectrometry (GC-MS) analysis of the crude extract of *C. gloeosporioides* revealed the presence of indole acetic acid (IAA) (0.80%) (Gupta, 2015). Thus, production of IAA by *C. gloeosporioides* might be responsible for enhanced root production in our study on *C. asiatica*. The production of IAA enhances root growth of the plants by stimulating plant cell elongation or cell division (Glick and Bashan, 1997). Rommert *et al.* (2002) found that a mycelial extract of the endophytic fungus, *Phialophora fortinii* produced IAA *in vitro* and increased *Larix decidua* shoot and root biomass. Several fungal endophytes (*Neotyphodium* spp. and *Balansia* spp.) forming mutualistic associations with grasses have been reported to produce IAA and related indole compounds in culture (De Battista *et al.*, 1990; Porter *et al.*, 1995; Yue *et al.*, 2000). The enhancement of growth by endophytes may thus be a result of the production of phytohormones by the fungal endophytes (Nassar *et al.*, 2005). The results of the present study evidently indicated the potential of the reintroduced endophytic fungus in improving the growth of an *in vitro* grown medicinal plant under a co-culture system.

14.5.4 Asiaticoside Content in Co-Cultivated and Control Plants

Co-cultivation with the endophytic fungus, *C. gloeosporioides* aided in enhancing the asiaticoside content in the co-cultivated plants. In the present study, asiaticoside content was evaluated, using a UV-visible spectrophotometer, in control and co-cultivated plants after 20 days. Maximum absorbance (λ max) for asiaticoside was determined by scanning of wavelength using the UV-visible spectrophotometer. Methanol was used as a reference sample, while 1 mg/ml of standard asiaticoside (Sigma, USA) was used for scanning at the wavelength (200–300 nm). The absorption spectra of asiaticoside showed maximum absorbance (λ max) at 215 nm (Figure 14.4). To plot a standard curve of asiaticoside, different concentrations of asiaticoside were used (5, 10, 20, 25, 30 μg/ml) for recording absorbance at 215 nm (Figure 14.5). The methanolic extract of control and co-cultivated *in vitro C. asiatica* plants were used for further spectral investigations.

The asiaticoside content was improved in co-cultivated micropropagated plants (4.08 mg/g) compared to control plants (2.63 mg/g) suggesting that *C. gloeosporioides* might have triggered the metabolic pathway leading to biosynthesis of asiaticoside. The reason for the increase in asiaticoside could be attributed to the increase in squalene content. In the GC-MS analysis it was found that there was an increase in squalene content in co-cultivated plants compared to control plants from 1.10% to

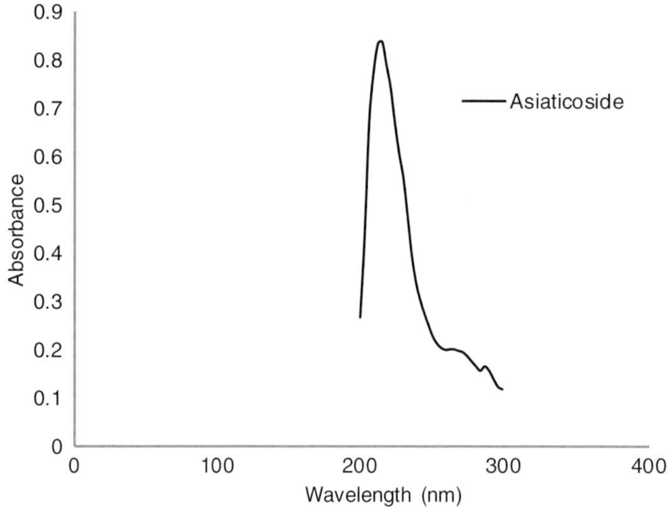

Figure 14.4 Absorption spectra of asiaticoside.

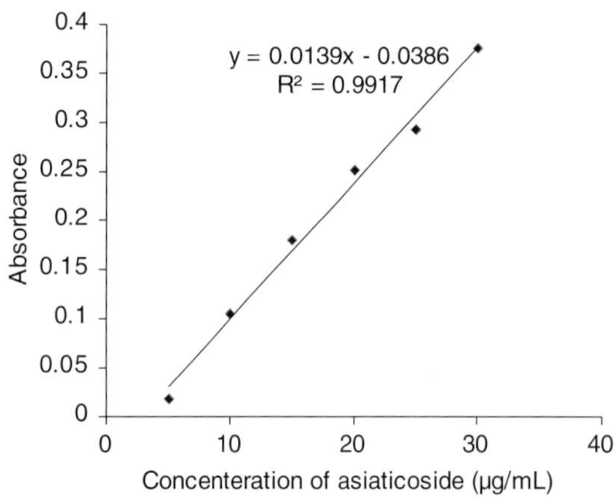

Figure 14.5 Standard curve of asiaticoside.

1.99% resulting in a 0.55 fold increase in squalene content in co-cultivated plants (Gupta, 2015). Squalene is the precursor in promoting triterpene production (Dubey *et al.*, 2003). The exchange of available substances (viz. elicitors produced by fungi and inhibitors produced by plant cells) via the separate membrane in the middle of the co-cultivated system might have regulated the fungal growth and enhanced asiaticoside accumulation in co-cultivated plants (Li and Tao, 2009). It might be possible that the endophytic infection of *C. gloeosporioides* resulted in specific-enhancement of the MEP (methylerythritol phosphate/ non-mevalonate pathway) pathway metabolic rate in co-cultivated *C. asiatica* plants leading to the formation of terpenes and finally enhanced asiaticoside (Yuan *et al.*, 2007).

Our study supports other investigations that have demonstrated terpenoid and other bioactive enhancement by endophytes. For example, Mucciarelli *et al.* (2003) evaluated growth and terpene production of *in vitro* generated peppermint plants (*Mentha piperita*) in response to inoculation with a foliar fungal endophyte PGP-HSF (plant growth promoter-hyaline sterile fungus) and found variation in the essential oil profile by the fungal infection. Jiang *et al.* (2003) reported that co-culture of endophytic *Fusarium* sp. and *Dracaena cochinchinensis* led to formation and accumulation of red resin in the inoculated sites. Prasad *et al.* (2004) reported an increase in bacoside content by several folds in *Bacopa monnieri* plants co-cultured with *Piriformospora indica* on growth regulator-free medium. Microbial elicitors derived from some fungal endophytes also promoted biomass and induced terpenoid (artemisinin) biosynthesis and production in plant suspension cells (Wang *et al.*, 2006). Yuan *et al.* (2007) showed that the weight of roots, seedlings and terpenoid production of *Euphorbia pekinensis* increased after they were inoculated with endophytic *Phomopsis* sp. *In vitro* and *ex vitro* co-cultures showed that the *Piriformospora indica* and *Sebacina vermifera* possessed pronounced plant growth stimulating capacity and induced plant secondary metabolite accumulation, such as spilanthol in *Spilanthes calva* and podophyllotoxins in *Linum album* (Rai *et al.*, 2001; Baldi *et al.*, 2008).

A co-culture of the suspension cells of *Taxus chinensis* var. *mairei* and its endophytic fungus *Fusarium mairei* was successfully established for paclitaxel production resulting in 38-fold higher productivity (Li *et al.*, 2009). *Fusarium mairei*, a paclitaxel-producing endophyte, also enhanced the paclitaxel production in *T. cuspidata* cell suspensions (Li and Tao, 2009). Furthermore, cells of *Catharanthus roseus* when co-cultured with its endophytic fungus gave a 48% higher alkaloid yield than the control group (Tang *et al.*, 2009). Bajaj *et al.* (2014) reported that co-cultivation of *Curcuma longa* and *Piriformospora indica* resulted in pronounced productivity and enhanced production of curcumin and volatile oil in farmer field conditions. These studies suggest that application of endophytic fungi may have induced a subset of secondary metabolite biosynthetic genes, which modulated expression of genes and accumulation of compounds. These results will further help to explore fungal endophytes as effective and beneficial fungal agents to enhance the target chemical production of medicinal plants (Saxena *et al.*, 2005; Yuan *et al.*, 2007).

14.6 Conclusions

The overall study of the microbe and the plant under a co-cultivation system demonstrates the potential of the endophytic fungus, *Colletotrichum gloeosporioides* to enhance/elicit the production of the biologically active secondary metabolite asiaticoside in micropopagated plants of *Centella asiatica*. These findings support the development of strategies for enhancing the production of bioactive secondary metabolites in medicinal plants using suitable fungal endophytes. The co-cultivation method is a promising and interesting approach for induction and augmentation of biologically active secondary metabolites in micropropagated plants. The approach can be further utilised for bulking up and enhancing plant secondary metabolite production under *in vitro*/bioreactor conditions.

Acknowledgements

We thank Dr. Manoj G. Kulkarni, Research Centre for Plant Growth and Development, University of KwaZulu-Natal, South Africa for assisting in preparing the final draft of this chapter. We also thank G. B. Pant University of Agriculture and Technology, India for providing financial assistance for this research.

References

Akiyama, K. and Hayashi, H. (2006). Strigolactones, chemical signals for fungal symbionts and parasitic weeds in plant roots. *Annals of Botany*, **97**, 925–931.

Andrews, J. H. and Hirano, S. S., eds. (1991). *Microbial Ecology of Leaves*. Brock/Springer Series. London: Springer Verlag.

Babykutty, S., Padikkala, J., Sathiadevan, P. P. *et al.* (2009). Apoptosis induction of *Centella asiatica* on human breast cancer cells. *African Journal of Traditional, Complementary and Alternative Medicines*, **6**, 9–16.

Bacon, C. W. and White, J. F. (2000). *Microbial Endophytes*. New York: Marcel Deker Inc.

Bajaj, R., Agarwal, A., Rajpal, K. *et al.* (2014). Co-cultivation of *Curcuma longa* with *Piriformospora indica* enhances the yield and active ingredients. *American Journal of Current Microbiology*, **2**, 6–17.

Baldi, A., Jain, A., Gupta, N., Srivastava, A. K. and Bisaria, V. S. (2008). Coculture of arbuscular mycorrhiza-like fungi (*Piriformospora indica* and *Sebacina vermifera*) with plant cells of *Linum album* for enhanced production of podophyllotoxins: a first report. *Biotechnology Letters*, **30**, 1671–1677.

Belesky, D. P. and Fedders, J. M. (1996). Does endophyte influence regrowth of tall fescue? *Annals of Botany*, 78, 499–505.

Boullard, B. (1957). Lamycotrophie chez les Pteridophytes. Sa frequence, ses characters, sa signification. *Botaniste*, **41,** 1–185.

Chen, Y., Han, T., Rui, Y. *et al.* (2005). Effects of total triterpenes of *Centella asiatica* on the corticosterone levels in serum and contents of monoamine in depression rat brain. *Zhong Yao Cai*, **28**, 492–496.

Chong, T. M., Abdullah, M. A., Lai, O. M., Nor'aini, F. M. and Lajis, N. H. (2005). Effective elicitation factors in *Morinda elliptica* cell suspension culture. *Process Biochemistry*, **40**, 3397–3405.

Clay, K. and Holah J. (1999). Fungal endophyte symbiosis and plant diversity in successional fields. *Science*, **285**, 1742–1744.

Dai, C. C., Yu, B. Y. and Li, X. (2008). Screening of endophytic fungi that promote the growth of *Euphorbia pekinensis*. *African Journal of Biotechnology*, **7**, 3505–3510.

Darnis, F., Orcel, L., de Saint Maur, P. P. and Mamov, P. (1979). Use of a titrated extract of *Centella asiatica* in chronic hepatic disorders. *Semaine Des Hopitaux*, **55**, 1749–1750.

de Bary, A. (1866). Morphologie und Physiologie der Pilze, Flechten, und Myxomyceten. *Hofmeister's Handbook of Physiological Botany*, Volume II. Leipzig, Germany: Engelmann.

De Battista, J. P., Bacon, C. W., Severson, R. F., Plattner, R. D. and Bouton, J. H. (1990). Indole acetic acid production by the fungal endophyte of tall fescue. *Agronomy Journal*, **82**, 878–880.

Ding, C., Wang, Q. B., Guo, S. and Wang, Z. Y. (2017). The improvement of bioactive secondary metabolites accumulation in *Rumex gmelini* Turcz through co-culture with endophytic fungi. *Brazilian Journal of Microbiology*, **49**, 362–369.

Dreyfuss, M. M. and Chapela, I. H. (1994). Potential of fungi in the discovery of novel, low-molecular weight pharmaceuticals. In *The Discovery of Natural Products with Therapeutic Potential*, ed. V. P. Gullo. London: Butterworth-Heinemann, pp. 49–80.

Dubey, V. S., Bhalla, R. and Luthra, R. (2003). An overview of the non-mevalonate pathway for terpenoid biosynthesis in plants. *Journal of Biosciences*, **28**, 637–646.

Freeman, E. M. (1904). The seed-fungus of *Lolium temulentum*, L., the Darnel. *Philosophical Transactions of Royal Society B*, **196**, 1–27.

Furumai, T., Yamakawa, T., Yoshida, R. and Igarashi, Y. (2003). Clethramycin, a new inhibitor of pollen tube growth with antifungal activity from *Streptomyces hygroscopicus* TP-A0623. I. Screening, taxonomy, fermentation, isolation and biological properties. *Journal of Antibiotics*, **56**, 700–704.

Ghisalberti, E. L. (2002). Anti-infective agents produced by the Hypomycetes genera *Trichoderma* and *Gliocladium*. *Current Medicinal Chemistry-Anti-Infective Agents*, **1**, 343–374.

Glick, B. R. and Bashan, Y. (1997). Genetic manipulation of plant growth-promoting bacteria to enhance biocontrol of phytopathogens. *Biotechnology Advances*, **15**, 353–378.

Guerin, P. (1898). Sur la presence d'un champignon dansl'ivraie. *Journal de Botanique*, **12**, 230–238.

Gunatilaka, A. A. L. (2006). Natural products from plant-associated microorganisms: Distribution, structural diversity, bioactivity and implication of their occurrence. *Journal of Natural Products*, **69**, 509–526.

Guo, B., Li, H. and Zhang, L. (1998). Isolation of the fungus producing vinblastine. *Journal of Yunnan University (Natural Science Edition)*, **20**, 214–215.

Gupta, S. (2015). *Characterization of Foliar Endophytic Fungi from* Centella asiatica *(L.) Urban as a Promising Source of Bioactive Metabolites*. PhD Thesis, G.B. Pant University of Agriculture & Technology, Pantnagar, India.

Gupta, S., Bhatt, P. and Chaturvedi P. (2018). Determination and quantification of asiaticoside in endophytic fungus from *Centella asiatica* (L.) Urban. *World Journal of Microbiology and Biotechnology*, **34**, 111.

Harper, J. K., Ford, E. J., Strobel, G. A. *et al.* (2003). Pestacin: a 1,3-dihydroisobenzofuran from

Pestalotiopsis microspora possessing antioxidant and antimycotic activities. *Tetrahedron*, **59**, 2471–2476.

Hill, N. S., Belesky, D. P. and Stringer, W. C. (1991). Competitiveness of tall fescue as influenced by *Acremonium coenophialum*. *Crop Science*, **30**, 156–161.

Hill, N. S., Pachon, J. G. and Bacon, C. W. (1996). *Acremonium coenophialum* mediated short- and long-term drought acclimation in tall fescue. *Crop Science*, **36**, 665–672.

Hussain, M. S., Rahman, M. A., Fareed, S. *et al.* (2012). Current approaches toward production of secondary plant metabolites. *Journal of Pharmacy and Bioallied Sciences*, **4**, 10–20.

Igarashi, Y., Iwashita, T., Fujita, T. *et al.* (2003). Clethramycin, a new inhibitor of pollen tube growth with antifungal activity from *Streptomyces hygroscopicus* TP-A0623. II. Physicochemical properties and structure determination. *Journal of Antibiotics*, **56**, 705–706.

Jiang, D. F., Ma, P., Yang, J. *et al.* (2003). Formation of blood resin in abiotic *Dracaena cochinchinensis* inoculated with *Fusarium* 9568D. *Ying Yong Sheng Tai XueBao (in Chinese)*, **14**, 477–478.

Junker, C., Draeger, S. and Schulz, B. (2012). A fine line endophytes or pathogens in *Arabidopsis thaliana*. *Fungal Ecology*, **5**, 657–662.

Karuppusamy, S. (2009). A review on trends in production of secondary metabolites from higher plants by *in vitro* tissue, organ and cell cultures. *Journal of Medicinal Plant Research*, **3**, 1222–1239.

Keith, C. (1998). Fungal endophytes of grasses: a defensive mutualism between plants and fungi. *Ecology*, **69**, 10–16.

Krings, M., Taylor, T. N., Hass, H. *et al.* (2007). Fungal endophytes in a 400-million-yr-old land plant: infection pathways, spatial distribution, and host responses. *New Phytologist*, **174**, 648–657.

Kumara, M. P., Zuehlke, S., Priti, V. *et al.* (2012). *Fusarium proliferatum*, an endophytic fungus from *Dysoxylum binectariferum* Hook, produces rohitukine, a chromane alkaloid possessing anti-cancer activity. *Antonie van Leeuwenhoek*, **101**, 323–329.

Kusari, S., Verma, V. C., Lamshoeft, M. and Spiteller, M. (2012). An endophytic fungus from *Azadirachta indica* A. Juss. that produces azadirachtin. *World Journal of Microbiology and Biotechnology*, **28**, 1287–1294.

Li, P. Q., Lou, J. F., Mou, Y. *et al.* (2012). Effects of oligosaccharide elicitors from endophytic *Fusarium oxysporum* Dzf17 on diosgenin accumulation in *Dioscorea zingiberensis* seedling cultures. *Journal of Medicinal Plants Research*, **6**, 5128–5134.

Li, Y. C., Tao, W. Y. and Cheng, L. (2009). Paclitaxel production using co-culture of *Taxus* suspension cells and paclitaxel-producing endophytic fungi in a co-bioreactor. *Applied Microbiology and Biotechnology*, **83**, 233–239.

Li, Y. L. and Tao, W. Y. (2009). Interactions of Taxol-producing endophytic fungus with its host (*Taxus* spp.) during Taxol accumulation. *Cell Biology International*, **33**, 106–112.

Malinowski, D. P. and Belesky, D. P. (2000). Adaptations of endophyte-infected cool-season grasses to environmental stresses: mechanisms of drought and mineral stress tolerance. *Crop Science*, **40**, 923–940.

Malinowsky, D. P., Brauer, D. K. and Belesky, D. P. (1999). *Neotyphodium coenophyialum* endophyte affects root morphology of tall fescue grown under phosphorous deficiency. *Journal of Agronomy and Crop Science*, **183**, 53–60.

Montecchio, G. P., Samaden, A., Carbone, S. *et al.* (1991). *Centella asiatica* triterpene fraction (CATTF) reduces the number of circulating endothelial cells in subjects with post phlebitic syndrome. *Haematologica*, **76**, 256–259.

Mucciarelli, M., Scannerini, S., Bertea, C. and Maffei, M. (2003). *In vitro* and *in vivo* peppermint (*Mentha piperita*) growth promotion by nonmycorrhizal fungal colonization. *New Phytologist*, **158**, 579–591.

Murashige, T. and Skoog, F. A. (1962). A revised medium for the rapid growth and bioassays with tobacco tissue cultures. *Physiologia Plantarum*, **15**, 473–497.

Murphy, B. R., Martin Nieto, L., Doohan, F. M. and Hodkinson, T. R. (2015). Profundae diversitas: the uncharted genetic diversity in a newly studied group of fungal root endophytes. *Mycology*, **6**, 139–150.

Nalini, K., Aroor, A. R., Karantu, K. S. and Rao, A. (1992). Effect of *Centella asiatica* fresh leaf aqueous extract on learning and memory and biogenic amine turnover in albino rats. *Fitoterapia*, **63**, 232–237.

Naoumkina, M. A., He, X. and Dixon, R. A. (2008). Elicitor-induced transcription factors for metabolic reprogramming of secondary metabolism in *Medicago truncatula*. *BMC Plant Biology*, **8**, 132.

Nassar, H. A., El-Tarabily, K. A. and Sivasithamparam, K. (2005). Promotion of plant growth by an auxin-producing isolate of the yeast *Williopsis saturnus* endophytic in maize (*Zea mays*) roots. *Journal of Biology and Fertility of Soils*, **42**, 97–108.

Nery-Silva, F. A., Fernandes, J. J., Juliatti, F. C. and Melo, B. (2007). Reação de germoplasma de mandioca a *Xanthomonas axonopodis* pv. *manihots*. *Semina: Ciênc. Agra*, **28**, 3–10.

Niere, B. (2001). *Significance of non-pathogenic isolates of Fusarium oxysporum Schlecht, Fries for the biological control of the burrowing nematode Radopholus similis (Cobb) Thorne on tissue-cultured bananas*. PhD Thesis, University of Bonn, Germany.

Ola A. R. B., Thomy, D., Lai, D., Brötz-Oesterhelt, H. and Proksch P. (2013). Inducing secondary metabolite production by the endophytic fungus *Fusarium tricinctum* through coculture with *Bacillus subtilis*. *Journal of Natural Products*, **76**, 2094–2099.

Peng, X., Zhao, Y., Liang, X. *et al.* (2006). Assessing the quality of RCTs on the effect of beta-elemene, one ingredient of a Chinese herb, against malignant tumors. *Contemporary Clinical Trials*, **27**, 70–82.

Plohman, B., Bader, G., Streich, S., Hiller, K. and Franz, G. (1994). Immuno-modulatory effects of triterpenoid saponins. *European Journal of Pharmaceutical Sciences*, **21**, 120.

Pocasangre, L. (2000). *Biological enhancement of tissue culture plantlets with endophytic fungi for the control of the burrowing nematode* Radopholus similis *and the Panama disease (*Fusarium oxysporum *f. sp.* cubense*)*. PhD Thesis, University of Bonn, Germany.

Porter, J. K. (1995). Analysis of endophyte toxins: fescue and other grasses toxic to livestock. *Journal Animal Science*, **73**, 871–880.

Prasad, R., Garg, A. P. and Varma, A. (2004). Interaction of medicinal plant growth-promoting rhizobacteria and symbiotic fungi. In *Basic Research and Applications: Mycorrhizae*, Microbiology Series, ed. G. Podila and A. Varma. Delhi: IK International, pp. 363–407.

Puri, S. C., Verma, V., Amna, T., Qazi, G. N. and Spiteller, M. (2005). An endophytic fungus from *Nothapodytes foetida* that produces camptothecin. *Journal of Natural Products*, **68**, 1717–1719.

Rai, M., Acharya, D. and Singh, A. (2001). Positive growth responses of the medicinal plants *Spilanthes calva* and *Withania somnifera* to inoculation by *Piriformospora indica* in a field trial. *Mycorrhiza*, **11**, 123–128.

Rajeswari, G., Murugan, M. and Mohan, V. R. (2013). GC-MS analysis of bioactive components of *Hugonia mystax* L. bark (Linaceae). *Journal of Pharmaceutical and Biomedical Sciences*, **29**, 818–824.

Raman, B. V., Samuel L. A., Pardhasaradhi, M. *et al.* (2012). Antibacterial, antioxidant activity and GC-MS analysis of *Eupatorium odoratum. Asian Journal of Pharmaceutical and Clinical Research*, **5**, 99–106.

Rommert, A. K., Oros-Sichler, M., Aust, H. J., Lange, T. and Schulz, B. (2002). Growth promoting effects of endophytic colonization of the roots of larch (*Larix decidua*) with *Cryptosporiopsis* sp. and *Phialophora* sp. *7th International Mycological Congress*, Oslo, Norway.

Sachin, N., Manjuntha, B. L., Mohana, K. P. *et al.* (2013). Do endophytic fungi possess pathway genes for plant secondary metabolites? *Current Science*, **104**, 178–182.

Saxena, P. K., Cole, I. B. and Murch, S. J. (2005). Approaches to quality plant based medicine: significance of chemical profiling. In *Applications of Plant Metabolic Engineering*, ed. R. Verpoorte, A. W. Alfermann and T. S. Johnson. Amsterdam: Springer, pp. 311–330.

Sherameti, I., Shahollari, B., Venus, Y. *et al.* (2005). The endophytic fungus *Piriformospora indica* stimulates the expression of nitrate reductase and the starch-degrading enzyme glucan-water dikinase in tobacco and *Arabidopsis* roots through a homeodomain transcription factor that binds to a conserved motif in their promoters. *Journal of Biological Chemistry*, **280**, 26241–26247.

Shukla, A., Rasik, A. K. and Dhawan, B. N. (1999). Asiaticoside-induced elevation of antioxidant levels in healing wounds. *Phytotherapy Research*, **13**, 50–54.

Staniek, A., Woerdenbag, H. J. and Kayser, O. (2008). Endophytes: exploiting biodiversity for the improvement of natural product-based drug discovery. *Journal of Plant Interaction*, **3**, 75–93.

Stierle, A., Strobel, G. A. and Stierle, D. (1993). Taxol and taxane production by *Taxomyces andreanae*, an endophytic fungus of Pacific yew. *Science*, **260**, 214–216.

Strobel, G. A. and Daisy, B. (2003). Bioprospecting for microbial endophytes and their natural products. *Microbiology and Molecular Biology Reviews*, **67**, 491–502.

Strobel, G. A., Yang, X., Sears, J. *et al.* (1996). Taxol from *Pestalotiopsis microspora*, an endophytic fungus of *Taxus wallichiana*. *Microbiology*, **142**, 435–440.

Strobel, G. A., Miller, R. V., Miller, C. *et al.* (1999). Cryptocandin, a potent antimycotic from the endophytic fungus *Cryptosporiopsis cf. quercina*. *Microbiology*, **145**, 1919–1926.

Strobel, G. A., Ford, A. E., Worapong, J. *et al.* (2002). Ispoestacin, an isobenzofuranone from *Pestalotiopsis microspora*, possessing antifungal and antioxidant activities. *Phytochemistry*, **60**, 179–183.

Tang, Z. H., Rao, L. Q., Peng, G. P. *et al.* (2009). Effects of endophytic fungus and its elicitors on cell status and alkaloid synthesis in cell suspension cultures of *Catharanthus roseus*. *Journal of Medicinal Plants Research*, **5**, 2192–2200.

Wang, J., Li, G., Lu, H. *et al.* (2000). Taxol from *Tubercularia* sp. strain TF5, an endophytic fungus of *Taxus mairei*. *FEMS Microbiology Letters*, **193**, 249–253.

Wang, J., Zheng, L., Tan, R. and Chin. J. (2006). The preparation of an elicitor from a fungal endophyte to enhance artemisinin production in hairy root cultures of *Artemisia annua* L. *Chinese Journal of Biotechnology*, **22**, 829–834.

Wang, J. W., Wu, J. H., Huang, W. Y. and Tan, R. X. (2006). Laccase production by *Monotospora* sp. an endophytic fungus in *Cynodon dactylon*. *Bioresource Technology*, **97**, 786–789.

Wang, Y., Dai, C. C., Cao, J. L. and Xu, D. S. (2012). Comparison of the effects of fungal endophyte *Gilmaniella* sp. and its elicitor on *Atractylodes lancea* plantlets. *World Journal of Microbiology and Biotechnology*, **28**, 575–584.

Yao, Y. Q., Ding, X., Jia, Y. C. *et al.* (2008). Anti-tumor effect of beta-elemene in glioblastoma cells depends on p38 MAPK activation. *Cancer Letters*, **264**, 127–134.

Yuan, Z. L., Dai, C. and Chen, L. (2007). Regulation and accumulation of secondary metabolites in plant-fungus symbiotic system. *African Journal of Biotechnology*, **6**, 1266–1271.

Yuan, Z. L., Zhang, C. L. and Lin, F. C. (2010). Role of diverse non-systemic fungal endophytes in plant performance and response to stress: progress and approaches. *Journal of Plant Growth Regulation*, **29**, 116–126.

Yue, C. C., Miller, J., White, J. and Richardson, M. (2000). Isolation and characterization of fungal inhibitors from *Epichloë festucae*. *Journal of Agricultural and Food Chemistry*, **48**, 4687–4692.

Zhang, L., Guo, B., Li, H. *et al.* (2000). Preliminary study on the isolation of endophytic fungus of *Catharanthus roseus* and its fermentation to produce products of therapeutic value. *Chinese Traditional and Herbal Drugs*, **31**, 805–807.

Zhang, Z. and Zhou, X. (2011). GC/MS analysis on benzene/alcohol extractives of *Manglietia glauca* leaves for biomedicine engineering. *Advanced Materials Research*, **213**, 475–47.

15

In Vitro Methods for Plant–Microbe Interaction and Biocontrol Studies in European Ash (*Fraxinus excelsior* L.)

Anindita Lahiri, Gerry C. Douglas, Brian R. Murphy and Trevor R. Hodkinson

Abstract

In vitro tissue culture systems are required for plant–microbe interaction studies on European ash, *Fraxinus excelsior*. Methods are needed for plant micropropagation and for physiological experimentation including pathogen/resistance testing and biocontrol studies. For example, systems are required for experiments on ash dieback disease, caused by the ascomycete fungus *Hymenoscyphus fraxineus*, that is killing ash plantations and natural populations across its native range. Methods are also needed to optimise the number of endophytes cultured from ash tissue and to taxonomically identify them. We present endophyte isolation protocols and media for ash, provide an optimised DNA barcoding procedure for endophyte identification and describe *in vitro* tissue culture methods suitable for ash–microbe interaction studies in both roots and shoots. Methods for both embryo culture and seed culture (with precutting) and for the bulking up of genotypes via single node culture are outlined. We also discuss the potential of tissue culture for establishing microbe/endophyte-free cultures.

15.1 Introduction

There has been growing interest in understanding the microbiome, and more specifically the endophytes, of trees for various applications including forestry, horticulture, plant protection and phytoremediation. The first published use of the term 'endophyte', but actually described as 'Entophytae', was in 1809 by Heinrich Friedrich Link, primarily for a group of fungi that are partly parasitic in nature (Link, 1809). The term endophyte was then also applied to bacteria and other organisms (Chanway, 1996; Hallmann, 1997). Petrini (1991) defined endophytes as 'organisms inhabiting plant organs that at some time in their life cycle can colonise internal plant tissues without causing apparent harm to their host'. We follow that definition in this chapter. According to sequence data obtained from National Center for Biotechnology Information (NCBI), fungal endophytes of plants mostly belong to the five primary classes (Hardoim *et al.*, 2015): Glomeromycota (40%), Ascomycota (31%), Basidiomycota (20%), Zygomycota (0.1%) and unidentified phyla (8%). Unidentified taxa demonstrate the huge diversity of fungal endophytes remaining to be discovered and further studied (Murphy *et al.*, 2015; Hodkinson, 2018).

Little is known about the endophytic community of European ash, *Fraxinus excelsior* L., but research has recently focused on this species because of the threat of ash dieback disease that is destroying ash populations and plantations across Europe (Kowalski, 2006; Kowalski and Holdenreider, 2009a, b). This chapter describes what is known about the microbiome and endophytic community of *F. excelsior* and outlines *in vitro* methods for the experimental study of endophyte, pathogen and ash tree interactions. Such methods are needed to reduce the risk of spreading the disease in open field sites of greenhouse environments. *In vitro* culture also allows precision so that experiments can include only the target organisms, in this case the dieback disease pathogen (*Hymenoscyphus fraxineus*), the endophyte species (or combinations of species) and the ash tree.

15.2 Ash Dieback Disease

Fraxinus excelsior is an important forest tree for timber, furniture, veneer, flooring, traditional medicine and sports equipment, including the hurley stick used in the Irish sport hurling (Thomasset *et al.*, 2011, 2014). Ash dieback disease is caused by a fungal pathogen *Hymenoscyphus fraxineus* (T. Kowalski) Baral, Queloz, Hosoya (Baral *et al.*, 2014). It was originally, and commonly, called *Chalara* but its generic name was changed to *Hymenoscyphus* after the full characterisation of its sexual and asexual stages, teleomorph and anamorph, respectively (Baral *et al.*, 2014). Ash dieback disease can affect all age groups of *F. excelsior* (Kowalski *et al.*, 2017). It often infects plants from fungal spores and leaves can be easily infected. Fungal spores are present in the previous year's ash litter and are able to disperse over tens of kilometres (Forestry Commission UK, 2015). Movement of diseased ash logs, following felling can also facilitate spread over large geographic areas (Forestry Commission UK, 2012).

The symptoms of the disease can be seen throughout the tree. Affected ash often first show leaf wilt in which the pinnae (leaflets) become dry and droopy because of loss of their turgor pressure (Kräutler *et al.*, 2012). The leaflet and petiole changes colour from green to light brown and then becomes black. The main veins are more resistant than minor ones and the fungus develops within them and can produce sclerotia on the rachis that can remain dormant (Douglas *et al.*, 2013). The disease causes a change in colour of the stem from green to brown in young material and then to black. It also affects the primary branches, and can lead to the formation of epicormic shoots, which can become numerous (Sansford, 2013; Gross *et al.*, 2014). In mature woody parts of the tree, the mycelium can grow inside the stem and trunk and, in turn, it blocks the xylem and damages the water transport system (Gross *et al.*, 2014). It can be observed when a young stem is cut longitudinally to reveal brown or black strips of necrotic damaged xylem tissue. The necrotic lesion on the stem is known as canker and can often look diamond or lens shaped. Once the disease develops, crown die back occurs which mostly affects the upper crown. Canker of ash can also be caused by other biological organisms such as bacteria *Pseudomonas syringae* pv. *savastanoi* (Smith, 1908; Young *et al.*, 1978) and other fungi such as *Nectria galligena* Bres. (Janse, 1981; Douglas *et al.*, 2013). The fungus can also be isolated from the roots of diseased ash where it weakens the root and can lead to the introduction of other fungal pathogens (Bakys, 2013). One of these is the honey fungus *Armillaria borealis* Marxm. & Korhonen, 1982, that is particularly damaging (Bakys, 2013).

15.3 Geography of Ash Dieback Disease

Ash die back disease was first observed in Poland in 1996 (Przybył, 2002; Kowalski and Łukomska, 2005). Major outbreaks have occurred in several European countries and the disease has spread rapidly. Altogether, at least 23 countries have so far reported the occurrence of ash dieback disease (Timmermann *et al.*, 2011). One of the earliest European regions to report significant levels of ash dieback was Scandinavia, where it was first observed in 2001. By 2004–06 the disease had spread throughout the Scandinavian ash distribution range leading to destruction of its populations (Barklund, 2005; Stenlid and Barklund, pers. comm). It was first recorded in Denmark in 2003 and in Norway ash dieback was first documented in the south-eastern parts in 2008 (Thomsen, 2005). According to the report of Forest Commission Survey of UK (Brome and Mitchell, 2017), the total number of cases of ash dieback has significantly increased between 2012 to 2017 in Scotland, England, Wales and Northern Ireland. The first reported case in the Republic of Ireland was in October 2012 in a forestry plantation site in County Leitrim and it has spread throughout the country (DAFM, n.d.). Thus, the disease has now spread throughout most of the natural distribution range of *F. excelsior*.

15.3.1 The Ash Microbiome

A few studies have been published on the ash microbiome in recent years. Those studies reveal our knowledge about endophytes in healthy trees and help us to compare our newly isolated endophytes with the list of organisms already revealed. Schlegel *et al.* (2016) studied ash endophytes as potential biocontrol agents. The study was conducted in *F. excelsior* and *F. ornus* L.. First, the influence of exudates from isolated *F. excelsior* endophytes was tested *in vitro* (on Petri dishes) against the ascospore germination of *H. fraxineus*. Strong inhibitory effects on ascospore germination were recorded from the exudates of *Paraconiothyrium* sp., *Boeremia exigua* (Desm.) Aveskamp, Gruyter & Verkley, *Kretzschmaria deusta* (Hoffm.) P.M.D. Martin, *Ampelomyces quisqualis* Ces. and Elsinoaceae sp.. Weak effects were observed for exudates of *Venturia* spp. and *Nemania serpens* (Pers.) Gray. Second, the protective effects of endophytes against the ash dieback pathogen were studied in the field (infected forest) using endophyte-free plants and plants pre-inoculated with endophytes. *Venturia* spp. dominated the endophytic community in the field-grown trees inoculated with endophytes, but no significant effect of endophyte inoculation was seen on plants after they had been infected with *H. fraxineus*. Thus, no evidence for field-based endophyte biocontrol of *H. fraxineus* was found.

Another study by Kosawang *et al.* (2018) was conducted on fungal communities from resistant *Fraxinus* for use as biocontrol agents. Fungal isolations were undertaken from five *H. fraxineus*, tolerant ash species (*F. chinensis* subsp. *rhynchophylla* (Hance) A.E. Murray, *F. lanuginosa* Koidz., *F. mandshurica* Rupr., *F. ornus* and *F. pennsylvanica* Marshall) and endophytes identified using DNA sequencing (internal transcribed spacer (ITS) 1 and 2). They isolated 196 fungal taxa belonging to 15 families, 9 orders and 40 species. Most of their endophytes were ascomycetes except a single basidiomycete, namely *Peniophora* sp.. They performed antagonistic activity assays by growing the endophyte with *H. fraxineus* on half-strength PDA (with ash leaf supplement). Endophyte species showing high antagonistic activity included *Boeremia exigua* (Desm.) Aveskamp, Gruyter & Verkley, *Epicoccum nigrum* Link, *Fusarium* sp, *Sclerostagonospora* sp. and *Setomelanomma holmii* M. Morelet. In our endophyte isolations from ash (Tables 15.1 and 15.2) we also found three of these species from leaf tissues, namely *Boeremia exigua*, *Epicoccum nigrum* and *Fusarium* sp..

15.4 *In Vitro* Culture of Ash for Plant–Microbe Interaction Studies

Tissue culture methods are required to improve ash micropropagation from seed and plantlets. They provide an enclosed system for the co-culture of pathogen, endophyte and plant and hence prevent unintentional release of the microorganisms

Table 15.1 Predominant endophytes from leaf and rachis tissue.

Sample name	Place	Provenance trial	Organism	Accession number GenBank	Identification score %
H1.a.H.M	Denmark	Bregentved	*Phoma* sp.	JX160059	99
K.H.M	France	Athis	*Diaporthe viticola*	KC145904	99
H1.aD.M	Denmark	Bregentved	*Phoma exigua* var. *exigua*	EU343168	96
T1.a.H.L	Czech Republic	Rabstejn	*Boeremia exigua*	KT004579	100
H1.b.H.R	Denmark	Bregentved	*Mycosphaerella coacervata*	EU167596	99
H1.b.D.M	Denmark	Bregentved	*Boeremia exigua*	KX618484	99
H1.a.H.L	Denmark	Bregentved	*Mycosphaerella coacervata*	EU167596	99
S1.a.H.R	Lithuania	Zeimelis	*Epicoccum nigrum*	MF509753	99
L1.H.M	Italy	Cadore	*Aureobasidium pullulans*	HG532077	98
K.H.R	France	Athis	*Diaporthe viticola*	KC145904	95

into the environment. The *in vitro* systems allow precise control of the target endophytes and pathogen and exclusion of other microbes. Hence, they represent ideal phenotyping systems in ash. Tissue culture methods are also needed for ash because germplasm needs conserving *ex situ*, as field collections are threatened by dieback disease. We therefore describe the methodology we have successfully used for the collection, isolation and culture of ash endophytes and describe *in vitro* systems for plant-microbe interaction studies in ash.

15.4.1 Plant Collection and Endophyte Culture

Endophytes can be isolated from different tissues of host plant such as leaf, leaf rachis and roots. Some endophytes have also been isolated from seed (Marshall *et al.*, 1999; O'Hanlon *et al.*, 2012). Isolation should be made within 24–48 hours of collection of the plant material before the decay of host tissue. For ash, we collected plant material and surface sterilised it with 5% sodium hypochlorite and

Table 15.2 List of predominant endophytes from root tissue.

Sample name	Place	Tree no.	Organism	Accession number GenBank	Identification score %
SITI-14-1b	*KRC	1	*Fusarium oxysporum*	MH055398	99
SITI-15-1b	KRC	1	*Cordyceps crassispora*	AB067714	100
SITI-15	KRC	1	*Neonectria candida*	LT821512	99
S1T1-11	KRC	1	*Acremonium alternatum*	NR144913	93
S4T2-6b	KRC	2	*Lecanicillium lecani*	FJ515768	100
S4T2-9-2b	KRC	2	*Dactylonectria macrdidyma*	MH06373	99
S4T2-10-b	KRC	2	*Dactylonectria macrdidyma*	MH06373	98
S5T3-2c	KRC	3	*Dactylonectria macrdidyma*	MF567498	99
S5T3-A-6a	KRC	3	*Fusarium culmorum*	MG274304	99
S6T3-D-4	KRC	3	*Ilynectria robusta*	LT821483	92

* Kinsealy Research Centre: KRC

70% ethanol prior to culture to kill epiphytic and contaminating microorganisms. After the surface-sterilisation process, the tissue was rinsed in sterile pure water to remove all surfactants. Surface-sterilised tissues were then dissected to prepare the explant that might, for example, be a leaf section, rachis section or root section. This was done immediately prior to plating. If the plant material is sterilised after dissection, the surfactants can more readily enter the plant cells and remain in the cut/damaged plant material and hence reduce the efficiency of endophyte isolation.

Surface-sterilised host tissues from leaf, rachis and roots were then cultured onto Petri dishes containing a suitable growth medium such as malt extract agar modified with Vegitone (MEA) (Sigma 38954). We have also successfully used half-strength MEA and half-strength MEA with additional leaves from healthy ash trees (MEA+F). For this MEA+F medium, ash leaves were collected from a healthy tree and cleaned with tap water three to four times, followed by cleaning with sterile pure water two times and then digested and autoclaved with the half MEA media (by including them in the sample bottle during autoclaving). We assume that MEA+F media will help to promote the endophyte growth differently than the MEA because it will contain extra metabolites from the ash leaf. In addition to MEA, we have found other commercial media to be good for fungal culture including corn meal agar, potato dextrose agar, Sabouraud maltose agar and Czapek Dox liquid medium (Murphy *et al.*, 2015). The separate use of a range of different media maximises the species richness of endophytes recovered because some endophytes do not grow on some media (Murphy *et al.*, 2015).

After plating out, the Petri dishes were sealed with Parafilm or masking tape and kept inside the incubator in dark condition at ±18°C for at least 4–21 days. We found that most endophytes start growing out within 5–7 days of culture. After appearance of mycelial growth, small subcultures were transferred onto a fresh medium in a Petri dish containing half-strength MEA media to maintain single isolates. Where more than one endophyte emerged from the same host tissue, each had to be separately subcultured. Pure cultures can be obtained from single spore sub-culture and can then be preserved in long-term storage. Two principal methods are used to preserve fungal isolates; preparing spore suspensions in ultra pure water at 0–4°C and cryogenic storage at −80°C. Cryogenic preservation is preferred for longer-term storage, and is accomplished by freezing the fungal cultures in a solution of glycerol (typically 10 or 20% glycerol). Multiple storage tubes for each pure fungal isolate are kept because each cryotube cannot be refrozen after thawing without significant damage to the sample.

15.4.2 Identification of Fungal Isolates

Fungi can be identified using a combination of morphological examination and DNA barcoding (Hawksworth, 2001; Nilsson *et al.*, 2008; Moore *et al.*, 2011). We extracted DNA from pure cultures of ash fungal endophytes using DNeasy Plant mini kits from Qiagen. Using sterile forceps and scalpels, approximately one-eighth of the fungal plate tissue was transferred into a 1.5 ml microcentrifuge tube with a sterile metal bead for homogenisation and tissue disruption. Disruption of the fungal tissue was done with bead homogenisation using a mixer mill (Retsch MM 300) for 1 min at 10 Hz. The mill block was then inverted and bead homogenisation run for a further 1 min. The rest of the DNA extraction followed the Qiagen manufacturer protocol. The final volume was 50 µL. It was stored at 20C until further use. We have also found that the CTAB method described in Hodkinson *et al.* (2007) is an efficient general method for fungal endophyte DNA extraction (Beekwilder *et al.*, 2019, Chapter 9).

DNA extracts were then used for PCR amplification of the nuclear ribosomal ITS using the ITS 5 and ITS 4 primers of White *et al.* (1990). The ITS region includes a small part of the 18S, all of the 5.8S and some of the 26S nrDNA. ITS 1F (Nilsson *et al.*, 2008; Begerow *et al.*, 2010) and ITS 4 also work well. After successful amplification, PCR products were cleaned using Exosap and processed for Sanger DNA sequencing (commercially by Macrogen or Source Biosciences; or using an in-house Applied Biosystems 3100xl Genetic Analyser). Sequence traces were edited in Geneious (v.6; Biomatters), or MEGA5 (Tamura *et al.*, 2011) and then used in a BLAST search to find the closest match (per cent similarity) in GenBank (NCBI).

We isolated 310 leaf endophytes on half-strength MEA and half-strength MEA+F media from leaf and leaf rachis tissues of healthy *F. excelsior* and its close relatives. Furthermore, we have isolated over 100 root endophytes on half-strength MEA media from three trees of *F. excelsior*. *Boeremia exigua* var. *pseudolilacis, Phoma sp., Mycosphaerella coecervata, Aureobasidium pullulans and Diaporthe viticola* were predominantly isolated from leaf tissues and *Acremonium alternatum, Dactylonectria sp., Fusarium oxysporium, Fusarium culmorum, Neonectria sp., Cordyceps crassispora* and *Lecanicillium lecani* from root tissues. Predominant fungi are listed in Tables 15.1 and 15.2.

The choice of potential endophyte for biocontrol experimentation, from the large list of potential isolates, can be based on a number of criteria. Information can be obtained about the isolate's antagonistic activity against *H. fraxineus in vitro* (as described in Kosawang *et al.*, 2018). Cultures with high spore formation may also be preferred (Murphy *et al.*, 2015). In addition, some basic information about the identified endophytes were also obtained from the online database (Mycobank, UNITE) and used as selection criteria. For example, known pathogens were avoided and known endophytes selected.

15.5 Methods for *In Vitro* Plant–Microbe Interaction Studies

In vitro co-culturing of fungal microbes with host plants for plant/endophyte and plant/endophyte/pathogen testing requires a source of plant material that can come from seed or from the micropropagation of existing stocks. For this process, we grew *in vitro* sterile (endophyte-free) plantlets from seeds and embryos of *F. excelsior*, which were then used for direct experimentation or propagated further by micropropagation. Although, several studies have found that endophytes can be vertically transmitted via seed, we found that no endophytes could be retrieved in endophyte isolation experiments from a total of 100 surface-sterilised ash seeds divided equally between half-strength PDA and half-strength MEA media and incubated at 18°C (Lahiri, pers. obs.).

15.5.1 Tissue Culture of Ash

Healthy undamaged fruits (samara) without holes in them that might indicate insect contamination by weevils such as *Lignyodes bisclwffi* (Blatchley), *L. helvolus*

(LeConte) and *L. horridulus* (Casey) were selected (Thomasset *et al.*, 2014). Fruits were depericarped and seeds surface sterilised with 7% calcium hypochlorite and rinsed with ultra pure sterile water. Seeds were then directly transferred to Petri dishes containing half-strength Murashige and Skoog (MS) media (Sigma M5524) with 0.6% agar and pH 5.7 (Mettler Toledo FE 20) and then kept in dark at ±18°C.

Old seeds, that are not from the same year, need to be kept in dry cold storage at 4–6°C. Those seeds need stratification with the same surface-sterilisation process. For this purpose, depericarped seeds were soaked in 0.3 M NaOH for 20 mins. The liquid was discarded and the seeds soaked in 0.2% calcium hypochlorite overnight at 4°C (to rehydrate the seeds which had become dry during storage). After discarding the liquid, the seeds were transferred to 2% calcium hypochlorite for 2 hours for sterilisation, and then thoroughly washed, five to six times, with sterile distilled water. They were then transferred to half-strength MS media with 0.6% agar without any additional sugars and kept in the dark for two weeks at ±18°C.

In vitro plantlets were also grown from germinated embryos cultured from surface-sterilised and stratified seeds. Dissection of embryos from the seed coats was carried out under a dissecting microscope (10× magnification; Jenco USA; ZM-500) with sterile forceps and scalpels. Sterile seeds were placed on a clean Petri dish under the dissecting microscope and cut along their middle. The half seeds containing the embryos were then carefully removed without damaging any tissues and placed on a Petri dish containing half-strength MS media, sealed with Parafilm on the edges and kept in ±18°C.

Germinated embryos and seedlings were transferred into individual test tubes of 25 × 150mm (Fisherbrand 14-961-34) with polypropylene closure of 25mm (Kimble 73660–25) containing full strength MS media with 1% of sucrose (BDH AnalaR: lot 236 K 18057286), pH 5.7, kept in light at ± 18°C. Plantlets began to root after a few weeks of transferring to test tubes without any hormone supplement.

Our studies (unpublished) have found embryo culture is the most effective for seedling initiation (83% germination) but seed cutting (62% germination) is the most successful for large-scale *in vitro* propagation because the plantlets generated by seed cutting are most robust and hence show the highest post establishment survival (survival recorded from embryo cultured plantlets was 40% comparted to 95% for seed established plantlets).

In vitro plantlets can be used directly or propagated further using single node culture. For micropropagation, single node explants were dissected from the plantlets and rooted on full strength MS media with 1% sucrose. They were then grown under 1600 μmol m^{-2} s^{-1} PAR to fill the culture tube once more before being propagated once more. In this way large stocks of identical genotype (ramets) were produced. Using such micropropagated stocks in plant–microbe interaction studies reduces variability caused by genotypic variation in a seed batch.

15.5.2 *In Vitro* Biocontrol and Plant–Microbe Interaction Testing

Some endophytes reside in the shoot and some in the roots of ash. Therefore, culture methods are required for both these types. We have found that agar culture system work well for shoot endophytes and perlite works well for root endophytes as outlined below.

To test the effects of the application of endophytes to ash plant tissue, we used *in vitro* plantlets growing on MS media with 1% sucrose and 0.6% agar (Sigma 9002-18-0). An endophyte inoculant solution was first obtained by diluting (9:1) with ultrapure water and 1 ml of the spore suspension was streaked out onto the agar surface to obtain single spore colonies, which were then subcultured individually. The surface of the sporulating subcultures were each washed with 5 ml of pure water and diluted to a spore concentration of 10^6 spores per millilitre. We used one of two methods to introduce the endophytes into the nodes of the in vitro plantlets: by either making a nick with scalpel at the node and then dusting spores from the pure culture plates onto leaf surface or by spraying the spore suspension onto the leaf. For non-sporulating cultures, a 'fluffed out' agar core (5 mm diameter) was added to the base of the plantlet upon establishment.

Root endophytes were tested on seedlings growing on MS plus 1% sucrose and perlite (Vitax: Perlite) without any agar. The perlite allows aeration required by the root fungi for growth through the medium. Plantlets were inoculated with either a spore solution (as outline above) or by fluffed out agar plug (as outline above). In this case the potential candidate root endophyte was introduced near the root tips (meristems) of *in vitro* plantlets.

An alternative approach for *in vitro* endophyte inoculation is to establish the experiment from seed and to coat the seeds with endophyte spore solution prior to germination. In this way the endophyte establishes with the seed as it germinates.

Biocontrol Experimentation

For plant/endophyte/pathogen testing, the culture system outlined above can be simply extended for a second microbe. This can be undertaken for both root and shoot endophytes and a pathogen. We have optimised the biocontrol experiments for *H. fraxineus* but the method is applicable to a wide range of pathogens so is not restricted to ash dieback disease studies.

We separately inoculated 10 replicates of *in vitro* grown ash plantlets with *H. fraxineus* for each of the tested root and shoot endophytes. Growth and disease development of the ash plantlets were recorded at several time intervals. Comparisons were made among plantlets with endophytes and plantlets without endophytes (for controls) in a fully randomised experimental design under 1600μmol.m^{-2} s^{-1} PAR light. Results not reported here.

15.6 Conclusion

Ash dieback is a potential threat to the ash population throughout Europe and is spreading rapidly by its spores that are able to travel over long distances. As common ash has high economic value and covers huge land areas in plantations and woodlands, it is vital to find ways or improving its disease resistance. The *in vitro* culture methods we have developed will aid in the study of disease resistance in ash. The *in vitro* antagonistic and *in vitro* biocontrol assays including endophyte and pathogen are performed in an enclosed and otherwise sterile environment so that the pathogen will not escape into the environment and so that the direct interaction of the microbes can be assessed without the influence of other organisms.

Acknowledgements

We thank Trinity College Dublin (TCD) for providing financial support through a PhD studentship and laboratory technicians Patricia Coughlan and Jacqueline Stone (TCD) for technical support.

References

Bakys, R. (2013). Dieback of *Fraxinus* in Baltic sea region. *Swedish University of Agricultural Sciences, Uppsala*, pp. 1–48.

Baral, H.-O., Queloz, V. and Hosoya, T. (2014). *Hymenoscyphus fraxineus*, the correct scientific name for the fungus causing ash dieback in Europe. *IMA Fungus*, **5**, 79–80.

Barklund, P. (2005). Ash dieback sweeping over south and central Sweden. *SkogsEko*, **3**,11–13. (In Swedish.)

Beekwilder, J., Murphy, B. R., Mathuna, E. M., Barry, A. and Hodkinson, T. R. (2019). Isolation, diversity and potential use of endophytes in the biomass and bioenergy crop *Miscanthus*. In *Endophytes for a Growing World*, ed. T. R. Hodkinson, F. M. Doohan, M. J. Saunders and B. R. Murphy. Cambridge: Cambridge University Press, Chapter 9.

Begerow, D., Nilsson, H., Unterseher, M. and Maier, W. (2010). Current state and perspectives of fungal DNA barcoding and rapid identification procedures. *Applied Microbiology and Biotechnology*, **87**, 99–108.

Brome A. and Mitchell, R. J. (2017). *Ecological Impacts of Ash Dieback and Mitigation Methods.* Research Note. London: Forestry Commission UK, pp. 1–16.

Chanway, C. P. (1996). Endophytes: they're not just fungi! *Canadian Journal Botany*, **74**, 321–322.

DAFM (n.d.). *Ash Dieback (Chalara).* www.agriculture.gov.ie/forestservice/treediseases/ashdiebackchalara/

Douglas, G. C., Pliura, A., Dufour, J. *et al.* (2013). Common ash (*Fraxinus excelsior* L.). In *Forest Tree Breeding in Europe: Current State of the Art and*

Perspectives, Vol. 2, ed. L. E. Pâques. *Springer, Managing Forest Ecosystems*, **25**, 403–462.

Forestry Commission UK (2015). Chalara dieback: managing ash trees and woodlands. www.forestry.gov.uk/forestry/infd-92pjkx

Gross, A., Holdenrieder, O., Pautasso, M., Queloz, V. and Sieber, T. N. (2014). *Hymenoscyphus pseudoalbidus*, the causal agent of European ash dieback. *Molecular Plant Pathology*, **15**, 5–21.

Hawksworth, D. L. (2001). The magnitude of fungal diversity: the 1.5 million species estimated revisited. *Mycological Research*, **105**, 1422–1432.

Hallmann, J., Quadt-Hallmann, A., Mahaffee, W. F. and Kloepper, J. W. (1997). Bacterial endophytes in agricultural crops. *Canadian Journal of Microbiology*, **43**, 895–914.

Hardoim, P. R., Overbeek, L. S., Berg, G. *et al.* (2015). The hidden world within plants: ecological and evolutionary considerations for defining functioning of microbial endophytes. *Microbiology and Molecular Biology Reviews*, **79**, 293–320.

Hodkinson, T. R. (2018). Evolution and taxonomy of the grasses (Poaceae): a model family for the study of species-rich groups. *Annual Plant Reviews Online*, doi: 10.1002/9781119312994.apr0622.

Hodkinson, T. R., Waldren, S., Parnell, J. A. N. *et al.* (2007). DNA banking for plant breeding, biotechnology and biodiversity evaluation. *Journal of Plant Research*, **120**, 17–29.

Janse, J. D. (1981). The bacterial disease of ash (*Fraxinus excelsior*), caused by *Pseudomonas syringae subsp. savastanoi pv. fraxini. European Journal of Forest Patholoìgy*, **11**, 306–315.

Kosawang, C., Amby, B. D., Bussaban, B. *et al.* (2018). Fungal communities associated with species of *Fraxinus* tolerant to ash dieback, and their potential for biological control. *Fungal Biology*, **122**, 110–120.

Kowalski, T. (2006). *Chalara fraxinea* sp. nov. associated with dieback of ash (*Fraxinus excelsior*) in Poland. *Forest Pathology*, **36**, 264–270.

Kowalski, T. and Holdenrieder, O. (2009a). Pathogenicity of *Chalara fraxinea*. *Forest Pathology*, **39**, 1–7.

Kowalski, T. and Holdenrieder, O. (2009b). The teleomorph of *Chalara fraxinea*, the causal agent of ash dieback. *Forest Pathology*, **39**, 304–308.

Kowalski, T. and Łukomska, A. (2005). Studies on *Fraxinus excelsior* L. dieback in Włoszczowa Forest Unit stands. *Acta Agrobotanica*, **59**, 429–440.

Kowalski, T., Bilański, P. and Kraj, W. (2017). Pathogenicity of fungi associated with ash dieback towards *Fraxinus excelsior. Plant Pathology*, **66**, 1228–1238.

Kräutler, K. and Kirisits, T. (2012). The ash dieback pathogen *Hymenoscyphus pseudoalbidus* is associated with leaf symptoms on ash species (*Fraxinus* spp.). *Journal of Agricultural Extension and Rural Development*, **4**, 261–265.

Link, H. F. (1809). Observationes in ordines plantarum naturales, disserta- tio prima, complectens anandrarum ordines Epiphytas, Mucedines, Gastromycos et Fungos. Berlin, Germany: Der Gesellschaft Naturforschender Freunde zu Berlin.

Marshall, D., Tunali, B. and Nelson, L. R. (1999). Occurrence of fungal endophytes in species of wild *Triticum*. *Crop Science*, **39**, 1507–1512.

Moore, D., Robson, G. D. and Trinci, A. P. J. (2011). *21st Century Guidebook to Fungi*. Cambridge: Cambridge University Press.

Murphy, B. R., Batke, S. P., Doohan, F. M. and Hodkinson, T. R. (2015). Media manipulations and the culture of beneficial fungal root endophytes. *International Journal of Biology*, **7**, 94–102.

Nilsson, R. H., Kristiansson, E., Ryberg, M., Hallenberg, N. and Larsson, K.-H. (2008). Intraspecific ITS variability in the Kingdom Fungi as expressed in the international sequence databases and its implications for molecular species identification. *Evolutionary Bioinformatics*, **4**, 193–200.

O'Hanlon, K. A., Knorr, K., Jørgensen, L. N., Nicolaisen, M. and Boelt, B. (2012). Exploring the potential of symbiotic fungal endophytes in cereal disease suppression. *Biological Control*, **63**, 69–78.

Petrini, O. (1991). Fungal endophytes of tree leaves. In *Microbial Ecology of Leaves*. ed. J. H. Andrews and S. S. Hirano. New York, NY: Springer Verlag, pp. 179–197.

Przybył, K. (2002). Fungi associated with necrotic apical parts of *Fraxinus excelsior* shoots. *Forest Pathology*, **32**, 387–394.

Sansford, C. E. (2013). Pest risk analysis for *Hymenoscyphus pseudoalbidus* (anamorph *Chalara fraxinea*) for the UK and the Republic of Ireland. *Forestry Commission*, 1-128.

Schlegel, M., Dubach, V., Buol, L. and Sieber, T. N. (2016). Effects of endophytic fungi on the ash dieback pathogen. *FEMS Microbiology Ecology*, **92**, 1–18.

Tamura K., Peterson D., Peterson N. *et al.* (2011). MEGA5: Molecular evolutionary genetic analysis using maximum likelihood, evolutionary distance, and maximum parsimony methods. *Molecular Biology and Evolution*, **28**, 2731–2739.

Thomasset, M., Fernández-Manjarrés, J. F., Douglas, G. C., Frascaria-Lacoste, N. and Hodkinson, T. R. (2011). Hybridisation, introgression and climate change: a case study for the tree genus Fraxinus (Oleaceae). In *Climate Change, Ecology and Systematics*, ed. T. R. Hodkinson, M. B. Jones, S. Waldren and J. A. N. Parnell. Cambridge: Cambridge University Press, pp. 320–344.

Thomasset, M., Hodkinson, T. R., Restoux, G. *et al.* (2014). Thank you for not flowering: conservation genetics and gene flow analysis of native and non-native populations of *Fraxinus* (Oleaceae) in Ireland. *Heredity*, **112**, 596–606.

Thomsen, I. M. (2005). Frost damage of ash. *Videnblade Skovbrug*, **8**, 2–1. (In Danish).

Timmermann, V., Borja, I., Hietala, A. M., Kirisits, T. and Solheim, H. (2011). Ash dieback: pathogen spread and diurnal patterns of ascospore dispersal, with special emphasis on Norway. *OEPP/EOPP Bulletin*, **41**, 14–20.

White, T. J., Bruns, T. D., Lee, S. and Taylor, J. (1990). Amplification and direct sequencing of fungal ribosomal RNA genes for phylogenetics. In *PCR Protocols: A Guide to Methods and Applications*, ed. M. A. Innis, D. H. Gelfand, J. J. Sninsky and T. J. White. New York: Academic Press, pp. 315–322.

Part V

Application and Commercialisation of Endophytes in Crop Production

16

The Science Required to Deliver *Epichloë* Endophytes to Commerce

LINDA J. JOHNSON AND JOHN R. CARADUS

Abstract

In some environments, the survival and production of ryegrass and fescue is heavily reliant on its mutualistic association with *Epichloë* endophytes. *Epichloë* endophytes produce a range of bioactive alkaloids, or secondary metabolites that can be effective in deterring insect pests, although some have also been shown to be toxic to grazing animals. These endophytes are being used in grassland farming systems in Australia, New Zealand, USA and some parts of South America. However, to achieve this outcome there has been considerable investment into developing a research pipeline for delivery of animal-safe endophyte strains that are still capable of deterring insect pests and providing protection against abiotic stresses. The pipeline starts with the discovery and isolation of endophytes from wild populations of ryegrass and fescue, characterisation of the known alkaloids they produce, use of genetic markers to determine the relationship between known well-characterised strains and new strains entering the collection, determination of their bioactivity against insect pests of economic significance, understanding issues of compatibility of a strain of interest with the elite germplasm into which it has been inoculated, determining ease of transmission to subsequent seed generations, and ensuring there will be no or minimal animal health and welfare issues associated with using the strain in grazing systems.

16.1 Introduction

Temperate grasses such as ryegrass and fescue have co-evolved over 40 million years with an obligate fungal endophyte *Epichloë* (Schardl *et al.*, 2004; Schardl *et al.*,

2008; Saikkonen *et al.*, 2016), which through the production of a range of secondary metabolites, referred to as alkaloids (Gallagher *et al.*, 1981; Lyons *et al.*, 1986; Rowan *et al.*, 1986; Reidell *et al.*, 1991), can provide deterrence to invertebrate pests and grazing animals (Clay, 1990; Clay *et al.*, 1993). This mutualistic association between grasses and *Epichloë* endophyte has proven to be crucial for the survival of ryegrass and fescue in many regions of the world where these grasses form the basis of intensive pastoral grazing systems. This has been clearly demonstrated in Australia, New Zealand and the USA (Fletcher, 2012; Johnson *et al.*, 2013a). However, these *Epichloë* endophytes bring both benefits and risks depending on the secondary metabolites they produce. Some secondary metabolites can be toxic or act as a deterrents to important insect pests, while others can also cause animal health and welfare issues, such as ryegrass staggers or exacerbation of heat stress. In the USA, this has been associated with the use of Kentucky 31 tall fescue, resulting in the expression of ergovaline (Bacon *et al.*, 1977, Schmidt *et al.*, 1982; Lyons *et al.*, 1986; Porter and Thompson, 1992; Paterson *et al.*, 1995; Ball *et al.*, 2003) and, in New Zealand, with the presence of a wild-type toxic endophyte strain that expresses lolitrem B and ergovaline (Gallagher *et al.*, 1984, Fletcher *et al.*, 1999) often referred to as the 'standard' strain.

Over the past 35 years, considerable investment has been made to understand how these *Epichloë* endophytes can be managed to provide the benefits of improved pest resistance/deterrence/tolerance while maintaining protection against abiotic stresses provided by endemic toxic strains of endophyte, without impacting animal welfare, health and production. This has required understanding the *Epichloë* life cycle; collecting a diverse range of *Epichloë* strains from differing environments and geographies; isolating and inoculating a fungus, which in nature has no free-living stage; understanding the fungal secondary metabolites (specifically alkaloids) associated with different *Epichloë* strains; associating the different chemical profiles with biotic and abiotic stress resistance offered to the host plant; providing proof of agronomic advantage resulting from incorporating novel *Epichloë* strains into elite grass germplasm; understanding the genetic control of compatibility between *Epichloë* strains and host genotypes; determining how to improve transmission and endophyte viability in seed; and finally linking research outputs with commercial imperatives to deliver a technology of value to end-user farmers.

16.2 Understanding the *Epichloë* Life Cycle

Epichloë includes species that can have either a sexual and asexual life cycle or just an asexual life cycle (Schardl *et al.*, 2004; Leuchtmann *et al.*, 2014). It is the latter types that have been used with temperate grasses to provide a technology that protects the host plant from abiotic and biotic stresses. *Epichloë* endophytes colonise only the aerial tissues of grasses by a mechanism known as intercalary growth, an infection process which is tightly controlled and synchronised with their grass host (Christensen *et al.*,

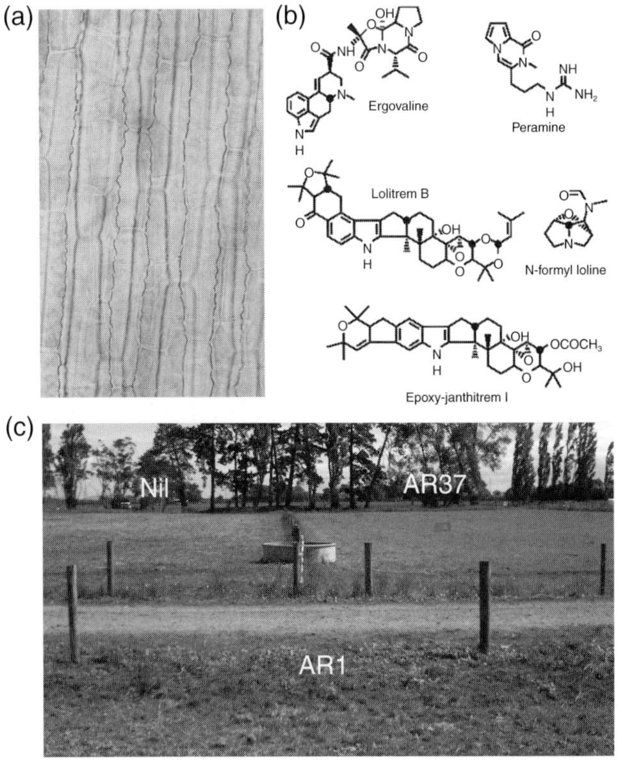

Figure 16.1 (a) An aniline blue-stained microscopy image showing the growth of *Epichloë festucae* in the outer leaf sheaths of perennial ryegrass. (b) Chemical structures of the five alkaloids produced by *Epichloë* endophytes. (c) The difference endophyte makes: pastures sown with perennial ryegrass inoculated without endophyte (Nil), or with AR37 endophyte (AR37) or with AR1 endophyte (AR1). (A black and white version of this figure will appear in some formats. For the colour version, please refer to the plate section.)

2008). Hyphae are confined to the intercellular spaces where long hyphal strands grow aligned parallel to the leaf axis (Figure 16.1a). The critical element of the *Epichloë* life cycle that has contributed significantly to its commercialisation success is that the fungus is only vertically transmitted, and maternally inherited, by growing into the embryo of the forming seed (Philipson and Christey, 1986). Therefore, as long as the seed is stored appropriately to maintain the viability of the endophyte this added value technology comes to the farmer as an 'endophytic seed'.

16.3 *Epichloë* Genetic Diversity

The centre of origin for *Epichloë* species found in ryegrass and fescue is likely to coincide with the origin of these host species in Europe, Eurasia and the Mediterranean

Basin (Borrill, 1976; Balfourier *et al.*, 1998). Germplasm collections across these regions have yielded considerable variation among *Epichloë* endophyte strains. Variability has been measured using genetic markers (Moon *et al.*, 1999; Young *et al.*, 2014; Sullivan *et al.*, 2016) and via the variation in alkaloid production (Takach and Young, 2014). The latter needs to be undertaken with the *Epichloë* strain *in planta* since the expression of secondary metabolites of interest only occurs when the endophyte is growing in the host plant. Karimi *et al.* (2012) showed strong host specificity of *Epichloë* endophytes related to both host species and their provenance.

16.4 Culturing and Inoculating *Epichloë*

In nature, the asexual *Epichloë* species have no free-living stage and it is only through intervention in the laboratory, through allowing the endophyte to grow from strongly surface disinfected leaf tissue onto agar, that they have been cultured outside of the host plant (Bacon and White, 1994). This allowed researchers to move desirable endophytes into elite breeding material and existing cultivars using artificial inoculation techniques (Latch and Christensen, 1985). It then became possible to link the best host plant genetics with the preferred endophyte strains to create a truly novel association with the potential to improve economic returns to farmers, while ensuring animal health and safety issues were minimised.

16.5 *Epichloë* Chemistry

Epichloë endophytes are all about chemistry. They produce a wide range of bioactive secondary metabolites some of which are alkaloids (a class of naturally occurring organic nitrogen-containing bases). Interestingly, the type of alkaloids produced depends on the strain of fungus present (Lane *et al.*, 2000; Schardl *et al.*, 2013), but the host plant has a major effect on the quantity of alkaloids expressed (Ball *et al.*, 1993; Latch, 1994; Easton *et al.*, 2002).

The chemistry of endophytes is diverse and complex with only a small number of alkaloid pathways identified. Furthermore, we only understand the impact of a few of these on plant resistance/tolerance to biotic and abiotic stresses. Those characterised fall into four main alkaloid categories (Figure 16.1b; Bush *et al.*, 1997; Tapper and Latch, 1999): the indole-diterpenoids, which include the tremorgens lolitrem B, epoxy-janthitrems, and paxilline; the ergot alkaloids, which include ergovaline, the causative agent of mammalian mycotoxicosis, ergotamine and other ergopeptides; the pyrrolizidines, which include lolines, a non-mammalian toxin with potent broad-spectrum anti-insect activity (Schardl *et al.*, 2007); and the pyrrolopyrazines which include peramine, a deterrent of Argentine stem weevil (Rowan *et al.*, 1990). Understanding the natural chemotype diversity and underpinning genetic diversity of these key alkaloid pathways between *Epichloë* species and strains has been a significant focus for research (Young *et al.*, 2005, 2006; Fleetwood *et al.*, 2007; Schardl *et al.*, 2007; Takach *et al.*, 2012; Saikia *et al.*, 2012; Schardl *et al.*, 2013).

There are many compounds produced by *Epichloë* endophytes that are yet to be identified and characterised. Analysis of secondary metabolite gene clusters from two assembled *E. festucae* genomes indicated these endophytes have a combined capacity to produce approximately 40 secondary metabolites, eight of which are only present in one or the other strain (Schardl *et al.*, 2013). Most of these predicted gene clusters contain either non-ribosomal peptide synthetases (NRPSs) or polyketide synthetases (PKSs). NRPSs and PKSs are both multifunctional enzymes required for the synthesis of a wide range of biologically active peptides, with PKSs being related to fatty acid synthases (Marahiel *et al.*, 1997; Cox, 2007). NRPSs are central to ergot alkaloid biosynthesis (Panaccione *et al.*, 2001; Fleetwood *et al.*, 2007), and the insect feeding deterrent alkaloid peramine requires a single NRPS gene for its production (Tanaka *et al.*, 2005). Other *Epichloë* secondary metabolites known to be produced by NRPSs are the iron-chelating siderophores epichloënin A (and B) and ferricrocin (Koulman *et al.*, 2012; Johnson *et al.*, 2013b; Forester *et al.*, 2017). *Epichloë* is also reported to synthesise multiple peptides (called epichloëcyclins) from a single gene (*gigA*) via ribosomal peptide synthesis (RiPS) (Johnson *et al.*, 2015). However, it is currently not known whether these RiPS produced peptides represent a new class of bioactive compounds or not.

16.6 Linking *Epichloë* Chemistry to Bioactivity

The central importance of *Epichloë* endophytes in temperate grasses lies in their *in planta* alkaloid production that can reduce both invertebrate pests and grazing vertebrates (Clay, 1989; Kuldau and Bacon, 2008). Early work on *Epichloë* endophytes identified two alkaloids that negatively affected vertebrate behaviour. Lolitrem B found in some *Epichloë* strains from perennial ryegrass causes the neuromuscular disorder ryegrass staggers (Fletcher and Harvey, 1981; Gallagher *et al.*, 1981). Ergovaline found in some *Epichloë* strains of both perennial ryegrass and fescue acts as a grazing deterrent (Stuedemann *et al.*, 1989), and where animals have no choice, exacerbates the symptoms of heat stress and in extreme cases causes a syndrome called fescue toxicosis (Bacon, 1995). Animal safety testing is undertaken on new grass associations of novel endophyte strains before commercial release. However, animal responses are also moderated by environment and to overcome this risk, testing is often carried out to manifest a 'worse case' scenario by increasing the stress on animals using controlled environments where temperatures can be elevated to 35–40°C and humidity to more than 75% (Fletcher *et al.*, 2000).

Studies in a number of countries have shown that *Epichloë* endophytes can provide protection against a range of insect pasture pests, including Argentine stem weevil (Prestidge and Gallagher, 1988), African black beetle (Bell *et al.*, 2011), root aphids (Popay and Thom, 2009), black cutworm (Baldauf *et al.*, 2011), pasture mealybug (Pennell *et al.*, 2005), field crickets (Barker *et al.*, 2015b), and white grubs including Japanese beetle (Scarabaeidae) (Potter *et al.*, 1992). Bird cherry-oat aphid (*Rhopalosiphum padi*) numbers are effectively reduced on fescue plants with some *Epichloë* endophytes (Clement *et al.*, 2007) and this reduction has been related to loline expression (Wilkinson *et al.*, 2000). Even with some ryegrass endophytes, which do not produce lolines, there can be a negative effect on bird cherry-oat aphid (Breen, 1993; Meister *et al.*, 2006). The motility of the lesion nematode *Pratylenchus scribneri* has been shown to be reduced by tall fescue endophytes due largely to lolines and some ergopeptides (Timper *et al.*, 2005; Bacetty *et al.*, 2007). However, there are some significant insect pests such as the larvae of redheaded (*Adoryphorus couloni*) and blackheaded (*Accrosidius tasmaniae*) cockchafers which cause significant damage to pastures in Australia that seem to be almost an intractable problem even for *Epichloë* endophytes (Watson, 2007).

In New Zealand, there are six recognised pasture pests and each of these can be affected by different secondary metabolites produced by *Epichloë* strains (Figure 16.2) (Popay and Rowan, 1994). Both adults and larvae of Argentine stem weevil (*Listronotus bonariensis*) are deterred by peramine but adults appear more sensitive than larvae (Rowan *et al.*, 1990). Loline, particularly high levels of N-formyl loline, produced by *Epichloë* endophytes found in fescues also had a toxic effect on Argentine stem weevil larvae (Patchett *et al.*, 2008; Jensen *et al.*, 2009; Popay *et al.*, 2009). Feeding and live-weight gain of African black beetle (*Heteronychus arator*) larvae have been negatively related to the loline concentration of feed (Barker *et al.*, 2015a). Certain ergopeptine alkaloids have also been shown to deter adult African black beetle (Ball *et al.*, 1997). Porina (*Wiseana cervinata* and *W. copularis*) are adversely affected by epoxy-janthitrems which are expressed by the AR37 endophyte strain. AR37 has been shown to be both deterrent and toxic to porina larvae in laboratory trials (Jensen and Popay, 2004). Grass grub (*Costelytra zealandica*) is best controlled by loline expressing endophytes (Patchett *et al.*, 2011). In AR37 pastures, grass grub numbers have been found to be lower in the Waikato (Popay and Thom, 2009) and Canterbury (A. J. Popay, unpublished data) than for AR1 and standard endophyte pastures. While there is no evidence that AR37 provides resistance to grass grub, lower grub numbers will create less pressure through using AR37 than AR1 or 'standard' endophyte. However, this does not mean that AR37 pastures will not be damaged by grass grub. Root aphid (*Aploneura lentisci*) numbers and activity can be reduced with loline-producing fescue endophytes (Jensen and Popay, 2007). Some endophytes such as AR1, which does not produce ergovaline or lolitrem B, have been found to increase root aphid numbers in comparison with

Figure 16.2 Major insect pests of pasture grasses in New Zealand. ASW refers to Argentine stem weevil. (A black and white version of this figure will appear in some formats. For the colour version, please refer to the plate section.)
Photos courtesy of AgResearch.

'standard' endophyte (Popay and Thom, 2009). The impact of pasture mealy bug (*Balanococcus poae*) appears to be reduced by the presence of 'standard' endophyte, AR1 and AR37 (Pennell *et al.*, 2005).

16.7 *Epichloë* and Tolerance to Abiotic Stresses

In the absence of biotic stresses, *Epichloë* endophytes can provide advantages for temperate grasses to tolerate abiotic stresses such as water and mineral deficiency (Arachevaleta *et al.*, 1989; Bacon, 1993; Malinowski *et al.*, 1997; Malinowski and Belesky, 2000; Kane, 2011). The mechanisms for improved tolerance due to endophyte presence is not entirely clear but there is some evidence it is linked to low stomatal conductance and increased cellular turgor pressure (Bacon, 1993). Several studies have also shown no beneficial effect of *Epichloë* endophytes in enhancing drought tolerance of grasses (Barker *et al.*, 1997; Cheplick *et al.*, 2000; Cheplick, 2004).

16.8 Compatibility and Transmission

Asexual *Epichloë* endophytes are efficiently vertically transmitted in host seeds and have a host range that is limited due to co-speciation (Schardl *et al.*, 1997). It has been challenging to move *Epichloë* strains across even closely related taxa since *Epichloë* species and even some strains within species, express a high degree of host specificity. Host genotypes can be selected for either increased (Easton *et al.*, 2007) or decreased compatibility (Easton, 2007, referencing a personal communication from M. J. Christensen), indicating that there is a strong genetic component of the host plant associated with determining the success of inoculating an endophyte strain into new grass germplasm. With the emergence of 'omics' technologies and sequence-based genotyping methods, it is now possible to study the biological processes that underlie the very low transmission and survival rates of endophytes in seeds that often accompany the introduction of novel endophytes into elite ryegrass cultivars. This will help confirm that ryegrass genotype plays an important role in the success of endophyte transmission (L. Johnson, unpublished data).

16.9 Seed Production and Storage

The agreed New Zealand industry target for the minimum percentage of seeds infected by a novel *Epichloë* endophyte strain is 70%; anything less than that should not be sold as an endophytic product. This standard has been adopted in other territories, such as Australia and the USA. Endophyte viability can decline at a much faster rate than seed viability (Rolston *et al.*, 1986; Welty *et al.*, 1987). Both genetic and environmental factors affect endophyte viability (Gundel *et al.*, 2011). To maintain high endophyte viability in seed it has been shown that storage at low temperature and humidity is preferable (Rolston *et al.*, 1986; Welty *et al.*, 1987; Rolston and Agee, 2007). While not commercially realistic for long-term storage of endophytic seed, it should be held at temperatures lower than 5°C and relative humidity less than 30% so that with stored seed moisture level is less than 8% (Hume *et al.*, 2013). The type of storage bag is also important. It has been shown that a thick 70 μm polyethylene or aluminium foil–polyethylene laminated bag serves as a barrier to ingress of moisture, and can maintain a low and stable seed moisture resulting in high viable endophyte levels in seed (Rolston *et al.*, 1986).

During seed production, management factors such as application rate of fungicides (Harvey *et al.*, 1982; Rolston *et al.*, 2002) and timing of nitrogen use (Stewart, 1986) need to be considered. Application of systemic fungicides to control leaf diseases in a seed crop partially kills or damages endophytes in the plant and reduces the storage life of endophytes in the seed. Safe fungicides for control of leaf diseases in endophytic tall fescue seed crops include contact, surface-acting fungicides and shallow systemic fungicides, e.g. the strobilurin group of fungicides (Rolston *et al.*, 2009).

16.9.1 Quality Control

Epichloë endophytes are a perishable technology and need to be produced and stored in ways outlined above to ensure a viable and effective product. To ensure that commercial endophyte strains are true to type and are viable in seed, a number of quality control tests have been developed. These include: grow out tests to measure endophyte viability in seed (Rolston and Agee, 2007), often associated with a tissue print-immunoblot technique that uses an antibody to detect endophyte viability in plant tissue (Gwinn *et al.*, 1991); squash tests to determine presence of endophyte in seed (Latch *et al.*, 1984; Card *et al.*, 2011), although this test does not determine whether the endophyte is alive or dead; ELISA tests to identify the presence of contaminant endophyte alkaloids such as lolitrems and ergot alkaloids in novel endophytes that do not express those secondary metabolites (Reddick Collins, 1988; Schnitzius *et al.*, 2001; Briggs *et al.*, 2007); ELISA assays to identify lolines in plants (Briggs *et al.*, 2017), and for the quantification of endophyte mycelia mass in plants (Faville *et al.*, 2015); polymerase chain reaction (PCR) methods, such as simple sequence repeats (SSRs) for routine endophyte detection (Groppe *et al.*, 1995) and discrimination between endophyte strains (Moon *et al.*, 1999; van Zijil *et al.*, 2003); and a patented method to measure levels of endophyte viability in plants (Johnson and Voisey, 2017).

16.10 Commercial Successes

Interestingly, the association of the *Epichloë* fungal endophyte with animal health and welfare disorders was identified in the USA and New Zealand at similar times (late 1970s and early 1980s) (Bacon *et al.*, 1977; Fletcher and Harvey, 1981). The initial reaction was to remove the endophyte from the plant to eliminate the animal disorders, but this reduced agronomic performance and plant persistence (Bouton *et al.*, 1993; West *et al.*, 1993; Popay *et al.*, 1999). Another more elegant solution was needed, and while requiring considerable investment through research and development it has provided significant outcomes of economic importance (Johnson *et al.*, 2013a). The first *Epichloë* endophytes successfully released commercially were MaxQ in the USA for use with tall fescue and AR1 in New Zealand for use with perennial ryegrass, in 2000 and 2001, respectively

In the USA, fescue toxicosis caused by *Epichloë* endophyte is estimated to be a US$1.0–1.5 billion per annum problem (Hoveland, 1993; Strickland *et al.*, 2011; Aiken and Strickland, 2013), affecting cattle grazing on 20 million ha of tall fescue pastures in south-eastern USA (Bouton, 2000). In Australia, ryegrass toxicosis caused by *Epichloë* endophyte was estimated to cost A$100 million in 2012 (Leury *et al.*, 2014). In New Zealand, ryegrass infected with standard endophyte can cause significant staggers in sheep (Fletcher *et al.*, 1999) resulting in 16% lower gross financial returns than a ryegrass system based on endophyte-free when compared in a systems grazing trial (Fletcher, 1999). It has been estimated that in New Zealand the use of AR37 endophyte has contributed NZ$3.6 billion to the economy through the life of its patent (ACIL Allen Consulting, 2017). This has been the result of not just excellent science but also the effective commercialisation of the technology and a remarkable uptake rate by farmers, particularly in New Zealand (Caradus *et al.*, 2013).

16.10.1 USA Case Study: MaxQ and MaxQ II

Infection of tall fescue pasture with *Epichloë coenophialium* in the USA has been associated with a complex of stock disorders referred to as fescue foot, fat necrosis and summer slump or fescue toxicity (Johnson, 1983; Bacon, 1995). This results in reduced animal live-weight gain and milk production, increased respiration and body temperature, and gangrene of extremities (Hoveland *et al.*, 1983; Bacon and Siegel, 1988; Schmidt and Osborne, 1993; Stuedemann and Thompson, 1993), primarily caused by the expression of ergovaline by some *Epichloë* endophyte strains (Hill *et al.*, 1994). Case study examples of *Epichloë* disorders include: significantly lower daily weight gains of cattle when animals consume Kentucky 31 over nil endophyte tall fescue pasture (Hoveland *et al.*, 1984; Danielson *et al.*, 1986; Peters *et al.*, 1992; Schmidt and Osborne, 1993; Paterson *et al.*, 1995); pregnancy rate reduced from over 85% with nil endophyte to below 60% for Kentucky 31 (Danielson *et al.*, 1986; Schmidt *et al.*, 1986; Essig *et al.*, 1993; Schmidt and Osborne, 1993; Coffey *et al.*, 2015); decreased reproductive performance of bulls (Looper *et al.*, 2009); increased abortion rate in gravid mares (Garrett *et al.*, 1980 Taylor *et al.*, 1985; Putnam *et al.*, 1991; Brendemuehl *et al.*, 1994); calving rate reduced from more than 85% to less than 70% (Boling, 1985; Schmidt *et al.*, 1986; Gay *et al.*, 1988; Essig *et al.*, 1989); delayed puberty (Porter and Thompson, 1992); intolerance to heat (Rhodes *et al.*, 1991); grazing time reduced by 20–90% (Bond *et al.*, 1984); reduced forage intake (Strahan *et al.*, 1987; Howard *et al.*, 1992) and grazing time (Bond *et al.*, 1984); reduced body condition score (Essig *et al.*, 1993); milk production reduced by 37–65% (Danielson *et al.*, 1986; Essig *et al.*, 1993; Brown *et al.*, 1996); and, reduced calf growth rates and weaning weights (Peters *et al.*, 1992; Essig *et al.*, 1993). Economic analysis has shown that cows grazing endophyte-free fescue returned a profit of US$98 per cow compared with a loss of US$54 per cow when cows and calves were grazed on Kentucky 31 (Essig *et al.*, 1993). Zhuang *et al.* (2005) found that cow–calf producer gross revenue diminished from US$198 per acre to US$92 per acre as toxic endophyte level increased from 0 to 100%. For small ruminants, such as sheep, similar negative effects of ergovaline-expressing endophytes have been observed (Aldrich *et al.*, 1993; Ditsch and Aiken, 2009).

However, persistence of tall fescue without an endophyte was invariably poorer in these studies (Pedersen *et al.*, 1990; Bouton *et al.*, 1993; Essig *et al.*, 1993) and hence alternative solutions were developed (Aitken and Strickland, 2013), one of which was to use non-toxic endophyte strains (Ball *et al.*, 2007). MaxQ was identified in the 1990s as a non-toxic *Epichloë* endophyte for the USA (Latch, 1997; Latch *et al.*, 2000) and was inoculated into the tall fescue cultivar Jesup (Bouton *et al.*, 1997; Bouton *et al.*, 2002). The endophyte strain has been protected through a patent, and was initially licensed exclusively for production and marketing as an added value product. Trials have clearly shown improved plant persistence without causing any adverse animal effects and have demonstrated improved on-farm performance of animals of MaxQ over toxic Kentucky 31 (Ball *et al.*, 2007; Johnson *et al.*, 2012; Coffey *et al.*, 2015). The benefits of MaxQ pastures over Kentucky 31 pasture for pasture-finished beef include: removing toxic alkaloids from meat and fat (Realini *et al.*, 2005); increasing pregnancy rate (+30%) (Burke *et al.*, 2004); increasing milk production (Watson *et al.*, 2004); increasing weaning weights (+15%) (Watson *et al.*, 2004); increasing grazing time and forage intake (Parish *et al.*, 2003b); increasing body condition score (Watson *et al.*, 2004); increasing average daily weight gains (+100%) (Watson *et al.*, 2004; Realini *et al.*, 2005; Beck *et al.*, 2008); reducing the number of days for steers to reach target weights (-60%) (Realini *et al.*, 2005); increasing live weight of feeder cattle per acre (+140%) (Parish *et al.*, 2003b; Franzluebbers *et al.*, 2009); and increasing carcass weights (+10%) (Realini *et al.*, 2005).

In the first seven years (2000 to 2007) of commercialisation MaxQ tall fescue endophyte had a significant level of farmer uptake with an estimated 121 000 ha established in at least 35 states (Ball *et al.*, 2007). Success has come because of the excellent field trial results in 25 grazing research experiments and over 150 on-farm demonstrations throughout the USA. Despite an outstanding start, the 121 000 ha is less than 1% of the total tall fescue area in the USA.

The benefits of MaxQ tall fescue are without question (Bouton *et al.*, 2000b; Parish *et al.*, 2003a, b). Bouton *et al.* (2001) conducted an extensive grazing study of beef cows, calves and feeder cattle grazing on tall fescue cv. Jesup, with either MaxQ, toxic wild-type (standard) endophyte or nil endophyte in central and north Georgia USA over a 2 year period. At weaning, heifer and steer calves grazing on MaxQ fescue weighed an extra 37 lb (16.8 kg) and 94 lb (42.6 kg) live weight, respectively (US$30 extra per heifer; $75 extra per steer), compared to calves grazing Jesup with standard endophyte. Similarly, at the completion of the stocker grazing phase, feeder cattle (heifers and steers) were 116 lb (52.6 kg) heavier (value US$93 extra per animal) and at the completion of the feedlot phase, live cattle (heifers and steers) that were backgrounded on MaxQ enhanced pastures were 108 lb (49.0 kg) heavier (after processing and grading, the carcass value was $108 higher per animal on average than animals grazing Jesup with standard endophyte).

Trials with sheep have shown similar advantages with using MaxQ over the standard endophyte in tall fescue (Bouton *et al.*, 2000a). For three spring grazing periods, lambs gained an average of 129 g/day on cultivars containing MaxQ which were equivalent to gains on nil endophyte tall fescue, but were 65% greater than for animals on standard endophyte tall fescue. Similar results were obtained during three successive autumn periods.

In a grazing study managed by the Noble Foundation and the University of Arkansas (Biermacher and Beck, 2013), two MaxQ containing tall fescue cultivars were compared with Kentucky 31, grown with either added nitrogen or white clover. Economic analysis showed a clear advantage to the MaxQ treatments over the Kentucky 31 treatment irrespective of whether clover or nitrogen was used. The mean net return for the MaxQ treatments was US$165 acre^{-1} (US$66.77 ha^{-1}) compared with minus US$20 acre^{-1} (US$8.09 ha^{-1}). They concluded that farmers who are interested in establishing tall fescue for use in a stocker cattle grazing enterprise would be better off with a novel endophyte variety compared to toxic Kentucky 31.

Duckett *et al.* (2007) grazed Angus-cross steers on either wild-type endophyte-infected (E+), MaxQ (AR542) endophyte, or endophyte-free (E-) tall fescue pastures at two locations. They took 24 animals per location to a research feedlot and fed them for 100 days. Average daily gain during grazing was 0.47 kg d^{-1}greater for cattle grazing AR542 than those grazing E+ pastures. In the feedlot, while overall average daily gain did not differ among treatments, hot carcass weight at slaughter was 40 kg greater per animal for AR542 than E+. The relative profitability of using MaxQ (net present values) compared with E+ was US$336.76 ha^{-1}; for E- it was US$177.08 ha^{-1}. The authors determined that the economic incentive for using MaxQ is derived from the savings in feeding costs once the animals reach the feedlot. Calves grazed on E+ weighed less and therefore had a higher overall feeding cost than calves grazed on either AR542 or E-.

While the initial release of the novel endophyte (AR542) was branded MaxQ, another strain with similar chemistry but better storage qualities (AR584) was subsequently released as MaxQ II (Hopkins and Bouton, 2007). The success of MaxQ and MaxQ II in the USA has led to its use in New Zealand and Australia, where it has been branded MaxP. It is now being trialled for use in parts of South America.

16.10.2 New Zealand Case Study: AR1 and AR37

In the 1980s and early 1990s, the primary pest considered to be associated with reducing ryegrass persistence in New Zealand pastures was Argentine stem weevil, which caused damage estimated at NZ$200 million per year (Prestidge *et al.*, 1991). The larvae of Argentine stem weevil cause the greatest damage through penetrating the stem and developing, through four instars, during which time they can mine and kill up to four tillers (Rowan and Gaynor, 1986). Perennial ryegrass persistence has been linked to the presence of the *Epichloë* endophyte (Prestidge *et al.*, 1982) and the production of the alkaloid peramine (Rowan *et al.*, 1990). However, the endophyte strain commonly found in New Zealand pastures at that time also produced lolitrem B associated with ryegrass staggers (Gallagher *et al.*, 1978; Fletcher and Harvey, 1981; Gallagher *et al.*, 1981, 1982, 1984) and ergovaline which has been associated exacerbating the symptoms of heat stress (Easton *et al.*, 1996).

In the early 1980s, it was postulated that it should be possible to produce an endophyte-infected ryegrass resistant to stem weevil that does not cause ryegrass staggers (Rowan and Gaynor, 1986) and this was found to be the case with the release of the 'Endosafe' strain in 1991 (Tapper and Latch, 1999). However, soon after release it was found that expression levels of ergovaline in some cultivars was high enough to be of concern and so the technology was withdrawn (Davies *et al.*, 1993). It took another decade of research and development before AR1, which expressed neither lolitrem B nor ergovaline, was released for the control of Argentine stem weevil (Fletcher, 1999; Tapper and Latch, 1999; Popay and Thom, 2009). It had the added bonus of also controlling mealy bug (Pennell *et al.*, 2005).

At its release, AR1 was hailed a technological breakthrough because it not only deterred Argentine stem weevil, but also improved animal performance compared with the 'standard' endophyte strain through preventing ryegrass staggers, not affecting heat stress, reducing fly strike, significantly increasing lamb live-weight gain by more than 10% (Fletcher, 1999) which in the summer/autumn period could exceed 65% (Fletcher, 2005) and significantly increasing milk solids by 9% (Bluett *et al.*, 2003). Trials with sheep showed a 22% increase in returns to farmers through using AR1 over 'standard' endophyte (Fletcher, 1999).

AR1 endophyte was a non-exclusive release and had a rapid uptake in New Zealand (Caradus *et al.*, 2013). AR1 is now licensed to more than 12 companies and marketed in more than 30 cultivars, and has been exported to Australia and Chile. It is currently being evaluated in the USA, Europe, Uruguay and Argentina. Uptake by New Zealand farmers since AR1's commercial release in 2001 was extraordinary, such that within 7 years AR1 was used in 70% of the proprietary perennial ryegrass seed sold (Caradus *et al.*, 2013).

However, AR1 did have some limitations. Through the 1990s, the predominant pest threat to perennial ryegrass was the Argentine stem weevil but it was controlled using the combination of AR1 and a wasp biocontrol agent (Goldson *et al.*, 1993). However, other insect pests became more prevalent including African black beetle and root aphid (Thom *et al.*, 2014) and the native insects, porina and grass grub. AR1 provided little or no protection against these pests. A new solution was required, particularly in those regions of New Zealand where African black beetle, root aphid and porina were rampant.

Field trials in the 1990s included a strain of endophyte called AR37, as a control to test the direct impact of an endophyte on plant growth without the confounding effect of improving plant growth through reducing the impact of pasture pests, because it did not express any of the then known alkaloids peramine, lolitrem B and ergovaline. However, it was soon noticed that the best performing ryegrasses were those containing this novel endophyte with no known alkaloids (Hume *et al.*, 2004), particularly in the northern parts of New Zealand (Hume *et al.*, 2009). In those trials, AR37 outperformed the 'standard' endophyte. It was subsequently shown to provide protection against a range of insect pests ensuring excellent ryegrass persistence (Popay and Hume, 2011). As a result it was decided to move AR37 into elite ryegrass germplasm for commercialisation. However, animal safety tests detected ryegrass staggers (Fletcher and Sutherland, 2009) and so the whole project was put on hold. The ryegrass staggers were considered to be associated with the presence of epoxy-janthitrems (Tapper and Lane, 2004).

In the mid-2000s when farmer complaints about the persistence of ryegrasses with AR1 began to increase, it was decided to take AR37 to market with the rider that it may cause ryegrass staggers (Fletcher, 2005). Fletcher and Sutherland (2009) concluded that the large agronomic advantages of AR37 outweigh the ryegrass staggers that may occur in sheep, making it a better option in terms of both pasture persistence and animal productivity and welfare than 'standard' endophyte. In controlled studies epoxy-janthitrems have not been reported to cause on-farm ryegrass staggers in cattle (Moate *et al.*, 2012). So the upside was that AR37 controlled not only Argentine stem weevil, but also African black beetle (Popay and Thom, 2009), and porina (Jensen and Popay, 2004). Later it was also shown to control root aphid and pasture mealy bug (Pennell *et al.*, 2005). While AR37 does not provide resistance to grass grub, lower grass grub numbers associated with AR37 indicate there will be less severe pressure from this pest in pastures with AR37 than with AR1 or 'standard' endophyte (Popay and Thom, 2009). As a result AR37 became the premier endophyte for use in areas with high pasture pest pressures. For a visual comparison of the performance of pastures infected with AR37, AR1 and nil endophyte see Figure 16.1c.

The cumulative area re-grassed with ryegrasses containing AR37 has been estimated to be in excess of 250 000 ha (extrapolated from 130 000 ha in 2013; Caradus *et al.*, 2013). On dairy land, this has resulted in producing an additional 8.4 million kg milksolids each year, with a cumulative value to date to dairy farmers, net of the re-grassing costs attributed to AR37, of NZ$125 million (Caradus *et al.*, 2013). In trials at AgResearch's Lincoln campus, growth during the summer and autumn for hogget/ lambs on pure ryegrass pasture averaged 29 g d^{-1} for the 'standard' endophyte, 74 g d^{-1} for AR1 and 93 g d^{-1} for AR37 (Fletcher and Sutherland, 2009).

16.11 Conclusion: Critical Elements for the Successful Commercialisation of *Epichloë* Endophytes

Before any significant investment, it is important to understand the market need and determine whether microbial endophytes might provide a workable solution. Discovery of novel strains with good efficacy against the target stress is required and often follows the effective characterisation of microbial strains from the species of interest. Characterisation will require the capabilities of mycology, biochemistry and molecular biology. Assuming the microbial endophyte is vertically transmitted then understanding the constraints and drivers of effective transmission will be required. Managing viability of the microbe in seed during harvest and storage will be needed to ensure the technology is alive when purchased and used by the farmer.

Apart from the scientific ability required to develop the technology, there is also the need to ensure that effective intellectual property protection is secured so that any commercial company involved in delivering the product has some certainty that they have control of the technology in the market place. This also provides the opportunity of obtaining a return for the value created by the technology for the investor in the research and development. This will ensure that the research group developing the technology has a committed commercial partner. Good branding in the market is needed and this must be supported by reliable information about the performance of the technology in different environments, seasons and geographies. Above all, success will be ensured by delivering a technology that works and farmers want to use and they can do this without it being disruptive to their traditional on-farm management systems.

References

ACIL Allen Consulting (2017). *New Zealand's Science System: Case Studies, Report to the Ministry of Business, Innovation and Employment*. Sydney, Australia: ACIL Allen Consulting.

Aiken, G. E. and Strickland, J. R. (2013). Forages and Pastures Symposium: Managing the tall fescue–fungal endophyte symbiosis for optimum forage-animal production. *Journal of Animal Science*, **91**, 2369–2378.

Aldrich, C. G., Rhodes, M. T., Miner, J. L., Kerley, M. S. and Paterson, J. A. (1993). The effects of endophyte-infected tall fescue consumption and use of a dopamine antagonist on intake, digestibility, body temperature, and blood constituents in sheep. *Journal of Animal Science*, **71**, 158–163.

Arachevaleta, M., Bacon, C. W., Hoveland, C. S. and Radcliffe, D. E. (1989). Effect of the tall fescue endophyte on plant response to environmental stress. *Agronomy Journal*, **81**, 83–90.

Bacetty, A. A., Snook, M. E., Glenn, A. E. *et al.* (2007). Nematotoxic effects of endophyte-infected tall fescue toxins and extracts in an in vitro bioassay using the nematode *Pratylenchus scribneri*. In *Research and Practice New Zealand Grassland Association 13: 6th International Symposium on Fungal Endophytes of Grasses*, ed. A. J. Popay and E. R. Thom. Palmerston, New Zealand: New Zealand Grassland Association, pp. 357–361.

Bacon, C. W. (1993). Abiotic stress tolerances (moisture, nutrients) and photosynthesis in endophyte infected tall fescue. *Agriculture Ecosystems and Environment*, **44**, 123–141.

Bacon, C. W. (1995). Toxic endophyte-infected tall fescue and range grasses: historic perspectives. *Journal of Animal Science*, **73**, 861–870.

Bacon, C. W. and Siegel, M. R. (1988). Endophyte parasitism of tall fescue. *Journal of Production Agriculture*, **1**, 45–55.

Bacon, C. W. and White, J. F. Jr. (1994). Stains, media, and procedures for analyzing endophytes. In *Biotechnology of Endophytic Fungi of Grasses*, ed. C. W. Bacon and J. F. Jr. White. Boca Raton, FL: CRC Press, pp. 47–56.

Bacon, C. W., Porter, J. K., Robbins, J. D. and Luttrell, E. S. (1977). *Epichloë typhina* from toxic tall fescue grasses. *Applied Environmental Microbiology*, **34**, 576–581.

Baldauf, M. W., Mace, W. J. and Richmond, D. S. (2011). Endophyte-mediated resistance to black cutworm as a function of plant cultivar and endophyte strain in tall fescue. *Environmental Entomology*, **40**, 639–647.

Balfourier, F., Charmet, G. and Ravel, C. (1998). Genetic differentiation within and between natural populations of perennial and annual ryegrass (*Lolium perenne* and *L. rigidum*). *Heredity*, **81**, 100–110.

Ball, D. M., Schmidt, S. P., Lacefield, G. D., Hoveland, C. S. and Young, W. C. III. (2003). Tall fescue/endophyte concepts. *Oregon Tall Fescue Commission Special Publication*, 1–03.

Ball, D. M., Lacefield, G. D., Agee, C. S. and Hoveland, C. S. (2007). Introduction and acceptance of novel endophyte tall fescue in USA. In *Research and Practice New Zealand Grassland Association 13: 6th International Symposium on Fungal Endophytes of Grasses*, ed. A. J. Popay and E. R. Thom. Palmerston, New Zealand: New Zealand Grassland Association, pp. 249–251.

Ball, O. J. P., Prestidge, R. A. and Sprosen, J. M. (1993). Effect of plant age and endophyte viability on peramine and lolitrem B concentration in perennial ryegrass seedlings. In *Proceedings of the Second International Symposium on Acremonium/Grass Interactions*, ed. D. E. Hume, G. C. M. Latch and H. S. Easton. Palmerston, New Zealand: Grasslands Research Centre, pp. 63–66.

Ball, O. J. P., Miles, C. O. and Prestidge, R. A. (1997). Ergopeptine alkaloids and *Neotyphodium lolii*-mediated resistance in perennial ryegrass against adult *Heteronychus arator* (Coleoptera: Scarabaeidae). *Journal of Economic. Entomology*, **90**, 1382–1391.

Barker, D. J., Hume, D. E. and Quigley, P. E. (1997). Negligible physiological responses to water deficit in endophyte-infected and uninfected perennial ryegrass. In *Neotyphodium/Grass Interactions*, ed. C. W. Bacon and N. S. Hill. New York: Plenus Press, pp. 137–139.

Barker, G. M., Patchett, B. J. and Cameron, N. E. (2015a). *Epichloë uncinata* infection and loline content afford Festulolium grasses protection from black beetle (*Heteronychus arator*). *New Zealand Journal of Agricultural Research*, **58**, 35–56.

Barker, G. M., Patchett, B. J. and Cameron, N. E. (2015b). *Epichloë uncinata* infection and loline content protect Festulolium grasses from crickets (Orthoptera: Gryllidae). *Journal of Economic Entomology*, **108**, 789–97.

Beck, P. A., Gunter, S. A., Lusby, K. S. *et al.* (2008). Animal performance and economic comparison of novel and toxic endophyte tall fescues to cool-season annuals. *Journal of Animal Science*, **86**, 2043–2055.

Bell, N., Townsend, R., Popay, A., Mercer, C. and Jackson, T. (2011). Black beetle: lessons from the past and options for the future. *New Zealand Grassland Association Pasture Persistence Symposium. Grassland Research and Practice Series*, **15**, 119–124

Biermacher, J. and Beck, P. (2013). Study reveals legume-fescue mixture economics. *The Samuel Roberts Noble Foundation Ag News and Views*, **31**, 4.

Bluett, S. J., Thom, E. R., Clark, D. A., Macdonald, K. A. and Minneé, E. M. K. (2003). Milksolids production from cows grazing perennial ryegrass containing AR1 or wild endophyte. *Proceedings of the New Zealand Grassland Association*, **65**, 83–90.

Boling, J. A. (1985). Endophytic fungus and tall fescue utilization by ruminants. *The Professional Animal Scientist*, **1**, 19–22.

Bond, J., Powell, J. B. and Weinland, B. T. (1984). Behaviour of steers grazing several varieties of tall fescue during summer conditions. *Agronomy Journal*, **76**, 707–709.

Borrill, M. (1976). Temperate grasses. In *Evolution of Crop Plants*, ed. N. Simmonds. London: Longman, pp. 137–142.

Bouton, J. H. (2000). The use of endophytic fungi for pasture improvement in the USA. In *Proceedings of the 4th International Neotyphodium/Grass Interactions Symposium. Soest, Germany*, ed. V. H. Paul and P. D. Dapprich. Soest, Germany: Universitat Paderborn, Gesamthochschulepp, pp. 163–168.

Bouton, J. H., Gates, R. N., Belesky, D. P. and Owsley, M. (1993). Yield and persistence of tall fescue in the southeastern coastal plain after removal of its endophyte. *Agronomy Journal*, **85**, 52–55.

Bouton, J. H., Duncan, R. R., Gates, R. N., Hoveland, C. S. and Wood, D. T. (1997). Registration of 'Jesup' tall fescue. *Crop Science*, **37**, 1011–1012.

Bouton, J. H., Hill, N. S. and Hoveland, C. S. *et al.* (2000a). Performance of tall fescue cultivars infected with non-toxic endophytes. In *Proceedings of the 4th International Neotyphodium/Grass Interactions Symposium. Soest, Germany*, ed. V. H. Paul and P. D. Dapprich. Soest, Germany: Universitat Paderborn, Gesamthochschulepp, pp. 179–185.

Bouton, J. H., Hill, N. S. and Hoveland, C. S. *et al.* (2000b). Infection of Tall Fescue Cultivars with Non-toxic Endophytes. In *Proceedings 55th Southern Forage Crop Improvement Conference*, Raleigh, NC, *June 12–14*, pp. 1–7.

Bouton, J. H., Hill, N. S. Hoveland, C. S. *et al.* (2001). *Georgia State Report to SERAIEG Group 8, Chapel Hill TN, USA*, November 5–6, 2001, pp. 9–20.

Bouton, J. H., Latch, G. C. M., Hill, N. S. *et al.* (2002). Reinfection of tall fescue cultivars with non-ergot alkaloid-producing endophytes. *Agronomy Journal*, **94**, 567–574.

Brendemuehl, J. P., Boosinger, T. R., Pugh, D. G. and Shelby, R. A. (1994).

Influence of endophyte-infected tall fescue on cyclicity, pregnancy rate and early embryonic loss in the mare. *Theriogenology* **42**, 489–500.

Breen, J. P. (1993). Enhanced resistance to three species of aphids (Homoptera: Aphididae) in *Acremonium* endophyte-infected turfgrasses. *Journal of Economic Entomology*, **86**, 1279–1286.

Briggs, L., Tapper, B., Sprosen, J., Mace, W. and Finch, S. (2017). Development of an enzyme-linked immunosorbent assay for the detection of lolines in pastures. *Food and Agricultural Immunology*, **28,** 1058–1070.

Briggs, L. R., Sprosen, J. M., Tapper, B. A. and Easton, H. S. (2007). AR1 quality assurance by ELISA. In *Research and Practice New Zealand Grassland Association 13: 6th International Symposium on Fungal Endophytes of Grasses*, ed. A. J. Popay and E. R. Thom. Palmerston, New Zealand: New Zealand Grassland Association, pp. 289–291.

Brown, M. A., Brown, A. H. Jr., Jackson, W. G. and Miesner, J. R. (1996). Milk production in Angus, Brahman, and reciprocal-cross cows grazing common Bermuda grass or endophyte-infected tall fescue. *Journal of Animal Science*, **74**, 2058–2066.

Burke, J. M., Brauer, D. K. and Looper, M. L. (2004). Use of novel endophyte-infected tall fescue for cow–calf production in Arkansas. *American Society of Animal Science Annual Meeting*, **82** (S. 1), 91.

Bush, L. P., Wilkinson, H. H. and Schardl, C. L. (1997). Bioprotective alkaloids of grass-fungal endophyte symbioses. *Plant Physiology*, **114**, 1–7.

Caradus, J. R., Lovatt, S. and Belgrave, B. (2013). Adoption of forage technologies. *Proceedings of New Zealand Grassland Association*, **75**, 39–44.

Card, S. D., Rolston, M. P., Park, Z., Cox, N. and Hume, D. E. (2011). Fungal endophyte detection in pasture grass seed utilising the infection layer and comparison to other detection techniques. *Seed Science & Technology*, **39**, 581–592.

Cheplick, G. P. (2004). Recovery from drought stress in *Lolium perenne* (Poaceae): are fungal endophytes detrimental? *American Journal of Botany*, 91, 1960–1968.

Cheplick, G. P., Perera, A. and Koulouris, K. (2000). Effect of drought on the growth of *Lolium perenne* genotypes with and without fungal endophytes. *Functional Ecology*, **14**, 657–667.

Clay, K. (1989). Clavicipitaceous endophyte of grasses: their potential as biocontrol agents. *Mycological Research*, **92**, 1–12.

Clay, K. (1990). Fungal endophytes of grasses. *Annual Review of Ecology and Systematics*, **21**, 275–297.

Clay, K., Marks, S. and Cheplick, G. P. (1993). Effects of insect herbivory and fungal endophyte infection on competitive interactions among grasses. *Ecology*, **74**, 1767–1777.

Clement, S. L., Elberson, L. R., Waldron, B. L. and Quisenberry, S. S. (2007). Variable performance of bird cherry-oat aphid on *Neotyphodium*-infected tall fescue from Tunisia. In *Research and Practice New Zealand Grassland Association 13: 6th International Symposium on Fungal Endophytes of Grasses*, ed. A. J. Popay and E. R. Thom. Palmerston, New Zealand: New Zealand Grassland Association, pp. 337–340.

Christensen, M. J., Bennett, R. J., Ansari, H. A. *et al.* (2008). *Epichloë* endophytes grow by intercalary hyphal extension in elongating grass leaves. *Fungal Genetics and Biology*, **45**, 84–93.

Coffey, K. P., Coblentz, W. K. and Caldwell, J. D. (2015). Pre- and postweaning performance by cows and calves that grazed toxic or nontoxic endophyte-infected tall fescue pastures. *The Professional Animal Scientist*, **31**, 577–587.

Cox, R. J. (2007). Polyketides, proteins and genes in fungi: programmed nano-machines begin to reveal their secrets. *Organic and Biomolecular Chemistry*, **5**, 2010–2026.

Danielson, D. A., Schmidt, S. P., King, C. C., Smith, L. A. and Webster, W. B. (1986). Fescue toxicity and reproduction in beef heifers. *Journal of Animal Science*, **63** (**S.1**), 296.

Davies, E., Lane, G. A., Latch, G. C. M. *et al.* (1993). Alkaloid concentrations in field-grown synthetic perennial ryegrass endophyte associations. In *Proceedings of the Second International Symposium on Acremonium/Grass Interactions*, ed. D. E. Hume, G. C. M. Latch and H. S. Easton. Palmerston, New Zealand: Grasslands Research Centre, pp. 72–76.

Ditsch, D. C. and Aiken, G. E. (2009). Endophyte infected tall fescue and small ruminant production: do we have a problem? Online. *Forage & Grazinglands*. doi: 10.1094/FG2009-1104-02-RS.

Duckett, S. K., Lacy, R. C., Andrae, J. G. *et al.* (2007). Animal performance, carcass quality and economics of cattle finished after grazing endophyte-infected, endophyte-free or nonergot alkaloid-producing endophyte-infected tall fescue. In *Research and Practice New Zealand Grassland Association 13: 6th International Symposium on Fungal Endophytes of Grasses*, ed. A. J. Popay and E. R. Thom. Palmerston, New Zealand: New Zealand Grassland Association, pp. 253–255.

Easton, H. S. (1999). Endophyte in New Zealand ryegrass pastures: an overview. Ryegrass endophyte: an essential New Zealand symbiosis. In *Ryegrass endophyte: an Essential New Zealand Symbiosis*, ed. D. R. Woodfield and C. Matthew. *Grassland Research and Practice Series* **7**, pp. 1–9.

Easton, H. S. (2007). Grasses and *Neotyphodium* endophytes: co-adaptation and adaptive breeding. *Euphytica*, **154**, 295–306.

Easton, H. S., Lane, G. A., Tapper, B. A. *et al.* (1996). Ryegrass endophyte-related heat stress in cattle. *Proceedings of the New Zealand Grassland Association*, **57**, 37–41.

Easton, H. S., Latch, G. C. M., Tapper, B. A. and Ball, O. J.-P. (2002). Ryegrass host genetic control of concentrations of endophyte-derived alkaloids. *Crop Science*, **42**, 51–57.

Easton, H. S., Lyons, T. B., Mace, W. J., Simpson, W. R. *et al.* (2007). Differential expression of loline alkaloids in perennial ryegrass infected with endophyte isolated from tall fescue. In *Research and Practice New Zealand Grassland Association 13: 6th International Symposium on Fungal Endophytes of Grasses*, ed. A. J. Popay and E. R. Thom. Palmerston, New Zealand: New Zealand Grassland Association, pp. 163–165.

Essig, H. W., Cantrell, C. E., Withers, F. T. Jr. *et al.* (1989). Performance and profitability of cow-calf systems grazing on EF and EI KY-31 fescue. (Preliminary report). *Proceeding of Tall Fescue Toxicosis Workshop*, November 13–14, Atlanta, GA, p. 61.

Essig, H. W., Aremu, B., Cantrell, C. E., Boyd, M. E. and Withers, F. T. Jr. (1993). Impacts of endophyte-infected fescue on cow/calf production. *The Professional Animal Scientist*, **9**, 64–69.

Faville, M. J., Briggs, L., Cao, M. *et al.* (2015). A QTL analysis of host plant effects on fungal endophyte biomass and alkaloid expression in perennial ryegrass. *Molecular Breeding*, **35**, 161.

Fleetwood, D. J., Scott, B., Lane, G. A., Tanaka, A. and Johnson, R. D. (2007). A complex ergovaline gene cluster in *Epichloë* endophytes of grasses. *Applied and Environmental Microbiology*, **73,** 2571–2579.

Fletcher, L. R. (1999). 'Non-toxic' endophytes in ryegrass and their effect on livestock health and production. In *Ryegrass endophyte: an Essential New Zealand Symbiosis*, ed. D. R. Woodfield and C. Matthew. *Grassland Research and Practice Series* **7**, pp. 133–139.

Fletcher, L. R. (2005). Managing ryegrass-endophyte toxicoses. In *Neotyphodium in Cool-Season Grasses*, ed. C. A., Roberts, C. P. West and D. E. Spiers. Ames, IA: Blackwell Publishing, pp. 229–241.

Fletcher, L. R. (2012). Novel endophytes in New Zealand grazing systems: the perfect solution or a compromise? In *Epichloae, Endophytes of Cool Season Grasses: Implications, Utilization and Biology. Proceedings of the 7th International Symposium on Fungal Endophytes of Grasses, Lexington, Kentucky, USA*, 28 June to 1 July 2010, ed. C. A. Young, G. E. Aiken, R. L. McCulley, J. R. Strickland and C. L. Schardl, pp. 5–13.

Fletcher, L. R. and Harvey, I. C. (1981). An association of a *Lolium* endophyte with ryegrass staggers. *New Zealand Veterinary Journal*, **29**, 185–186.

Fletcher, L. R. and Sutherland, B. L. (2009). Sheep responses to grazing ryegrass with AR37 endophyte. *Proceedings of New Zealand Grassland Association*, **71**, 127–133.

Fletcher, L. R., Sutherland, B. L. and Fletcher, C. G. (1999). The impact of endophyte on the health and productivity of sheep grazing ryegrass-based pasture. *Grassland Research and Practice Series*, **7**, 11–17.

Fletcher, L. R., Fletcher, C. G. and Sutherland, B. L. (2000). The health and performance of sheep grazing a non-toxic tall fescue endophyte association. In *Proceedings of the 4th International Neotyphodium/Grass Interactions Symposium. Soest, Germany*, ed. V. H. Paul and P. D. Dapprich. Soest, Germany: Universitat Paderborn, Gesamthochschulepp, pp. 459–464.

Forester, N. T., Lane, G. A., Steringa, M., Lamont, I. L. and Johnson, L. J. (2017). Contrasting roles of fungal siderophores in maintaining iron homeostasis in *Epichloë festucae*. *Fungal Genetics and Biology*, **111**, 60–72.

Franzluebbers, A. J., Seman, D. H. and Stuedemann, J. A. (2009). Tall fescue persists and cattle perform well on a novel-endophyte association in the southern Piedmont USA. *Forage & Grasslands*, **7**, doi: 10.1094/FG-2009-0227-01-RS.

Gallagher, R. T., Keogh, R. G., Latch, G. C. M. and Reid, C. S. W. (1978). The role of fungal tremorgens in ryegrass staggers. *New Zealand Journal of Agricultural Research*, **20**, 431–440.

Gallagher, R. T., White, E. P. and Mortimer, P. H. (1981). Ryegrass staggers: isolation of potent neurotoxins lolitrem A and lolitrem B from staggers producing pastures. *New Zealand Veterinary Journal*, **29**, 189–190.

Gallagher, R. T., Campbell, A. G., Hawkes, A. D. *et al.* (1982). Ryegrass staggers: the presence of lolitrem neurotoxins in ryegrass seed. *New Zealand Veterinary Journal*, **30**, 183–184.

Gallagher, R. T., Hawkes, A. D., Steyn, P. S. and Vleggaar, R. (1984). Tremorgenic neurotoxins from perennial ryegrass causing ryegrass staggers disorder of livestock: Structure elucidation of lolitrem B. *Journal of Chemistry Society Chemistry Communications*, **1984**, 614–616.

Garrett, L. W., Heimann, E. D., Pfander, W. H. and Wilson, L. L. (1980). Reproductive problems of pregnant mares grazing fescue pastures. *Journal of Animal Science* **51**(S. 1), 237.

Gay, N., Boling, J. A., Dew, R. and Miksch, D. E. (1988). Effects of endophyte-infected tall fescue on beef cow-calf performance. *Applied Agricultural Research*, **3**, 182–186.

Goldson, S. L., McNeill, M. R., Proffitt, J. R. *et al.* (1993). Systematic mass rearing and release of *Microctonus hyperodae* (Hym.: Braconidae, Euphorinae), a parasitoid of the Argentine stem weevil *Listronotus bonariensis* (Col.: Curculionidae) and records of its establishment in New Zealand. *Entomophaga*, **38**, 527–536.

Groppe, K., Sanders, I., Wiemken, A. and Boller, T. (1995). A microsatellite marker for studying the ecology and diversity of fungal endophytes (*Epichloë* spp.) in grasses. *Applied and Environmental Microbiology*, **61**, 3943–3949.

Gwinn, K. D., Collinsshepard, M. H. and Reddick, B. B. (1991). Tissue print-immunoblot, an accurate method for the detection of *Acremonium coenophialium* in tall fescue. *Phytopathology*, **81**, 747–748.

Harvey, I., Fletcher, L. and Emms, L. (1982). Effects of several fungicides on the *Lolium* endophyte in ryegrass plants, seeds, and in culture. *New Zealand journal of Agricultural Research*, **25**, 601–606.

Hill, N. S., Thompson, F. N., Dawe, D. L. and Stuedemann, J. A. (1994). Antibody binding of circulating ergopeptine alkaloids in cattle grazing tall fescue. *American Journal of Veterinary Research*, **55**, 419–424.

Hopkins, A. A. and Bouton, J. H. (2007). A new tall fescue-novel endophyte combination for the south-central USA. In *Research and Practice New Zealand Grassland Association 13: 6th International Symposium on Fungal Endophytes of Grasses*, ed. A. J. Popay and E. R. Thom. Palmerston, New Zealand: New Zealand Grassland Association, pp. 257–258.

Hoveland, C. S. (1993). Importance and economic significance of the *Acremonium* endophytes to performance of animals and grass plant. *Agricultural Ecosystem & Environment*, **44**, 3–12.

Hoveland, C. S., Schmidt, S. P., King, Jr, C.C. *et al.* (1983). Steer performance and association of *Acremonium coenophialum* fungal endophyte on tall fescue pasture. *Agronomy Journal*, **75**, 821–824.

Hoveland, C. S., Schmidt, S. P., King, C. C. Jr. and Clark, E. M. (1984). Association of fungal endophyte with seasonal gains of beef steers grazing tall fescue pasture. In *Proceedings of European Grassland Federation*, ed. H. Riley and A. D. Skelvag. Zurich, Switzerland: European Grassland Federation, pp. 382–386.

Howard, M. D., Muntifering, R. B., Bradley, N. W., Mitchell, G. E. Jr. and Lowry, S. R. (1992). Voluntary intake and ingestive behaviour of steers grazing Johnstone or endophyte-infected Kentucky-31 tall fescue. *Journal of Animal Science*, **70**, 1227–1237.

Hume, D. E., Popay, A. J., Cooper, J. P. *et al.* (2004). Effect of a novel endophyte on the productivity of perennial ryegrass (*Lolium perenne*) in New Zealand. In *Proceedings of 5th International Symposium on Neotyphodium/ Grass Interactions*, ed. R. Kallenbach, C. Rosekrans Jr. and T. R. Lock, Poster #313.

Hume, D. E., Cooper, B. M. and Panckhurst, K. A. (2009). The role of endophyte in determining the persistence and productivity of ryegrass, tall fescue and meadow fescue in Northland. *Proceedings of the New Zealand Grassland Association*, **71**, 145–150.

Hume, D. E., Card, S. D. and Rolston, M. P. (2013). Effects of storage conditions on endophyte and seed viability in pasture grasses. *Proceedings of the 22nd International Grassland Congress*, Sydney, pp. 405–408.

Gundel, P., Rudgers, J. and Ghersa, C. (2011). Incorporating the process of vertical transmission into understanding of host–symbiont dynamics. *Oikos*, **120**, 1121–1128.

Jensen, J. G. and Popay, A. J. (2004). Perennial ryegrass infected with AR37 endophyte reduces survival of porina larvae. *New Zealand Plant Protection*, **57**, 323–328.

Jensen, J. G. and Popay, A. J. (2007). Reductions in root aphid population by non-toxic endophyte strains in tall fescue. In *Research and Practice New Zealand Grassland Association 13: 6th International Symposium on Fungal Endophytes of Grasses*, ed. A. J. Popay and E. R. Thom. Palmerston, New Zealand: New Zealand Grassland Association, pp. 341–344.

Jensen, J. G., Popay, A. J. and Tapper, B. A. (2009). Argentine stem weevil adults are affected by meadow fescue endophyte and its loline alkaloids. *New Zealand Plant Protection*, **62**, 12–18.

Johnson, R. D. and Voisey, C. R. (2017). *Method for Detection of Viable Endophyte in Plants*. US Patent number 9631242.

Johnson, R. D., Lane, G. A., Koulman, A. *et al.* (2015). A novel family of cyclic oligopeptides derived from ribosomal peptide synthesis of an in planta-induced gene, gigA, in *Epichloë* endophytes of grasses. *Fungal Genetics & Biology*, **85,** 14–24.

Johnson, J. M., Aiken, G. E., Phillips, T. D. *et al.* (2012). Steer and pasture responses for a novel endophyte tall fescue developed for the upper transition zone. *Journal of Animal Science*, **90**, 2402–2409.

Johnson, L. J., de Bonth, A. C. M., Briggs, L. R. *et al.* (2013a). The exploitation of epichloae endophytes for agricultural benefit. *Fungal Diversity*, **60**, 171–188.

Johnson, L. J., Koulman, A., Christensen, M. *et al.* (2013b). An extracellular siderophore is required to maintain the mutualistic interaction of *Epichloë festucae* with *Lolium Perenne*. *PLoS Pathogens*, **9**: e1003332.

Johnson, T., ed. (1983). *Proceedings of Tall Fescue Toxicosis Workshop.* Cooperative Extension Service. Athens, Georgia: University of Georgia College of Agriculture, pp. 1–26.

Kane, K. H. (2011). Effects of endophyte infection on drought stress tolerance of *Lolium perenne* accessions from the Mediterranean region. *Environmental and Experimental Botany*, **71**, 337–344.

Karimi, S., Mirlohi, A., Sabzalian, M. R., Tabatabaei, B. E. S. and Sharifnabi, B. (2012). Molecular evidence for *Neotyphodium* fungal endophyte variation and specificity within host grass species. *Mycologia*, **104**, 1281–1290.

Koulman, A., Lee, T. V., Fraser, K. *et al.* (2012). Identification of extracellular siderophores and a related peptide from the endophytic fungus *Epichloë festucae* in culture and endophyte-infected *Lolium perenne*. *Phytochemistry*, **75,** 128–139.

Lane, G. A., Christensen, M. J. and Miles, C. O. (2000). Coevolution of fungal endophytes with grasses: the significance of secondary metabolites. In *Microbial Endophytes*, ed. C. W. Bacon and J. F. White. New York: Marcel Dekker, pp. 341–388.

Latch, G. C. M. (1994). Influence of *Acremonium* endophytes on perennial ryegrass improvement. *New Zealand Journal of Agricultural Research*, **37**, 311–318.

Latch, G. C. M. (1997). An overview of *Neotyphodium*/Grass interactions. In *Neotyphodium/Grass Interactions*, ed. C. W. Bacon and N. S. Hill. New York: Plenus Press, New York, pp. 1–11.

Latch, G. C. M. and Christensen, M. J. (1985). Artificial infections of grasses with endophytes. *Annals of Applied Biology*, **107**, 17–24.

Latch, G. C. M., Christensen, M. J. and Samuels, G. J. (1984). Five endophytes of *Lolium* and *Festuca* in New Zealand. *Mycotaxon*, **20**, 535–550.

Latch, G. C. M., Christensen, M. J., Tapper, B. A. *et al.* (2000). *Tall fescue Endophytes*. U.S. Patent 6 111 170. Date issued: 29 August 2000.

Leuchtmann, A., Bacon, C. W., Schardl, C. L., White, J. F. and Tadych, M. (2014). Nomenclatural realignment of *Neotyphodium* species with genus *Epichloë*. *Mycologia*, **106,** 202–215.

Leury, B., Henry, M., Kemp. S. and Webb-Ware, J. (2014). *Developing Increased Understanding, Awareness and Potential Mitigation Strategies for Perennial Ryegrass Toxicosis in Sheep Production Systems*. Project B.AHE.0039. North Sydney, Australia: Meat and Livestock Australia Limited.

Looper, M. L., Rorie, R. W., Person, C. N. *et al.* (2009). Influence of toxic endophyte-infected fescue on sperm characteristics and endocrine factors of yearling Brahman-influenced bulls. *Journal of Animal Science*, **87**, 1184–1191.

Lyons, P. C., Plattner, R. D. and Bacon, C. W. (1986). Occurrence of peptide and clavine ergot alkaloids in tall fescue grass. *Science*, **232**, 487–489.

Kuldau, G. and Bacon, C. (2008). Clavicipitaceous endophytes: Their ability to enhance resistance of grasses to multiple stresses. *Biological Control*, **46**, 57–71.

Marahiel, M. A. (1997). Protein templates for the biosynthesis of peptide antibiotics. *Chemistry and Biology*, **4,** 561–567.

Malinowski, D. P. and Belesky, D. P. (2000). Adaptations of endophyte-infected cool-season grasses to environmental stresses: mechanisms of drought and mineral stress tolerance. *Crop Science*, **40**, 923–940.

Malinowski, D., Leuchtmann, A., Schmidt, D. and Nosberger, J. (1997). Growth and water status in meadow fescue (*Festuca pratensis*) is affected by *Neotyphodium* and *Phialophora* endophytes. *Agronomy Journal*, **89**, 673–678.

Meister, B., Krauss, J., Härri, S. A., Schneider, M. V. and Müller, C. B. (2006). Fungal endosymbionts affect aphid population size by reduction of adult life span and fecundity. *Basic and Applied Ecology*, **7**, 244–252.

Moate, P. J., Williams, S. R. O., Grainger, C. *et al.* (2012). Effects of wild-type,

AR1 and AR37 endophyte-infected perennial ryegrass on dairy production in Victoria, Australia. *Animal Production Science*, **52**, 1117–1130.

Moon, C. D., Tapper, B. A. and Scott, B. (1999). Identification of Epichloë endophytes in planta by a microsatellite-based PCR fingerprinting assay with automated analysis. *Applied and Environmental Microbiology*, **65**, 1268–1279.

Panaccione, D. G., Johnson, R. D., Wang, J. *et al.* (2001). Elimination of ergovaline from a grass-*Neotyphodium* endophyte symbiosis by genetic modification of the endophyte. *Proceedings of the National Academy of Sciences of the United States of America*, **98,** 12820–12825.

Parish, J. A., McCann, M. A., Watson, R. H. *et al.* (2003a). Use of non-ergot alkaloid-producing endophytes for alleviating tall fescue toxicosis in sheep. *Journal of Animal Science*, **81**, 1316–1322.

Parish, J. A., McCann, M. A., Watson, R. H. *et al.* (2003b). Use of non-ergot alkaloid-producing endophytes for alleviating tall fescue toxicosis in stocker cattle. *Journal of Animal Science*, **81**, 2856–2868.

Patchett, B. J., Chapman, R. B., Fletcher, L. R. and Gooneratne, S. R. (2008). Endophyte infected *Festuca pratensis* containing loline alkaloids deter feeding by *Listronotus bonariensis*. *New Zealand Plant Protection*, **61**, 205–209.

Patchett, B. J., Gooneratne, S. R. and Chapman, R. B. (2011). Effects of loline-producing endophyte-infected meadow fescue ecotypes on New Zealand grass grub (*Costelytra zealandica*). *New Zealand Journal of Agricultural Research*, **54**, 303–313.

Paterson, J., Forcherio, C., Larson, B., Samford, M. and Kerley, M. (1995). The effects of fescue toxicosis on beef cattle productivity. *Journal of Animal Science*, **73**, 889–98.

Pedersen, J. F., Lacefield, G. D. and Ball, D. M. (1990). A review of the agronomic characteristics of endophyte-free and endophyte infected tall fescue. *Applied Agricultural Research*, **5**, 188–194.

Pennell C., Popay, A. J., Ball, O. J. P., Hume, D. E. and Baird, D. B. (2005). Occurrence and impact of pasture mealybug (*Balanococcus poae*) and root aphid (*Aploneura lentisci*) on ryegrass (*Lolium* spp.) with and without infection by *Neotyphodium* fungal endophytes. *New Zealand Journal of Agricultural Research*, **48**, 329–337.

Peters, C. W., Grigsby, K. N., Aldrich, C. G. *et al.* (1992). Performance, forage utilization, and ergovaline consumption by beef cows grazing endophyte fungus-infected tall fescue, endophyte fungus-free tall fescue, or orchardgrass pastures. *Journal of Animal Science*, **70**, 1550–1561.

Philipson, M. N. and Christey, M. C. (1986). The relationship of host and endophyte during flowering, seed formation, and germination of *Lolium perenne*. *New Zealand Journal of Botany*, **24**, 125–134.

Popay, A. J. and Hume, D. H. (2011). Endophytes improve ryegrass persistence by controlling insects. *New Zealand Grassland Association Pasture Persistence Symposium. Grassland Research and Practice Series*, **15**, 149–156.

Popay, A. J. and Rowan, D. D. (1994). Endophytic fungi as mediators of plant insect interactions. In *Insect-Plant Interactions*, ed. E. A. Bernays. Boca Raton, FL: CRC Press, pp. 83–103.

Popay, A. J. and Thom, E. R. (2009). Endophyte effects on major insect pests in Waikato dairy pasture. *Proceedings of the New Zealand Grassland Association*, **71**, 121–127.

Popay, A. J., Hume, D. E., Baltus, J. G. *et al.* (1999). Field performance of perennial ryegrass (*Lolium perenne*) infected with toxin-free fungal endophytes (*Neotyphodium* spp.). In *Ryegrass endophyte: an Essential New Zealand Symbiosis*, ed. D. R. Woodfield and C. Matthew. *Grassland Research and Practice Series* **7**, 113–122.

Popay, A. J., Tapper, B. A. and Podmore, C. (2009). Endophyte-infected meadow fescue and loline alkaloids affect Argentine stem weevil larvae. *New Zealand Plant Protection*, **62**, 19–27.

Porter, J. K. and Thompson, F. N. Jr. (1992). Effects of fescue toxicosis on reproduction in livestock. *Journal of Animal Science*, **70**, 1594–1603.

Potter, D. A., Patterson, C. G. and Redmond, C. T. (1992). Influence of turfgrass species and tall fescue endophyte on feeding ecology of Japanese beetle and southern masked chafer grubs (Coleoptera: Scarabaeidae). *Journal of Economic Entomology*, **85**, 900–909.

Prestidge, R. A. and Gallagher R. T. (1988). Endophyte fungus confers resistance to ryegrass: Argentine stem weevil larval studies. *Ecological Entomology*, **13**, 429–435.

Prestidge, R. A., Pottinger, R.P. and Barker, G. M. (1982). An association of *Lolium* endophyte with ryegrass resistance to Argentine stem weevil. *Proceedings of the 35th New Zealand Weed and Pest Control Conference*, 119–122.

Prestidge, R. A., Barker, G. M. and Pottinger, R. P. (1991). The economic cost of Argentine stem weevil in pastures in New Zealand. *Proceedings of the New Zealand Weed and Pest Control Conference*, **44**, 165–170.

Putnam, M. R., Bransby, D. I., Schumacher, J. *et al.* (1991). Effects of the fungal endophyte *Acremonium coenophialum* in fescue on pregnant mares and foal viability. *American Journal of Veterinary Research* **52**, 2071–2074.

Realini, C. E., Duckett, S. K., Hill, N. S. *et al.* (2005). Effect of endophyte type on carcass traits, meat quality, and fatty acid composition of beef cattle grazing tall fescue. *Journal of Animal Science*, **8,** 430–439

Reddick, B. and Collins, M. (1988). An improved method for detection of *Acremonium coenophialum* in tall fescue plants. *Phytopathology*, **78**, 418–420.

Reidell, W., Kieckhefer, R., Petroski, R. and Powell, R. (1991). Naturally occurring and synthetic loline alkaloid derivatives: insect feeding behaviour modification and toxicity. *Journal of Entomological Science*, **26**, 122–129.

Rhodes, M. T., Paterson, J. A., Kerley, M. S., Garner, H. E. and Laughlin, M. H. (1991). Reduced blood flow to peripheral and core body temperatures in sheep and cattle induced by endophyte infected tall fescue. *Journal of Animal Science*, **69**, 2033–2043.

Rolston, M. and Agee, C. (2007). Delivering quality seed to specification-the USA and NZ novel endophyte experience. In *Research and Practice New Zealand Grassland Association 13: 6th International Symposium on Fungal Endophytes of Grasses*, ed. A. J. Popay and E. R. Thom. Palmerston, New Zealand: New Zealand Grassland Association, pp. 229–232.

Rolston, M., Hare, M., Moore, K. and Christensen, M. (1986). Viability of *Lolium* endophyte fungus in seed stored at different moisture contents and temperatures. *New Zealand Journal of Experimental Agriculture*, **14**, 297–300.

Rolston, M. P., Archie, W. J. and Simpson, W. (2002). Tolerance of AR1 *Neotyphodium* endophyte to fungicides used in perennial ryegrass seed production. *Proceedings New Zealand Plant Protection Conference*, **55**, 322–325.

Rolston, M. P., McCloy, B. L., Harvey, I. C. and Chynoweth, R. W. (2009). Ryegrass (*Lolium perenne*) seed yield response to fungicides: a summary of 12 years of field research. *New Zealand Plant Protection*, **62**, 343–348.

Rowan, D. D. and Gaynor, D. L. (1986). Isolation of feeding deterrents against Argentine stem weevil from ryegrass infected with the endophyte *Acremonium loliae*. *Journal of Chemical Ecology*, **12**, 647–658.

Rowan, D. D., Hunt, M. B. and Gaynor, D. L. (1986). Peramine, a novel insect feeding deterrent from ryegrass infected with the endophyte *Acremonium loliae*. *Journal of Chemistry Society Chemistry Communications*, **1986**, 935–936.

Rowan, D. D., Dymock, J. J. and Brimble, M. A. (1990). Effect of fungal metabolite peramine and analogs on feeding and development of Argentine stem weevil (*Listronotus bonariensis*). *Journal of Chemical Ecology*, **16**, 1683–1695.

Saikia, S., Takemoto, D., Tapper, B. A. *et al.* (2012). Functional analysis of an indole-diterpene gene cluster for lolitrem B biosynthesis in the grass endosymbiont *Epichloë festucae*. *FEBS Lett*, **586**, 2563–2569.

Saikkonen, K., Young, C. A., Helander, M. and. Schardl, C. L. (2016). Endophytic Epichloë species and their grass hosts: from evolution to applications. *Plant Molecular Biology*, **90**, 665–675.

Schardl, C. L., Leuchtmann, A., Chung, K. R., Penny, D. and Siegel, M. R. (1997). Coevolution by common descent of fungal symbionts (*Epichloë spp.*) and grass hosts. *Molecular Biology & Evolution*, **14**, 133–143.

Schardl, C. L., Leuchtmann, A. and Spiering, M. J. (2004). Symbioses of grasses with seedborne fungal endophytes. *Annual Review of Plant Biology*, **55**, 315–340.

Schardl, C. L., Grossman, R. B., Nagabhyru, P., Faulkner, J. R. and Mallik, U. P. (2007). Loline alkaloids: currencies of mutualism. *Phytochemistry*, **68,** 980–996.

Schardl, C. L., Craven, K. D., Speakman S. *et al.* (2008). A novel test for host-symbiont codivergence indicates ancient origin of fungal endophytes in grasses. *Systematic Biology*, **57**, 483–498.

Schardl, C. J., Young, C. A., Hesse, U. *et al.* (2013). Plant-symbiotic fungi as chemical engineers: multi-genome analysis of the Clavicipitaceae reveals dynamics of alkaloid loci. *PLoS Genetics* **9**, e1003323.

Schmidt, S. P. and Osborn, T. G. (1993). Effects of endophyte-infected tall fescue on animal performance. *Agriculture Ecosystems and Environment*, **44**, 233–262.

Schmidt, S. P., Hoveland, C. S., Clark, E. M. *et al.* (1982). Association of an endophytic fungus with fescue toxicity in steers fed Kentucky 31 tall fescue seed or hay. *Journal of Animal Science*, **55**, 1259.

Schmidt, S. P., Danielson, A. D., Holliman, J. A., Grimes, H. W. and Webster, W. B. (1986). Fescue fungus suppresses growth and reproduction in replacement beef heifers. *Highlights of Agricultural Research Alabama*

Agricultural Experiment Station, Auburn University, Auburn, **33**, 15.

Schnitzius, J. M., Hill, N. S., Thompson, C. S. and Craig, A. M. (2001). Semiquantitative determination of ergot alkaloids in seed, straw, and digesta samples using a competitive enzyme-linked immunosorbent assay. *Journal of Veterinary Diagnostic Investigation* **13**, 230–237.

Stewart, A. (1986). Effect on the Lolium endophyte of nitrogen applied to perennial ryegrass seed crops. *New Zealand Journal of Experimental Agriculture*, **14**, 393–397.

Strahan, S. R., Hemken, R. W., Jackson, J. A. Jr. *et al.* (1987). Performance of lactating dairy cows fed tall fescue forage. *Journal of Dairy Science*, **70**, 1228–1234.

Strickland, J. R., Looper, M. L., Matthews, J. C. *et al.* (2011). St. Anthony's fire in livestock; causes, mechanisms and potential solutions. *Journal of Animal Science*, **89**, 1603–1626.

Stuedemann, J. A. and Thompson, F. N. (1993). Management strategies and potential opportunities to reduce the effects of endophyte-infested tall fescue on animal performance. In *Proceedings of the Second International Symposium on Acremonium/Grass Interactions*, ed. D. E. Hume, G. C. M. Latch and H. S. Easton. Palmerston, New Zealand: Grasslands Research Centre, pp. 103–114.

Stuedemann, J. A., Breedlove, D. L., Pond, K. R. *et al.* (1989). Effect of endophyte (*Acremonium coenophialum*) infection of tall fescue and paddock exchange on intake and performance of grazing steers. *Proceedings of the XVI International Grassland Congress*, pp. 1243–1244.

Sullivan, T. J., Bultman, T. L. and Schoolcraft, J. (2016). Primers to amplify SNP markers in *Epichloë canadensis* (Clavicipitaceae). *Applications in Plant Sciences*, **4** (3), apps.1500078.

Takach, J. E. and Young, C. A. (2014). Alkaloid genotype diversity of tall fescue endophytes. *Crop Science*, **54**, 667–678.

Takach, J. E., Mittal, S., Swoboda, G. A. *et al.* (2012). Genotypic and chemotypic diversity of Neotyphodium endophytes in tall fescue from Greece. *Applied and Environmental Microbiology*, **78**, 5501–5510.

Tanaka, A., Tapper, B. A., Popay, A., Parker, E. J. and Scott, B. (2005). A symbiosis expressed non-ribosomal peptide synthetase from a mutualistic fungal endophyte of perennial ryegrass confers protection to the symbiotum from insect herbivory. *Molecular Microbiology*, **57,** 1036–1050.

Tapper, B. A. and Lane, G. A. (2004). Janthitrems found in a *Neotyphodium* endophyte of perennial ryegrass. In *Proceedings of 5th International Symposium on Neotyphodium/Grass interactions*, ed. R. Kallenbach, C. Jr. Rosekrans and T. R. Lock, #301.

Tapper, B. A. and Latch, G. C. M. (1999). Selection against toxin production in endophyte-infected perennial ryegrass. In *Ryegrass Endophyte: an Essential New Zealand Symbiosis*, ed. D. R. Woodfield and C. Matthew. *Grassland Research and Practice Series* **7**, Napier, New Zealand, pp. 107–111.

Taylor, M. C., Loch, W. E. and Ellersieck, M. (1985). Toxicity in pregnant pony mares grazing Kentucky-31 fescue pastures. *Nutrition Report International* **31**, 787.

Thom E. R., Popay A. J., Waugh C. D. and Minneé E. M. (2014). Impact of novel endophytes in perennial ryegrass on herbage production and insect pests from pastures under dairy cow grazing in northern New Zealand. *Grass Forage Science*, **69**, 191–204.

Timper, P., Gates, R. N. and Bouton J. H. (2005). Response of *Pratylenchus* spp. In tall fescue infected with different strains of the fungal endophyte *Neotyphodium coenophialum. Nematology*, **7**, 105–110.

van Zijll de Jong, E., Guthridge, K. M., Spangenberg, G. C. and Forster, J. W. (2003). Development and characterization of EST-derived simple sequence repeat (SSR) markers for pasture grass endophytes. *Genome*, **46,** 277–290.

Watson, B. M. (2007). The effect of endophyte in perennial ryegrass and tall fescue on red and blackheaded pasture cockchafers. In *Research and Practice New Zealand Grassland Association 13: 6th International Symposium on Fungal Endophytes of Grasses*, ed. A. J. Popay and E. R. Thom. Palmerston, New Zealand: New Zealand Grassland Association, pp. 347–352.

Watson, R. H., McCann, M. A., Parish, J. A. *et al.* (2004). Productivity of cow-calf pairs grazing tall fescue pastures infected with either the wild-type endophyte or a non-ergot alkaloid-producing endophyte strain, AR542. *Journal of Animal Science*, **82**, 3388–3393.

Welty, R., Azevedo, M. and Cooper, T. (1987). Influence of moisture content, temperature, and length of storage on seed germination and survival of endophytic fungi in seeds of tall fescue and perennial ryegrass. *Phytopathology*, **77**, 893–900.

West, C. P., Izekor, E., Turner, K. E. and Elmi, A. A. (1993). Endophyte effects on growth and persistence of tall fescue along a water-supply gradient. *Agronomy Journal*, **85**, 264–270.

Wilkinson, H. H., Siegel, M. R., Blankenship, J. D. *et al.* (2000). Contribution of fungal loline alkaloids to protection from aphids in a grass–endophyte mutualism. *Molecular Plant–Microbe Interactions*, **13**, 1027–1033.

Young, C. A., Bryant, M. K., Christensen, M. J. *et al.* (2005). Molecular cloning and genetic analysis of a symbiosis-expressed gene cluster for lolitrem biosynthesis from a mutualistic endophyte of perennial ryegrass. *Molecular Genetics & Genomics*, **274,** 13–29.

Young, C. A., Felitti, S., Shields, K. *et al.* (2006). A complex gene cluster for indole-diterpene biosynthesis in the grass endophyte *Neotyphodium lolii. Fungal Genetics & Biology*, **43,** 679–693.

Young, C. A., Charlton, N. D., Takach, J. E. *et al.* (2014). Characterization of *Epichloë coenophiala* within the US: Are all tall fescue endophytes created equal? *Chemistry and Biology*, **2**, 95.

Zhuang, J., Marchant, M. A., Schardl, C. L. and Butler, C. M. (2005). Economic analysis of replacing endophyte-infected with endophyte-free tall fescue pastures. *Agronomy Journal*, **97**, 711–716.

17

Plant Growth-Promoting Bacteria Field Trials in Europe

KAREN O'HANLON

Abstract

There are increasingly restrictive EU regulations surrounding the use of chemicals in farming due to increased information linking environmental behaviour and ecotoxicity, such as the effect of the insecticide class neonicotinoids on bees. For this reason there will be a continued move towards a reduction of chemical use in farming throughout the EU. In this context, the use of plant growth-promoting bacteria (PGPB) offers an attractive alternative to chemical fertilisers. A range of commercially available PGPB products, consisting of different species of bacteria, offer a potential alternative to chemicals. This includes: *rhizo power*® by nadicom who provide bacteria-based organic fertilisers. In this chapter, we outline detailed findings of four field trials undertaken using PGPB in Europe on alfalfa, broccoli, faba beans and tomatoes. All trials using these products yielded positive results such as faster germination, higher yields, increased chlorophyll in leaves and improved tap root formation. Larger plants resulting in significantly higher yield of organic tomatoes were also observed in the tomato field trial. Longer tap root and deeper green colour was observed in the broccoli trial and a reduction in chocolate spot was shown when treated with *Bacillus subtilis*. Statistically significant higher yields were shown in treated alfalfa crops using *Ensifer meliloti*.

17.1 Introduction

The large number of research groups publishing on PGPB shows the great interest in identifying organisms for the development of products for agricultural use.

Currently, bacterial strains are used routinely and effectively in many developing countries for plant growth promotion (Timmusk, 2017). In Europe, where agricultural chemicals remain relatively inexpensive compared to the cost of food, the use of PGPB occupies a small but growing niche in the development of organic agriculture (Timmusk, 2017). The microbial inoculum has several advantages over artificial agricultural chemicals. They are environmentally friendly renewable sources of nutrients and phytohormones and they restore soil fertility, ultimately improving plant growth. In addition to fighting agricultural pathogens, microbes can also alleviate abiotic stresses.

Notwithstanding the long-time use of chemicals in agriculture, PGPB use is in its infancy in Europe but is slowly being recognised in the non-organic sector. The growth of this area is hampered however by: (1) the specificity of microbial strains to particular environmental regions or plants; (2) inconsistencies among regulatory agencies worldwide regarding the release of PGPB products commercially; and (3) shortage of data on the mechanisms of the interaction between the bacteria and other microbes and fungi in the soil (Glick, 2012). Furthermore, the difficulty in uptake by non-organic farmers in the developed world is hampered by the historic reliance on chemicals in monoculture patterned farming and the expectation of immediate biocidal effects shown when using chemicals.

PGPB as antifungal agents are highly targeted to particular species. In contrast, chemicals have a broad-spectrum mode of action; for example, fungicides target all or most fungi. This can result in variable quality and efficacy under field conditions. Currently, there are also challenges with microbial product registration. However, due to the probable lower toxicity and ecotoxicity of these environmentally available strains of bacteria, one can envisage that the registration process will become quicker once the regulatory bodies establish a clear framework among all EU member states for registering these products (Timmusk, 2017).

Despite the problems with uptake and acceptance of PGPB use in agriculture in Europe, many companies are commercialising these organisms as products for use as soil conditioners and fertilisers. These beneficial claims circumvent the need to register the products as plant protection products (PPPs). Other long established agri-chemical companies are also registering these under the PPP legislative framework. Bayer Crop Science, for instance, have commercialised Serenade ASO in Europe (*Bacilus subtilis*) as an antifungal agent containing the spore-forming microorganism.

The PGPB field is also growing as other smaller commercial companies are entering from various sectors. One such company, nadicom GmbH, began as a German bacterial identification contract company for pharmaceuticals with a large bank of bacterial strains at its disposal. Progressive seed companies have purchased nitrogen-fixing bacterial strains to coat legume seeds for some time now. Because of these partnerships, soya, lupine, clover, alfalfa and bean seeds have also been successfully coated and used in Germany and Italy since 2012. No regulations on the use of rhizobia strains is enforced in Germany or Italy currently. However, the EU Biofertiliser Regulation is due to be published in 2018/2019 and this will include rhizobia bacteria. Therefore, nitrogen-fixing rhizobia bacterial strains will be regulated under this biofertiliser regulation.

In order to approve the use of PGPB by commercial growers and farmers in the non-organic sector, efficacy trials need to demonstrate the benefits of their use to improve yields. The environmental sustainability and long-term soil conditioning benefits are unquestionable. However, farmers are mainly interested in the potential percentage yield increases. One of the major problems with the long-term benefits is that many farmers occupy rented land and cannot justify the costs of using PGPB for the long-term improvement in soil quality when they will not be using the land the following season. This is one challenge that needs to be addressed by integrated pest management (IPM), including the Framework Directive on the Sustainable Use of Pesticides (EU DG Environment) and member states should incentivise farmers to think of the long-term sustainability of land rather than short-term measures.

17.2 Challenges Faced in the Uptake of PGPB in the Agricultural Sector

17.2.1 Lack of Harmonisation in the Process Between Member States

Regulations in the EU with respect to the lack of quality restrictions concerning biofertilisers has left a situation where national or regional rules are applied; these are often variable and inconsistent with each other. For example, rhizobia does not currently need to be registered with a competent authority (CA) in Germany, Italy and Ireland, however, in France, registration is necessary. These products are usually registered first by the organic associations in member states as this is their niche market. Mutual recognition requires that one member state recognises that it has been approved for use in another member state and on their approval, they will register it also. However, problems can arise when one member state is not required to register the PGPB with the CA and the product can be marketed and sold freely without notification to the CA, yet another member state cannot mutually recognise it and register it because it does not have the CA paperwork from that member state.

17.2.2 The Registration Process

Regulatory processes and documentation for product registration are very complex and require significant levels of expertise to understand. The procedure for registration of PPPs, firstly requires that the active substance is authorised by the EU Commission DG SANCO (Directorate General for Health and Consumer Affairs) and subsequently, the formulated product is still a matter of national authorisation.

The process can take several years, as deadlines are often not met and numerous steps are difficult to comprehend. Fees are high and differ considerably from one country to another. Plant growth products are utilised in a number of countries and are not regulated or subjected to a level of regulation below that of PPP. It is important to determine which regulatory framework naturally occurring environmentally available soil organisms come under. It is usually the PPP framework if they claim to have plant protection properties and the fertiliser framework if they fertilise the soil and do not claim any PPP properties. However, some argue that there should be a completely separate framework developed for PGPB to encompass the range of bacteria and fungi we are currently aware of and leave scope to include the huge numbers yet to be discovered.

Policy in the EU aimed at developing the agricultural sector emphasises the necessity for a reduction in the use of chemical agents. The EU is required by Regulation (EC) No 1107/2009 to establish a list of substances identified as 'candidates for substitution'. The list identifies active substances with certain properties. For PPPs containing these active substances, member states will be required to evaluate whether they can be replaced (substituted) by other adequate solutions (chemical and non-chemical). The analysis is conducted by comparing the agreed and peer reviewed endpoints against the relevant seven conditions specified in Annex II, point 4 of the Regulation. The information is grouped in a comprehensive database that will be updated on a regular basis. The current draft list contains 77 candidates for substitution (October 2017). Likewise, products can be removed from the market by member states; for example, Ireland revoked the licence of 187 products between January 2017 and 2018. Removal from the list can occur for a number of reasons, most notably, the banning of the pesticide in the EU, failure to pass the risk assessment or the removal of the product from the market by the applicant.

Other legislative measures put in place in 2009 include IPM, and the Framework Directive on the Sustainable Use of Pesticides (EU DG Environment). The idea behind IPM is to combine different crop management practices to overcome the shortcomings of the individual system. Likewise, member states have been encouraged to use rural development programmes (funded under the Common Agricultural Policy) to provide financial incentives to farmers to begin implementation. The new legislation gives a specific status to non-chemical and natural alternatives to conventional chemical pesticides and requires them to be given priority wherever possible. Biopesticides should generally qualify as low-risk active substances under the legislation. Low-risk substances are granted initial approval for 15 years rather than the standard 10. A reduced dossier can be submitted for low-risk substances, but this must include a demonstration of sufficient efficacy. One requirement for low-risk substances, still to be elaborated, is that their half-life in the soil should be less than 60 days. This may cause problems for some microbial biopesticides, such as rhizosphere-competent antagonists of soil-borne plant pathogens (Timmusk, 2017).

17.3 Properties of PGPB Currently Known to Enhance Crop Production

17.3.1 Nitrification

Bacterial nitrification is a biological process in which energy is extracted by sequential oxidation of nitrogen that occurs as ammonia. Complete oxidation of nitrate is carried out by two metabolically distinct groups of bacteria: (1) ammonia-oxidising bacteria, such as *Nitrosomonas, Nitrosospira, Nitrosovibrio, Nitrosolobus* and *Nitrosococcus*, transform ammonia to nitrite, and (2) nitrifying bacteria that transform nitrite is to nitrate, such as *Nitrobacter, Nitrococcus, Nitrospira* and *Nitrospina*. Nitrification is important for soil and ecosystem health because it completes the mineralisation of organic nitrogen started with the ammonification process (nitrogen fixation) (Cohen and Mazzola, 2006; Cohen *et al.*, 2010; Ardisson *et al.*, 2014). It is believed that nitrification is the principal source of nitric oxide (NO) emitted from the soil. The ability to perform nitrification is considered as an important trait in the selection of beneficial bacteria which are capable of improving plant growth and crop yield (Prasad *et al.*, 2015).

17.3.2 Denitrification

Denitrification is defined as a microbial respiratory process during which soluble nitrogen oxides are used as alternative electron acceptors when O_2 is not available for aerobic respiration. It involves sequential reduction of NO^{3-} into dinitrogen in four steps coupled with energy conservation (NO to NO_2, NO_2 to NO, NO to N_2O and N_2O to N_2). Denitrification completes the N cycle and usually balances the total biological N fixation in the global N cycle (Hofstra and Bouwman, 2005; Philippot *et al.*, 2007). In the rhizosphere, oxygen concentration could be lowered because of root and microorganism respiration. In addition, organic compounds released by the plant's roots can be used as electron donors in the denitrification process. Denitrifying bacteria may prevent nitrogen accumulation to toxic levels, reduce nitrate contents in groundwater and maintain a balance between soil and atmospheric nitrogen avoiding serious problems that could occur if no alternative mechanism is available to return nitrogen to the atmosphere (Gupta *et al.*, 2000; Antoun and Prévost, 2005; Philippot *et al.*, 2007).

17.3.3 Siderophore Production

Iron participates in the formation of several types of vegetable proteins such as ferredoxin, cytochrome and leghemoglobin (Fukuyama, 2004; Liu *et al.*, 2014).

This element is relatively insoluble in soil solution. Plants can use microbial siderophores (from fungi and bacteria) which are synthesised under iron-starved conditions. Siderophores are defined as low molecular weight compounds (500–1500 Da) possessing high affinity for ferric iron. They are mainly produced by bacteria (Kümmerli *et al.*, 2014), fungi (Renshaw *et al.*, 2002) and graminaceous plants (Hider and Kong, 2010) to scavenge iron from the environment.

17.3.4 Bacterial Phytohormones

Phytohormones or 'plant growth hormones' including auxins are naturally occurring organic substances that exert, at low concentrations, a major influence on plant growth and upregulation of physiological process. Auxins are produced by plants (Bari and Jones, 2009), algae (Kiseleva *et al.*, 2012) and by microorganisms (Narayanasamy, 2013). Among microbes, PGPB can also modulate phytohormone levels in plant tissues affecting the hormonal balance of the host plant (Figueiredo *et al.*, 2016). Some of the most common phytohormones affecting plant growth are auxins, giberellins, cytokinins, ethylene and ABA. The bacterial ACC deaminase enzyme is key to regulating ethylene in plants and believed to be important in PGPB (Glick, 2012).

17.3.5 Production of Antimicrobial Agents

The mechanisms by which antagonistic microorganisms protect plants are generally attributed to parasitism, competition and the production of secondary metabolites with toxic effects on pathogens (Harman *et al.*, 2012). Among naturally occurring diversified microorganisms, many fungi such as *Trichoderma* sp., *Chaetomium* sp., *Coniothyrium minitans* and *Gliocladium* sp., bacteria including *Pseudomonas* sp. and *Bacillus* sp., and bacterial actinomycetes such as *Streptomyces* sp., have been screened and developed to control the soil-borne pathogens such as, *Pythium* sp., *Phytophthora* sp., *Fusarium* sp., *Rhizoctonia* sp. and *Sclerotium* sp. (Pan *et al.*, 2001; Pal and Gardener, 2006; Whipps *et al.*, 2008). Some antagonistic microorganisms, such as *Trichoderma* sp., *Ampelomyces quisqualis*, *Bacillus* sp. and *Ulocladium* sp., have also been developed for foliar disease control of powdery mildew and Botrytis rot (Paulitz and Belanger, 2001). *Pseudomonas oculosa* was developed for the control of powdery mildew (Paulitz and Belanger, 2001). *Bacillus amyloliquefaciens* has been developed to control soil-borne and foliar diseases, anthracnose and bacterial pustule of soybean (Prathuangwong and Kasem, 2003). The PGPB microorganism can also produce bacteriocin that inhibits closely related bacterial species and induces systemic resistance in plants (Khan *et al.*, 2006).

17.4 PGPB Field Trials in Europe

We conducted a series of plant growth-promoting bacteria field trials in Europe and report a summary of the findings here. From a commercial grower's perspective, the most important results to consider in these trials are yields, shelf life, crop uniformity, disease resistance and impact on soil fertility. The long-term improved fertility is important to farmers who own their own land, however, commercial growers often rent land for one or two seasons. This short-term view of soil fertility is an obstacle to overcome and perhaps the onus is on the land owner to ensure the long-term soil fertility of the land.

The trials conducted involved four treatments: (1) *rhizo power® Bacillus subtilis* applied as an antifungal agent for chocolate spot on faba beans (variety Fuego) in Ireland; (2) *rhizo power® Ensifer meliloti* on alfalfa in Ireland; (3) *rhizo power®* Increased Crop on broccoli (variety Parthenon) in Ireland; and (4) *rhizo power®* mix (*Bacillus subtilis*, Azotobacter and rhizobia) on industrial and organic tomatoes in Italy.

rhizo power® Bacillus subtilis is used for the treatment of pathogens. *Bacillus subtilis* stimulates plant-induced systemic resistance by bacterial metabolites produced in the vicinity of plant roots, this is the key mechanism in the biocontrol action of Gram-positive endospore-forming bacteria. It produces a broad-spectrum iturin antibiotic that disrupts the cell wall formation of pathogens, and it triggers an advantageous induced systemic resistance (ISR) in plants, whereby a plant's natural immune system is activated to fight plant diseases (Doornbos *et al.*, 2012). As a fungicide, *Bacillus subtilis* functions by stopping spore germination, disrupting the cell membrane and inhibiting growth.

rhizo power® Ensifer meliloti contains *E. meliloti* a rhizobium that enhances nitrogen fixation, siderophore formation, phytohormone production, indirectly promotes growth, and enhances biocontrol of pathogens and induces plant resistance.

rhizo power® Increased Crop is based on several different bacteria including: (1) *Pseudomonas fluorescens* (2 strains), which conveys a faster and higher growth of the shoot, gives higher yields, more seeds, and protects the plant against the toxic nature of cadmium; (2) *P. jesenii* (2 strains), are phosphorus (P), nitrogen (N) and copper (Cu) solubilising and supply the plant with these elements. It can increase shoot and root biomass, provide effective metal sequestering and promote growth bioinoculant for plants in metal stressed soil; (3) *Bacillus subtilis*, provides antifungal activity towards *Fusarium*, increases auxin and cytokinin production, aids phosphate solubilisation, reduces plant (fungal) disease, produces siderophores, and confers a more effective promotion of plant development when combined with nitrogen-fixing rhizobia; (4) *B. simplex* (=*muralis*), provides antifungal activity towards *Fusarium*, auxin production, produces siderophores, and confers a more effective promotion of plant development when combined with nitrogen-fixing rhizobia; (5) *B. licheniformis*, produces high amounts of physiologically active gibberellins, aids stem elongation, and reduces plant disease by inducing some mechanism of systemic resistance.

rhizo power® mix contains (1) Azotobacter, which utilises atmospheric nitrogen for cell protein synthesis (Jnwali *et al.*, 2015). This cell protein is then mineralised in soil after the death of *Azotobacter* cells thereby contributing towards the N availability to the crop plants. The agri-economically important species of *Azotobacter* include *A. chroococcum*, *A. vinelandii*, *A. salinestris*, and *A. beijerinckii*. Like other PGPB, *Azotobacter* has also been reported to enhance germination, growth and yield of vegetable crops (Dhumal, 1992; Verma and Shende, 1993); (2) rhizobia, enhances nitrogen fixation, siderophore formation, increases phytohormone production, indirectly promotes growth, enhances biocontrol of pathogens and induces plant resistance; and (3) *Bacillus subtilis*, confers the following characteristics for use as a biofertiliser: (a) tryptophan-dependent synthesis of indole-3-acetic acid for auxin production and root development (Idris *et al.*, 2007); (b) release of volatiles, such as 2,3-butanediol and acetoin to enhance plant growth (Borriss, 2011); (c) phytase production to enhance phosphorous availability (Idris *et al.*, 2002); and (d) nitrogen-fixing, mineral solubilisation and secretion of macromolecule-degrading enzymes (Borriss, 2011).

17.4.1 Treatments for Bean, Alfalfa, Broccoli and Tomato Field Trials

Bean trials included an untreated control and plants treated with *Bacillus subtilis* sprayed onto the emerged crop. The full treatment included three applications of Signum @ 0.75kg ha^{-1}. The beans were grown on a field following a previous crop of winter wheat. No beans had been grown in the past 10 years: nitrogen index 1 (very low), phosphorus index 3 (good), potassium index 3 (good), pH 6.8, organic matter 4.6%. Beans were sown in 24 × 2.5 m plots at 35 seeds m^{-2}, with plough + 1 pass (power harrow plus wintersteiger plot drill to 10 cm depth). Herbicide was applied pre-sowing as glyphosate + pre-emergence Nirvana at 3.5 l ha^{-1} and Lingo at 1.7 l ha^{-1}. Establishment was excellent (> 90% in all plots). Fertiliser was applied at 20 kg ha^{-1} P + 60 kg ha^{-1} K. Insecticides were applied as follows: Aphox – 1 May for bean weevil control. Weed and insect control was excellent. Blocks 2 and 3 suffered slightly from drought stress in late July. All blocks suffered from drought stress in early/mid-August. Statistical analysis was conducted in Genstat with $p < 0.05$ for significant differences between treatments.

Alfalfa seed was treated with *Ensifer meliloti* or left as untreated. The area of each plot was 9 × 1.5 m, and planted area for each plot: 6.3 × 1.1 m. Five replicates were included. The previous crop was spring barley. Soil characteristics included: nitrogen index 1 (very low), phosphorus index 2, potassium index 2, pH 6.7, and organic matter 4.8%. Sowing method was plough + 1 pass (power harrow plus wintersteiger plot drill). Herbicide application occurred pre-sowing with glyphosate. Fertiliser was applied at 20 kg ha^{-1} P + 60 kg ha^{-1} K. Weed control was manual. In 2016, one cut was carried out (harvesting date: 27 September, 2016). In 2017, four cuts were carried out. Yield was determined by weighing the entire plot (fresh weight yield). A subsample was retrieved that was subsequently weighed, dried and further weighed to calculate the dry matter. Two-way ANOVA was performed using harvest (cut) and seed treatment as factors and accounting for blocking effects. The analysis was completed for fresh weight (t ha^{-1}) and dry weight (t ha^{-1}) as variates, separately.

Broccoli was treated with *Bacillus subtilis, Pseudomonas fluorescens and P. jesenii* applied over ready-made drills prior to seedling planting with an application rate of 500 ml 10 l^{-1} water sprayed in field at a rate of 1000 l ha^{-1}. The soil was fertile, and 2 ha was treated and 2 ha untreated. Broccoli was grown in propagation trays from seed for 6 weeks before being planted into the field. An area of 2 ha treated plants and 2 ha untreated plants were sown. Soil samples were taken at 4 weeks post planting in field and 12 weeks post planting for bacteria analysis. Heads were picked at harvest randomly by the grower by walking 30 m into the drill and selecting ten heads together, of both treated and untreated crops.

Tomatoes were treated with *Bacillus subtilis, Azotobacter* and rhizobia, or left as untreated controls. The organic trials were in four blocks; total: 2 tests (treated and control) × 4 repetitions = 8 plots. The non-organic trials had two variables: normal fertiliser and reduced fertiliser with four blocks; total: 2 tests (treated and untreated control) × 2 variables × 4 repetitions = 16 plots. The preparation of seeds and sowing was carried out at a nursery. A 10% solution (1:10) of bacteria at 10^9 cfu ml^{-1} was diluted with chlorine-free water. Propagation trays containing seedlings were placed in a trough containing the bacterial mix for approximately 40 seconds for the roots to soak up the bacterial mix. The seedlings were transplanted to the field in mid-May and harvested mid-August. Normal field agronomic management was applied. Mechanical transplantation, fertilisation, irrigation, plant protection and treatments were carried out. Tomatoes were harvested manually (80% of ripe berries). Data collection was undertaken on the plant and berries before harvesting was conducted and a score of 1 (worst) to 5 (best value) was assigned for the following: plant health, fruit cover, berry consistency, uniformity of colour, resistance to berry burns, resistance to splits berries, resistance to over-maturation. Yield, plant and juice were measured. Measurements also included: red berry weight (kg): to calculate the marketable production in tonnes ha^{-1}; green berry weight (kg): to calculate the immature production in t ha^{-1}; berry counting: to calculate rotten production in t ha^{-1}; weight of 100 berries: average berry weight (g); average length of branches (in cm measured on ten plants per parcel); Brix value to measure the sugar content of the tomatoes; and major and minor defects on a sample of 100 berries picked from each parcel.

17.4.2 Results

Bacillus subtilis on Bean Leaves for the Treatment of Chocolate Spot

The leaf area index, number of pods, flowers, and height differences between treated with *Bacillus subtilis,* untreated bean crop and full treatment crop showed no difference. The percentage of disease observed at the top, middle and bottom of the plant

Table 17.1 The effect of *Bacillus subtilis* on bean leaves (var. Fuego) for the treatment of chocolate spot: percentage disease on leaves at bottom, middle and top of plant by visual assessment.

	Bottom	Middle	Top
10 May			
Full Treatment	0.80	0.20	0
Treated	3.53	0.76	0
Untreated	4.33	0.50	0.03
25 June			
Full Treatment	1.20	0.80	0.50
Treated	4.80	2.20	1.60
Untreated	6.30	3.10	2.10
27 July			
Full Treatment	4.50	2.50	1.60
Treated	9.50	6.30	3.90
Untreated	11.2	7.40	5.30

is presented in Table 17.1. Full treatment was significantly better than untreated and *Bacillus* treatment at each date. *Bacillus* treatment was not significantly worse than full treatment but not significantly better than untreated. Bean yields for each treatment were untreated 6.74 t ha^{-1}, treated 7.15 t ha^{-1} and full fungicide treatment 7.42 t ha^{-1}. Treated crops showed a 4% reduction in yield while untreated showed a 9% reduction in yield in comparison to full fungicide treated plants.

Ensifer-meliloti-Coated Alfalfa Seeds for Yield Analysis

Means for yield of treated and untreated trials are presented in Table 17.2 for fresh and dry weight (t ha^{-1}). ANOVA results revealed that both treatment and harvest and the interaction between those factors were statistically significant ($P< 0.001$) for yield measured as fresh or dry weight. Significant differences in yield between treated and untreated seed plots were found for the first three cuts. No significant differences were detected for the last two cuts. Differences in yield between harvests were found to be significant. (Figure 1.1).

Bacillus and Pseudomonas for the Growth of Broccoli

Approximately 5×10^{10} bacteria ml^{-1} in weeks 4 and 12 was measured in treated soil samples, whereas the total count of bacteria in the untreated soil was below

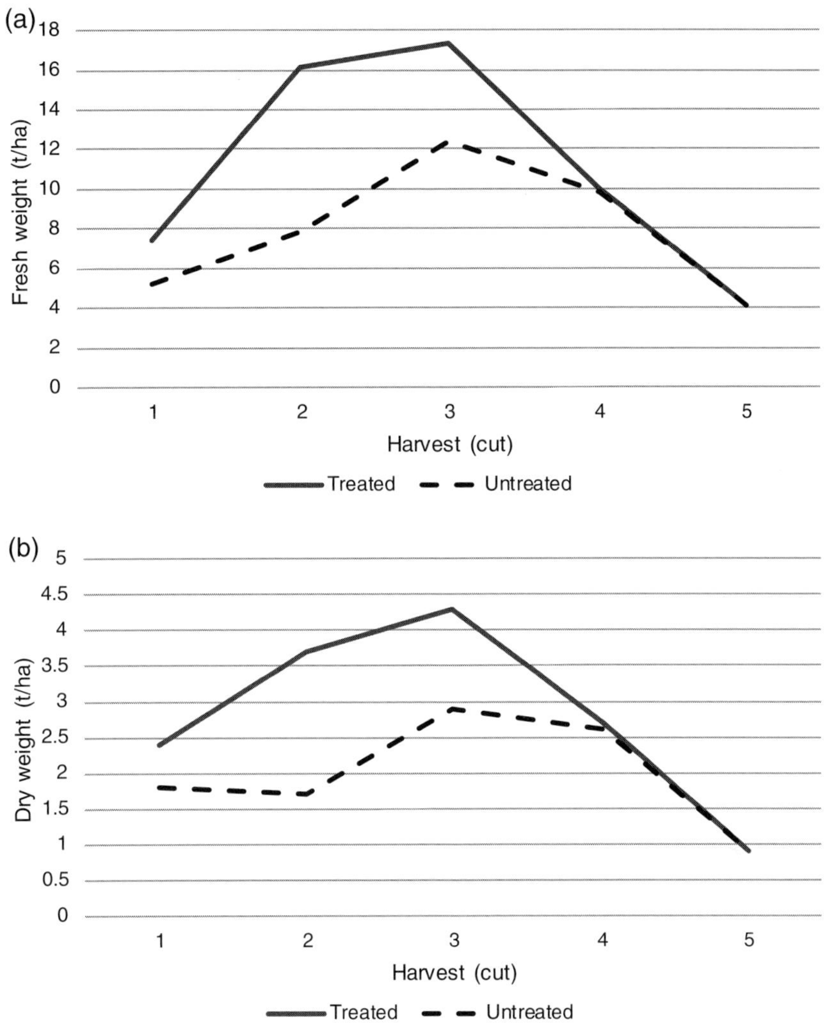

Figure 17.1 Yield means plotted against sequential harvests for rhizobia (*Ensifer meliloti*) coated alfalfa seeds: (a) fresh weight; (b) dry weight.

10^7 bacteria ml^{-1} at 4 and 12 weeks post spraying. A 100-fold increase in fungi from untreated compared to treated, possibly mycorrhiza, was measured at week 4. The fungi were not measured at week 12.

The plant size or head size was not statistically significantly bigger at P = 0.01 or 0.05 for treated compared to untreated plant (Table 17.3). There was no difference in broccoli head sizes by observation by the agronomist. Treated broccoli was greener when viewed in the field. Roots and leaves were bigger after treatment. However, the head size, uniformity of crop, and shelf life are the most important

Table 17.2 Effect of Rhizobia (*Ensifer meliloti*) coated alfalfa seeds on yield.
Fresh weights shown before diagonal and dry weight after diagonal, measured in (t ha^{-1}).

	2016	2017			
		1st cut	2nd cut	3rd cut	4th cut
Treated	7.4/2.4	16.1/3.7	17.3/4.3	10.0/2.7	4.1/0.9
Untreated	5.2/1.8	7.9/1.7	12.4/2.9	9.8/2.6	4.1/0.9

Table 17.3 Effect *of Bacillus and Pseudomonas* treatment on the growth of broccoli (var. Parthenon) shown by comparing weights of broccoli head and plant at harvest.

Head number	Treated		Untreated	
	Head (g)	Plant (g)	Head (g)	Plant (g)
1	500	2860	290	1830
2	670	2400	300	1820
3	450	2280	550	1930
4	530	2640	550	2770
5	390	1920	490	2290
6	470	2580	310	1900
7	410	2150	530	1900
8	430	2010	600	1890
9	570	2550	630	2010
10	410	1610	520	2270
Standard Deviation	87	381	128	300
Mean	483	2300	477	2061

factors for growers. The best result for growers is when head sizes are 500 g plus. There was no increase in head size of plants treated with PGPB in comparison to normal fertilisers. Analysis of observations by the agronomist on comparison of ten broccoli heads revealed that PGPB treated plants had larger roots, less mildew, less branching, greater number of plants with hollow stalks, darker green colouration, and greater number of roots breaking off.

Bacillus subtilis, *Azotobacter* and Rhizobia for Yield Analysis of Processing Tomatoes (Non-Organic and Organic)

No statistically significant difference between treated and non-treated tomatoes were shown in industrial non-organic tomatoes tests (Table 17.4). Plants treated with fertilisers and lower level of fertilisation were analysed and no difference occurred. However, when less fertiliser waas applied to the soil, *Bacillus subtilis*, Azotobacter and rhizobia appear to work better as is evident in the % marketable yield (96.4%) compared to when soil was chemically fertilised with NPK (92.4%).

There was a significant increase, $P < 0.05$, between treated and untreated soluble solids, marketable and total crop yields in treated crops compared to untreated crops in organic tomatoes (Table 17.5). The most significant value is 'soluble solids, the Brix value' which was 12% higher (value per ha) when treated with PGPB. The farmer is paid according to this value. The acidity and lycopene level was slightly higher by 2–3%. The commercial production was also higher by about 10% when treated with these PGPB.

17.5 Conclusions Regarding the Use of PGPB for Plant Protection and Yield

Approximately 25% of the world's crop yield is lost every year, mainly due to diseases caused by fungi, other pathogens and pests. PPP, mainly chemicals, are on the market to fight these diseases. Disease control with beneficial microbes is an alternative which allows more sustainable crop production, as the mode of action of microbes are different to chemicals. The use of microbes as PPPs is growing, and their prevalence is likely to increase because of political and public pressure. As knowledge surrounding the specific interactions between microbes and plants becomes more widely studied, the benefits of their targeted specificity can be used to replace the broad-spectrum mode of action of the chemical pesticides.

The world population is growing, and the amount of food needed by 2050 is estimated to be double from current values (FAO *et al.*, 2017), whereas the area of available new agricultural land is decreasing, therefore we must increase crop yields in a sustainable way. A selection of crop trials have been summarised in this chapter, showing that bacteria can be used for sustainable and cost-effective fertilisation. It is recognised that the range of soils, even within small countries, will influence the crop production outcome and efficacy of PGPB. It is therefore important to use bacteria which are adapted to the particular soil types (clay,

Table 17.4 Effect of *Bacillus subtilis, Azotobacter* and rhizobia treatment on industrial non-organic tomatoes. Summary of tomato plant yield and analysis results.

		Yield						
		Soluble solids (kg ha^{-1})	**Marketable (t ha^{-1})**	**Unripe (t/ha)**	**Rotten (t ha^{-1})**	**Total (t ha^{-1})**	**Marketable total (%)**	**Unripe total (%)**
Treated	**Conventional fertilisers**	5474	118.0	7.0	2.4	127.3	92.7	5.4
Treated	**Reduced fertilisers**	5714	122.4	2.2	2.8	127.4	96.0	1.8
Not treated	**Conventional fertilisers**	5659	122.2	7.9	2.6	132.7	92.1	5.8
Not treated	**Reduced fertilisers**	5866	123.3	2.0	2.0	127.4	96.9	1.5
Treated		5594	120.2	4.6	2.6	127.4	94.3	3.6
Not treated		5763	122.8	5.0	2.3	130.0	94.5	3.6
	Conventional fertilisers	5567	120.1	**7.4**	2.5	130.0	**92.4**	5.6
	Reduced fertilisers	5790	122.9	**2.1**	2.4	127.4	**96.4**	1.7
Mean		5678	121.5	4.8	2.5	128.7	94.2	3.6
Coefficient of var. (%)		8.5	8.3	41.4	21.7	8.1	2.0	45.7
Sig. factor A		n.s	n.s	n.s	n.s	n.s	n.s	n.s
Sig. factor. B		n.s	n.s	**	n.s	n.s	**	**
Sig. factor AB		n.s	n.s	n.s	n.s	n.s	n.s	n.s

Significance: ** P=0.01; * P=0.05; n.s. = not significant; (-) = not calculated; Scott-Knott's test

		Plant characteristics				Analysis of juice			
Rotten total (%)	**Total (%)**	**Crop Cycle (days)**	**Health (P:5-1)**	**Fruit coverage (P:5-1)**	**Branch length**	**Soluble solids (Brix)**	**pH**	**Colour Gardener/ (a/b)**	**Lycopene**
1.9	100	89	3.0	2.0	115.6	4.64	4.49	1.52	13.15
2.2	100	88	3.0	2.0	106.0	4.67	4.52	1.40	10.84
2.1	100	90	3.0	2.0	115.4	4.66	4.46	1.53	12.71
1.6	100	88	3.0	2.0	101.0	4.76	4.48	1.53	12.71
2.1	100	89	3.0	2.0	110.8	4.66	4.50	**1.46**	11.99
1.9	100	89	3.0	2.0	108.2	4.71	4.47	**1.53**	12.71
2.0	100	**90**	3.0	2.0	**115.5**	4.65	4.47	**1.52**	12.93
1.9	100	**88**	3.0	2.0	**103.5**	4.71	4.50	**1.46**	11.78
2.0	100.0	88.8	3.0	2.0	109.5	4.68	4.49	1.49	12.35
27.9	–	1.0	–	–	4.6	2.3	0.7	2.6	6.1
n.s	–	n.s	–	–	n.s	n.s	n.s	**	n.s
n.s	–	*	–	–	**	n.s	n.s	*	n.s
n.s	–	n.s	–	–	n.s	n.s	n.s	**	n.s

Table 17.5 Effect of *Bacillus subtilis, Azotobacter* and rhizobia treatment on industrial organic tomatoes. Summary of tomato plant yield and analysis results.

	Yield						
	Soluble solids (kg ha^{-1})	**Marketable (t ha^{-1})**	**Unripe (t ha^{-1})**	**Rotten (t ha^{-1})**	**Total (t/ha)**	**Marketable total (%)**	**Unripe total (%)**
Treated	5,527	120.8	1.1	16.3	138.2	87.1	0.8
Not treated	4,855	108.0	1.1	16.6	125.6	85.7	0.8
Mean	5,191	114.4	1.1	16.5	131.9	86.4	0.8
Coeff. of var. (%)	3.68	4.68	18.49	23.28	3.57	2.69	20.62
Sig.	*	*	n.s	n.s	*	n.s	n.s

Notes: Significance: **P = 0.01; *P = 0.05; n.s. = not significant; (-) = not calculated

sand, loam, etc.) The soil fertility and ecosystem of the starting soil is perhaps the greatest factor at play here and best results are likely to be shown in organic farming where the lack of chemical fertilisation will result in the benefits of PGPB taking effect more efficiently. This was shown in the organic tomato trial presented here, where the treated plants were healthier and had a higher yield. The alfalfa trial highlighted the lack of *Ensifer meliloti* (nitrogen production) in the soil in the untreated plots. The implication of this result is for the general application of rhizobia for legumes to ensure higher sustainable yields. Going forward, we have sufficient information on many PGPB and a substantial library of bacterial strains allowing for the substitution of PGPB bacteria depending on soil and crop type in the future as required by growers.

Acknowledgements

We thank John Carroll, Teagasc, Oakpark, Co. Carlow, Ireland for the *Bacillus subtilis* for the treatment of chocolate spot and *Ensifer meliloti* on the alfalfa crop trials. Paul Brophy Produce, Naas, Co. Kildare, Ireland for the *Bacillus* and *Pseudomonas* on broccoli trial and Stuard, Parma, Italy for the treatment of organic tomatoes with *Bacillus, Azotobacter* and rhizobia trial.

		Plant characteristics				Analysis of juice			
Rotten total (%)	**Total (%)**	**Crop Cycle (days)**	**Health (P:5-1)**	**Fruit coverage (P:5-1)**	**Branch length (cm)**	**Soluble solids (Brix)**	**pH**	**Colour Gardener (a/b)**	**Lycopene**
12.2	100.0	96	2.6	2.6	85.2	4.57	4.37	1.44	11.41
13.5	100.0	96	2.1	2.1	85.2	4.49	4.41	1.43	11.28
12.8	100.0	96	2.4	2.4	85.2	4.53	4.39	1.43	11.34
18.58	-	-	-	-	0.95	2.10	1.02	3.92	9.56
n.s	-	-	-	-	n.s	n.s	n.s	n.s	n.s

References

Antoun, H. and Prévost, D. (2005). Ecology of plant growth-promoting rhizobacteria. In *PGPR: Biocontrol and Biofertilization*, ed. Z. A. Siddiqui. Netherlands: Springer, pp. 1–38.

Ardisson, G. B., Tosin, M., Barbale, M. and Degli-Innocenti, F. (2014). Biodegradation of plastics in soil and effects on nitrification activity: a laboratory approach. *Frontiers in Microbiology*, **5**, 710.

Bari, R. and Jones, J. D. G. (2009). Role of plant hormones in plant defence responses. *Plant Molecular Biology*, **69**, 473–488.

Borriss, R., Chen, X. H., Rueckert, C. *et al.* (2011). Relationship of *Bacillus amyloliquefaciens* clades associated with strains DSM7^T and FZB42^T: a proposal for *Bacillus amyloliquefaciens* subsp. *amyloliquefaciens* subsp. nov. and *Bacillus amyloliquefaciens* subsp. *plantarum* subsp. nov. based on complete genome sequence comparisons. *International Journal of Systematic Evolutionary Microbiology*, **61**, 1786–1801.

Cohen, M. F. and Mazzola, M. (2006). Resident bacteria, nitric oxide emission and particle size modulate the effect of *Brassica napus* seed meal on disease incited by *Rhizoctonia solani* and *Pythium* spp. *Plant and Soil*, **286**, 75–86.

Cohen, M. F., Lamattina, L. and Yamasaki, H. (2010). Nitric oxide signaling by plant-associated bacteria. In *Nitric Oxide in Plant Physiology*, ed. S. Hayat. Weinheim, Germany: Wiley-VCH, pp. 161–172.

Dhumal, K. N. (1992). Effect of *Azotobacter* on germination, growth and yield of some vegetables. *Journal of Maharashtra Agricultural University*, **17**, 500.

Doornbos, R. F., van Loon, L. C. and Bakker, P. A. (2012). Impact of root exudates and plant defense signaling on bacterial communities in the rhizosphere. A review. *Agronomy for Sustainable Development*, **32**, 227–243.

Khan, M. R., Fischer, S., Egan, D. and Doohan, F. M. (2006). Biological control of Fusarium seedling blight disease of wheat and barley. *Phytopathology*, **96**, 386–394.

FAO, IFAD, UNICEF, WFP and WHO (2017). *The State of Food Security and Nutrition in the World 2017. Building resilience for peace and food security*. Rome, FAO.

Figueiredo, M. V. B., Bonifacio, A., Rodrigues, A. C. and Araujo, F. F. (2016). Plant growth-promoting rhizo-bacteria: key mechanisms of action. In *Microbial-Mediated Induced Systemic Resistance in Plants*, ed. D. K. Choudhary and A. Varma. Singapore: Springer Science, pp. 23–37.

Fukuyama, K. (2004). Structure and function of plant-type ferredoxins. *Photosynthesis Research*, **81**, 289–230.

Glick, B. R. (2012). Applications and mechanisms: bacteria promoting growth. *Scientifica*, **2012**, 1–15.

Gupta, A., Gupal, M. and Tilak, K. V. (2000). Mechanism of plant growth promotion by rhizobacteria. *Indian Journal of Experimental Biology*, **38**, 856–862.

Harman, G. E., Herrera-Estrella, A. H., Horwitz, B. A. and Lorito, M. (2012). *Trichoderma* – from basic biology to biotechnology. *Microbiology*, **158**, 1–2.

Hider, R. C. and Kong, X. (2010). Chemistry and biology of siderophores. *Royal Society of Chemistry*, **27**, 637–657.

Hofstra, N. and Bouwman, A. F. (2005). Denitrification in agricultural soils: summarizing published data and estimating global annual rates. *Nutrient Cycling in Agroecosystems*, **72**, 267–278.

Idris, E. E., Iglesias, D. J., Talon, M. *et al.* (2007). Tryptophan dependent production of indole-3-acetic acid (IAA) affects level of plant growth promotion by *Bacillus amyloliquefaciens* FZB42. *Molecular Plant–Microbe Interactions*, **20**, 619–626.

Idris, E. E. S., Makarewicz, O., Farouk, A. *et al.* (2002). Extracellular phytase activity of *Bacillus amyloliquefaciens* FZB 45 contributes to its plant growth-promoting effect. *Microbiology*, **148**, 2097–2109.

Jnawali, A. D., Ojha, R. B. and Marahatta, S. (2015). Role of *Azotobacter* in soil fertility and sustainability – a review. *Advances in Plants and Agriculture Research*, **2**, 64–69.

Kiseleva, A. A., Tarachovskaya, E. R. and Shishova, M. F. (2012). Biosynthesis of phytohormones in algae. *Russian Journal of Plant Physiology*, **59**, 595–610.

Kümmerli, R., Schiessl, K. T., Waldvogel, T., McNeill, K. and Ackermann, M. (2014). Habitat structure and the evolution of diffusible siderophores in bacteria. *Ecology Letters*, **17**, 1536–1544.

Liu, J., Chakraborty, S., Hosseinzadeh, P. *et al.* (2014). Metalloproteins containing cytochrome, iron–sulfur, or copper redox centers. *Chemical Reviews*, **114**, 4366–4469.

Narayanasamy, P. (2013). Mechanisms of action of fungal biological control agents. In *Biological Management of Diseases of Crops, Progress in Biological Control*, ed. P. Narayanasamy. Dordrecht, The Netherlands: Springer Science, pp. 99–200.

Pal, K. K. and Gardener, B. M. (2006). Biological control of plant pathogens.

Plant Health Instructor. doi: 10.1094/PHI-A-2006-1117-02.

Pan, S., Roy, A. and Hazra, S. (2001). *In vitro* variability of biocontrol potential among some isolates of *Gliocladium virens*. *Advances in Plant Science*, **14**, 301–303.

Paulitz, T. M. and Belanger, R. R. (2001). Biological control in greenhouse system. *Annual Review of Phytopathology*, **39**, 103–133.

Philippot, L., Hallin, S. and Schloter, M. (2007). Ecology of denitrifying prokaryotes in agricultural soil. *Advances in Agronomy*, **96**, 249–305.

Prasad, R., Kumar, M. and Varma, A. (2015). Role of PGPR in soil fertility and plant health. In *Plant-Growth-Promoting Rhizobacteria (PGPR) and Medicinal Plants. Soil Biology*, Vol. 42, ed. D. Egamberdieva, S. Shrivastava and A. Varma. Switzerland: Springer International Publishing, pp. 247–260.

Prathuangwong, S. and Kasem, S. (2003). Potential of new antagonists for controlling soybean bacterial pustule and reducing bactericide application. In *Proceedings of the 7th International Conference of Plant Pathology*, 2003, Christchurch, New Zealand, pp. 2–11.

Renshaw, J. C., Robson, G. D., Trinci, A. P. J. *et al.* (2002). Fungal siderophores: structures, functions and applications. *Mycological Research*, **106**, 1123–1142.

Timmusk, S., Behers, L., Muthoni, J., Muraya, A. and Aronsson, A. C. (2017). Perspectives and challenges of microbial application for crop improvement. *Frontiers in Plant Sciences*, **8,** 49.

Verma, O. P. and Shende, S. T. (1993). *Azotobacter* a biofertilizer for vegetable crops. *Biofertilizer Newsletter*, **1**, 6–10.

Whipps, J. M., Sreenivasaprasad, S., Muthumeenakshi, S., Rogers, C. W. and Challen, M. P. (2008). Use of *Coniothyrium minitans* as a biocontrol agent and some molecular aspects of sclerotial myco-parasitism. *European Journal of Plant Pathology*, **121**, 323–330.

18

Prospecting Crop Wild Relatives for Beneficial Endophytes

BRIAN R. MURPHY, FIONA M. DOOHAN AND TREVOR R. HODKINSON

Abstract

The wild relatives of agricultural crops represent a largely untapped source of beneficial microbial endophytes that have potential for agricultural applications. Much of the research into the effects of endophytes on crop species has focused on a relatively small selection of well-characterised bacterial or fungal strains. However, many of these strains can have inconsistent and even unpredictable agronomic effects depending on the complex relationship between host, endophyte, microbiota and environment. We argue that a more focused approach to endophyte selection and application to crop production can generate more predictable results. We show that the appropriate identification of novel fungal endophyte strains from defined source host populations along with the consideration of the target crop species, cultivar and site can improve the chances of a successful endophyte-induced benefit. We discuss the implications for agriculture and suggest further research that will provide more robust support for this approach.

18.1 Introduction

While the value of crop wild relatives (CWR) and landraces as genetic resources for agricultural crop species is understood (Zhang *et al.*, 2017) the equally valuable microbial resources of these CWRs are largely unknown and as a result underutilised (Schlaeppi and Bulgarelli, 2015; Murphy *et al.*, 2018). This is mainly due to the unreliability and inconsistency of the beneficial effects of microbial inoculants in field crops (Selosse and Rousset, 2011; Kusari *et al.*, 2014; Nadeem *et al.*, 2014; Heijden *et al.*, 2015; Owen *et al.*, 2015), and the fact that many of the microbes have been recovered from agricultural field crops rather than CWRs. The complex and dynamic nature of the crop-growing environment makes it difficult to successfully translate encouraging results from 'pot to plot' (Nelissen *et al.*, 2014). Wild relatives of crops represent a valuable repository of symbiotic microorganisms which can be beneficial for modern crops and research is beginning to reveal their diversity and potential use in agriculture (Mousa *et al.*, 2015; Murphy *et al.*, 2015b,d; Wiewióra *et al.*, 2015; Yokoya *et al.*, 2017).

The few microbial products which have been successfully developed as field crop inoculants have focused on a relatively small selection of well-characterised strains which have not necessarily been derived from a CWR (Varma *et al.*, 2012; Coleman-Derr and Tringe, 2014; Johnson *et al.*, 2014; O'Callaghan, 2016). Many of these strains are endophytic (Kabaluk *et al.*, 2010), and these microorganisms have in some instances shown promise as beneficial crop treatments, both for increasing yield (Kedar *et al.*, 2014; Owen *et al.*, 2015; Tkacz and Poole, 2015) and for improving plant stress tolerance (Waqas *et al.*, 2012; Coleman-Derr and Tringe, 2014; Chadha *et al.*, 2015). Endophytes are microorganisms (bacteria, fungi and unicellular eukaryotes) which live at least part of their life cycle inter- or intracellularly inside plants with or without inducing pathogenic symptoms and are often symbiotic (endosyms) with potential to improve crop performance (Schulz and Boyle, 2006; Kuldau and Bacon, 2008; Rodriguez *et al.*, 2009; Achatz *et al.*, 2010; O'Hanlon *et al.*, 2012; Murphy *et al.*, 2013; Saikkonen *et al.*, 2013; Busby *et al.*, 2015). This can include competent, facultative, obligate, opportunistic and passenger endophytes (Card *et al.*, 2016). Endophytes can have several functions and/or may change function during their life cycle (Murphy *et al.*, 2013; Hodkinson and Murphy, 2019, Chapter 1).

While bacterial and fungal endophytes can enhance abiotic and biotic plant stress tolerance in a controlled environment (Waller *et al.*, 2005; Rodriguez *et al.*, 2008; Belesky and West, 2009; Singh *et al.*, 2011; Johnson *et al.*, 2014; Gill *et al.*, 2016), these beneficial microbes can have unpredictable effects in the field due to the complexity of the interactions between host, inoculant, plant-associated microbiota and environment (Kogel *et al.*, 2006; Lugtenberg and Kamilova, 2009; Atkinson and Urwin, 2012; Murphy *et al.*, 2014; Lareen *et al.*, 2016), and also because they are often facultative in behaviour and need the correct environment for them to be effective (Kabaluk *et al.*, 2010). One of the most successful microbial associations that has been commercially used in pastoral agriculture for nearly 20 years involves *Epichloë*, an obligate mutualist fungus that moves systemically through the plant and is vertically transferred through seed (Johnson *et al.*, 2013). *Epichloë* has co-evolved with several pooid grasses and shows host specificity (Hodkinson, 2018; Johnson and Caradus, 2019, Chapter 16). However, this is a reasonably unique association that has not been found in other sown crop species.

As with other types of microbial inoculants, most experimental work that has shown endophyte-induced beneficial effects for crops has been conducted using a small number of well-tested endophyte species, such as *Serendipita indica* (= *Piriformospora indica*), *Epicoccum* spp. and *Chaetomium* spp. (Dhingra *et al.*, 2003; Istifadah and McGee, 2006; Sherameti *et al.*, 2008; Sun *et al.*, 2010; Fávaro *et al.*, 2012; Johnson *et al.*, 2014; Mansotra *et al.*, 2015). However, more novel endophytes recovered from CWRs have been less well studied and used (Ames *et al.*, 2015; Bulgarelli *et al.*, 2015; Shehata *et al.*, 2016). We, and others, maintain that endophytes derived from CWRs represent a largely unknown, untapped reservoir with vast potential for improving agronomic traits in crop cultivars (Clement *et al.*, 1997; Yuan *et al.*, 2010; Murphy *et al.*, 2015a,c, 2017b).

Endophytes from wild plant populations have proved to be extremely diverse (Kumar and Hyde, 2004; Yuan *et al.*, 2010; Weiss *et al.*, 2011; Murphy *et al.*, 2015d) and this diversity represents an enormous resource for the testing and development of strains with potential for agriculture. However, this very diversity and volume begs the question as to how best to design projects that will return the most useful set of strains for crop treatment. In this chapter, we outline a framework methodology for producing a tailored and more focused endophyte inoculant for a particular environmental setting. We show how we have used a set of basic principles to guide a more efficient research pipeline from endophyte recovery to successful agricultural application.

18.2 Fundamental Concept

Natural populations of plants harbour a vast and diverse range of endophytic microorganisms (Bayman, 2006; Rodriguez *et al.*, 2009; Porras-Alfaro and Bayman, 2011; Sun and Guo, 2012; Murphy *et al.*, 2015d; Wemheuer and Wemheuer, 2017), but many of these are unculturable using standard cultivation methods (Sun and Guo, 2012; Prior *et al.*, 2014; Hegazi *et al.*, 2017) and fall outside the scope of this chapter. In contrast to their CWRs, the microbial communities that associate with crop cultivars are relatively depauperate (Kiers *et al.*, 2007; Minz *et al.*, 2013; Lugtenberg *et al.*, 2016; Pérez-Jaramillo *et al.*, 2016). Just as a diverse and functional human gut microbiome is essential for good health, the same is true for the plant microbiome (Singh and Trivedi, 2017). The elite crop cultivars do not possess the full set of microbiota associated with their wild relatives, with a consequent reduction in health and stress resistance (Wissuwa *et al.*, 2009; Bulgarelli *et al.*, 2013). We consider the discovery and application of endophytic microorganisms to modern agricultural crops to be an attempt to restore this core beneficial microbiota which has been lost from elite cultivars through selection and intensive cultivation during domestication.

There is evidence that crop landraces that are less domesticated than modern elite cultivars are more likely to have maintained a diverse microbial association.

Hetrick *et al.* (1992) showed that wheat landraces gained more benefit from mycorrhizal symbioses than modern cultivars, and a similar landrace-associated benefit has been demonstrated for other important crops, such as maize (Xing *et al.*, 2012). Germida and Siciliano (2001) revealed that the rhizosphere bacterial community of ancient landraces was more diverse than that of two modern cultivars of wheat. These landrace resources may also contain endophytes that can be beneficial for modern crops (Shehata *et al.*, 2016)

However, indiscriminately introducing a complete set of culturable CWR-recovered endophytes into a related crop species would be inadvisable at best; pathogenic endophytes within the group and competitive microbial interactions would likely reduce crop fitness (Bulgarelli *et al.*, 2013). So the goal should be to recover and identify a core CWR set of culturable endophytes that will be beneficial for the crop and that are compatible in an inoculant consortium (Gopal *et al.*, 2013; Vandenkoornhuyse *et al.*, 2015; de Souza *et al.*, 2016; Busby *et al.*, 2017). To achieve this goal (while still hitting the 'sweet spot' of crop-beneficial interactions between endophyte(s), host and environment) requires an approach that is focused on understanding the eventual application from the very start. Another important aspect influencing this goal is how quickly the beneficial microbe community can be delivered to the crop cultivar. This is often microbe specific and in some cases a microbial application to an established crop might be the best application option. For endophytes where application as a seed treatment is the best delivery method, we have developed a system which ensures the inoculant is in place to immediately colonise the plant soon after germination.

18.3 Application

18.3.1 Understanding the Application

The most important part of any potential application of endophyte technology is to understand what we want to achieve as the final outcome. For successful application in the field, previous and promising research results need to be transferable to a viable commercial product where new intellectually property can be protected. Many factors need to be managed before this can happen, but here we will focus on the few that have delivered the best outcomes from our own work.

First, it helps if the individual endophyte strains to be used in the product are not closely related to strains used in any other competing products already on the market. Novelty is a strength here as it will allow an easier route to intellectual property and patent protection. As we discussed earlier, selecting previously unstudied CWRs as the endophyte source will, more often than not, result in at least some novel strain discovery. As well as being easier to protect, the novel strains are less likely to have any associated indications of human and/or plant pathogenicity; an important consideration when facing lengthy and costly regulatory requirements.

Second, endophyte inoculants should be applied to crop-growing sites that have similar environmental descriptors to the endophyte source site. Plants can dynamically recruit particular sets of microbes that are of functional benefit to them in a defined environment, controlled by plant exudates and molecular cues (Bulgarelli *et al.*, 2012; Lundberg *et al.*, 2012; Schlaeppi *et al.*, 2014). Therefore, it follows that a CWR growing on a site that experiences a particular stress will recruit soil microbes that will help alleviate that stress, in a habitat-adapted symbiosis (Rodriguez *et al.*, 2008; Redman *et al.*, 2011; Soares *et al.*, 2016; Rho *et al.*, 2017). Comparisons of endophytes recovered from control versus stressed conditions may reveal which endophytes have been differentially recruited under that particular stress (Haichar *et al.*, 2008; Berg and Smalla, 2009; Deepika and Kothamasi, 2014). These stress related endophytes should ideally be applied to a related crop species growing under a similar stress, and this can be tested and confirmed in a controlled environment before conducting field trials. We develop this theme further in Section 4.1 'Prospecting' below.

Third, the endophyte inoculant must be consistent in the benefits it brings to a crop. Inconsistency in field performance has been a real and disillusioning factor in the use of microbial crop inoculants (Nadeem *et al.*, 2014; Owen *et al.*, 2015). Unfortunately, microbe based products are sometimes brought to market with very little prior testing in agricultural field conditions, and those that have field validation are sometimes treated as a panacea for ailments in all crops in all growing environments. Even a previously almost universally beneficial endophyte such as the model species *Serendipita indica* (= *Piriformospora indica*) has been shown to reduce yield in a growing environment not normally encountered by the endophyte in its native range (Murphy *et al.*, 2014). Inconsistent and even negative responses of crops to microbial inoculants reduces end-user confidence in the technology.

Fourth, if the endophytes are intended for commercial use, the product must be marketable and economically viable to produce, store and apply. Questions of marketability are beyond the scope of this article but are extremely important in the successful commercialisation of the endophyte inoculant: questions include 'who is the customer? how much extra are they willing to pay for the treatment? who are the main competitors? what measurable benefits will the treatment deliver?' Of more immediate concern is whether the product is economically viable. An endophyte-based treatment may be extremely effective but will be unmarketable if it is too expensive to produce and difficult to store while maintaining microbial vigour. Inoculant culture methods and crop delivery formulation will largely determine the cost to the end-user (see Section 4.1 'Prospecting' for further discussion).

Finally, before final distribution to growers, the endophyte inoculant must be modifiable and adaptable to changing conditions. This can be a real problem if only one endophyte strain is used in the inoculant. If the endophyte strain is reliable in every environment and with every crop cultivar then this is not an issue; but this is rarely ever the case. While it may be easier to negotiate the regulatory minefield with only one strain, a consortium of different strains may offer a more flexible and adaptable formulation, and also offer a degree of redundancy in the dynamic environment of a field crop (Brenner *et al.*, 2008; Vandegrift *et al.*, 2015; Lindemann *et al.*, 2016; Murphy *et al.*, 2017b). A farmer may face an unusually wet spring for example, and a consortium designed for a drier environment can be 'tweaked' by adding an individual strain originally recovered from a moist environment. One of our recent studies demonstrates the importance of matching source and sink, where we found a strong correlation between endophyte-induced yield increase and low rainfall; the endophytes in this consortium of four strains were all sourced from particularly dry sites (Murphy *et al.*, 2017b). We had previously tested these strains as individual inoculants on drought-stressed barley in a controlled environment, with favourable results (Murphy *et al.*, 2015c).

18.3.2 Understanding the Crop

Understanding the crop-growing environment is only one aspect of the application that needs to be considered; of equal importance is understanding the target crop cultivar(s) and the agricultural practices normally applied to the crop. The beneficial effect of the endophyte inoculant may be cultivar specific (Hasan, 2002; Assuero and Tognetti, 2010; Murphy *et al.*, 2017b), so it is necessary to test the effect on all cultivars in commercial use before application. Cultivar parentage and history may also affect the degree of response to a particular stress (McMullen *et al.*, 2012; Ali *et al.*, 2014; Bedada *et al.*, 2014) and this may alter the outcome of the plant–endophyte relationship.

The history of the field growing site needs to be considered so that an agronomist can select the optimal inoculant for the site; what crop types have previously been grown there? have particular crop rotations been carried out? what inputs has the site received etc. These and other site-specific factors will have changed the soil characteristics, particularly with regard to soil microorganisms, so that there may be a particularly heavy or light disease load and various levels of chemical residues (Sinha, 1997; Nandwani, 2016; Zaidi and Khan, 2017). Some soils are naturally more suppressive of disease (Weller and Cook, 1983; Duffy and Simon, 1996; Yang *et al.*, 2011; Gopal *et al.*, 2013) and this usually indicates the presence of particular groups of soil microbes, which may influence the effectiveness of any endophyte inoculant through competitive processes (Combès *et al.*, 2012; Chagas *et al.*, 2013; Zhou *et al.*, 2014; Kemen *et al.*, 2015). If there are indications of a disease-suppressive soil then this may affect the choice of inoculant delivery methods.

The choice of plant tissue to inoculate depends on the type of tissue from which the endophyte was recovered and the mode of action, though this is only a rough guide for effective usage. For instance, endophytes recovered from seeds of a CWR can significantly increase yield in barley when applied as a seed dressing

(Murphy *et al.*, pers. obs.). It may be that previous work has shown that the endophyte boosts early crop growth and here it might be most profitable to apply the inoculant as a seed treatment (Achatz *et al.*, 2010; Murphy *et al.*, 2017a; Zaidi and Khan, 2017). On the other hand, if the endophyte suppresses development of a late season pathogen, then a foliage treatment may be more effective (Mousa *et al.*, 2015). Furthermore, if the microorganism normally infects the roots via horizontal transmission then a soil treatment would be preferred (Behie and Bidochka, 2014).

Another, often overlooked, aspect of crop inoculation is the timing of application. Cook *et al.* (1999) found that disease of winter wheat caused by *Septoria tritici* and *Erysiphe graminis* f. sp. *tritici* was most effectively controlled on the final leaf by treatment at or immediately after its emergence. The endophyte may be more effective if it colonises the plant tissue before any potential pathogen, and the order of arrival of endophyte and pathogen may even change the effect on the plant from beneficial to detrimental (Adame-Álvarez *et al.*, 2014).

How long the inoculant stays viable is a crucial aspect of inoculant delivery timing. A survey of different products by Owen *et al.* (2015) revealed that a number of microbial inoculants (primarily bacterial) have a very short shelf life, with a severely reduced window of effectiveness. The choice of carrier substrate for the inoculant is important as the substrate should provide a stable environment for the microbial fractions and prolong product shelf life (Malusá *et al.*, 2012).

Plants, and most other complex organisms, have distinct circadian rhythms (McClung, 2006) and this may have some significance for the timing of inoculant application in the cycle. For instance, the inoculant may be able to gain access to the plant through the stomata more effectively during the early part of the day before the plant is water stressed (Tuteja and Sopory, 2008). The same inoculant can have different effects on the plant depending on the timing of application during the growing season (Kecskes *et al.*, 2008).

As potential end users may be using the endophyte inoculant as part of a comprehensive integrated pest management (IPM) system, endophyte tolerance of normal agricultural inputs such as fungicide, fertilisers and pesticides would make the inoculant product more attractive to a broader range of farmers (McDonald, 2015). We have demonstrated that selected endophytes are tolerant of both fungicidal seed dressings (Murphy *et al.*, 2017a) and the fungicides applied to the crop during the growing season (Murphy *et al.*, 2017b).

In summary, there are many and varied complex contributory factors in determining the effect of endophyte inoculants on the crop (Figure 18.1) and these must be fully analysed and addressed to ensure successful application; the importance of comprehensive testing for all these aspects of the target crop cultivar(s) and site-specific inputs cannot be over-emphasised.

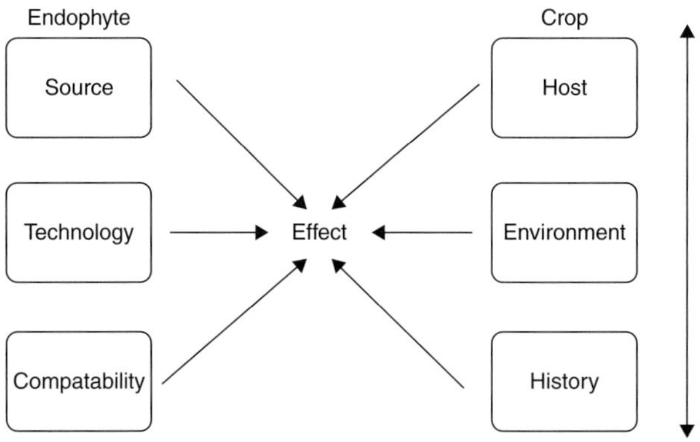

Figure 18.1 Contributory factors in determining the effect of endophyte inoculants on the crop yield. The vertical double-headed arrow on the 'Crop' side indicates dynamic and temporal interactions between the elements.

18.3.3 Prospecting

The first principle in selecting endophyte source sites is to know exactly what kind of application will eventually be targeted. This is directly linked to the choice of endophyte source site. For example, if the targeted crop application is to use an endophyte inoculant to alleviate drought stress then our evidence (Murphy *et al.*, 2015a) suggests that it is best to select healthy host plants from a source site that has low soil moisture; a site where we would expect the plants to have recruited those endophytes that enable them to cope with this stress.

In our own research, we have followed a set of standard principles when sampling plants for fungal endophytes:

- The sampling site should contain a minimum population of ten individual plants, not ramets.
- Plants on sampling site should be healthy and free of disease.
- If sampling for biocontrol, the site should preferably be in an area of potentially high disease pressure for the target crop species.
- If sampling for abiotic stress tolerance, there should be some such stress present at the site, e.g. very dry, saline or poor soil which must be measurable.
- The site should be relatively undisturbed and natural, i.e. not recently worked.
- The site should not be exposed to contaminants, such as close to a busy roadside.
- The plant species should be identifiable using a recognised and reliable key.
- The plants should be actively growing and not in a senescent phase.
- The plants should be sampled when the site is in its 'normal' condition, i.e. not after any heavy rain, frost or unusual weather event, nor after any disturbance.

Whole healthy plants should be collected and processed in the laboratory as soon as possible before any damage or necrosis can affect the efficiency of endophyte recovery. Established methods for processing plant tissue and culturing endophytes can be followed to build up a collection of candidate isolates for screening (for example Murphy *et al.*, 2015b). Candidate fungal isolates can now be identified using a recognised barcoding gene region such as the internal transcribed spacer (ITS) (Kõljalg *et al.*, 2013) and morphological features (Cannon and Kirk, 2007). Once the isolates have been identified a further screening step can be made to select the final strains for testing. The endophyte strain should ideally show vigorous growth on a range of substrates, produce early and copious spores, and be pure (i.e. subcultured from a single spore). Endophyte cultures that do not easily produce spores are of limited use in large-scale agricultural applications. As mentioned earlier, it is also important that the selected strains are not closely related to any known human pathogen.

The final set of selected strains should now be tested, individually and in various combinations, under the targeted stress factor in a series of replicated controlled environment trials. Only those individual strains and consortia which show consistently significant improvements in the desired trait should be considered for field testing. At this point the selected strains can be subcultured and prepared for medium and/or long-term storage (Ryan *et al.*, 2000), and added to a central strain collection. The benefit of a local, institutionalised strain collection becomes apparent when selecting strains for an inoculant consortium, where previously validated strains from other applied stress factors can be chosen as members of a compatible consortium. Endophyte strains that have shown different modes of action are likely to be compatible when used in combination as they will often not be competitive (for resource and/or niche).

The final inoculant is now ready for the target crop in the target environment, and a delivery formulation can be developed. This on its own is a significant challenge and many microbial biocontrol products have failed at this point. Many different inoculant delivery methods have been researched, and formulation selection will be based on experimentation and previous experience (Jakobs-Schönwandt *et al.*, 2014).

18.3.4 Validation of Concept

Throughout this chapter, we have made reference to our experimental work that has supported the successful development of a beneficial barley crop inoculant. We now summarise the field validation steps for proof of concept for our approach (see also Murphy *et al.*, 2017a).

We conducted tests over two seasons on three contrasting field sites with three barley cultivars (Mickle, Planet and Propino). Seeds were either untreated or dressed with a consortium of four endophyte strains isolated from the barley

CWR, *Hordeum murinum*. The selected strains had increased barley grain yield under various biotic and abiotic stresses in earlier controlled environment experiments. As we were testing for the effects of the endophytes under different nitrogen (N) inputs, we applied three levels of N: normal N input for the site, 50% N and 0 N. On the field site with the lowest overall N input, the endophyte treatment with 50% N restored yield for one cultivar (Planet) to that associated with untreated plants receiving the full N input. On the same site and with the same cultivar, endophyte treatment increased yield by 15% under normal N, and by a mean 12% for all three cultivars with 50% N input. Over both seasons and on all three sites, the endophyte treatment increased yield for the cultivar Planet by a mean 9%. For the endophyte-associated increase in grain yield, a strong correlation was found between increased yield and low seasonal rainfall. Furthermore, the endophytes were just as effective when used with regular foliar fungicidal treatment. A recently completed field trial on a dry site returned significant endophyte-induced yield increases for all three cultivars (Murphy *et al.*, pers. obs.). These results suggest that fungal endophytes can contribute to improving barley yield grown in low rainfall areas and under a range of fertiliser input regimes, provided that endophyte treatments are applied to compatible crop cultivars and sites.

The endophyte seed dressing was relatively cheap to produce and apply, and so is attractive to potential investors. In other tests (results not published), it was shown that the endophyte spores can be economically produced on an industrial scale, and these strains have been repeatedly subcultured with no apparent loss of competence. Even as a simple carrier-based seed dressing, the endophyte inoculant remains effective as a seed treatment when stored at room temperature for at least six months after dressing the seed (Murphy *et al.*, 2017b), which is a more than acceptable shelf life for the seeds of annual cereal crops.

So, here the beneficial effect of the endophyte inoculant was environment, cultivar and site specific – exactly what we suspected! Only at this stage, after over 5 years of continuous work, could we say with any confidence that these endophytes can be successfully applied to a specific crop in a particular environment. We recovered endophytes from a wild barley relative growing on water-stressed sites and this is the type of environment where they were most effective when subsequently applied to barley cultivars growing on agricultural sites.

18.4 Implications for Agriculture

The fungal endophyte methods we have used can in principle be applied to any crop type and any growing environment. But of course, as clearly demonstrated here, this requires a great deal of time and effort. What we and others have learnt may help to reduce the time and effort required in future work, but similar positive outcomes cannot be assumed. What is certain is that we will require an increase in food production to cater for the growing population, and this increase must be sustainable (Dobermann and Nelson, 2013). While traditional, genetic modification and biotech based agricultural practices may go some way to addressing the required increase in production (Richards, 2006; ISAAA, 2012; James, 2013), the development of biological treatments for crops will certainly enhance our ability to meet future needs (Duhamel and Vandenkoornhuyse, 2013; Rai *et al.*, 2014; Lugtenberg *et al.*, 2016). If food producers use endophyte inoculants on the correctly matched crops and environments then they can expect increased profit margins, new marketing opportunities, and resource optimisation. This will require companies who sell these solutions to farmers to have a direct and respected advisory input into farm-scale decisions.

18.5 Future Research Directions

The use of endophyte-based technology in sustainable crop production is in its infancy and further research is necessary to both develop new products and to fully understand the crop-endophyte relationship. Knowing exactly how the relationship is established, maintained and lost in the dynamic and complex environment of a field crop is a formidable task. However, we are beginning to elucidate the mechanisms involved (Porras-Alfaro and Bayman, 2011; Ansari *et al.*, 2013; Kemen *et al.*, 2015; Vandenkoornhuyse *et al.*, 2015; Abreu and Taga, 2016; Deng and Cao, 2016; Müller *et al.*, 2016; Rosier *et al.*, 2016;). In general, future research would best focus on providing further validation of the mechanisms behind plant–endophyte relationships, thus providing a more robust platform for product development. Of most importance in this endeavour will be to prove the technology in the field over many seasons and many different environments, and to fully translate promising controlled environment results to agricultural field sites.

Perhaps the greatest challenge to crop production in the future will be the effects of global climate change, where events of drought, flooding, sea-level change, and storms threaten many agricultural production areas and where a change in climate brings new pests and diseases (Hodkinson *et al.*, 2011; FAO, 2015). If agriculture is to adapt to these changes, then future experiments in this field may best be guided by examining the effects of this technology in realistic future local climate change scenarios. Experiments should make maximum use of the molecular and technical tools available to improve the depth and focus of analyses. In short, we should design experiments for likely future climate and future technologies.

Regardless of future climate change, we contend that endophyte research should focus primarily on choosing the most productive path to achieving viable new products. However, basic research into modes of action, and genetic and molecular characterisation of interaction dynamics should also be encouraged as valuable support for the primary focus. Perhaps we can use newer software tools to develop model agricultural systems *in silico* that can test out many hypotheses and predict/model holistic interactions without having to conduct time-consuming and expensive experiments (Heng *et al.*, 2007)

Importance should be placed on building up extensive compatibility libraries of competent endophyte strains to enable the selection of consortia for a best possible fit in any agricultural situation. These repositories would be maintained in internationally recognised collections to complement seed banks of CWRs, landraces and cultivars. We need to conduct research that will discover the core microbiota in different environments, and to understand plant microbiome assembly factors and dynamics. Finally, rather than just cataloguing and describing environmentally determined core sets of endophytes it might be even more useful to reinforce this with an accompanying description of the primary functional benefit to the plant and the functional interactions between each member.

18.6 Conclusions

In this chapter, we have shown that the focused selection of novel endophyte strains from a CWR of barley can significantly improve the chances of discovering a commercially viable endophyte inoculant. The benefits of considering the choice of a target crop cultivar and growing environment from the very beginning of a research project have been highlighted, and this can lead to an effective experimental design and subsequent application. Our published work has reported a significant endophyte-induced yield increase under different chemical fertiliser inputs and in a range of stressed environments. The potential reductions in agricultural chemical use will have immediate and lasting beneficial impacts on the environment as well as reducing economic costs. While we have described the methods of prospecting for, and the application of, the endophytes we have not addressed the equally complex and lengthy process of bringing such products to market. We are currently in the process of commercialising the application of endophytes in a project known as e-Seed (Enterprise Ireland reference number CF-2017-0625-P, 'e-Seed: a sustainable cereal seed endophyte treatment to reduce agricultural chemical costs').

Acknowledgements

We thank: Goldcrop Limited, Cork, Ireland for the generous supply and dressing of barley seeds, and for advice on suitable cultivars to use; Crop Plot Trials and the Irish Department of Agriculture, Food and the Marine for carrying out the field trials; Helena Murphy for proofreading and the de-cluttering of technical terms; laboratory technicians at Trinity College Dublin and University College Dublin for providing supplies and technical support. Science Foundation Ireland funded the 2016 field trials through a Technology Innovation Development Award (Grant Number R15449). Enterprise Ireland are supporting further research and commercialisation through a Commercialisation Fund (Grant number CF 2017 0625-P). Trinity College Dublin provided financial support for the 2015 field trial through a PhD studentship grant.

References

Abreu, N. A. and Taga, M. E. (2016). Decoding molecular interactions in microbial communities. *FEMS Microbiology Reviews*, **40**, 648–663.

Achatz, B., Rüden, S., Andrade, D. *et al.* (2010). Root colonization by *Piriformospora indica* enhances grain yield in barley under diverse nutrient regimes by accelerating plant development. *Plant and Soil*, **333**, 59–70.

Adame-Álvarez, R.-M., Mendiola-Soto, J. and Heil, M. (2014). Order of arrival shifts endophyte-pathogen interactions in bean from resistance induction to disease facilitation. *FEMS Microbiology Letters*, **355**, 100–107.

Ali, S. S., Gunupuru, L. R., Kumar, G. B. S. *et al.* (2014). Plant disease resistance is augmented in uzu barley lines modified in the brassinosteroid receptor BRI1. *BMC Plant Biology*, **14**, 227.

Ames, N., Dreiseitl, A., Steffenson, B. and Muehlbauer, G. (2015). Mining wild barley for powdery mildew resistance. *Plant Pathology*, **6**, 1396–1406.

Ansari, M. W., Trivedi, D. K., Sahoo, R. K., Gill, S. S. and Tuteja, N. (2013). A critical review on fungi mediated plant responses with special emphasis to *Piriformospora indica* on improved production and protection of crops. *Plant Physiology and Biochemistry*, **70**, 403–410.

Assuero, S. G. and Tognetti, J. A. (2010). Tillering regulation by endogenous and environmental factors and its agricultural management. *The American Journal of Plant Science and Biotechnology*, **2010**, 1–14.

Atkinson, N. J. and Urwin, P. E. (2012). The interaction of plant biotic and abiotic stresses: from genes to the field. *Journal of Experimental Botany*, **63**, 3523–3543.

Bayman, P. (2006). Diversity, scale and variation of endophytic fungi in leaves of tropical plants. In *Microbial Ecology of Aerial Plant Surfaces*, ed. M. Bailey. Wallingford, UK: CABI, pp. 37–50.

Bedada, G., Westerbergh, A., Müller, T. *et al.* (2014). Transcriptome sequencing of two wild barley (*Hordeum spontaneum* L.) ecotypes differentially adapted to drought stress reveals ecotype-specific transcripts. *BMC Genomics*, **15**, 995.

Behie, S. W. and Bidochka, M. J. (2014). Ubiquity of insect-derived nitrogen transfer to plants by endophytic insect-pathogenic fungi: an additional branch of the soil nitrogen cycle. *Applied and Environmental Microbiology*, **80**, 1553–1560.

Belesky, D. P. and West, C. P. (2009). Abiotic stresses and endophyte effects. In *Tall Fescue for the 21st Century*, ed. H. Fribourg, D. Hannaway and C. West. Madison, WI: American Society of Agronomy, pp. 49–64.

Berg, G. and Smalla, K. (2009). Plant species and soil type cooperatively shape the structure and function of microbial communities in the rhizosphere. *FEMS Microbiology Ecology*, **68**, 1–13.

Brenner, K., You, L. and Arnold, F. H. (2008). Engineering microbial consortia: a new frontier in synthetic biology. *Trends in Biotechnology*, **26**, 483–489.

Bulgarelli, D., Rott, M., Schlaeppi, K. *et al.* (2012). Revealing structure and assembly cues for Arabidopsis root-inhabiting bacterial microbiota. *Nature*, **488**, 91–95.

Bulgarelli, D., Schlaeppi, K., Spaepen, S., van Themaat, E. V. L. and Schulze-Lefert, P. (2013). Structure and functions of the bacterial microbiota of plants. *Annual Review of Plant Biology*, **64**, 807–838.

Bulgarelli, D., Garrido-Oter, R., McHardy, A. C. and Schulze-Lefert, P. (2015). Structure and function of the bacterial root microbiota in wild and domesticated barley. *Cell Host and Microbe*, **17**, 392–403.

Busby, P. E., Ridout, M. and Newcombe, G. (2015). Fungal endophytes: modifiers of plant disease. *Plant Molecular Biology*, **90**, 645–655.

Busby, P. E., Soman, C., Wagner, M. R. *et al.* (2017). Research priorities for harnessing plant microbiomes in sustainable agriculture. *PLoS Biology*, **15**, 1–14.

Cannon, P. F. and Kirk, P. M. (2007). *Fungal Families of the World*. London: CABI.

Card, S., Johnson, L., Teasdale, S. and Caradus, J. (2016). Deciphering endophyte behaviour: The link between endophyte biology and efficacious biological control agents. *FEMS Microbiology Ecology*, **92**, 1–19.

Chadha, N., Mishra, M., Rajpal, K. *et al.* (2015). An ecological role of fungal endophytes to ameliorate plants under biotic stress. *Archives of Microbiology*, **197**, 869–881.

Chagas, F. O., Dias, L. G. and Pupo, M. T. (2013). A mixed culture of endophytic fungi increases production of antifungal polyketides. *Journal of Chemical Ecology*, **39**, 1335–1342.

Clement, S. L., Wilson, A. D., Lester, D. G. and Davitt, C. M. (1997). Fungal endophytes of wild barley and their effects on *Diuraphis noxia* population development. *Entomologia Experimentalis et Applicata*, **82**, 275–281.

Coleman-Derr, D. and Tringe, S. G. (2014). Building the crops of tomorrow: advantages of symbiont-based approaches to improving abiotic stress tolerance. *Frontiers in Microbiology*, **5**, 1–6.

Combès, A., Ndoye, I., Bance, C. *et al.* (2012). Chemical communication between the endophytic fungus *Paraconiothyrium variabile* and the phytopathogen *Fusarium oxysporum*. *PloS One*, **7**, e47313.

Cook, R. J., Hims, M. J. and Vaughan, T. B. (1999). Effects of fungicide spray timing on winter wheat disease control. *Plant Pathology*, **48**, 33–50.

de Souza, R. S. C., Okura, V. K., Armanhi, J. S. L. *et al.* (2016). Unlocking the bacterial and fungal communities assemblages of sugarcane microbiome. *Nature Scientific Reports*, **6**, 1–15.

Deepika, S. and Kothamasi, D. (2014). Soil moisture-a regulator of arbuscular mycorrhizal fungal community assembly and symbiotic phosphorus uptake. *Mycorrhiza*, **25**, 67–75.

Deng, Z. and Cao, L. (2016). Fungal endophytes and their interactions with plants in phytoremediation: A review. *Chemosphere*, **168**, 1–7.

Dhingra, O. D., Mizubuti, E. S. G. and Santana, F. M. (2003). *Chaetomium globosum* for reducing primary inoculum of *Diaporthe phaseolorum* f. sp. *meridionalis* in soil-surface soybean stubble in field conditions. *Biological Control*, **26**, 302–310.

Dobermann, A. and Nelson, R. (2013). *Solutions for Sustainable Agriculture and Food Systems: Technical Report for Post-2015 Development Agenda.* New York: Sustainable Development Solutions Network.

Duffy, B. K., Simon, A. and Weller D. M. (1996). Combination of *Trichoderma koningii* with fluorescent pseudomonads for control of take-all on wheat. *Phytopathology*, **86**, 188–194.

Duhamel, M. and Vandenkoornhuyse, P. (2013). Sustainable agriculture: possible trajectories from mutualistic symbiosis and plant neodomestication. *Trends in Plant Science*, **18**, 597–600.

FAO (2015). Keeping plant pests and diseases at bay: experts focus on global measures. Rome: FAO. www.fao.org/news/story/en/item/280489/icode/

Fávaro, L. C. D. L., Sebastianes, F. L. D. S. and Araújo, W. L. (2012). *Epicoccum nigrum* P16, a sugarcane endophyte, produces antifungal compounds and induces root growth. *PloS One*, **7**, e36826.

Germida, J. J. and Siciliano, S. D. (2001). Taxonomic diversity of bacteria associated with the roots of modern, recent and ancient wheat cultivars. *Biological Fertilisation of Soils*, **33**, 410–415.

Gill, S. S., Gill, R., Trivedi, D. K. *et al.* (2016). *Piriformospora indica*: potential and significance in plant stress tolerance. *Frontiers in Microbiology*, **7**, 1–20.

Gopal, M., Gupta, A. and Thomas, G. V. (2013). Bespoke microbiome therapy to manage plant diseases. *Frontiers in Microbiology*, **4,** 355.

Haichar, F. el Z., Marol, C., Berge, O. *et al.* (2008). Plant host habitat and root exudates shape soil bacterial community structure. *The ISME Journal*, **2**, 1221–1230.

Hasan, H. A. H. (2002). Gibberellin and auxin-indole production by plant root-fungi and their biosynthesis under salinity-calcium interaction. *Acta Microbiologica et Immunologica Hungarica*, **49**, 105–118.

Hegazi, N. A., Sarhan, M. S., Fayez, M. *et al.* (2017). Plant-fed versus chemicals-fed rhizobacteria of Lucerne: plant-only teabags culture media not only increase culturability of rhizobacteria but also recover a previously uncultured *Lysobacter* sp., *Novosphingobium* sp. and *Pedobacter* sp. *PLoS One*, **12**, e0180424.

Heijden van der, M. G., Martin, F. M., Selosse, M.-A. and Sanders, I. R. (2015). Mycorrhizal ecology and evolution: the past, the present, and the future. *New Phytologist*, **205**, 1406–1423.

Heng, L. K., Asseng, S., Mejahed, K. and Rusan, M. (2007). Optimizing wheat productivity in two rain-fed environments of the West Asia-North Africa region using a simulation model. *European Journal of Agronomy*, **26**, 121–129.

Hetrick, B. A. D., Wilson, G. W. T. and Cox, T. S. (1992). Mycorrhizal dependence of modern wheat varieties, landraces, and ancestors. *Canadian Journal of Botany*, **70**, 2032–2040.

Hodkinson, T. R. (2018). Evolution and taxonomy of the grasses (Poaceae): a model family for the study of species-rich groups. *Annual Plant Reviews Online*, doi: 10.1002/9781119312994.apr0622.

Hodkinson, T. R., Jones, M. B., Waldren, S. and Parnell J. A. N., eds. (2011). *Climate Change Ecology and Systematics*. Cambridge: Cambrige University Press.

Hodkinson, T. R. and Murphy, B. R. (2019). Endophytes for a growing world. In *Endophytes for a Growing World*, ed. T. R. Hodkinson, F. M. Doohan, M. J. Saunders and B. R. Murphy. Cambridge: Cambridge University Press, Chapter 1.

ISAAA (2012). *Beyond Promises: Top 10 Facts about Biotech/GM crops in 2012*. Los Baños, Philippines: ISAAA.

Istifadah, N. and McGee, P. A. (2006). Endophytic *Chaetomium globosum* reduces development of tan spot in wheat caused by *Pyrenophora tritici-repentis*. *Australasian Plant Pathology*, **35**, 411.

Jakobs-Schönwandt, D., Döring, M. and Patel, A. (2014). *Application Techniques of Endophytes*. Bielefeld, Germany: F. H. Bielefeld, University of Applied Sciences, pp. 89.

James, C. (2013). Brief 39. Global status of commercialized biotech/GM crops: 2008. *ISAAA Briefs*, **46,** 317.

Johnson, L. J., De Bonth, A. C. M., Briggs, L. R. *et al.* (2013). The exploitation of *Epichloë* endophytes for agricultural benefit. *Fungal Diversity*, **60,** 171–188.

Johnson, J. M., Alex, T. and Oelmüller, R. (2014). *Piriformospora indica*: the versatile and multifunctional root endophytic fungus for enhanced yield and tolerance to biotic and abiotic stress in crop plants. *Journal of Tropical Agriculture*, **52**, 103–122.

Johnson, L. J. and Caradus, J. R. (2019). The science required to deliver *Epichloë* endophytes to commerce. In *Endophytes for a Growing World*, ed. T. R. Hodkinson, F. M. Doohan, M. J. Saunders and B. R. Murphy. Cambridge: Cambridge University Press, Chapter 16.

Kabaluk, J. T., Svircev, A. M., Goettel, M. S. and Woo, S. G. (2010). *The Use and Regulation of Microbial Pesticides in Representative Jursidictions Worldwide*. Lincoln, New Zealand: IOBC Global.

Kecskes, M. L., Kennedy, I., Choudhury, A. *et al.* (2008). Efficient nutrient use in rice production in Vietnam achieved using inoculant biofertilisers. *ACIAR project 130*. Hanoi: ACIAR, pp. 49–59.

Kedar, A., Rathod, D., Yadav, A., Agarkar, G. and Rai, M. (2014). Endophytic *Phoma* sp. isolated from medicinal plants promote the growth of *Zea mays*. *Nusantara Bioscience*, **6**, 132–139.

Kemen, E., Kemen, A. C., Agler, M. T. and Kemen, E. (2015). Tansley review: host-microbe and microbe-microbe interactions in the evolution of obligate plant parasitism. *New Phytologist*, **214**, 526–532.

Kiers, E. T., Hutton, M. G. and Denison, R. F. (2007). Human selection and the relaxation of legume defences against ineffective rhizobia. *Proceedings of the Royal Society of London B*, **274**, 3119–3126.

Kogel, K.-H., Franken, P. and Hückelhoven, R. (2006). Endophyte or parasite-what decides? *Current Opinion in Plant Biology*, **9**, 358–363.

Kõljalg, U., Nilsson, R. H., Abarenkov, K. *et al.* (2013). Towards a unified paradigm for sequence-based identification of fungi. *Molecular Ecology*, **22**, 5271–5277.

Kuldau, G. and Bacon, C. (2008). Clavicipitaceous endophytes: their ability to enhance resistance of grasses to multiple stresses. *Biological Control*, **46**, 57–71.

Kumar, D. S. S. and Hyde, K. D. (2004). Biodiversity and tissue-recurrence of endophytic fungi in *Tripterygium wilfordii*. *Fungal Diversity*, **17**, 69–90.

Kusari, S., Singh, S. and Jayabaskaran, C. (2014). Biotechnological potential of plant-associated endophytic fungi: hope versus hype. *Trends in Biotechnology*, **32**, 297–303.

Lareen, A., Burton, F. and Schäfer, P. (2016). Plant root-microbe communication in shaping root microbiomes. *Plant Molecular Biology*, **90**, 575–587.

Lindemann, S. R., Bernstein, H. C., Song, H.-S. *et al.* (2016). Engineering microbial consortia for controllable outputs. *The ISME Journal*, **10**, 2077–2084.

Lugtenberg, B. and Kamilova, F. (2009). Plant-growth-promoting Rhizobacteria. *Annual Review of Microbiology*, **63**, 541–556.

Lugtenberg, B., Caradus, J. and Johnson, L. (2016). Fungal endophytes for sustainable crop production. *FEMS Microbiology Ecology*, **92**, 1–37.

Lundberg, D. S., Lebeis, S. L., Paredes, S. H. *et al.* (2012). Defining the core *Arabidopsis thaliana* root microbiome. *Nature*, **488**, 86–90.

Malusá, E., Sas-Paszt, L. and Ciesielska, J. (2012). Technologies for beneficial microorganisms inocula used as biofertilizers. *The Scientific World Journal*, **2012**, 491206.

Mansotra, P., Sharma, P., Sirari, A. and Sharma, S. (2015). Impact of *Piriformospora indica*, *Pseudomonas* species and *Mesorhizobium cicer* on growth of chickpea (*Cicer arietinum* L.). *Journal of Applied and Natural Science*, **7**, 373–380.

McClung, C. R. (2006). Plant circadian rhythms. *The Plant Cell*, **18**, 792–803.

McDonald, B. A. (2015). How can research on pathogen population biology suggest disease management strategies? The example of barley scald (*Rhynchosporium commune*). *Plant Pathology*, **64**, 1005–1013.

McMullen, M., Bergstrom, G., De Wolf, E. *et al.* (2012). Fusarium head blight disease cycle, symptoms, and impact on grain yield and quality frequency and magnitude of epidemics since 1997. *Plant Disease*, **96**, 1712–1728.

Minz, D., Ofek, M. and Hadar, Y. (2013). Plant rhizosphere microbial communities. In *The Prokaryotes*, ed. E. Rosenberg, E. F. DeLong, S. Lory, E. Stackebrandt and F. Thompson. Berlin: Springe, pp. 56–84.

Mousa, W. K., Shearer, C. R., Limay-Rios, V., Zhou, T. and Raizada, M. N. (2015). Bacterial endophytes from wild maize suppress *Fusarium graminearum* in modern maize and inhibit mycotoxin accumulation. *Frontiers in Plant Science*, **6**, 805.

Müller, D. B., Vogel, C., Bai, Y. and Vorholt, J. A. (2016). The plant microbiota: systems-level insights and perspectives. *Annual Review of Genetics*, **50**, 211–234.

Murphy, B. R., Doohan, F. M. and Hodkinson, T. R. (2013). Fungal endophytes of barley roots. *The Journal of Agricultural Science*, **152**, 602–615.

Murphy, B. R., Doohan, F. M. and Hodkinson, T. R. (2014). Yield increase induced by the fungal root endophyte *Piriformospora indica* in barley grown at low temperature is nutrient limited. *Symbiosis*, **62**, 29–39.

Murphy, B. R., Doohan, F. M. and Hodkinson, T. R. (2015a). Fungal root endophytes of a wild barley species increase yield in a nutrient-stressed barley cultivar. *Symbiosis*, **65**, 1–7.

Murphy, B. R., Doohan, F. M. and Hodkinson, T. R. (2015b). Persistent fungal root endophytes isolated from a wild barley species suppress seed-borne infections in a barley cultivar. *Biocontrol*, **60**, 281–292.

Murphy, B. R., Martin Nieto, L., Doohan, F. M. and Hodkinson, T. R. (2015c). Fungal endophytes enhance agronomically important traits in severely drought-stressed barley. *Journal of Agronomy and Crop Science*, **201**, 419–427.

Murphy, B. R., Martin Nieto, L., Doohan, F. M. and Hodkinson, T. R. (2015d). Profundae diversitas: the uncharted genetic diversity in a newly studied group of fungal root endophytes. *Mycology*, **6**, 139–150.

Murphy, B. R., Doohan, F. M. and Hodkinson, T. R. (2017a). A seed dressing combining fungal endophyte spores and fungicides improves seedling survival and early growth in barley and oat. *Symbiosis*, **71**, 69–76.

Murphy, B. R., Hodkinson, T. R. and Doohan, F. M. (2017b). A fungal endophyte consortium counterbalances the negative effects of reduced nitrogen input on the yield of field-grown spring barley. *The Journal of Agricultural Science*, **155**, 1324–1331.

Murphy, B. R., Doohan, F. M. and Hodkinson, T. R. (2018). From concept to commerce: developing a successful fungal endophyte inoculant for agricultural crops. *Journal of Fungi*, **4**, 24.

Nadeem, S. M., Ahmad, M., Zahir, Z. A., Javaid, A. and Ashraf, M. (2014). The role of mycorrhizae and plant growth promoting rhizobacteria (PGPR) in improving crop productivity under stressful environments. *Biotechnology Advances*, **32**, 429–448.

Nandwani, D. (2016). *Organic Farming for Sustainable Agriculture*. Berlin: Springer.

Nelissen, H., Moloney, M. and Inzé, D. (2014). Translational research: from pot to plot. *Plant Biotechnology Journal*, **12**, 277–285.

O'Callaghan, M. (2016). Microbial inoculation of seed for improved crop performance: issues and opportunities. *Applied Microbiology and Biotechnology*, **100**, 5729–5746.

O'Hanlon, K. A., Knorr, K., Jørgensen, L. N., Nicolaisen, M. and Boelt, B. (2012). Exploring the potential of symbiotic fungal endophytes in cereal disease suppression. *Biological Control*, **63**, 69–78.

Owen, D., Williams, A. P., Griffith, G. W. and Withers, P. J. A. (2015). Use of commercial bio-inoculants to increase agricultural production through improved phosphrous acquisition. *Applied Soil Ecology*, **86**, 41–54.

Pérez-Jaramillo, J. E., Mendes, R. and Raaijmakers, J. M. (2016). Impact of plant domestication on rhizosphere microbiome assembly and functions. *Plant Molecular Biology*, **90**, 635–644.

Porras-Alfaro, A. and Bayman, P. (2011). Hidden fungi,emergent properties: endophytes and microbiomes. *Annual Review of Phytopathology*, **49**, 291–315.

Prior, R., Görges, K., Yurkov, A. and Begerow, D. (2014). New isolation method for endophytes based on enzyme digestion. *Mycological Progress*, **13**, 849–856.

Rai, M., Rathod, D., Agarkar, G. *et al.* (2014). Fungal growth promotor endophytes: a pragmatic approach towards sustainable food and agriculture. *Symbiosis*, **62**, 63–79.

Redman, R. S., Kim, Y. O., Woodward, C. J. D. A. *et al.* (2011). Increased fitness of rice plants to abiotic stress via habitat adapted symbiosis: a strategy for mitigating impacts of climate change. *PloS One*, **6**, e14823.

Rho, H., Hsieh, M., Kandel, S. L., Cantillo, J., Doty, S. L. and Kim, S.-H. (2017). Do endophytes promote growth of host plants under stress? A meta-analysis on plant stress mitigation by endophytes. *Microbial Ecology*, **75**, 407–418.

Richards, R. A. (2006). Physiological traits used in the breeding of new cultivars for water-scarce environments. *Agricultural Water Management*, **80**, 197–211.

Rodriguez, R. J., Henson, J., Van Volkenburgh, E. *et al.* (2008). Stress tolerance in plants via habitat-adapted symbiosis. *The ISME Journal*, **2**, 404–416.

Rodriguez, R. J., White, J. F., Arnold, A. E. and Redman, R. S. (2009). Fungal endophytes: diversity and functional roles. *The New Phytologist*, **182**, 314–330.

Rosier, A., Bishnoi, U., Lakshmanan, V., Sherrier, D. J. and Bais, H. P. (2016). A perspective on inter-kingdom signaling in plant-beneficial microbe interactions. *Plant Molecular Biology*, **90**, 537–548.

Ryan, M. J., Smith, D. and Lane, B. (2000). A decision-based key to determine the most appropriate protocol for the preservation of fungi. *World Journal of Microbiology and Biotechnology*, **16**, 183–186.

Saikkonen, K., Gundel, P. E. and Helander, M. (2013). Chemical ecology mediated by fungal endophytes in grasses. *Journal of Chemical Ecology*, **39**, 962–968.

Schlaeppi, K. and Bulgarelli, D. (2015). The plant microbiome at work. *Molecular Plant-Microbe Interactions*, **28**, 212–217.

Schlaeppi, K., Dombrowski, N., Oter, R. G., Ver Loren van Themaat, E. and Schulze-Lefert, P. (2014). Quantitative divergence of the bacterial root microbiota in *Arabidopsis thaliana* relatives. *Proceedings of the National Academy of Sciences of the United States of America*, **111**, 585–592.

Schulz, B. and Boyle, C. (2006). What are endophytes? In *Microbial Root Endophytes, Soil Biology*, Vol. 9, ed. B. Schulz, C. Boyle and T. N. Sieber. Berlin: Springer-Verlag, pp. 1–14.

Selosse, M.-A. and Rousset, F. (2011). The plant-fungal marketplace. *Science*, **333**, 828–829.

Shehata, H. R., Lyons, E. M., Jordan, K. S. and Raizada, M. N. (2016). Bacterial endophytes from wild and ancient maize are able to suppress the fungal pathogen *Sclerotinia homoeocarpa*. *Journal of Applied Microbiology*, **120**, 756–769.

Sherameti, I., Tripathi, S., Varma, A. and Oelmüller, R. (2008). The root-colonizing endophyte *Pirifomospora indica* confers drought tolerance in *Arabidopsis* by stimulating the expression of drought stress-related genes in leaves. *Molecular Plant-Microbe Interactions, MPMI*, **21**, 799–807.

Singh, B. K. and Trivedi, P. (2017). Microbiome and the future for food and nutrient security. *Microbial Biotechnology*, **10**, 50–53.

Singh, L. P., Gill, S. S. and Tuteja, N. (2011). Unraveling the role of fungal symbionts in plant abiotic stress tolerance. *Plant Signaling and Behavior*, **6**, 175–91.

Sinha, R. K. (1997). Embarking on the second green revolution for sustainable agriculture in India: a judicious mix of traditional wisdom and modern knowledge in ecological farming. *Journal of Agricultural and Environmental Ethics*, **10**, 183–197.

Soares, M. A, Li, H. Y., Kowalski, K. P. *et al.* (2016). Evaluation of the functional roles of fungal endophytes of *Phragmites australis* from high saline and low saline habitats. *Biological Invasions*, **18**, 2689–2702.

Sun, C., Johnson, J. M., Cai, D. *et al.* (2010). *Piriformospora indica* confers drought tolerance in Chinese cabbage leaves by stimulating antioxidant enzymes, the expression of drought-related genes and the plastid-localized CAS protein. *Journal of Plant Physiology*, **167**, 1009–1017.

Sun, X. and Guo, L. (2012). Endophytic fungal diversity: review of traditional and molecular techniques. *Mycology*, **3**, 65–76.

Tkacz, A. and Poole, P. (2015). Role of root microbiota in plant productivity. *Journal of Experimental Botany*, **66**, 2167–2175.

Tuteja, N. and Sopory, S. K. (2008). Chemical signaling under abiotic stress environment in plants. *Plant Signaling and Behavior*, **3**, 525–536.

Vandegrift, R., Roy, B. A., Pfeifer-Meister, L., Johnson, B. R. and Bridgham, S. D. (2015). The herbaceous landlord: integrating the effects of symbiont consortia within a single host. *PeerJ*, **3**, e1379.

Vandenkoornhuyse, P., Quaiser, A., Duhamel, M., Le Van, A. and Dufresne, A. (2015). Tansley Review: the importance of the microbiome of the plant holobiont. *New Phytologist*, **206**, 1196–1206.

Varma, A., Bajaj, R., Agarwal, A. *et al.* (2012). *Memoirs of 'Rootonic': The Magic Fungus*. Noida, India: Amity Institute of Microbial Technology.

Waller, F., Achatz, B., Baltruschat, H. *et al.* (2005). The endophytic fungus *Piriformospora indica* reprograms barley to salt-stress tolerance, disease resistance, and higher yield. *Proceedings of the National Academy of Sciences of the United States of America*, **102**, 13386–13391.

Waqas, M., Khan, A. L., Kamran, M. *et al.* (2012). Endophytic fungi produce gibberellins and indoleacetic acid and promotes host-plant growth during stress. *Molecules (Basel, Switzerland)*, **17**, 10754–10773.

Weiss, M., Sýkorová, Z., Garnica, S. *et al.* (2011). Sebacinales everywhere: previously overlooked ubiquitous fungal endophytes. *PloS One*, **6**, e16793.

Weller, D. M. and Cook, R. J. (1983). Supression of take-all of wheat by seed treatmetns with fluorescent pseudomonads. *Phytopathology*, **73**, 463–469.

Wemheuer B. and Wemheuer F. (2017). Assessing Bacterial and Fungal Diversity in the Plant Endosphere. In *Metagenomics, Methods in Molecular Biology*, vol 1539, ed. W. Streit and R. Daniel. New York, NY: Humana Press, pp. 75–84.

Wiewióra, B., Żurek, G. and Żurek, M. (2015). Endophyte-mediated disease resistance in wild populations of perennial ryegrass (*Lolium perenne*). *Fungal Ecology*, **15**, 1–8.

Wissuwa, M., Mazzola, M. and Picard, C. (2009). Novel approaches in plant breeding for rhizosphere-related traits. *Plant and Soil*, **321**, 409–430.

Xing, X., Koch, A. M., Jones, A. M. *et al.* (2012). Mutualism breakdown in breadfruit domestication. *Proceedings of the Royal Society B*, **22**, 1122–1130.

Yang, M. M., Mavrodi, D. V., Mavrodi, O. V. *et al.* (2011). Biological control of take-all by fluorescent *Pseudomonas* spp. from Chinese wheat fields. *Phytopathology*, **101**, 1481–1491.

Yokoya, K., Postel, S., Fang, R. and Sarasan, V. (2017). Endophytic fungal diversity of *Fragaria vesca*, a crop wild relative of strawberry, along environmental gradients within a small geographical area. *PeerJ*, **5**, e2860.

Yuan, Z.-L., Zhang, C.-L., Lin, F.-C. and Kubicek, C. P. (2010). Identity, diversity, and molecular phylogeny of the endophytic mycobiota in the roots of rare wild rice (*Oryza granulate*) from a nature reserve in Yunnan, China. *Applied and Environmental Microbiology*, **76**, 1642–1652.

Zaidi, A. and Khan, M. S. (2017). *Microbial Strategies for Vegetable Production.* Berlin: Springer International Publishing.

Zhang, H., Mittal, N., Leamy, L. J., Barazani, O. and Song, B.-H. (2017). Back into the wild-Apply untapped genetic diversity of wild relatives for crop improvement. *Evolutionary Applications*, **10**, 5–24.

Zhou, Y., Bradshaw, R. E., Johnson, R. D., Hume, D. E., Simpson, W. R. and Schmid, J. (2014). Detection and quantification of three distinct *Neotyphodium lolii* endophytes in *Lolium perenne* by real time PCR of secondary metabolite genes. *Fungal Biology*, **118**, 316–324.

Index

Page numbers in **bold** refer to tables; those in *italic* to figures.

β-tubulin, 72
γ-actin (actA), 72
1-(4-ethylphenyl)-ethanone, 240
16S rRNA, 171, 173, 176, **176**, 178, *215*, 221, 238
1-aminocyclopropane-1-carboxylate (ACC) deaminase, 216
2,3-butanediol, 378
2,4-diacetylphloroglucinol (2,4-DAPG), 241
2,4-dichlorophenoxyacetic acid (2,4-D), 243
2-(methylthio) benzothiazole, 240
2-butenoic acid, 240
4-methoxyindole, 240

abscisic acid, 33–34, 49, 376
ACC deaminase, 10, 217, **236**, 236–237, 376
Accrosidius, 348
Acephala, 144
acetoin, 378
acid phosphatases, 151
Acido, 96
Acidobacteria, 234, 239
Aconitum, 287, 297, 299, 308
Acremonium, 10, 53, 71, 99, 101, 285, 292, 333, 335
Actinoallomurus, 168, 182, 185
Actinobacteria, 96, 167–173, *174*, *175*, **176**, 177–179, *181*, **181**, 181–182, **182**, 213, *215*, 219, 234, 239
Actinomyces, 241
Actinoplanes, 234
adhesion aids, 56
Adoryphorus, 348
Aegicerus, *215*, 219
Aegle, 282–283, 302
Aereobasidium, 259
African black beetle, 193, 348, 355–356
Agaricomycetes, 96
Agaricomycotina, 125
Agrobacterium, 223, 234, 313
agroecosystems, 9, 12, 89, 240, 274
Alcaligenes, 212, 230, 236
Aleppo pine, 266
alfalfa, 57, 147, *148*, 148–149, 151, 371–372, 377–378, 380, *381*, **382**, 386
alkaline phosphatase, 151
alkaloids, 10, 35, 37, 192–193, 282, 299, 313, 343–344, *345*, 345–348, 351, 353, 356
Alphaproteobacteria, 213
Alternaria, 37, 198, 201, 259–260, 282–286, 291, 294, 297, 301, 306, 308
Alzheimer's disease, 291
AMF, *see* arbuscular mycorrhizal fungi
ammonia-oxidizing bacteria, 375
Ammophila, 106, 108
Amoora, 287, 297, 303
Ampelomyces, 331, 376
amplicon sequencing, 29–31, 221, 239, 241
Ancylobacter, 96
Andropogon, 96
angiosperms, 125, 144
antagonism, 11, 39, 90, 156, 205, **236**, 306, 313, 315
antibiosis, 11, 39, 73–75, 94, **98**, **99**, **102**, 109–112, 157, 192, 216
antibiotic resistance, 234, **236**
antibiotics, 9, 75, 192, 208, 225, 228, 277, 307, 311
anticancer, 9, 277–278, 281, 290, 297, 299–300, 302–305, 313
antidiabetic, 278
antifungal, 43–44, 75, 77–78, 90, 172, *181*, **181**, 186, 219, 226, 247, 278, 323–324, 327, 377, 403
anti-inflammatory, 278, 291, 296–297

antimalarial, 220
antimicrobial, 9, 26, 30, 37, 73, 75, 78, 185, 240, 295, 297
anti-obesity, 278
antioxidant, 9, 110, 156, 277–278, **286**, 291, 295–296, 313
apigenin, 242
apoplast, 33, 71, 218, 225, 238
AR1 endophyte, *345*, 348–349, 351, 354–356, 359–360, 366, 368
AR37 endophyte, *345*, 348–349, 352, 354, 356, 362, 364, 366
AR542 endophyte, 354, 370
AR584 endophyte, 354
Arabidopsis, 34, 110, 126, 149, 153, 215, 217, 238
Arachis, 294
arbuscular mycorrhizal fungi, 11, 96, 120, 122, 159, 202
Argentine stem weevil, 346, 348, *349*, 354–356, 363–364, 367
Armillaria, 330
aromatics, 9
Arthrobacter, 96
artichoke, 57
Ascomycota, 11, 53, 68, 70, 85, 87, 90, 96, 120, 131, 188, 196, *200*, 329
ash, 10–11, 40, 328–335, 337–340
  dieback disease, 10–11, 40, 328–330, 337
asiaticoside, 310–311, 315–316, 319, *320*, 320–321
Aspergillus, 43, 105, 196, 266, 274, 285–286, 290–291, 295, 303, 314
Atkinsonella, 96
Aureobasidium, 258, 260, 271, 332, 335
auxin, 10, 33–34, 149, 153, 376–378
  IAA indole acetic acid, 16, 33–34, 153, 173, 181, **182**, 182–183, 217, 237, 319
Avr4, 33
azadirachtin, 312, 324
Azospirillum, 10, 212, 223
Azotobacter, 377–379, 383, **384**, **386**, 386–389

Bacilli, 96, 213
Bacillus, 101, 114, 213, 215, 228–229, 231, 234–236, 241, 245, 325, 371–372, 376–380, **380**, **382**, 382–383, **384**, **386**, 386–388
Bacopa, 319, 321
Bacterial nitrification, 375
Bacteroidetes, 234, 239
Balansia, 319
Balansiaceous, 192
banana, 60, 71, 84, 168, 183, 219, 318
barcoding, 12, 188, 193–196, **197**, 200, 202, 205–206, 312, 328, 334, 338, 398
barley, 5, 11, 28, 34, 38, 40, 93–95, 97, **98**, **104**, **108**, 104–112, 126, 128, *128*, 191–192, 234, 252, 378, 395, 398–399, 401
  yellow dwarf (*BYDV*), 94
Bartalinia, 282–283, 302
Basidiomycota, 11, 70, 96, 120, 125, 188, 196, *200*, 329
Bassiacridin, 73, 90
Bassianin, 73
Bassianolide, 73, 75, 91
BCA, *see* biocontrol agent
bead systems, 59
bean, 57–58, 74, 229, 372, 378–379, **380**
Beauveracin, 73
Beauveria, 10, 42, 53–54, 56–58, 60, 68–69, 71, 73, 76–78, *79*, *80*, 84–87, 89–92, 206
Beauveriolides, 73
benzopyranones, 35
benzothiadiazole, 26
benzothiazole, 240
Berberine, 286, 296, 307
Berberis, 296
Betaproteobacteria, 213
bioactive compounds, 277–282, **283**, 287, 291, 295, 297, 299, 312, 347
biocontrol, 3, 10–11, 25, 27, 30, 37, *38*, 38–39, 42, 54, 58, 67–70, 73, 78, 82–83, 86, 93–97, **98**, **108**, 106–112, 156, **157**, 157–159, 162, 168, 177, 181–182, **182**, 191–192, 208, 215–216, 218, 234, **236**, 241, 243, 328, 331, 335, 337–338, 355, 360, 388–389, 397–398
biocontrol agent, 10, 25, 27, 30, 37, 54, 69, 93, 95–97, **98**, 106, 108, 159, 168, 181, 192, 208, 243, 331
biodiversity, 8–9, 204, 233, 237, 240, 247, 252, 271, 280, 294, 307, 312, 326, 339
bioenergy, 9, 188–189, 202, 213
bioethanol, 189
biofertiliser, 41, 373, 378
biofilm, 216, 218
biofuel, 189, 233
bioinsecticides, 83, 91
biomass, 9, 32, 59, 110, 121, 126, 129, 132, 147, 150, 154, 188–191, 202, 213, 216, 234, **236**, 236–237, 240, 243, 315, 319, 321, 377
biomolecules, 8, 13, 15, 275
biopesticides, 67–70, 80–81, 94, 243
bioprospect, 3
Bipolaris, 46, 98–101, 104–106, 108–109, 111, 115–116
bird cherry-oat aphid, 348
Biscogniauxia, 257, 261, 272
BIV, 295
black cutworm, 348
Blastomyces, 285, 292

blastospores, 54, 80
Blumeria, 35, 48, 94, 98, 100–101, 105–106, 108–111
Boeremia, 331–332, 335
Botryosphaeria, 259, 261, 267
Botryosphaeriaceae, 258, 274
Botrytis, 48, 76–77, *80*, 88, 157, 261, 283, 285, 292, 376
Bouteloua, 279
Bradyrhizobiaceae, 234
Brassica napus, 232–235, *235*, **236**, 238, 240–241
Brassicaceae, 49, 125, 233
Brevibacillus, 216
broccoli, 371, 377–378, 380–381, **382**, 386
brown rust, 94
bryophytes, 125
Bucephalogonia, 212, 225
Burkholderia, 88, 96, 117, 212, 224, 234, 246
Burkholderiaceae, 234
BVDV, 295

C_4 grasses, 96
Cadophora, 132, *133*, 133–134, 144, *145*, *148*, 151, 154–158, *158*, 160, 163
Camptotheca, 284, 289, 303–304, 306
camptothecin, 281, 284, *288*, 288–289, 299–300, 303, 305–306, 312, 326
carbohydrates, 9, 72, 189, 313
carbon sequestration, 9
carboxymethyl cellulose, 56
Cardinium, 212
catalase, 156, 158
Catharanthus, 212, 282, 300, 303–305, 308, 321, 327
Cd, 154, **155**, 155–156, 237, 249
Cd-oxalate complexes, 156
Centella asiatica, 310–311, 315, *317*, 317–318, 321–323, 325
Ceriporia, 286, 292
CERK receptor-like protein kinase, 33
Cf4 resistance gene, 33
Chaetomium, 98, 101, 105, 109, 117–118, 153, 198, 376, 392, 404–405
Chaetosphaeriales, 144
Chalara, 329, 339–340
chemoattractants, 211
Chilean Collection of Microbial Genetic Resources, 76
chitin, 33, 122, 173
Chloroflexi, 234, 239
ciclosporin, 73
cinnamoyl-CoA reductase, 215
Cirsium, 96
Citrobacter, 213
Cladosporium, 10, 33, 38, 53, 71, 75, 92, *145*, 162, 196, 203, 261, 283, 285, 287, 292, 297, 308–309
Clavicipitaceae, 192, 369
clavicipitaceous, 3, 53, 71, 96, 257
climate change, 3, 8, 15, 26, 189, 223, 233, 250, 256, 259, 268, 270–271, 274, 400, 408
climate change index, 256
Clonostachys, 10, 38, 42, 53, 71, 97, 99–102, 111, 115
Clostridium, 216
cockchafer, 348
Coelomycetes, 297, 300
coevolved, 10
coffee, 57, 65, 68, 90, 118, 192
Colletotrichum, 28, 44, 63, 236, 285, 292, 296, 301, 310–311, *317*, 316–318, 321
Comamonadaceae, 235, 241
Common Agricultural Policy, 374
community ecology, 165
conidia, 10, 54, 69, 76, 80, 82, 86–87, 89–90, 109, 111, 115, 117
Coniothyrium, 376, 389
conservation tillage, 237, 239, *239*, 241
consortia, 13, 42
copiotrophic bacterial families, 240
Coptis, 286, 296
Cordycepin, 73
Cordyceps, 62, 206, 333, 335
cork oak, 10–11, 250–252, 254, *255*, 255–259, 266
corn borer, 10, 53
Coronamycin, 220
Corynephorus, 99, 106
Coryneum, 259, 261
Coscinium, 286, 296, 307
Costelytra, 348, 366
cotton, 57, 86, **103**, 107
CRISPR-Cas9, 26
crop inoculant, 398
crop resistance, 11
crop rotation, 13, 26, 232, 234, 237, 240–241, *242*, 247
crop wild relative, 13, 95, 390, 401
crown gall, 97
cryogenic preservation, 334
cryptocandin, 312
cucumber, 74, 85, 87–88, 114, 138, 168
Cucumis, 74, 138, 151, 153, 156–158, 161
Cucumis sativus, 74, 138, 153, 161
cultivation-dependent methods, 29
cultivation-independent methods, 29, 228
Curcuma, 287, 297, 308, 321–322
curcumin, 287, 297
Curtobacterium, 213
CWR, *see* crop wild relative

Cyanobacteria, 235
cyclopentonone and 2-pyranone derivatives, 219
cytochalasines, 35
Cytophagaceae, 240
Czapek Dox, 188, 194–195

Dactylis, 12
Dactylonectria, 196, 333, 335
dark septate endophyte, 3, *7*, 11, 96, 120, 122, 143–144, *145*, 160–164
denitrification, 375, 388
Destruxins, 73
Devosia, 234
Diapensia, 96
Diaporthaceae, 196
Diaporthe, 196–197, 199, 202, 284, 299, 332, 335, 404
diazotroph, 3, 11, 173, 213
diffusion pathways, 125
Dimethomorph, 58
Dioscorea, 314, 324
Diplodia, 251, 257–258, 262
Discula, 257, 262
diversity, 12–13, *14*, 29, 31, 35, 70, 73, 95, 132, 167–169, 171, 178, 182, 188–189, 194–195, 200, 202, 208–209, 212, 215, 220–221, *222*, 222–223, 233, 237, 239–240, 250, 252, 259, 280, 299, 311–312, 329, 345–346, 391–392
DNA barcoding, 193, 196
Dothideomycetes, 96, 188, 196, 201
Dracaena, 321, 324
drought, 18–19, 32, 89, 95, 115, 117, 125, 153, 164, 205, 237, 251, 256–257, 259, 266–267, 269, 272, 324, 349, 360, 365, 378, 395, 397, 400, 402, 407, 409
drug discovery, 159, 277–278, 299, 326
drug resistance, 9
DSE, *7*, 11, 96, 122, 131–132, *133*, 133–134, 143–145, *145*, *146*, 146–147, *148*, 148–151, *152*, **155**, **157**, *158*, 152–161, 164
Dysoxylum, 287, 297, 303, 324

ecomycins, 219, 228
ecosystem services, 9, 254
Ecp6 effector, 33
EEPF, *see* endophytic entomopathogenic fungi
EFEs entomopathogenic fungal endophytes, 71–75, 78, 81–84
effectors, 29, 32–33, 35
elicitors, 9, 75, 299, 310, 312, 314, 320–321
ELISA tests, 351
embryo culture, 328, 336
emodin, *293*, 295, 299
endomicrobiome, 28
endopaths, 3, 6, *7*, 11, 13
endophytic entomopathogenic fungi, *53*, *55*, 53–60
endophytologist, 3, 7, 9
endosymbionts, 3
endosyms, 3, *7*, 9–10, 12–13
enhancers, 9
enniatines, 35
Ensifer, 371, 377–378, 380, *381*, **382**, 386
Enterobacter, 10, 117, 211, 213, 216, 236, 249
enterobacteria, 218
Enterococcus, 213, 219
entomopathogen, 87, 90
entomopathogenic, 10, 52–56, 59, 67–69, 71, 83, 86–87, 89–92, 192
Entrophosphora, 290
environment, 26–27, 37, 40–42, 68, 78, 83, 96, 149–150, 156, 177, 190, 210–212, 218, 220, 222–223, 233, 251, 256, 258, 266, 290, 332, 338, 347, 376, 390–391, 393–396, 398–401
Epichloë, 10, 14, 36, 40–41, 53, 96, 192, *345*, 343–353, 355, 357, 391
epichloëcyclins, 347
Epichloënin A (and B), 347
Epicoccum, 201, 259, 263, 331–332, 392
epiphyte, 4, 95, 194
epoxy-janthitrems, 346, 348, 356
ergopeptides, 346, 348
ergot alkaloids, 192–193, 351
ergotamine, 346
ergovaline, 22, 192, 344, 346, 348, 352, 355–356, 362, 366
ericoid mycorrhiza, 125
ERM, *see* extraradical mycelium
erosion, 9, 233–234, 237, 241
Erysiphe, 118, 396
Escherichia coli, 218
ethylene, 33–34, 75, 88, 153, 213, 216, 225, 376
EU Commission DG SANCO, 373
Eucalyptus, 211, 225
Euphorbia, 319, 321
Eurotiomycetes, 188, 196
Eutypella, 284, 289
Exophiala, 144, 153–155, 161, 164
extraradical mycelium, 124, 128, **130**

faba beans, 371, 377
fatty acid methyl ester (FAME) profiling, 235
ferns, 125, 312
Ferricrocin, 347
fertiliser, 8, 13, 26–27, 58, 121, 134, 188, 232, 237, 241, 243, 371–372, 374, 379, 382–384, 396, 399, 401
fescue, 20, 205, 322–325, 343, 345, 347–348, 350–354, 357–370

Festuca, see fescue
field crickets, 348
field trials, 107–108, **236**, 236–237, 371, 376, 378, 394, 402
fipronil insecticide, 242
Firmicutes, *215*, 234, 239–240
Flavobacteriaceae, 241
Flavobacterium, 234, 237
flavonoids, 35, 248, 313–314
food security, 8, 15, 387
forage, 10, 12, 190, 352–353, 357, 360, 366
formulation, 25, 41, 52, 54–55, 57, 59–60, 82, 86, 91, 394–395, 398
fossil record, 279
Framework Directive on the Sustainable Use of Pesticides, 373–374
Fraxinus, 40, 328–331, 335
functional genomic, 13
fungicides, 19, 26, 38, 57–58, 82, 91, 350, 368, 372, 396, 407
Funneliformis, 126, 160
funnelling approach, 13
Fusarium, 28–29, 38–39, 71, 74–75, 77, *80*, 94, 98–99, 101–103, 105–106, 108–110, 156–158, 191, 196, 219, 282–283, 287, 290, 297, 299, 301–303, 305–307, 314, 318, 321, 324–325, 331, 333, 335, 376–377

Gaeumannomyces, 99, 103, 106, 108, 114, 116–117, 144, 155, 159, 168, 184, 196, 198, 202, 232, 234
Gemmatimonadetes, 234
gene editing, 26
genetic engineering, 26
genome-wide association study (GWAS), 216
Geomyces, 287, 297
Gibberella, 287, 297
gibberellins, 21, 33–34, 46, 50, 377, 409
Ginkgo, 283, 287, 299, 306
globalisation, 9
Glomeromycota, 96, 329
Glomeromycotina, 11, 120, 122
Glomus, 123, 126, 135, 137–138
gluthatione reductase, 158
Gossypium, 73
granules, 52, *53*, 55, *55*, 58–60, 136
grapevine, 73, 212, 215, 217, 223, 231
grasses, 10, 12, 53, 71, 91, 96, 106, 112, 114, 192, 204, 319, 324–325, 339, 343–344, 347, 349, *349*, 359–360, 362, 364–365, 368, 391, 405–406
grassland, 12, 96, 119, 203, 279, 343
green fluorescent protein (GFP)-labelled strains, 211
Green Revolution, 9
greenhouse gas, 8–9, 233
Guarea, 279

habitat loss, 8
Harpophora, 150, 157–158, 163–164
Hartig net, 144, 190
haustorium, 71
heavy metal, 11, 161, 216, 234, **236**
Heinrich Friedrich Link, 5, 329
Helotiales, 144, 151, 164
hemagglutinin gene, 217
hemocele, 69
Herbaspirillum, 190, 206, 213, 227
Herbicide, 61, 245, 378
Heterobasidion, 157, *158*, 163
Heterochomium, 156–157
Heteronychus, 348, 358
hexoses, 122
HIV, 220, 295
Hodgkin's disease, 282
holobiont, 3, 52
Homalodisca, 212, 217
Hordeum, 22, 97, 117, 191, 399, 402
horizontal transmission, 3, 71–72, 208, 210–211, 396
host-induced gene silencing
  HIGS, 26
HSV, 295
Huperzia, 285, 291, 301, 304, 306–309
huperzine, 281, *293*, 301, 304, 306–307, 309
Huperzine A (HupA), 285, 291, 307–308
Hydrastis, 296
Hymenoscyphus, 40, 328–329, 331, 335, 337–340
Hymenoscyphus fraxineus, 40, 328, 338
hypericin, 281, 286, *293*, 295, 299, 303–304
Hypericum, 287, 295, 303
hyperparasitism, 192
Hypocreales, 53, 68, 85, 87, 90, 144, 151, 196
Hypoxylon, 292

IAA, *see* auxin
in vitro testing, 37, 107
indole-diterpenoids, 346
indolosesquiterpene, 220, 225
induced resistance, 11, 37, 39, 75, **99**, 109, 111, 115
induced systemic resistance, 73, 75
influenza A virus, 295
insecticides, 9, 82, 244
integrated pest management, 52, 58, 60, 93–94, 373, 396
IP protection, 357
IPM, *see* integrated pest management
iron, 157, 163, 191, 249, 347, 362, 376, 388
irrigation, 8, 26, 39, 169, 218, 379

Isaria, 10, 53, 56, 81
ISR, *see* induced systemic resistance

jasmonate, 34, 75, 230, 313
jasmonic acid (JA) defense signaling, 215
Jatropha, 216
Juncus, 96
Juniperus, 285, 291, 303–304

Kaistobacter, 234
kakadumycin compounds, 220
Kennedia, 219
Kineococcus, 168
Klebsiella, 213
Knoellia, 96
Koch's Postulates, 29

landraces, 26, 390, 392, 401, 405
land-use change, 8
Lasiodiplodia, 282–283, 305
latent pathogens, 28, *38*, 95, 258, 278
latent saprotrophs, 95
leaf blotch, 94
Lecanicillium, 10, 53, 73–75, 86–88, 103, 333, 335
lectins, 33
Leptodontidium, 144, *145*, 147, *148*, 148–149, 151, *152*, 155–157, 159
lesion nematode, 348
lignins, 158
Lignyodes, 335
Linum, 321–322
lipopeptides, 219
Listronotus, 348, 363, 366, 368
liverworts, 312
locking nucleic acids (LNA), 221
lolines, 192, 346, 348, 351, 360
lolitrem B, 192, 344, 348, 355–356, 358, 362
lolitrems, 192, 351
Lolium, 21, *145*, 149, 155, 265, 270, 279, 301, 311, 323, 358, 360, 362, 364–368, 409–410
Lupinus, 252
luteolin, 242
Lygeum, 99, 106

Magnaporthaceae, 196
Magnaporthe, 150, 157–158, 196
maize, 10, 71, 97, 117, 126, 155, 159, 161, 164, 168, 177, 183, 243, 325, 393
malting, 94
mammalian mycotoxicosis, 346
MAMP, *see* microbe associated molecular patterns
mangrove, 219
Manilkara, 279
MaxQ, 351–354
MaxQ II, 352, 354
Medicine, 9, 303, 308
Mediterranean, 250, 252
melanin, 144, 156
Melia, 287, 299, 304
melon, 57, 161
Mentha, 321, 325
metabarcoding, 30
metabolites, 3, 9, 31, 35–36, 41–42, 73, 75, 78, 92, 109, 211, 215, 217–218, 278, 300–302, 305, 311–312, 323–324, 333, 377
metagenomics, 12, 29–31, 159, 200, 202, 221, 234, 240
metal hyperaccumulators, 145
metal transport protein (MTP), 156
meta-omics, 232
Metarhizium, 10, 42, 53–54, 56–58, 60, 68–69, 71–78, *79*, *80*, 85–91, 282–283
meta-transcriptomics, 31
Methylobacterium, 212–213, 216, 225, 227, 235
Methylophilaceae, 241
microarrays, 149
Microbacterium, 213, 236
microbe associated molecular patterns, 33, 35
microbiome, 3, 6, 12–13, *14*, 25, 27, *30*, 30–31, 36–38, 82, 95, 97, 202, 209, 211–213, 215, 222–223, 237–238, 241, 243, 267, 310, 328–329, 331, 392, 401
microbiota, 31, 43, 68, 117, 135, 213, 218, 238–241, 243–245, 390–392, 401, 403, 406, 408
microbispora, 167–168, 173, 177–178, 182–184
Micrococcus, 213, 237
Microdochium, 96, 100–101, 105–106, 111, 149–150, 153, 160, 196–199
Micromonospora, 104, 177
micronutrient, 11
microsclerotia, 54, 131, 143–144, *145*
mineral solubilisation, 378
Miscanthus, 188–191, 193–194, **195**, 195–196, **197**, *200*, 200–202, *215*
Monochaetia, 282
monoculture, 234, 241, 282, 313, 372
Morinda, 282–283, 305, 314, 322
mosses, 125, 312
mouldboard plough based conventional tillage, 239
Mucilaginibacter, 96
Musa, 71, 84
mutualism, 3–4, 7, 25, 28, 143, 147, *148*, 148–149, 154, 324
mutualism-parasitism, *5*
mycoacaricides, 56, 85
Mycobank, 335
mycoinsecticides, 55, 87
mycoparasitism, 74, 78, 109, 111, 202

mycopathogen, 74
mycorrhizal fungi, 3, 6, *7*, 10–12, 32, 42, 95, 120, 122, 124–126, *131*, 131–132, 136, 138–139, 144–145, 150, 153–156, 159, 190, 266, 322, 381, 393
Mycosphaerella, 29, 219, 223, 286, 296, 332, 335
mycotoxins, 41, 87

Nectria, 330
Nectriaceae, 196
Nemania, 331
Neonectria, 333, 335
Neotyphodium, 190, 193, 205, 265, 319, 324, 358–362, 364–370
net blotch, 94, 113
neuroprotective, 278, 296
Neurospora, 191, 207, 284, 290, 306
neutralism, 149
Nicotiana, 216, 229
Nigrospora, 37, 289, 300
nitrification, 375
nitrifying bacteria, 375
Nitrobacter, 375
Nitrococcus, 375
nitrogen, 5, 8, 10, 72, 120, 147, 151, 189–190, 213, 216, 234, 237, 240, 242, 346, 350, 354, 372, 375, 377–378, 386, 399
nitrogen fixation, 10, 168, 237, 242, 375, 377–378
nitrogen fixing, 190, 213, 216, 243
nitrogen fixing rhizobia, 373, 377
Nitrosococcus, 375
Nitrosolobus, 375
Nitrosomonas, 375
Nitrosospira, 375
Nitrosovibrio, 375
Nitrospina, 375
Nitrospira, 375
Nocardioides, 96
Nocardiopsis, 219, 227
Nod D receptors, 242
Nodulisporium, 284, 290
non-ribosomal peptide synthetases (NRPSs), 347
nrITS, *146*, 188, **195**, 194–197, **197**
nrLSU, 188, 195, **195**, 194–197, **197**
nutrient media, 30
nutrient uptake, 5, 32, 110, 147, 150–151, *152*, 152–153, 167, 190, 216, 240, 266, 311

oilseed rape, 12, 57–59, 232–233, 244–247, 249
oligotrophic bacterial families, 240
oomycetes, 28, 35, 74, 257
Oosporein, 73, 92
Ophiosphaerella, 196–198
organic agriculture, 27, 94, 112, 117, 141, 372
orthophosphate, 121, 126
Oryza, 47, 157–158, 168, 170, 185–186, 213, *215*, 296, 301
Ostrinia, 10, 15, 53
Oxalobacteraceae, 124, 240–241
Oxyria, 96

Pacific yew, 281, 306, 326
paclitaxel, 312, 321, 324
Paecilomyces, 46, 62, 71, 75, 85, 259, 286, 292, 307
Paenibacillus, 96, 138, 216, 228, 236
Panicum, 96, 213, *215*, 225
Pantoea, 105, 113, 184, 213, 219, 223, 225, 229
Paracoccus, 96
parasite, 74, 95, 220, 405
parasitism, 3–4, *5*, 7, 11, 25, 28, 39, 73–74, 94, 109, 143, 147, *148*, 148–149, 154, 156, 358, 376, 389, 405
Passalora, 33
passenger endophytes, 6, 391
pasture mealybug, 348, 366
pathogen, 4–5, 10–11, 27–29, 32–33, 35, 37, 39–41, 73–75, 77–78, 94, 96, **98**, 107, 109–112, 150, 156–157, 172, 181, 191–193, 216–217, 219, 235, 240, 250–251, 256–257, 259, 266, 313, 315, 328–329, 331, 335, 337–338, 396, 398
pathogenesis related gene, 111
pathogenesis related proteins, 75, 110
pathosystem, 4
paxilline, 346
Pb, **155**, 159, 237, 249
PCR-based DGGE (denaturing gradient gel electrophoresis), 169
Pedobacter, 96, 404
Penicillin, 277
Penicillium, 105, 196, 199, 258, 263, 266, 274, 277, 282, 287, 297, 299, 302, 304
Peniophora, 331
peptides, 9, 219, 226, 347
peramines, 192
Periconia, 96, 132, *133*, 133–134, 286, 296, 307
peroxidase, 111, 158
Pestalotiopsis, 282–283, 287, 299, 302, 306, 324, 326–327
pesticide, 10, 13, 26–27, 57, 67, 69, 78, 83, 94, 188, 192–193, 232, 237, 241–243, 374, 383, 396
Pezizomycotina, 188, 196
PGP, *7*, 9, 11–12, 14, 150–151, 234–235, 241, 321, *see also* plant growth promoting
PGPB, 242, 371–376, 378, 382–383, *see also* plant growth promoting bacteria
pharmaceutical, 3, 9, 36, 222, 278, 282, 302, 326

Phaseolus, 84, 213, *215*
Phellodendron, 286, 296, 301
phenolics, 10, 313
phenotype, 6
phenotyping, 13, 332
phenylpropanoids, 35, 313
Phialocephala, 138, 144, 147, *148*, 148–151, 154, 157, 160–161, 163–164, 285, 291
Phialophora, 139, 144, *145*, *148*, 148–149, 151, *152*, 152–153, 155, *158*, 319, 326, 365
Phlegmariurus, 285–286, 292, 308
Phoma, 99, 162, 263, 282–283, 303, 332, 335, 405
Phomopsis, 199, 286, 296, 301, 321
phosphatase, 124, 127
phosphate, 10–11, 120–121, *127*, 123–129, **130**, 132, *133*, 137–139, 141, 167, 170, 173, 181, **182**, 182–183, **236**, 236–237, 290, 320, 377
phosphate transporter gene, 126
phosphorus, 8, 120, 378
phyllosphere, 27, 40, 95, 116, 209, 212, 218, 224, 226, 230
phytase, 378, 388
phytoalexins, 35–36
phytoanticipins, 35–36
phytohormones, 11, 33, 143, 153, 213, 259, 319, 376, 388
phytopathogens, 11, 67–68, 74–75, 78, 82–83, 143, 156, 163, 217–218, 323
Phytophthora, 27, 35, 157, *158*, 163, 240, 257, 266, 268, 376
phytoremediation, 208, 329, 404
phytostimulation, 10, 216
Pi, inorganic P, 121–124, 127–128, 131, 134
Pierce's disease (grapevine), 217
Pinus, 65, 81, 138, 147, 151, 161–162
Piper, 286, 296, 301, 307
Piperine, 286, 296, 301–302, 307
Piriformospora, 5, 34, 97, 113–115, 118–119, 126, 135–136, 139, 141–142, 161, 193, 207, 319, 321–322, 326, 392, 394, 402, 404–406, 409
Pitef1, fungal gene, 129, 136
Pithomyces, 282
Planctomycetes, 234
Planococcus ficus, 83
plant breeding, 9, 26, 204, 339, 409
plant growth promoting, *7*, 9, 32, 138, 183, 224, 229, 233, 237, 243–244, 325, 376, 386
plant growth-promoting bacteria (PGPB), 371
plant pathogenic bacteria, 208
plant protection products (PPPs), 372
Plantago, 96
Plasmodiophora, 185, 241
Plasmopara, 73, 215
Pleosporales, 144, 147, 151
podophyllotoxin, 281, 285, *288*, 290
Podophyllum, 285, 290, 301, 305
pollution, 8–9, 83, 153, 160, 163, 223
Polygonum, 246, 294
polyketide synthetases (PKSs), 347
polyketides, 35, 304, 403
Populus, 43, 155, 215, *215*
Porina, 348
Portulaca, 168
powdery mildew, 48, 82, 87, 94, 109, 376
PR, *see* pathogenesis related
Pratylenchus, 348, 358, 370
probenazole, 26
Prosopis, 97
protein nucleic acids (PNA), 221
Proteobacteria, 96, *215*, 234, 239
proteomic, 13
Pseudomonadaceae, 240–241
Pseudomonas, 97, 100, 102–105, 108, 111, 113–114, 117, 124, 187, 190, 213, 215–216, 219, 223, 226, 228–229, 235–236, 241, 243–245, 248, 330, 339, 376–377, 379–380, **382**, 386, 406, 410
pteridophytes, 125
Puccinia, 47, 94, 191
pumpkin, 57
Pyrenophora, 94, 98, 100–101, 103–109, 111, 405
pyriproxyfen insecticide, 242
pyrrolizidines, 346
pyrrolopyrazines, 346
Pythium, 48, 74, 85, 157, 217, 376, 387

Quercus suber, 250–251, 254, 256–257, 260, 268–269, 271–272
  decline, 257
quinones, 35

Ralstonia, 186, 216
Ramularia, 28, 48, 94, 112
Rathayibacter, 237
reactive oxygen species, 110, 289
REBECA policy, 41
Regulatory processes, 373
resilience, 9, 117, 387
Resveratrol, 286, 294–295, 301, 306
Rhizobia, 10–11, 95, 212, 242, 373, 377–379, *381*, **382**, 382–383, **384**, 386, **386**
Rhizobiaceae, 235, 241
Rhizobium, 96, 234, 242, 244–245
Rhizoctonia, 73, 82, 167–168, 172, 219, 236, 376, 387
rhizodeposition, 240
Rhizoglomus, 123–124, 141
Rhizophagus, 124, 140
rhizosphere, 13, 27, 32, 39–40, 95, 97, **103**, 114, 118, 120, 123–124, 127, 134, 137, 150, 160,

169, 183, 190, 211–213, 223, 229, 232, 234, *235*, 238–239, *239*, 241, *242*, 244–245, *253*, 375, 387, 393, 403, 409
Rhodanobacter, 96
Rhodococcus, 96
Rhodoplanes, 234
Rhodothorula, 188, 196, 199
Rhopalosiphum, 348
Rhynchosporium, 94, 100, 105–106, 111
ribosomal peptide synthesis (RiPS), 347
ribosomal polymerase B, 72
rice, 87, 150, 158, 163–164, 167–170, 173, *174*, *175*, **175**, **176**, 176–178, *179*, *180*, **181**, 180–182, **182**, 184–186, 213, 224, 230, **236**, 259, 405, 408
rohitukine, 287, 297, *298*, 312
root aphids, 348
root derived polysaccharides, 211
ROS, *see* reactive oxygen species
Roselinia, 196
Rumex, 96, 314, 323
ryegrass, 21, 36, 44, *145*, 147, *148*, 148–149, 204, 279, 343, 345, *345*, 347–348, 350–351, 354–356, 358–359, 361–362, 364, 366–370, 409

Saccharopolyspora, 167, 176, 182–183
Saccharum, 188, 191, 196, 201, 205, 213, *215*
salicylic acid, 33, 75, 153, 215
salinity, 70, 153, 257, 404
Salix, 154–155, 160–161
Salmonella, 218, 223, 227, 229
saprophytes, 81–82
SAR, *see* systemic acquired resistance
Schizachyrium, 96
Sclerostagonospora, 331
Sclerotinia, 76–77, *80*, 82, 219, 236, 247
Sebacinales, 11, 119–120, 122, 125, 134, 139, 409
secondary metabolite, 3, 9, 36, 53, 73, 153, 157, 168, 192, 217, 278, 297, 299–300, 304, 310–311, 313, 315, 321–322, 325, 344, 347, 410
seed dressing, 19, 81, 395, 399, 407
seed treatment, 46, 58, 62, 72, 84, 87, 118, 379, 393, 396, 399
Seimatoantlerium, 282
Septoria, 28, 115, 396
Serendipita, 33, 97–99, 109, 122, 125–126, *127*, 129, 193, 392, 394, *see also Piriformospora*
Serendipitaceae, 126
Serratia, 213, 236–237, 244
Setomelanomma, 331
siderophore production, 173, *181*, **182**, 375
silvicultural, 3, 9, 252
soil fertility, 9, 233, 372, 377, 383, 388
soil management, 9, 169
soil structure, 125, 140, 234, 240, 244
Solanum, 65, 97, 127, 157
Sordaria, 259, 264, 269
Sordariales, 144
Sordariomycetes, 188, 196
Sorghastrum, 96
Sorghum, 81, 154
Sphingobacteriaceae, 241
Sphingobacteriia, 213
Sphingomonas, 96, 213, 235
Spilanthes, 321, 326
spore formation, 13, 59, 335
Sporormia, 282
Stagonospora, 191, 206
Staphylococcus, 213, 219, 307
Stenotrophomonas, 216, 236
Stereum, 259, 265
steroids, 35, 313
stomata, 18, 124, *210*, 212, 218, 396
Streptomyces, 103, 167–169, 173, **175**, 175–178, 182–186, 216, 219, 225, 227, 234, 241, 323–324, 376
superoxide dismutase, 156, 158
Sustainable Development Goals, 8, 15
symbionts, 4, 7, 12, 28, 42, 71, 147, 160, 250, 322
systemic acquired resistance, 75
systemic resistance, 10, 73–75, 91, 110, 158, 168, 216–217, 376–377

take-all fungus, 241
Talaromyces, 284, 287, 289, 299, 305, 308
tall fescue, 344, 348, 351–354, 366, 368
taxol, 279, 281, *288*, 299, 301–308, 312
Taxomyces, 279, 281, 283, 306, 326
Taxus, 279, 281, 283, 301–302, 304, 306–308, 321, 324, 326–327
TE, *see* trace element
TEF, *see* translation elongation factor
terminal restriction fragments (TRFs), 172
Terminalia, 282–283, 302
terpenoids, 9–10, 35, 313, 315
tetralones, 35
Theobroma, 65, 90, 114, 259, 266–267
Thielavia, 287, 295, 304
Thiram, 58
tillage, 13, 26, 232, 237–238, *239*, 239–241, *242*, 245, 248
tissue print-immunoblot technique, 351
tobacco, *148*, 153, 162, 212, 216, 219, 226, 228, 240, 326
tobacco-specific nitrosamines (TSNAs), 219
tomato, 33–34, 57–59, 67, 71, 76, 78, *79*, *80*, 97, 127, *127*, *128*, 128–129, **130**, *131*, 133, 147, 156, 168, 259, 371, 377–379, 383, **384**, 386, **386**

toxicity, 11, 69, 75, 154, 156, 192, 243–244, 289, 352, 361, 368, 372
trace element, 154
Trametes, 285, 291, 306
translation elongation factor, 72, 129, 136, 195
tremorgens lolitrem B, 346
TRF, *see* terminal restriction fragments
T-RFLP profile analysis, 167
Trichoderma, 43, 48, 91, 97, 101, 103–104, 259, 265, 267, 284–286, 290, 292, 299, 305, 323, 376, 388
Trichothecium, 282
Triticum, 97, 183, *215*
Truncatella, 282
Tubercularia, 282–283, 327
Tuta absoluta, 83
Type IV pili gene, 217

Ulocladium, 376
United Nations, 8–12, 15, 28–29, 32–36, 38–39, 113–114
Ustilago, 101, 105–106
UV radiation, 37, 57, 82

Venturia, 38, 47, 331
Verrucomicrobia, 234, 239–240
vertical transmission, 3, 10, 12, 14, 208, *210*, 210–211, 335, 345, 350, 357, 364
Verticillium, 71, 74, 85, 87, 91, 156–157, 162, 235–236, 244
vinblastine, 281–282, *288*, 312
Vinca, 287, 297, 308
vincristine, 281–282, *288*, 303, 308, 312
virus, 9, 73–74, 87, 94, 119, 295
viruses, 4
Vitis vinifera, 73, 215, *215*, 224, 286, 294, 301
VOCs, *see* volatile organic compounds
volatile organic compounds, 11, 94, 143, *148*, 153, 162, 278, 313, 378

wheat, 10, 28, 31, 34, *38*, 57–58, 97, 105, 111, 113–115, 117–118, 168–169, 183–184, 229, 234, 240–241, 245, 247, 252, 269, 378, 387, 393, 396, 403–405, 409–410
white grubs, 348
whitefly, 83
Wiseana, 348

xanthan, 56, 58
Xanthomonadaceae, 235, 241
Xanthomonas, 73, 226, 236, 325
xanthones, 35
xenohormesis hypothesis, 281
Xylariales, 144, 151
Xyllela, 212

yellow Sigatoka disease, banana, 219
yield losses, 69, 192

Zizyphus, 97
zucchini yellow mosaic virus, 74
Zygomycota, 70, 329
Zymoseptoria, 27–28